Psychoendocrinology

Edited by

F. Robert Brush
PSYCHOLOGICAL SCIENCES
PURDUE UNIVERSITY

Seymour Levine
PSYCHIATRY AND BEHAVIORAL SCIENCES
STANFORD UNIVERSITY

Academic Press, Inc.

Harcourt Brace Jovanovich, Publishers

SAN DIEGO   NEW YORK   BERKELEY   BOSTON

LONDON   SYDNEY   TOKYO   TORONTO

This book is printed on acid-free paper. ∞

Academic Press, Inc.
San Diego, California 92101

United Kingdom Edition published by
Academic Press Limited
24–28 Oval Road, London NW1 7DX

Library of Congress Cataloging-in-Publication Data

Psychoendocrinology / F. Robert Brush and Seymour Levine, eds.
      p.   cm.
    Prepared with support from the National Institute of Mental
Health.
    Includes bibliographies and index.
    ISBN   0-12-137952-3   (alk. paper)
    1. Psychoneuroendocrinology.   I. Brush, F. Robert.   II. Levine,
Seymour.   III. National Institute of Mental Health (U.S.)
    [DNLM: 1. Brain--physiology.   2. Hormones--physiology.
3. Psychophysiology.   4. Steroids--physiology.   WK 102 P974]
QP356.45.P79   1989
599'.0188--dc19
DNLM/DLC
for Library of Congress   89-408
                                                              CIP

Printed in the United States of America
89   90   91   92   9   8   7   6   5   4   3   2   1

Dedication

This book is dedicated to Professor Frank Beach, who was Professor Emeritus of Psychology at the University of California, Berkeley. It is an acknowledgment of his contributions to the understanding of the relationships between hormones and behavior and a recognition of his inspiration to so many of his students.

Frank Beach was a scholar, teacher, and perhaps most importantly, a revered friend of many in the field of psychoendocrinology. All of us who are concerned with the biological bases of behavior and, in particular, the role that hormones play in regulating behavior are, in one way or another, students of Frank. He once commented at a large meeting on reproductive behavior that had such a meeting been held in the late 1930s, it would have had three participants: William Young, Josephine Ball, and himself. No other field of research owes its inception to the efforts of so few and, in particular, to Frank. He was among the first to explore the concept that chemical substances originating from structures far removed from the brain could influence as complex a process as sexual behavior. Throughout his long and illustrious career he made fundamental contributions, both experimentally and theoretically, without which the field would not have advanced. It was through his vision, creativity, and devotion that the discipline he represented has reached its current state of development. Most of the major figures who are currently publishing and making contributions to our understanding of the biological bases of reproductive behavior are either his students or students of his students. Undoubtedly, the third generation of students will continue his tradition.

People who knew Frank were overwhelmed by his scholarship; his knowledge of the literature was encyclopedic. Whenever one had the opportunity to present data in his presence, there was always a small amount of anxiety that one would make some generalization to which he would know the exception. Above all, he maintained throughout his career a commitment to the study of behavior. In recent years he expressed concern that the glamour of molecular biology would divert attention from the fundamental question of what regulates behavior. In some quarters it appears that his concern was a valid one.

All those associated with him realize the privilege it was to have known him and to have shared his knowledge, humor, and dedication to his work. For many of us, Frank Beach was, and always will be, an inspiration and a model to which one might aspire. He never lost his excitement and enthusiasm for all things scientific and, indeed, for life itself.

Contents

Contributors

N. T. ADLER   Department of Psychology, University of Pennsylvania, 3815 Walnut Street, Philadelphia, Pennsylvania 19104

T. O. ALLEN   728 Dodds Lane, Gladwyne, Pennsylvania 19035

F. ROBERT BRUSH   Department of Psychological Sciences, Purdue University, West Lafayette, Indiana 47907

CHRISTOPHER COE   Department of Psychology, University of Wisconsin, Madison, Wisconsin 53706

PAUL E. GOLD   Department of Psychology, University of Virginia, Charlottesville, Virginia 22901

J. HERBERT   Department of Anatomy, University of Cambridge, Downing Street, Cambridge CB2 3DY, England

T. F. LEE   Department of Zoology, Biological Science Building, University of Alberta, Edmonton, Alberta, Canada T6G 2E9

SEYMOUR LEVINE   Department of Psychiatry and Behavioral Sciences, Stanford University School of Medicine, Stanford, California 94305

B. S. MCEWEN   Laboratory of Neuroendocrinology, The Rockefeller University, 1230 York Avenue, New York, New York 10021

JAMES L. MCGAUGH   Center of the Neurobiology of Learning and Memory and Department of Psychobiology, University of California, Irvine, California 92717

D. W. PFAFF   Laboratory of Neurobiology and Behavior, The Rockefeller University, 1230 York Avenue, New York, New York 10021

H. RIGTER   The Health Council of the Netherlands, P.O. Box 90517, 2509 LM's-Cravenhage, The Netherlands

S. SCHWARTZ-GIBLIN   Laboratory of Neurobiology and Behavior, The Rockefeller University, 1230 York Avenue, New York, New York 10021

CAROLYN NAGASE SHAIN   Department of Anesthesiology, Yale University School of Medicine, New Haven, Connecticut 06520

JUDITH M. STERN   Department of Psychology, Rutgers-The State University of New Jersey, New Brunswick, New Jersey 08903

LAWRENCE C. H. WANG Department of Zoology, Biological Science Building, University of Alberta, Edmonton, Alberta, Canada T6G 2E9

SANDRA G. WIENER Department of Psychiatry and Behavioral Sciences, Stanford University School of Medicine, Stanford, California 94305

TJ. B. VAN WIMERSMA GREIDANUS Rudolf Magnus Institute of Pharmacology, University of Utrecht, Medical Faculty, Vondellaan 6, 3521 GD Utrecht, The Netherlands

Preface

In 1948 the late Professor Frank Beach, to whom this volume is dedicated, published the first volume of "Hormones and Behavior." In 1972 another book with the same title was published, and in its preface was stated, "In one sense, therefore, this book represents a progress report in the area of hormonal control, not only of sexual behavior but of many other behaviors which now have been shown to be influenced in some way by a multitude of hormonal determinants." It is difficult to find an appropriate description for the events that have occurred in the area of hormones and behavior since 1972. There has been an explosion of information that in many ways has dramatically altered our view of the action of hormones and the relationship between hormones and the central nervous system (CNS). The discovery of many new hormones, which are synthesized by cells in the CNS, has validated the view of the late Professor Geoffrey Harris that the brain is one of the major endocrine organs.

Advances in the field of biology and the development of highly refined new measurement techniques for hormones have resulted in new insights and have made it necessary to once again reevaluate the relationship between hormones and behavior. The work reported and reviewed by all of the chapters in this volume reflects these developments. It is also now apparent that specific environmental events and behaviors directly address such influences. Another theme that emerges from many of the chapters in this book is that hormones do not exert their actions simultaneously or directly, but rather that they exert their influence in concert with many other biological events in a variety of different systems. There appears to be a running conversation in the CNS among hormones, neurotransmitters, neuromodulators, and probably other, as yet unknown, elements of CNS function.

One of the principal objectives in assembling and editing the chapters for this volume was to present material that covers a wide spectrum of procedures, processes, interpretations, and species to facilitate a broad and integrated understanding of the area that has been designated "psychoendocrinology." No volume on this topic can be inclusive without becoming overbearing. However, the chapters in this volume cover their specific areas of interest in depth and reveal many aspects of the complexity of hormonal regulation. This book is intended for advanced students and specialists in psychobiology, neuroscience, neuroendocrinology, psychology, and psychiatry but might also be of interest to anyone concerned with understanding the biological bases of behavior.

To the contributors, each of whom we hold in highest esteem, both personally and professionally, we wish to express our gratitude for their effort and patience.

This book was prepared with the support to S. Levine of a U.S. Public Health Service Research Scientist Award (MH-19936) and to F. R. Brush of Research Grant MH-39230, both from the National Institute of Mental Health.

F. Robert Brush
Seymour Levine

Partitioning of Neuroendocrine Steroids and Peptides between Vascular and Cerebral Compartments

J. Herbert

I. Introduction

The response of any animal (or human) to an adaptively important stimulus from the environment is formulated through its behavior, its endocrine activity, and its autonomic nervous system. To say this is to say nothing new or surprising, for this division of effectors has been apparent for many years. However, largely as the result of the traditional boundaries around different disciplines, these sets of responses have often been studied separately, by different people deriving from equally different backgrounds and thus using significantly different approaches. On some occasions this division of effort has resulted in individuals in one subject being unaware of what the others were doing and thus in danger of interpretations which do not take into account sufficiently the complexity of an animal's total response or the possibility of interactions between the constituent components. There are signs, however, that this situation is being remedied, partly because of a more flexible approach, but also as the result of recent findings in neuroendocrinology. In particular, the description of systems within the brain, endocrine glands, and somatic tissues which seem to contain the same (or similar) peptides has forced us to consider whether a more integrated view of their function might be appropriate. A number of attempts to synthesize currently available knowledge have appeared, notably from those concerned with drinking behavior and fluid balance (Mogenson *et al.*, 1980; Swanson and Mogenson, 1981), for this area offers some of the best examples of coordinated behavioral and endocrine activity to date. Because many peptides have been found in high concentration in the limbic system, it is natural to wonder whether they may play a part in this system's function, which is central to the organization of the response triad described previously. Several lines of evidence, to be discussed more fully later in this chapter, indeed suggest that some neuropeptides can regulate the categories of behavior traditionally associated with the limbic system, for example, eating, drinking, sexual activity, and maternal behavior. This seems to apply particularly to peptides which are called here "neuroendocrine peptides," not as a watertight definition separating them from other sorts of peptides, but to indicate that many of those active in these sorts of behavior are also concerned with more usual neuroendocrine function, such as the regulation of the pituitary (Everitt *et al.*, 1983). If this is accepted, then it follows that since many of the same functions are also acted upon by the steroid hormones, the interaction between these two classes of compound becomes interesting.

Behavior, hormones, and autonomic activity not only form part of a total response, they also interact with each other. An example might be the hormonal changes which follow some forms of sexual interaction. Copulation by a female rat induces the corpus luteum to secrete progesterone, which in turn inhibits further sexual responsiveness (Marrone *et al.*, 1979). Secretion of noradrenaline from the adrenal during "stress" may act upon the pituitary to release adrenocorticotrophic hormone (ACTH), thus activating a second set of endocrine responses (Mezey *et al.*, 1984). The existence of such interactions means that we have to know about the way that events taking place in the vascular compartment (for example, the secretion of a hormone) influence those in the brain (such as acting on a behavioral response, or a system controlling an autonomic or endocrine activity). This leads us to consider the ways that levels in one compartment are reflected in the other, and whether, in cases where the same substance derives from both, there is evidence of functional linkage between the two systems. For steroids the situation is simpler than for peptides. Steroid hormones are only produced, so far as we know, from peripheral endocrine glands. In their case the question is how alterations in their level in the blood, in either the long or the short term, are reflected in the brain's extracellular environment, since it is from the latter that neurons receive their hormonal information. For peptides, the story is more complicated. In some cases at least, both central neurons and peripheral glands produce the same peptide, so the intra- and extracerebral compartments have, potentially, different sources of supply. The question then is: do the two sources coordinate, and is there any direct communication between them?

Much of the information to be discussed here is based upon measuring levels of hormones in the cerebrospinal fluid (CSF), though what interests us, in fact, is the hormone levels in the brain's extracellular fluid (ECF). The reasons for such measurements on CSF are entirely pragmatic; it is the nearest thing to ECF that can be obtained in an endocrine context. It must be reasonably certain, therefore, that hormone levels in the one give a satisfactory index of those in the other. Many studies, both anatomical and physiological, show that there is no effective barrier between the two except, perhaps, around the tanycyte ependyma (Bradbury, 1979). Though this means that hormones in the ECF will pass readily into the CSF (and vice versa), there may be regional differences in concentration in the CSF related, for example, to those in the neighboring ECF, which will be missed in samples drawn from, say, the cisterna magna. Furthermore, clearance from the CSF may result in sink effects, so that concentration gradients exist within the ventricles which are not apparent from sampling at one point. Nevertheless, while these reservations need to be remembered, it remains true that CSF levels can be assumed to give at least a first approximate guide to those in the ECF, though the whole question awaits development of methods of assaying hormones in the ECF itself. Finally, it is worth noting that many limbic structures, on which this chapter is focused, lie very close to the ventricles, a further reassurance that CSF levels of steroids and peptides may give a realistic indication of those in the fluid surrounding the neurons themselves.

II. Steroid Hormones

The advent of highly sensitive assays has permitted the operating characteristics of the steroid hormones to be specified in some detail. For our purposes, it is important to note that their secretion (and therefore blood levels) is not constant, but shows the features of at least three sets of rhythms (Hastings and Herbert, 1986). The pulsatile release from their glands of origin depends, of course, on antecedent pulses from the pituitary and hypothalamus; the significant point is that blood levels are changing relatively rapidly all the time. Is this change detected by the brain and, if so, in what form? Many steroids also show circadian alterations in blood levels; this seems to be related to circadian changes in the frequency or amplitude of the circhoral pulses. The important point here is that the mean level of the steroid is changing because of underlying alterations in the pulse characters; how does the brain perceive these circadian changes? Annual rhythms in the reproductive steroids typify those species showing breeding seasons; are there more exact correlations between levels in blood and brain in the case of these very slow changes than in the more rapid ones during the day? It must be emphasized that a steroid-sensitive neuron will respond to a steroid according to the concentration of that hormone in the fluid surrounding it; many *in vitro* studies have shown this without question (McEwen, 1981). The fluid surrounding such a neuron is the extracellular fluid of the brain, not the blood. We need to know, therefore, how the variations of steroid levels in the blood already described are reflected in the brain's ECF (or, since it is in free communication with it, the CSF).

There is increasing evidence that the temporal characteristics of changes in steroid levels are also important. For example, the duration of the estrogen surge, as well as its level, is critical for provoking the secretion of luteinizing hormone (LH) (Karsch *et al.,* 1973). Equally significant may be the duration of a hormone's absence. Withdrawal of progesterone for a defined period is required to enable a female rat to respond to estradiol by showing lordosis (Lisk, 1978). The occurrence of an estradiol surge in a female monkey not only releases LH but renders the LH discharge system unresponsive to a second estrogen surge for a finite period during which hormone levels are basal (reviews by Feder, 1981; Hastings and Herbert, 1986). All these endocrine events are measured in the blood, yet the response to them occurs in the CNS. If there are serious discrepancies between the change in absolute levels or levels over time in the two compartments, we may be inferring false parameters by taking account of only those detectable peripherally and assuming that they also occur centrally. Furthermore, even if we take the trouble to determine corresponding changes in the blood and brain ECF for one steroid, can we be sure that the rules derived are applicable to other steroids? The data reviewed in this chapter show clearly that (1) we cannot infer in a simple way the parameters of steroid levels in the cerebral compartment from those in the blood, and (2) there are large differences in the relationship between blood and brain levels of different steroids.

Studying the relationship between hormones in the two compartments is more than simply studying the entry of steroids into the brain. Of course, exact knowledge of a steroid's ability to access the brain's ECF from the blood is a prerequisite, but it is only part of the information we require. It is important to recognize that the great majority of investigations on this subject have been concerned with this parameter (Pardridge and Meitus, 1979; Marynick *et al.*, 1980). But these do not tell us, nor can we infer from them, how the levels of a hormone in the CSF will alter in response to those in the blood. If, for example, clearance of the hormone differs in the two compartments, then so will relative levels over time, especially if blood titers are changing rapidly. There is no substitute for measuring CSF (to infer ECF) levels under different endocrine conditions. We can anticipate the data discussed in the following sections by pointing to the differences between cortisol and testosterone. Both enter the brain at about the same proportion of blood levels, but the clearance of the former is much slower than that of the latter and presumably accounts for the striking difference in blood/CSF ratios of the two at high serum concentrations. We have to remind ourselves continually that it is the levels of a steroid in the CSF (ECF) and the way that these change over time which are the important parameters so far as a neuron's response is concerned.

III. Cortisol

Cortisol shares two properties with many other steroids: it is lipid soluble; and a large fraction of that in the serum is bound to a plasma protein (Fig. 1) which, in the case of cortisol, is a globulin called transcortin or corticoid-binding globulin (CBG) (Westphal, 1970; Siiteri *et al.*, 1982). Serum albumin is also a larger-capacity (but lower-affinity) source of binding of cortisol, a fact which has led to uncertainties about the definition of "bound" hormone in the context of its passage into the CSF. The problem is to what extent the fraction bound to albumin actually acts as a deterrent to the hormone's entry into the brain, and if it acts as such, under what circumstances. Thus, Pardridge and Meitus (1979) report that albumin-bound steroid is available for transport across the blood–brain barrier. However, an experiment by them in rats, in which cortisol was preequilibrated with albumin before infusion, showed that "first-pass" extraction by the brain was significantly reduced, suggesting that albumin binding might regulate entry of cortisol into the brain in some circumstances, though it is difficult to compare first-pass techniques with longer term steady-state observations on relative levels of steroids in blood and CSF.

Repeated measurements of blood cortisol over the very short term in man have shown that circhoral pulses are very characteristic (Sachar *et al.*, 1973). Levels in blood may alter several-fold during each pulse, and thus the

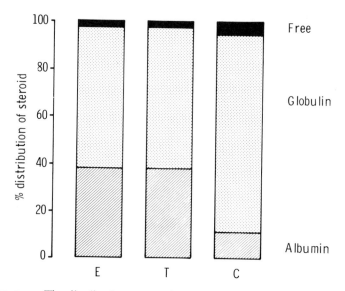

FIGURE 1. The distribution among three compartments in the blood (albumin, globulin, and "free" fraction) of the steroid hormones estradiol (E), testosterone (T), and cortisol (C). Redrawn from data in Siiteri *et al.* (1982).

binding sites on the plasma proteins may be alternately filled and emptied, with corresponding changes in the "free" (or unbound) fraction. But there is no information which allows us to know how these circhoral pulses in plasma cortisol are reflected in the brain ECF or the CSF. As we shall see, corresponding measurements over the longer term make us suspect that the shape of circhoral pulses in the CSF (if they exist) might be very different from those in the blood. Since variations in the frequency of circhoral pulses underlie the daily rhythm in mean blood cortisol levels, and since it is these which seem abnormal in a proportion of those with affective disorders (Carroll *et al.*, 1976), a more exact understanding of the way that circhoral pulses in the blood are related to those in the fluid surrounding the brain is overdue.

Simultaneous study of circadian rhythms in cortisol in blood and CSF has shown unequivocally that the two differ (Fig. 2). In primates, serum cortisol is at its highest during the early part of the daily light cycle and falls during the day to reach its nadir in the evening and early part of the dark phase, before rising once more during the latter part of the night to the morning peak. The major difference in the cerebral compartment occurs during the afternoon fall, which occurs much more slowly in the CSF than in the blood. Peak levels therefore persist in the CSF for longer than in the

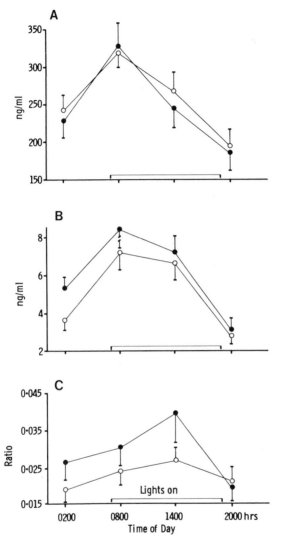

FIGURE 2. Circadian rhythms in cortisol levels in (A) serum, (B) CSF, and (C) CSF/serum ratio of male (●, $n = 4$) and female (○, $n = 7$) rhesus monkeys to show differences between compartments. From Umberkoman-Wiita et al. (1981).

blood, and thus the CSF/serum ratio increases at this time (Fig. 2) (Umberkoman-Wiita et al., 1981). By the evening, the two compartments are consonant again. Abnormal circadian rhythms can be generated by, for example, giving a single injection of 5-OH-tryptophan, a serotonin precursor that causes a pulse of ACTH (and hence cortisol). This results in a correspondingly abnormal rhythm in the CSF, reinforcing the view that cortisol-sensitive neurons are also exposed to rhythms of abnormal shapes in conditions when those in the blood are disturbed. Injecting cortisol itself results in a proportionately larger amount of the hormone entering the CSF; that is, the CSF/serum ratio increases (Uete *et al.*, 1970; Umberkoman-Wiita *et al.*, 1981). This suggests that pulselike alterations in the blood may have exaggerated results on CSF cortisol and hence on the fluid to which neurons are exposed. These studies indicate that changes in circadian parameters in blood cortisol may result in nonparametric fluctuations in the CSF (but see Perlow *et al.*, 1981, for a dissenting view).

Further studies on the CSF of monkeys in which blood cortisol was persistently elevated have suggested at least a partial explanation for these findings. In such conditions, CSF cortisol increases proportionately more than serum cortisol (the CSF/serum ratio rises) (Fig. 3). In both normal and hypercortisolemic animals, there is a significant correlation between serum and CSF cortisol levels in individual samples, but the slope of the regression is increased when blood levels are chronically high (Martensz *et al.*, 1983). The obvious explanation for this is that the "free" fraction of cortisol in the serum is increased, as indeed it is, so that proportionately more cortisol enters the CSF. This, however, is not the whole explanation, for the slope of the regression between free serum cortisol and that in the CSF (which is all free), is greater than one. Studying the decline in CSF cortisol after a single injection of the hormone into the blood reveals that clearance from the cerebral compartment is very slow, much slower than from the blood (Fig. 4) (Peterson and Chaikoff, 1963; Walker *et al.*, 1971).

The increasing amount of free cortisol in the blood thus accounts for the proportionately larger amount which enters the CSF after an acute rise in blood levels. But relatively slow clearance from the CSF means that such an event, whether part of the normal circadian variation or as part of the "stress" response, will have a lingering or carry-over effect on the brain, and chronically high levels may result in accumulation of cortisol in the CSF (Martensz *et al.*, 1983). The importance of these effects should not be overlooked in the context of a hormone which responds so easily to environmental events and which may have potent actions on the brain.

One such environmental event, an animal's position in its social group, may also modulate the actual partitioning of cortisol between blood and CSF. Dominant male talapoin monkeys have more cortisol in their CSF at a given serum concentration than do the most subordinate members of their group, and variations in cortisol in the dominants' blood have a diminished effect on CSF cortisol levels (Fig. 5) (Herbert *et al.*, 1985).

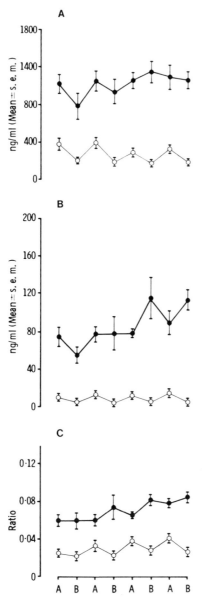

F I G U R E 3. (A) Serum and (B) CSF cortisol levels and (C) CSF/serum ratio before and during chronic treatment with ACTH (1–24) in rhesus monkeys to show disproportionate effects in the CSF compartment during chronic hypercortisolemia. ○, Control (n = 6); ●, ACTH (n = 6); A, Sample at 1000 hr; B, sample at 1630 hr. From Martensz *et al.* (1983).

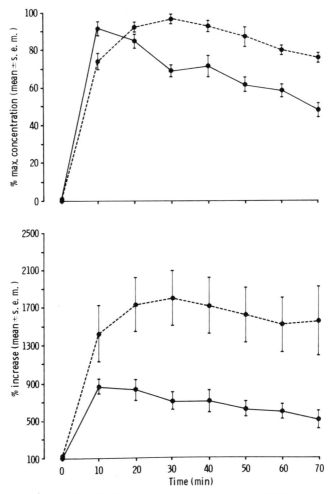

FIGURE 4.   Levels of cortisol in serum (●——●, $n = 6$) and CSF (●----●, $n = 6$) of rhesus monkeys following a bolus intravenous injection of 500 μg cortisol to show slower clearance rate from the CSF compartment. From Martensz *et al.* (1983).

IV. Gonadal Steroids

In some species gonadal steroids in the blood are bound to a globulin, sex hormone-binding globulin (SHBG), which therefore limits the free fraction of steroid available for passage into the brain in a manner analogous to the role of transcortin and cortisol (Fig. 1) (Westphal, 1970). In others, however, including rats, hamsters, and sheep, SHBG is absent, so that all

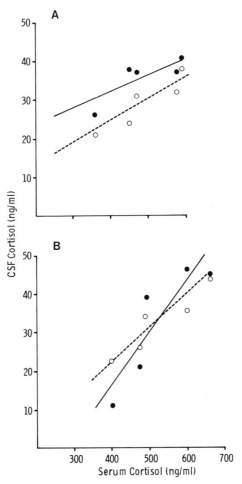

FIGURE 5. The relationship between CSF and serum cortisol levels in socially dominant (A) and subordinate (B) male talapoin monkeys living in groups, showing the predicted (-----o) and actual (———●) effects of social rank on distribution of this hormone. From Herbert *et al.* (1985).

serum binding is to albumin. Comparative studies on the relation between blood and CSF levels of gonadal hormones in a species with or without significant SHBG activity are lacking, but would be interesting. The proportion of free steroid can vary markedly even between species possessing SHGB (Siiteri *et al.*, 1982).

Testosterone has major effects on both gonadotropic function and be-

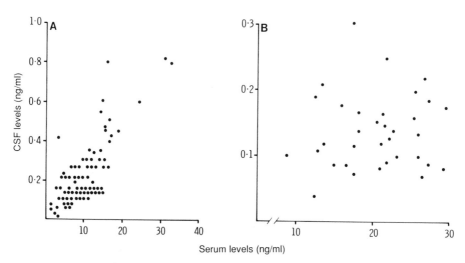

FIGURE 6. The relationship between serum and CSF levels of (A) testosterone and (B) dihydrotestosterone (DHT) in castrated male rhesus monkeys given increasing amounts of either hormone. There is a predictable relationship in the case of testosterone, but not for DHT. From Dubey *et al.* (1984) and Abbott *et al.* (1985).

havior in males, but surprisingly little is known about the way blood levels relate to those in the CSF. This steroid differs from others in that the circulating form (testosterone) acts as a precursor for the formation of active metabolites [dihydrotestosterone) (DHT) and estradiol] in the target tissues themselves (Baird *et al.*, 1968). Concentrations of testosterone in the CSF therefore represent the amounts of substrate available for the converting process within the neurons. Since DHT and estradiol are also formed peripherally from precursor testosterone, a second question of interest is how far this pool can gain access to the brain to supplement that made *in situ*.

Testosterone enters the CSF readily from the blood in monkeys and in rats (David and Anand Kumar, 1974; Marynick *et al.*, 1976; Pardridge and Meitus, 1979). Studies on the amount of hormone in the CSF as levels in the blood rise have demonstrated that there is a predictable relation between testosterone in the two compartments (Fig. 6), though it is possible that this may break down when blood levels are very high (for example, during the nocturnal surge in monkeys). Under steady-state conditions, CSF testosterone is directly proportional to free hormone in the blood. Unlike cortisol, testosterone is cleared from the CSF more rapidly than from the blood (Dubey *et al.*, 1984), which may account for the CSF/blood relation at high peripheral concentrations. Like cortisol, there is no evidence for protein binding of the steroid in the CSF (Dubey *et al.*, 1984).

The negative feedback action of testosterone on LH is well established and offers an opportunity to try to correlate CSF testosterone levels with a physiological parameter. However, higher levels of testosterone are needed to initiate feedback in castrated monkeys than to maintain it, which complicates such experiments (Thornton and Abbott, 1981; Dubey *et al.*, 1984). Nevertheless, it has been shown that, as blood levels are gradually increased, there comes a point at which LH output abruptly declines (Fig. 7), and this coincides with a sharp increase in CSF testosterone, presumably the result of the capacity of SHBG being exceeded (Dubey *et al.*, 1984). There is a need to determine whether giving testosterone in a manner which mimics the normal circadian rhythm is more effective than steady-state regimes, and how such rhythms in the blood are reflected in the CSF. However, the peripheral parameters determining the CSF levels of testosterone seem quite straightforward compared to cortisol or, as we shall see, to DHT.

It is becoming apparent that the regulation of entry of DHT into the brain is quite different from that of testosterone, which, because of their metabolic relationship, is interesting. Studies in which a single dose of labeled DHT has been given intravenously show a curious discrepancy: measurements on the CSF have revealed little DHT (Marynick *et al.*, 1976), whereas those concerned with mapping DHT-concentrating neurons apparently yielded enough useful label (Sheridan and Weaker, 1982). A study in which CSF levels were measured in castrated monkeys given repeated daily injections of DHT has confirmed that, although CSF levels rise when those in the serum are very high, penetration is minimal (Abbott *et al.*, 1985). More importantly, unlike testosterone, there is no correlation between either total or free DHT in the blood and that in the CSF (Fig. 8), in which levels are very low. These results do not support the suggestion that low entry of DHT into the CSF is caused by high serum binding, since even when the free fraction was raised to 4–5%, CSF levels remained low. There is no information about the clearance of DHT from the brain.

These results can also be given a functional interpretation. The LH levels in castrated male monkeys were not correlated with the CSF DHT levels (Fig. 8), but with those in the serum (particularly the free fraction). In contrast to testosterone, injecting DHT parenterally does not restore the sexual behavior of castrated rats, though it has somewhat more effect in monkeys (Macdonald *et al.*, 1970). Implanting DHT directly into the amygdala, however, does improve the castrated rat's sexual performance (Baum *et al.*, 1982). These results suggest that much of the ineffectiveness of DHT given peripherally may be due to its relative inability to gain access to the brain. The important point, therefore, is that the supply of DHT is regulated

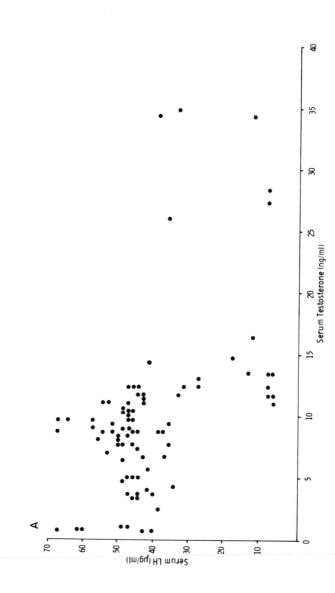

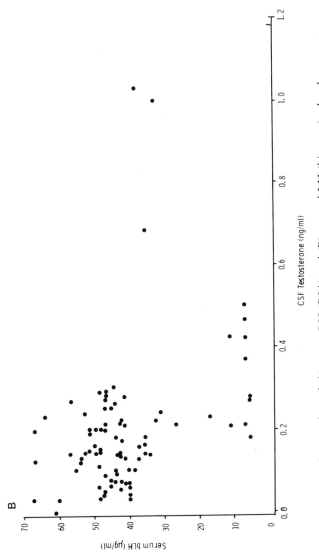

FIGURE 7. Suppression of (A) serum LH (RIA) and (B) serum bLH (bioassay) related to testosterone levels in either CSF or serum in castrated male rhesus monkeys given increasing amounts of the steroid. From Dubey *et al.* (1984).

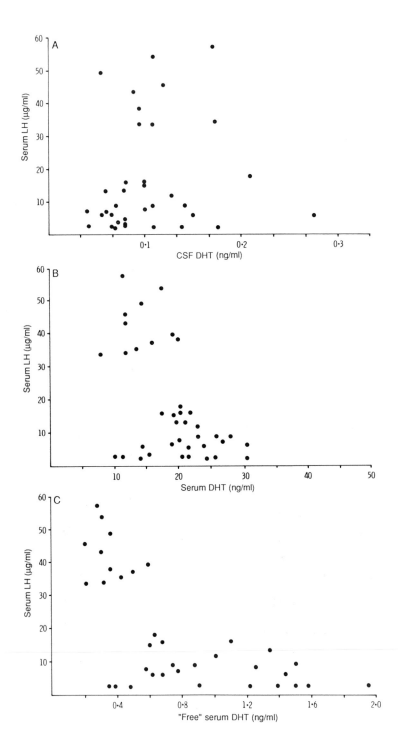

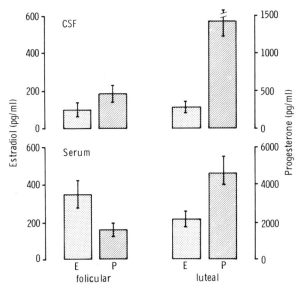

F I G U R E 9. CSF and serum levels of estradiol (E) and progesterone (P) during the follicular or luteal phases of the menstrual cycle of the female rhesus monkey. Redrawn from Anand Kumar *et al.* (1980).

by that of testosterone which, as we have seen, enters the CSF in amounts related to blood levels. Exactly how DHT is impeded is obscure; differences between it and testosterone in binding or lipid solubility do not seem adequate explanations, and it remains possible (though unproven) that some special mechanism limits the entry of DHT into the CSF (Abbott *et al.*, 1985).

There is even less information on the two major ovarian steroids, estradiol and progesterone. It is well established that labeled estradiol enters the CSF from the blood (Anand Kumar and Thomas, 1968). Measurements of CSF levels during short-term infusions show that the same proportion is found relative to blood levels as for testosterone (i.e., 2–4%) (Marynick *et al.*, 1976). A study of CSF levels during the menstrual cycle of the monkey confirmed this (Fig. 9), but also confirmed that CSF levels did not fluctuate

F I G U R E 8. Relationship between serum LH and levels of DHT in (A) CSF, (B) total serum DHT, or (C) "free" serum DHT in castrated male rhesus monkeys given increasing amounts of DHT. Serum, but not CSF, DHT levels are reflected in the levels of LH. From Abbott *et al.* (1985).

in parallel with those in the blood (Anand Kumar *et al.*, 1980). It seems possible, therefore, that the two compartments are dissociated under some circumstances. However, another study in humans showed that, in common with testosterone, CSF estradiol correlated quite well with free levels of the hormone in the blood (Backstrom *et al.*, 1976). Progesterone is also found in the CSF and seems to be closely related to blood levels (Fig. 9) (Anand Kumar *et al.*, 1980). Though Marynick *et al.* (1976) were not able to find authentic progesterone in the CSF of female monkeys during a short-term intravenous infusion of labeled hormone, others found about 10% of plasma levels by direct radioimmunoassay, and these correlated quite well with free progesterone in the blood (Backstrom *et al.*, 1976). There is no good evidence on the clearance of progesterone from the CSF, but one of its principal metabolites (17-OH-progesterone) may be cleared slowly, since it remains elevated in the CSF after a bolus intravenous injection even after blood levels have fallen (David and Anand Kumar, 1974; Marynick *et al.*, 1980). It has been reported that the high blood levels characteristic of late pregnancy are not reflected in the CSF, for reasons unknown, but this has been held to explain the lack of evident central effects in women at this time (Lurie and Weiss, 1967). We have no information about the fluctuations in CSF estradiol or progesterone relative to short-term pulses in the blood; knowledge of CSF levels during the preovulatory estradiol surge would be particularly welcome.

V. Neuroendocrine Peptides

The use of the term *neuroendocrine* peptide has already been described (see Section I). An essential part of this category is that its members themselves act as hormones, either by regulating the pituitary or as peripheral hormones in their own right. A method of classifying these peptides is proposed which relies on their anatomical distribution and on the way they are represented in the vascular and cerebral compartments. It will emerge that there may be four major divisions into which they fall, which will each be represented by an archetypical example, namely: (1) LHRH, (2) oxytocin (vasopressin), (3) β-endorphin, and (4) prolactin. There are clear indications that the difference between the partitioning of these peptides has functional as well as systematic significance. It is important to note that the discussion which follows ignores instances where a peptide can be found in peripheral structures. Here we are concerned with the distribution of the neuropeptides between the cerebral and vascular compartments.

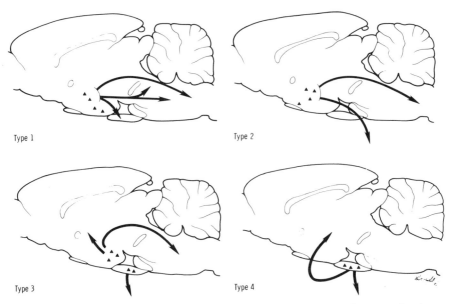

Type 1

Type 2

Type 3

Type 4

F I G U R E 10. Diagrammatic representation of the four types of neuroendocrine peptides. The sites (either neurons or endocrine cells) producing the peptide are indicated by ▲▲▲, and the pathways, either neural or vascular, by which the peptide reaches its target are shown by arrows. See text for full explanation.

A. Type 1 Peptides: LHRH

Peptides like LHRH originate in cell bodies lying in well-defined clusters in the basal forebrain (Fig. 10). These neurons project, in general, to two sites: to various parts of the limbic system and associated parts of the brain stem, and to the external layer of the median eminence, where they are discharged into the portal system and reach the pituitary. A wide variety of peptides acting on the anterior pituitary fall into this group, including not only luteinizing hormone-releasing hormone (LHRH), but also corticotropin-releasing factor (CRF), thyrotropin-releasing factor (TRF), and growth hormone-releasing factor (GRF). It should be noted that vasopressin (AVP), which is considered more closely as a type 2 peptide, has some of the features of a type 1 peptide, a point to be taken up again later.

There now seems to be general agreement that the cell bodies of LHRH neurons are found in the rostral hypothalamus and preoptic area (Barry *et al.*, 1973; Naik, 1976; Jennes and Stumpf, 1980), though some authors report them in more posterior regions of the ventrobasal hypothalamus as well (e.g., Silverman *et al.*, 1982). They seem not be restricted to any anatomically recognized cell groups, though the medial preoptic area is a major site at which they are found; but LHRH neurons are also seen in the septum and diagonal band (Silverman, 1976; King et al., 1982). Since we are here concerned with their relation to the pituitary, an important point is whether LHRH neurons are to be found in the arcuate nucleus, well established as a site of steroid feedback regulating the pituitary. Earlier studies affirmed their presence (Silverman, 1976), but later ones, probably using better antibodies to demonstrate them immunohistochemically, deny it (Merchenthaler et al., 1980). LHRH, like other neuroendocrine peptides, derives from a larger molecular weight precursor (Seeburg and Adelman, 1984). Recent studies suggest that this precursor, rather than the processed peptide, may be present in arcuate cell bodies, and can be demonstrated there by appropriate antibodies (Rubin et al., 1987). This may help resolve these disagreements in the literature. Like other type 1 peptides, LHRH neurons project to the limbic system (see references and above) and to the hypothalamus, septum, and amygdala (and, in the case of LHRH, to the olfactory bulbs). A second projection reaches the brainstem, either by passing through the hypothalamus close to the third ventricle or with the fibers of the medial forebrain bundle, or by traveling through the stria medullaris, the habenulae, and the fasciculus retroflexus. Part of the first pathway terminates in the median eminence and hence forms the "peripheral" component of this system (Hoffman and Gibbs, 1982). The route of the second is similar to that taken by oxytocin-containing fibers (see below), here classified as type 2 neuroendocrine peptides.

It is important to know whether LHRH (and peptides like it) is or can be released simultaneously both centrally and into the portal vessels. The effects upon the pituitary of these peptides are well known and do not need documentation here. A more contentious issue is whether these peptides have a "central" function which can be in any way related to their peripheral one. In the case of LHRH, which has such a potent action on the reproductive organs as its peripheral function, the idea that the same peptide might play a role in regulating sexual behavior is an attractive one. A number of studies show that injecting LHRH into the cerebral ventricles or the midbrain of the female rat facilitates the stimulating action of estrogen on the reflexive display of lordosis, the posture taken by an estrous female in response to a mounting attempt by the male (Beyer et al., 1982; Tennent et al., 1982; see also Schwartz-Gibblin, McEwen and Pfaff, this volume, Chapter 2). Conversely, antibodies or antagonists to LHRH reduce lordosis (Sirinathsinghji et al.,

1983). These findings give grounds for postulating a coordinated role for LHRH in timing both fertility and sexual behavior; in many animals, both are strictly limited and occur together during the fertile period. Other type 1 peptides may also be postulated to have similar coordinated roles: for example, CRF may induce behavioral "activation" (Sutton *et al.*, 1982) and behavioral responses that resemble those observed to occur during stress (Thatcher-Britton and Koob, 1986; Berridge and Dunn, 1986), which would fit its peripheral role. GRF has been reported to induce fearlike responses (Tannenbaum, 1984), and the secretion of growth hormone in such situations is well known. In all these cases (including that of LHRH) the central functions of these peptides is still uncertain, and their exact behavioral role (if any) is still undefined.

A common theme in this chapter is concern with partitioning of neuroendocrine peptides and steroids between peripheral and central compartments. In the case of LHRH and other type 1 peptides, the distribution of the peptide within the central compartment seems to be limited (like any other neural system) by the distribution of its intracerebral fiber system. Potential interaction between central and peripheral compartments is prevented by the manner by which the latter is limited to the portal vessels (Fig. 10). Not only does this ensure that maximum amounts reach the cells of the anterior pituitary (the usual explanation for the existence of this system), but amounts of peptides reaching the peripheral bloodstream from this source may be too small to have any effect upon the central compartment, even if the peptide were able to cross the blood–brain barrier in effective quantities (this may be equally significant). A number of peptides that are concerned with anterior pituitary function, and which are released into the portal vessels, have also been detected in the blood. However, in most cases, no correlation has been found between the values for a given peptide in the two compartments. Thus, the diurnal rhythms of both corticotropin-releasing factor (CRF) and somatostatin in blood and CSF are different, suggesting that the sources for the peptides in the two compartments are also different (Barreca *et al.* 1988; Kalin *et al.* 1987; Garrick *et al.* 1987). In some cases, the peripheral peptide may derive from peripheral tissue; in others, it may be that the central neural system addressing the pituitary may be distinct from that distributed to other parts of the brain. Thus it is the anatomical arrangement of the peripheral component of these peptides which is the major factor in determining their distribution between blood and brain after release, and this also distinguishes them from type 2 peptides, now to be considered.

B. Type 2 Peptides: Oxytocin and Vasopressin

Type 2 peptides are closely related to type 1 insofar as their distribution within the brain is similar (particularly the terminal areas to which they project). However, their peripherally projecting axons are not anatomically delimited by discharging directly into the portal system and hence to a local action on the anterior pituitary. Instead, they are released into the peripheral vascular system in sufficient concentrations to act as systemic hormones. They therefore gain access to the blood–brain barrier and the potential of passing into the cerebral compartment in physiologically important amounts. The crucial distinction between type 1 and 2 peptides is, therefore, the nature of their peripheral organization (Fig. 10).

Oxytocin and vasopressin neurons lie in the anterior hypothalamus, though scattered cells have been described at other sites (van Leeuwen and Caffe, 1983). Unlike LHRH and other type 1 peptides, the majority of these neurons lie within anatomically discrete nuclei: the supraoptic, paraventricular and suprachiasmatic areas (Vandesande and Dierickx, 1975; Sofroniew et al., 1979), though this is not relevant to their classification. The parvicellular oxytocin and vasopressin neurons project to many of the intracerebral areas also receiving type 1 terminals, though a more prominent descending projection to the brainstem autonomic nuclei and spinal cord seems to exist (Sofroniew and Schrell, 1981; Sawchenko and Swanson, 1982). Vasopressin-containing cells also project to the external layer of the median eminence, and thus resemble type 1 peptides. It is suggested that there are distinct populations of neurons (in the paraventricular nuclei at least) projecting to the posterior pituitary, the median eminence, and the brainstem and spinal cord (Swanson and Kuypers, 1980; Swanson et al., 1980). There may, therefore, be a subcomponent of the vasopressin system which is type 1, rather than type 2. This will only have functional significance if it can be shown that this subsystem can operate independently of the rest (as, for instance, in certain circumstances controlling ACTH release).

As for LHRH, there is some evidence, for a related central peripheral action of oxytocin. In the female, peripheral oxytocin plays a well-known role in milk ejection, and a possible one in paturition, both processes closely associated with the display of maternal behavior. Oxytocin injected into the ventricles has been claimed to induce maternal behavior in nulliparous rats (Pederson and Prange, 1979), which is difficult to achieve in any other way (Rosenblatt et al., 1979), though other peptides can have similar, if lesser, effects (Pederson et al., 1982); but these findings have not always been repli-

cated. Support for this behavioral action of oxytocin comes from ovariectomized ewes, in which icv infusions of oxytocin were shown to potentiate
maternal responses, but only if they were also treated with estrogen (Kendrick *et al.*, 1987). There is a growing list of contexts in which the behavioral
effects of peptides depend on coincident steroid action.

The literature on the central effects of vasopressin is abundant and
contentious. There are numerous reports of its effect on memory (de Weid,
1977), though many of these have been criticized on technical grounds (Gash
and Thomas, 1983). Since these behavioral changes have been reported to
follow parenteral injection as well as direct infusion into the brain (de Wied,
1977), they are not evidence for a compartmentalized coordinated action by
the same peptide. However, it has also been claimed that many of these
effects are due to "arousal" (Sahgal *et al.*, 1982). Arousal is a generalized
behavioral response, and one which is difficult to define, so that a plausible
behavioral role for vasopressin which can be directly linked to its peripheral
effects remains elusive. A more obvious link is between its central autonomic
projections and its peripheral vasoactive actions (which form part of the
somatic component of arousal). However, the principal concern here is to
establish if coordination could occur across the blood–brain barrier, bearing
in mind the relatively high concentrations of these peptides in the blood.

The evidence seems unequivocal. Both peptides are present in the CSF
in amounts which rival those in the blood. But infusing either into the blood
has no effect on levels in the CSF (Jones and Robinson, 1982; Wang *et al.*,
1982). There seems to be a highly effective mechanism which prevents
blood-borne type 2 peptides from gaining access to the cerebral compartment
and hence to the receptors addressed by the same peptides present in anatomically discrete neural systems (Fig. 10). Any coordination between the activities or the same peptides in the two compartments, therefore, will have to
come from synchronization of their neurons of origin rather than from cross-
talk between the two compartments themselves. Levels in the two compartments seem to be regulated separately, since homeostatic challenges, such as
sodium infusions (icv) or hemorrhage, increase plasma AVP leaving that in
the CSF unchanged (Ota *et al.* 1988; Eriksson *et al.* 1987).

C. Type 3 Peptides: β-Endorphin

Peptides of this group, which includes the proopiomelanocortin
(POMC) family, have an anatomical arrangement which is very different
from types 1 and 2. The central and peripheral systems have separate cells of
origin, each forming the peptide but not connecting anatomically with each

other. Thus, β-endorphin neurons are found in the hypothalamic arcuate nucleus (Sofroniew, 1979), from which projections run to limbic, diencephalic, and midbrain regions (Finley et al., 1981). However, there is no peripheral projection from this source; instead, the same peptide is found in cells of the anterior pituitary, from which it is released into the circulation (Guillemin et al., 1977). In common with type 2 peptides, therefore, there is both a central and a peripheral component (Fig. 10); unlike type 2, these come from separate sources and might, therefore, act independently of each other. Although in the discussion which follows attention is fixed on β-endorphin, it is clear that this peptide is part of a family which includes ACTH, melanocyte-stimulating hormone (MSH), and β-lipotropin (LPH) (Eipper and Mains, 1980), and it is therefore relevant to these peptides as well.

Immunohistochemical techniques have successfully separated this system from those containing the other two opiates presently known, enkephalin and dynorphin (Dupont et al., 1980). Unlike the others, β-endorphin neurons are mostly confined to the arcuate nucleus and the surrounding area, though there have been suggestions that magnocellular neurons may also contain it, but these have not been confirmed by appropriate absorption techniques (see Everitt et al., 1983). From the arcuate nucleus, fibers run to the areas already listed as well as to the ventral striatum (nucleus accumbens) (Finley et al., 1981), particularly noteworthy in view of this structure's close anatomical relation to the limbic system (Kelley et al., 1982) and its postulated role in behavior (Koob et al., 1978). Otherwise projections from β-endorphin-containing neurons resemble quite closely those from type 1 (LHRH) and type 2 (oxytocin) systems (Sofroniew, 1979). One exception is the median eminence; some authors report high levels of β-endorphin here also (Dupont et al., 1980), but others much less (Wilkes et al., 1980). There seems to be general agreement that β-endorphin has little direct effect on the cells of the pituitary (Wardlaw et al., 1980), so that if it is released in physiologically important amounts in the median eminence, it presumably acts on neighboring nerve terminals rather than on the gland itself (unlike LHRH). Because the external layer of the median eminence is outside the blood–brain barrier, the interactions here between peripherally derived and central β-endorphin are interesting, since this part of the brain is accessible to both.

Despite attempts to do so, no coherent account of the function of blood-borne β-endorphin has so far been provided, though it can release prolactin from the pituitary itself, not by a direct action, but by one on the external layer of the median eminence (Gudelsky and Porter, 1979; Catlin et *al.*, 1980). This demonstrates that peripherally derived β-endorphin can access this part of the central neuroendocrine system. But can peripheral β-endorphin reach other parts of the CNS and thus cause correlated behavioral effects?

Specific behavioral effects in response to intracerebrally administered β-endorphin need to be separated from generalized opiate effects. If β-endorphin is given into the cerebral ventricles, especially in high concentrations, it will act on receptors to which it might not normally gain access and thus cause a mixed syndrome. The position of the neurons making β-endorphin suggests that it may be concerned with reproductive events. This belief is encouraged by observations of LH discharge after opiate blockade (Cicero *et al.*, 1979; van Vugt and Meites, 1980) and of inhibited sexual behavior following intraventricular β-endorphin (Meyerson and Terenius, 1977), or direct infusion into the pre-optic area of the hypothalamus (Hughes *et al.*, 1987, 1988). States of reproductive inhibition, such as those occurring during pregnancy in rats, during the nonbreeding season in hamsters, or in the subordinate males in a group of monkeys, are associated with elevated β-endorphin levels in the hypothalamus or CSF, respectively (Fig. 11) (Wardlaw and Frantz, 1983; Roberts *et al.*, 1985; Martensz *et al.*, 1986), which suggests

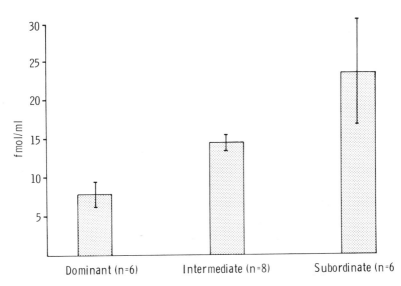

F I G U R E 11.   Levels of β-endorphin in the CSF of male talapoin monkeys of different social rank. Drawn from data in Martensz *et al.* (1986).

that a variety of environmental events may be able to act on β-endorphin neurons to suppress reproduction.

The weight of evidence, though mixed, is against peripheral β-endorphin being able to enter the CSF. Giving β-endorphin systemically is generally ineffective on central neural function (Foley *et al.*, 1979), and elevated blood levels in humans are not associated with raised CSF levels or pain thresholds (Nakao *et al.*, 1980; Mohs *et al.*, 1982). However, there are studies which have shown passage of the peptide into the CSF (e.g., Pezalla *et al.*, 1978), so this interpretation is not unanimous. It should be noted that small modifications to its molecule, making it more lipophilic, can greatly enhance the transfer of β-endorphin into the CSF (Kastin *et al.*, 1980). There is no evidence for coordinated central and peripheral activation of β-endorphin-containing systems as yet; the existence of an effective barrier between them may ensure that their operations remain secluded from each other (Barna *et al.*, 1988).

A second member of this family of peptides, ACTH, is also excluded from the CSF. Studies in monkeys have shown that ACTH (1-24) given systemically is not detected in the CSF, though serum levels increased four-fold. The serum levels of ACTH in the untreated, adrenalectomized monkey reach about 10 times their normal values; yet ACTH in the CSF does not increase (Fig. 12) (Beckford *et al.*, 1985). Clinical results also suggest that CSF and blood ACTH levels are dissociated, and that elevated peripheral levels do not cause those in the CSF to increase (Daly *et al.*, 1974; Hoffman *et al.*, 1974). Unlike β-endorphin, there may be no authentic ACTH (1-39) in the CSF (Smith *et al.*, 1982). There thus seems to be a highly effective means of preventing this member of the POMC family from reaching the brain's ECF, despite claims that ACTH can reach the brain by retrograde transport up the portal vessels (Bergland *et al.*, 1980). Passage into the CSF of some other pituitary peptides is, however, very different, as we shall see. There are a number of other peptides whose status as "neuroendocrine" seems plausible, though they are not usually considered releasing factors in the classical sense. Angiotensin II and atrial natriuretic peptide (ANP) are two; both are found in the brain as well as in the blood, and both can release hormones from the anterior pituitary after icv injection. Both, however, are formed in considerable amounts from peripheral sources (i.e., the blood or the heart, respectively). In this they rather resemble the more conventional type 3 peptides described above. It also seems that neither peptide can cross the blood–brain barrier (Doczi *et al.*, 1988; Masuda *et al.* 1988; Levin *et al.* 1987), another feature comparable to type 3 peptides.

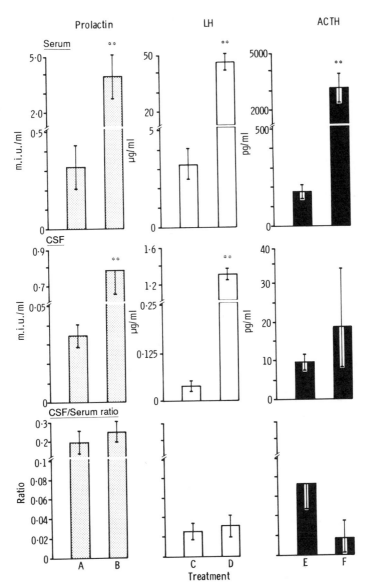

F I G U R E 12. CSF levels in male rhesus monkeys of three anterior pituitary peptides before and following chronic elevation of serum levels of the same peptides. A, B: Serum prolactin elevated by giving daily sulpiride injections. C, D: Serum LH elevated following castration. E, F: Serum ACTH elevated by daily injections of ACTH (1–24). CSF levels of the first two peptides (prolactin and LH) respond to changes in the serum, but that of ACTH does not. From Dubey *et al.* (1983).

D. Type 4 Peptides: Prolactin

Type 4 peptides are those that are released from the anterior pituitary, but for which the presence of an intracranial component seems open to question. The peptides of this group, which includes prolactin, LH, and growth hormone, have appreciably larger molecular weights than those of the others already considered. However, they differ from the latter in that it appears they can enter the CSF from the blood and thus act as peripherally derived neuropeptides. There are thus two criteria, one dependent on the anatomical position of their cells of origin, the other on their passage into the CSF, which distinguishes this group from the peptides of type 3 (Fig. 10).

It must be acknowledged that there is evidence that at least two of this group (prolactin and LH) can be found within neurons of the CNS. Immunohistochemistry and (in the case of prolactin) *in situ* hydridization are said to identify a population of cells in the mediobasal hypothalamus containing these peptides (Emmanuele *et al.*, 1979; Toubeau *et al.*, 1979). There is, of course, no reason why any neuron should not form these (or any other) peptides, though the technical problems of identifying with certainty large peptides using these methods should not be underestimated. Should more critical evidence substantiate these views, then the distinction between this group and those of type 3 will lie in their relative accessibility to the CSF from the blood. A further problem will also be generated—explaining how the intracerebral receptors distinguish peptide derived by local release from that arriving by way of the extracerebral compartment. One possibility is that these receptors, whose characters are hardly known, only respond to high concentrations and thus to blood levels at the extreme end of the physiological range. But these speculations remain in abeyance until the existence of intraneuronal systems containing type 4 peptides is established or disproved.

Prolactin, somewhat surprisingly for a peptide of its size, enters the brain relatively easily. Levels in the CSF are about 10–20% of those in the blood and rise in concert with those in the vascular compartment during hyperprolactinemia (Fig. 12) (Dubey *et al.*, 1983). Infusing heterologous prolactin intravenously shows that the hormone passes into the CSF, but chromatographic separation demonstrates that this is limited to monomeric "little" prolactin (Martensz and Herbert, 1982). Larger-molecular-weight forms ("big" prolactin) are excluded. The same experiments exclude retrograde flow through the pituitary portal veins as a route of privileged access to the brain, since prolactin given intravenously enters the CSF as easily as that from the animal's own pituitary. Prolactin derived from an intravenous injection would not have access to the so-called privileged route via the portal system.

The exclusion of larger forms of prolactin from the CSF suggests that its entry is not entirely passive, which more detailed study of the dynamics of this process seems to confirm (Martensz and Herbert, 1982). Repeated sampling of the CSF during continuous infusion of prolactin shows that levels in the CSF rise only slowly. Furthermore, a bolus of prolactin given intravenously also causes a progressive rise in CSF levels which may continue even while those in the blood are falling. Though it remains to be proven, some selective mechanism may be operating on the uptake of prolactin into the CSF (which may also apply to other peptide hormones showing similar uptake characteristics, such as growth hormone, LH, and chorionic peptides) (Belchetz et al., 1982; Assies et al., 1978).

There is evidence for direct actions of prolactin on the brain. Infusing prolactin into the cerebral ventricles has shown, both in rats and in monkeys, that it can suppress its own secretion from the anterior pituitary, probably by acting on the dopamine-containing neurons in the mediobasal hypothalamus (Nicholson et al., 1980; Herbert and Martensz, 1983). The terminals (but not the cell bodies) of these neurons (which lie in the median eminence) would be accessible to prolactin from the blood without the need for it to pass into the CSF. However, high levels of prolactin can diminish sexual activity in the males of several species (Svare et al., 1979; Bailey and Herbert, 1982; Franks et al., 1978), and this effect is not relieved by giving them testosterone, estradiol, or DHT, though prolactin may impair the metabolism of testosterone in parts of the limbic system (Bailey et al., 1984). Heightened prolactin levels in the blood may thus be one of the mechanisms controlling sexual responsiveness, and these levels are sensitive to changes in both an animal's physical and social environment.

The evidence suggests that both growth hormone and LH (Fig. 12) also reach the CSF from the blood, though in amounts which are proportionately less than in the case of prolactin (Belchetz et al., 1982; Dubey et al., 1983). So far, no behavioral sequelae seem to have been reported for either peptide in the CSF. Thus, while these members of type 4 share the common property of access to the CSF from the anterior pituitary, there are differences between them which may prove to be functionally important.

VI. Conclusions

It has been necessary to consider the steroids and the neuroendocrine peptides separately, and even to discuss the latter as a series of subtypes, to try to propose a scheme which attempts to draw attention to (1) the functional importance of compartmentalization in neuroendocrinology, and (2) how this may suggest one way of classifying the hormones and peptides concerned. But, to repeat the truism stated at the start of this chapter, hormones (or hormonal systems) do not operate in isolation, and to leave the discussion as if they did would be also to repeat the dangers of a fragmented view suggested in the same paragraphs. A complete consideration of this topic is well beyond the confines of this chapter; all that is possible is the briefest account of some of the possible implications of the way that the subjects of this chapter interact with each other in the context of their partitioning between blood and CSF.

We are beginning to become aware of surprising differences in the partitioning of the steroids. To take those considered here, the changes in the CSF resulting from those in the blood are distinct for testosterone, DHT, and cortisol. Whereas fluctuations in serum testosterone seem to be mirrored fairly faithfully in the CSF, the fluid around the brain seems hardly aware of those in DHT. We badly need more information about the principal ovarian steroids, estradiol and progesterone; they may resemble testosterone rather than DHT, though we cannot be sure. Cortisol has such a prolonged half-life in the CSF that surges in the blood are likely to result in elevations in the CSF greatly outlasting those in the plasma. Since alterations in its pulsatile release are typical of "stress," the precise effect on the brain's extracellular environment is important. The major point, however, is that we cannot assume the nature of the relationship between the blood and the CSF for any steroid from the parameters of any other; in particular, simply measuring their lipid solubility, free fraction, and entry into the CSF, while yielding essential information, is not enough. We also need to know much more about the effects on brain function of changing relative concentrations of steroid in the CSF over time, in such a way as to mimic those occurring in response to physiological fluctuations in the blood.

There are clearly effective mechanisms to keep separate the cerebral and vascular effects of the majority of the neuroendocrine peptides (except those of type 4), though the ways of achieving this differ, as described in this chapter. In some cases, peptides which have a neuroendocrine function (e.g., somatostatin, CRF) are distributed widely outside the hypothalamus; it should be made clear that this discussion refers only to the component that is implicated in the control of the pituitary. There is no reason to suppose that the same peptide located elsewhere (e.g., in the dorsal roots) has any common function. There is also no reason to think that the principles applying to more restricted peptides (e.g., LHRH) should not apply to the neuroendocrine systems containing these more widely distributed ones. Some peptides, as has already been pointed out, are represented in more than one category, for example, vasopressin. The question here is whether this is a case of two separate components of the neuroendocrine system which, for obscure reasons, happen to use the same peptide, or whether there is a functional relation to be discerned in such an arrangement. In the case of vasopressin, both systemic hormone and that releasing ACTH would be released after, say, hemorrhage.

Coexistence of two or more peptides in the same terminal, now generally recognized in the CNS, complicates but does not destroy the scheme proposed here. If the constraints are anatomical (e.g., both peptides are released together into the portal vessels) then both will behave in the same way, since their partitioning is determined by their site of release. If they were to enter the bloodstream together, then each would interact with the intracerebral compartment according to its passage across the blood–brain barrier and clearance from the brain. In this case, there might be an interesting divorce between their representation in the CSF.

Acknowledgments

I thank my colleagues, whose work is here described, for their collaboration, stimulation, and forbearance. The work of my laboratory is supported by a program grant from the Medical Research Council, by project grants from the MRC and AFRC, and by visiting fellowships from the Wellcome Trust.

This chapter is dedicated to the memory of Josef Meller.

References

Abbott, D. A., Batty, K. A., Dubey, A. K., Herbert, J., and Shiers, H. M. (1985). The passage of 5α-dihydrotestosterone from serum into cerebrospinal fluid and LH negative feedback in castrated rhesus monkeys. *J. Endocrinol.* **104**, 325–330.

Anand Kumar, T. C., and Thomas, G. H. (1968). Metabolites of 3-H-oestradiol-17β in the cerebrospinal fluid of the rhesus monkey. *Nature (London)* **219**, 628.

Anand Kumar, T. C., David, G. F. X., and Puri, V. (1980). Levels of estradiol and progesterone in the cerebrospinal fluid of the rhesus monkey during the menstrual cycle. *J. Med. Primatol.* **9**, 222–232.

Assies, J., Schellekens, A. P. M., and Touber, J. L. (1978). Protein hormones in cerebrospinal fluid: Evidence for retrograde transport of prolactin from the pituitary to the brain in man. *Clin. Endocrinol. (Oxford)* **8**, 487–491.

Backstrom, T., Carstensen, H., and Sodergard, R. (1976). Concentration of estradiol, testosterone and progesterone in cerebrospinal fluid compared to plasma unbound and total concentrations. *J. Steroid Biochem.* **7**, 469–472.

Bailey, D. J., and Herbert, J. (1982). Impaired copulatory behavior of male rats with hyperprolactinaemia induced by domperidone or pituitary grafts. *Neuroendocrinology* **35**, 186–193.

Bailey, D. J., Dolan, A. L., Pharoah, P. D. P., and Herbert, J. (1984). Role of gonadal and adrenal steroids in the impairment of the male rat's sexual behavior by hyperprolactinaemia. *Neuroendocrinology* **39**, 555–562.

Baird, D., Horton, R., Longcope, C., and Tait, J. F. (1968). Steroid pre-hormones. *Perspect. Biol. Med.* **11**, 348–421.

Barna, I., Sweep, C. G., Veldhuis, H. D., and Wiegant, V. M. (1988). Differential effects of cisterna magna cannulation on beta-endorphin levels in rat plasma and cerebrospinal fluid. *Acta Endocrinol. Kbh.* **117**, 517–524.

Barreca, T., Francheschini, R., Siani, C., Perria, C., Francaviglia, N., Cataldi, A., and Rolandi, E. (1988). Diurnal changes of plasma and cerebrospinal fluid somatostatin and 24-h growth hormone secretory pattern in man. A study in hydrocephalic patients. *Acta Endocrinol. Kbh.* **117**, 130–134.

Barry, J., Dubois, M. P., and Poulain, P. (1973). LRF-producing cells of the mammalian hypothalamus: A fluorescent antibody study. *Z. Zellforsch. Mikrosk. Anat.* **148**, 351–366.

Baum, M. J., Tobet, S. A., Starr, M. S., and Bradshaw, W. G. (1982). Implantation of dihydrotestosterone into the lateral septum or medial amygdala facilitates copulation in castrated male rats given estradiol systemically. *Horm. Behav.* **16**, 208–223.

Beckford, U., Herbert, J., Jones, M. T., Martensz, N. D., Nicholson, S. A., Gillham, B., and Hamer, J. D. (1985). Relationship between adrenocorticotrophin bioactivity in blood and cerebrospinal fluid of rhesus monkeys. *J. Endocrinol.* **104**, 331–338.

Belchetz, P., Ridley, and Baker, H. F. (1982). Studies on the accessibility of prolactin and growth hormone to brain: Effect of opiate agonists on hormone levels in serial, simultaneous plasma and cerebrospinal fluid samples in the rhesus monkey. *Brain Res.* **239**, 310–314.

Bergland, R. M., Blume, H., Hamilton, A., Monica, P., and Paterson, R. (1980). Adrenocorticotropic hormone may be transported directly from the pituitary to the brain. *Science* **210**, 541–543.

Berridge, C. W., and Dunn, A. J. (1986). Corticotropin-releasing factor elicits naloxone sensitive stress-like alterations in exploratory behavior in mice. *Regul. Peptides* **16**, 83–93.

Beyer, C., Gomora, P., Canchola, E., and Sandorai, Y. (1982). Pharmacological evidence that LHRH action on lordosis behavior is modulated through a rise in c-AMP. *Horm. Behav.* **16**, 107–112.

Bradbury, M. (1979). "The Concept of a Blood-Brain Barrier." Wiley, Chichester.

Carroll, B. J., Curtis, G. C., and Mendels, J. (1976). Cerebrospinal fluid and plasma free cortisol concentrations in depression. *Psychol. Med.* **6,** 235–247.

Catlin, D. H., Poland, R. E., Gorelick, D. A., Gerner, R. H., Hui, K. K., Rubin, R. T., and Li, C. H. (1980). Intravenous infusion of β-endorphin increases serum prolactin, but not growth hormone or cortisol, in depressed subjects and withdrawing methadone addicts. *J. Clin. Endocrinol. Metab.* **50,** 1021–1025.

Cicero, T. J., Schainker, B. A., and Meyer, E. R. (1979). Endogenous opioids participate in the regulation of the hypothalamic-pituitary-luteinising hormone. *Endocrinology (Baltimore)* **104,** 1286–1291.

Daly, J. R., Fleischer, M. R., Chambers, D. J., Bitensky, L., and Chayen, J. (1974). Application for the cytochemical bioassay for corticotrophin to clinical and physiological studies in man. *Clin. Endocrinol. (Oxford)* **3,** 335–345.

David, G. F. X., and Anand Kumar, T. C. (1974). Transfer of steroidal hormones from blood to the cerebrospinal fluid in the rhesus monkey. *Neuroendocrinology* **14,** 114–120.

de Wied, D. (1977). Peptides and behavior. *Life Sci.* **20,** 195–204.

Doczi, T., Joo, F., Vecsernyes, M., and Bodosi, M. (1988). Increased concentration of atrial natriuretic factor in the cerebrospinal fluid of patients with aneurysmal subarachnoid hemorrhage and raised intracranial pressure. *Neurosurgery* **23,** 16–19.

Dubey, A. K., Herbert, J., Martensz, N. D., Beckford, U., and Jones, M. T. (1983). Differential penetration of three pituitary peptide hormones into the cerebrospinal fluid of rhesus monkeys. *Life Sci.* **32,** 1857–1863.

Dubey, A. K., Herbert, J., Abbott, D. A., and Martensz, N. D. (1984). Serum and CSF concentrations of testosterone and LH related to negative feedback in male rhesus monkeys. *Neuroendocrinology* **39,** 176–185.

Dupont, A., Lepine, J., Langelier, P., Merand, Y., Rouleau, D., Vaudry, H., Gros, C., and Barden, N. (1980). Differential distribution of β-endorphin and enkephalins in rat and bovine brain. *Regul. Pept.* **1,** 43–52.

Eipper, B. A., and Mains, R. E. (1980). Structure and biosynthesis of pro-adrenocorticotropin/endorphin and related peptides. *Endocr. Rev.* **1,** 1–27.

Emmanuele, N., Connick, E., Baker, G., Kirstens, L., and Lawrence, A. M. (1979). Hypothalamic luteinizing hormone: mediator of short-loop feedback. *Clin. Res.* **27,** 678A.

Eriksson, S., Simon-Oppermann, C., Simon, E., and Gray, D. A. (1987). Interaction of changes in the third ventricular CSF tonicity, central and systemic AVP concentrations and water intake. *Acta Physiol. Scand.* **130,** 575–583.

Everitt, B. J., Herbert, J., and Keverne, E. B. (1983). The neuroendocrine anatomy of the limbic system: a discussion with special reference to steroid responsive neurons, neuropeptides and monoaminergic systems. *In* "Progress in Anatomy" (V. Navaratnam and R. J. Harrison, eds.), Vol. 3, pp. 235–260. Cambridge Univ. Press, London and New York.

Feder, H. H. (1981). Experimental analysis of hormone actions on the hypothalamus, anterior pituitary, and ovary. *In* "Neuroendocrinology of Reproduction" (N. T. Adler, ed.), pp. 243–278. Plenum, New York.

Finley, J. C. W., Lindstrom, P., and Petrusz, P. (1981). Immunocytochemical localization of β-endorphin-containing neurons in the rat brain. *Neuroendocrinology* **33**, 28–42.

Foley, K. M., Kourides, I. A., Inturrisi, C. E., Kaiko, R. F., Zaroulis, C. G., Posner, J. B., Houde, R. W., and Li, C. H. (1979). β-endorphin: analgesic and hormonal effects in humans. *Proc. Natl. Acad. Sci. U.S.A.* **76**, 5377–5381.

Franks, S., Jacobs, H. S., Martin, N., and Nabarro, J. D. N. (1978). Hyperprolactinaemia and impotence. *Clin. Endocrinol. (Oxford)* **8**, 277–287.

Garrick, N. A., Hill, J. L., Szela, F. G., Tomai, T. P., Gold, P. W., and Murphy, D. L. (1987). Corticotropin-releasing factor: a marked circadian rhythm in primate cerebrospinal fluid peaks in the evening and is inversely related to the cortisol circadian rhythm. *Endocrinology* **121**, 1329–1334.

Gash, D. M., and Thomas, G. J. (1983). What is the importance of vasopressin in memory processes? *Trends Neurosci.* **6**, 197–198.

Gudelsky, G. A., and Porter, J. C. (1979). Morphine and opioid induced inhibition of the release of dopamine from tuberoinfundibular neurons. *Life Sci.* **25**, 1697–1699.

Guillemin, R., Vargo, T., Rossier, J., Minick, S., Ling, N., Rivier, C., Vale, W. O., and Bloom, F. E. (1977). β-endorphin and adrenocorticotropin are secreted concomitantly by the pituitary gland. *Science* **197**, 1367–1369.

Hastings, M. H., and Herbert, J. (1986). Endocrine rhythms. *In* "Neuroendocrinology" (S. L. Lightman and B. J. Everitt, eds.). Blackwell, Oxford.

Herbert, J., and Martensz, N. D. (1983). The effects of intraventricular prolactin infusions on pituitary responsiveness to thyrotropin-releasing hormone, 5-hydroxytryptophan or morphine in rhesus monkeys. *Brain Res.* **258**, 251–262.

Herbert, J., Keverne, E. B., and Yodyingyuad, U. (1985). Modulation by social status of the relationship between CSF and serum cortisol levels in male talapoin monkeys. *Neuroendocrinology* **42**, 436–442.

Hoffman, G. E., and Gibbs (1982). LHRH pathways in the rat brain: 'Deafferentiation' spares a sub-chiasmatic LHRH projection to the median eminence. *Neuroscience* **7**, 1979–1993.

Hoffman, J. E., Baumgartner, J., and Gold, E. M. (1974). Dissociation of plasma and spinal fluid ACTH in Nelson's syndrome. *JAMA, J. Am. Med. Assoc.* **228**, 491–492.

Hughes, A. M., Everitt, B. J., and Herbert, J. (1987). Selective effects of β-endorphin infused into the hypothalamus, preoptic area and bed nucleus of the Stria terminalis on sexual and ingestive behaviour of male rats. *Neuroscience* **23**, 1063–1073.

Hughes, A. M., Everitt, B. J., and Herbert, J. (1988). The effects of simultaneous or separate infusions of some pro-opiomelanocortin-derived peptides (β-endorphin, melanocyte stimulating hormone, and corticotropin-like intermediate polypeptide) and their acetylated derivatives upon sexual and ingestive behavior of male rats. *Neuroscience* **27**, 689–698.

Jennes, L., and Stumpf, W. E. (1980). LHRH-systems in the brain of the golden hamster. *Cell Tissue Res.* **209**, 239–256.

Jones, P. M., and Robinson, I. C. A. F. (1982). Differential clearance of neurophysin and neurohypophysial peptides from the cerebrospinal fluid in conscious gurons in the paraventricular nucleus of the hypothalamus that project to the medulla or to the spinal cord in the rat. *J. Comp. Neurol.* **205**, 260–272.

Kalin, N. H., Shelton, S. E., Barksdale, C. M., and Brownfield, M. S. (1987). A diurnal rhythm in cerebrospinal fluid corticotropin-releasing hormone different from the rhythm of pituitary–adrenal activity. *Brain Res.* **426**, 385–391.

Karsch, F. J., Weick, R. F., Butler, W. R., Dierscke, D. J., Krey, L. C., Weiss, G. T., Hotchkiss, J., Yamaji, Y., and Knobil, E. (1973). Induced LH surges in the rhesus monkey: strength-duration characteristics of the estrogen stimulus. *Endocrinology (Baltimore)* **92**, 1740–1747.

Kastin, A. J., Jemison, M. T., and Coy, D. H. (1980). Analgesia after peripheral administration of enkephalin and endorphin analogues. *Pharmacol., Biochem. Behav.* **11**, 713–716.

Kelley, A. E., Domesick, V. B., and Nauta, W. J. H. (1982). The amygdalostriatal projection in the rat—an anatomical study by anterograde and retrograde tracing methods. *Neuroscience* **7**, 615–630.

Kendrick, K. M., Keverne, E. B., and Baldwin, B. A. (1987). Intracerebroventricular oxytocin stimulates maternal behaviour in the sheep. *Neuroendocrinology* **46**, 56–61.

King, J. C., Tobet, S. A., Snavely, F. L., and Arimura, A. A. (1982). LHRH immunopositive cells and their projections to the median eminence and organum vasculosum of the lamina terminalis. *J. Comp. Neurol.* **209**, 287–300.

Koob, G. F., Riley, S. J., Smith, S. C., and Robbins, T. W. (1978). Effects of 6-hydroxydopamine lesions of the nucleus accumbens septi and olfactory tubercle on feeding, locomotor activity, and amphetamine anorexia in the rat. *J. Comp. Physiol. Psychol.* **92**, 917–927.

Levin, E. R., Frank, H. J., Weber, M. A., Ismail, M., and Mills, S. (1987). Studies of the penetration of the blood–brain barrier by atrial natriuretic peptide. *Biochem Biophys. Res. Commun.* **147**, 1226–1231.

Lisk, R. D. (1978). The regulation of sexual "heat." *In* "Biological Determinants of Sexual Behavior" (J. B. Hutchinson, ed.), pp. 425–466. Wiley, New York.

Lurie, A. O., and Weiss, J. B. (1967). Progesterone in cerebrospinal fluid during human pregnancy. *Nature (London)* **215**, 1178.

Macdonald, P., Beyer, C., Newton, F., Brien, B., Baker, R., Tan, H. S., Sampson, C., Kitching, P., Greenhill, R., and Pritchard, D. (1970). Failure of 5a-dihydrotestosterone to initiate sexual behavior in the castrated male rat. *Nature (London)* **227**, 964–965.

McEwen, B. S. (1981). Cellular biochemistry of hormone action in brain and pituitary. *In* "Neuroendocrinology of Reproduction" (N. T. Adler, ed.), pp. 485–518. Plenum, New York.

Marrone, B. L., Rodriguez-Sierra, J. F., and Feder, H. H. (1979). Intrahypothalamic implants of progesterone inhibit lordosis behavior in ovariectomised, estrogen-treated rats. *Neuroendocrinology* **28**, 92–98.

Martensz, N. D., and Herbert, J. (1982). Relationship between prolactin in the serum and cerebrospinal fluid of ovariectomised female rhesus monkeys. *Neuroscience* **7**, 2801–2812.

Martensz, N. D., Herbert, J., and Stacey, P. M. (1983). Factors regulating levels of cortisol in cerebrospinal fluid of monkeys during acute and chronic hypercortisolemia. *Neuroendocrinology* **36**, 39–48.

Martensz, N. D., Vellucci, S. V., Keverne, E. B., and Herbert, J. (1986). β-endorphin levels in the CSF of male talapoin monkeys in social groups related to dominance status and the LH response to naloxone. *Neuroscience* **18**, 651–658.

Marynick, S. P., Havens, W. W., Ebert, M. H., and Loriaux, D. L. (1976). Studies on the transfer of steroid hormones across the blood-cerebrospinal fluid barrier in the rhesus monkey. *Endocrinology (Baltimore)* **99**, 400–405.

Marynick, S. P., Wood, J. H., and Loriaux, D. L. (1980). Cerebrospinal fluid steroid hormones. *In* "Neurobiology of Cerebrospinal Fluid" (J. H. Wood, ed.), pp. 605–611. Plenum, New York.

Masuda, T., Ando, K., and Marumo, F. (1988). The existence of low concentrations of atrial natriuretic peptide (ANP) in canine cerebrospinal fluid which does not correlate with plasma ANP levels. *Neurosci. Lett.* **88**, 93–99.

Merchenthaler, I., Kovacs, G., Lovasz, G., and Setalo, G. (1980). The preoptic-infundibular LHRH tract of the rat. *Brain Res.* **198**, 63–74.

Meyerson, B. J., and Terenius, L. (1977). β-endorphin and male sexual behavior. *Eur. J. Pharmacol.* **42**, 191–192.

Mezey, E., Reisine, T. D., Brownstein, M. J., Palkovits, M., and Axelrod, J. (1984). β-adrenergic mechanism of insulin-induced adrenocorticotrophin release from the anterior pituitary. *Science* **226**, 1085–1086.

Mogenson, G. J., Jones, D. L., and Yim (1980). From motivation to action: functional interface between the limbic system and the motor system. *Prog. Neurobiol.* **14**, 69–97.

Mohs, K. C., Davis, B. M., Rosenberg, G. S., Davis, K. L., and Krieger, D. T. (1982). Naloxone does not affect pain sensitivity, mood or cognition in patients with high levels of beta-endorphin in plasma. *Life Sci.* **30**, 1827–1833.

Naik, D. V. (1976). Immunohistochemical localization of LHRH neurons in the mammalian hypothalamus. *In* "Neuroendocrine Regulation of Fertility" (T. C. Anand Kumar, ed.), pp. 80–91. Karger, Basel.

Nakao, K., Nakai, Y., Oki, S., Matsubara, S., Konishi, T., Nishitani, H., and Imura,

H. (1980). Immunoreactive β-endorphin in human cerebrospinal fluid. *J. Clin. Endocrinol. Metab.* **50,** 230–233.

Nicholson, G., Greeley, G. H., Humm, J., Youngblood, W. W., and Kizer, J. S. (1980). Prolactin in cerebrospinal fluid: a probable site of prolactin autoregulation. *Brain Res.* **190,** 447–457.

Ota, K., Kimura, T., Matsui, K., Iitake, K., Shoji, M., Inoue, M., and Yoshinaga, K. (1988). Effects of hemorrhage on vasopressin and Met-Enk releases in blood and cerebrospinal fluid in dogs. *Am. J. Physiol.* **255,** R731–736.

Pardridge, W. M., and Meitus, C. J. (1979). Transport of steroid hormones through the blood-brain barrier. *J. Clin. Invest.* **64,** 145–154.

Pederson, C. A., and Prange, A. J. (1979). Induction of maternal behavior in virgin female rats after intracerebroventricular administration of oxytocin. *Proc. Natl. Acad. Sci. U.S.A.* **76,** 6661–6665.

Pederson, C. A., Ascher, J. A., Monroe, Y. L., and Prange, A. J. (1982). Oxytocin induces maternal behavior in virgin female rats. *Science* **216,** 648–649.

Perlow, M. J., Reppert, S. M., Boyar, R. M., and Klein, C. (1981). Daily rhythms in cortisol and melatonin in primate cerebrospinal fluid. *Neuroendocrinology* **32,** 193–196.

Peterson, N. A., and Chaikoff, I. L. (1963). Uptake of intravenously-injected [H-14-C] cortisol by adult rat brain. *J. Neurochem.* **10,** 17–23.

Pezalla, P. D., Lis, M., Seidah, N. G., and Chrétien, M. (1978). Lipotropin, malanotropin and endorphin: *in vivo* catabolism and entry into cerebrospinal fluid. *J. Can. Sci. Neurol.* **5,** 183–188.

Roberts, A. C., Hastings, M. H., Martensz, N. D., and Herbert, J. (1985). Changes in photoperiod alter the daily rhythms of pineal melatonin content and hypothalamic β-endorphin content and the LH response to naloxone in the male Syrian hamster. *Endocrinology (Baltimore)* **117,** 141–148.

Rosenblatt, J. S., Siegel, H. I., and Mayer, A. D. (1979). Progress in the study of maternal behavior in the rat: Hormonal, non-hormonal, sensory and developmental aspects. *Adv. Study Behav.* **19,** 225–311.

Sachar, E. J., Hellman, L., Roffwarg, H. P., Halpern, F. S., Fukushima, D. K., and Gallagher, T. F. (1973). Disrupted 24-hr patterns of cortisol secretion in psychiatric depression. *Arch. Gen. Psychiatry* **28,** 19–24.

Sahgal, A., Keith, R., Wright, C., and Edwardson, J. A. (1982). Failure of vasopressin to enhance memory in a passive avoidance task in rats. *Neurosci. Lett.* **28,** 87–92.

Sawchenko, P. E., and Swanson, L. W. (1982). Immunohistochemical identification of neurons in the paraventricular nucleus of the hypothalamus that project to the medulla or to the spinal cord in the rat. *J. Comp. Neurol.* **205,** 260–272.

Seeburg, P. H., and Adelman, J. P. (1984). Characterisation of cDNA for precursor of human luteinising hormone releasing hormone. *Nature (London)* **311,** 666–668.

Sheridan, P. J., and Weaker, F. J. (1982). Androgen receptor systems in the brain stem of the primate. *Brain Res.* **235,** 225–232.

Siiteri, P. K., Murai, J. T., Hammond, G. L., Nisker, H. A., Raymoure, W. J., and Kuhn, R. W. (1982). The serum transport of steroid hormones. *Recent Prog. Horm. Res.,* 457–504.

Silverman, A. J. (1976). Distribution of luteinizing hormone-releasing hormone (LHRH) in the guinea pig brain. *Endocrinology (Baltimore)* **99,** 30–41.

Silverman, A. J., Antunes, J. L., Abrams, G. M., Nilaver, G., Thau, R., Robinson, J. A., Ferin, M., and Krey, L. C. (1982). The luteinizing hormone-releasing hormone pathways in rhesus (*Macaca mulatta*) and pigtailed (*Macaca nemestrina*) monkeys: new observations on thick, unembedded secretions. *J. Comp. Neurol.* **211,** 309–317.

Sirinathsinghji, D. J. S., Whittington, P. E., Audsley, A., and Fraser, H. M. (1983). β-endorphin regulates lordosis in female rats by modulating LHRH release. *Nature (London)* **301,** 62–64.

Smith, A. I., Keith, A. B., Edwardson, J. A., Biggins, J. A., and McDermott, J. R. (1982). Characterization of corticotrophic-like immunoreactive peptides in rat brain using high performance liquid chromatography. *Neurosci. Lett.* **30,** 133–138.

Sofroniew, M. V. (1979). Immunoreactive β-endorphin and ACTH in the same neurons of the hypothalamic arcuate nucleus in the rat. *Am. J. Anat.* **154,** 283–289.

Sofroniew, M. V., and Schrell, U. (1981). Evidence for a direct projection from oxytocin and vasopressin neurons in the hypothalamic paraventricular nucleus to the medulla oblongata: immunohistochemical visualization of both the horseradish peroxidase transported and the peptide produced by the same neurons. *Neurosci. Lett.* **22,** 211–217.

Sofroniew, M. V., Weindl, A., Schinko, I., and Wetzstein, R. (1979). The distribution of vasopressin-, oxytocin-, and neurophysin-producing neurons in the guinea pig brain. *Cell Tissue Res.* **196,** 367–384.

Song, T., Nikolics, K., Serburg, P. H., and Goldsmith, P. C. (1987). GnRH-prohormone-containing neurons in the primate brain: immunostaining for GnRH-associated peptide. *Peptides* **8,** 335–346.

Sutton, R. E., Koob, G. F., Le Moal, M., Rivier, J., and Vale, W. (1982). Corticotropin releasing factor produces behavioral activation in rats. *Nature (London)* **297,** 331–333.

Svare, B., Bartke, A., Doherty, P., Mason, P., Michael, S. D., and Smith, M. S. (1979). Hyperprolactinaemia suppresses copulatory behavior in male rats and mice. *Biol. Reprod.* **21,** 529–535.

Swanson, L. W., and Kuypers, H. G. J. M. (1980). The paraventricular nucleus of the hypothalamus: cytoarchitectonic subdivisions and organization of projections to the pituitary, dorsal vagal complex, and spinal cord as demonstrated by retrograde fluorescence double-labelling methods. *J. Comp. Neurol.* **194,** 555–570.

Swanson, L. W., and Mogenson, C. J. (1981). Neural mechanisms for the functional coupling of autonomic, endocrine and somatomotor responses in adaptive behavior. *Brain Res. Rev.* **3,** 1–34.

Swanson, L. W., Sawchenko, P. E., Wiegand, S. J., and Price, J. L. (1980). Separate neurons in the paraventricular nucleus project to the median eminence and to the medulla or spinal cord. *Brain Res.* **198,** 190–195.

Tannenbaum, G. S. (1984). Growth hormone-releasing factor: Direct effects on growth hormone, glucose, and behavior *via* the brain. *Science* **226,** 464–466.

Tennent, B. J., Smith, E. R., and Dorsa, D. M. (1982). Comparison of some CNS effects of luteinizing-hormone releasing hormone and progesterone. *Horm. Behav.* **16,** 76–86.

Thatcher-Britton, K., and Koob, G. F. (1986). Alcohol reverses the proconflict effect of corticotropin-releasing factor. *Regul. Peptides* **16,** 315–320.

Thornton, J. E., and Abbott, D. A. (1981). Initiation and maintenance of gonadotrophin suppression with testosterone propionate in long-term castrated male rhesus. *Biol. Reprod.* **24,** Suppl. 1, 120A.

Toubeau, G., Desclin, J., Parmentier, M., and Pasteels, J. M. (1979). Compared localization of prolactin-like and ACTH immunoreactivity within the brain of the rat. *Neuroendocrinology* **29,** 384–394.

Uete, T., Nishimura, S., Ohya, H., Shimomura, T., and Tatebayashi, Y. (1970). Corticosteroid levels in blood and cerebrospinal fluid in various diseases. *J. Clin. Endocrinol. Metab.* **30,** 208–214.

Umberkoman-Wiita, B., Hansen, S., Herbert, J., and Moore, G. F. (1981). Circadian rhythms in serum and CSF cortisol of rhesus monkeys and their modulation by timed injections of l-5-hydroxytryptophan. *Brain Res.* **222,** 235–252.

Vandesande, F., and Dierickx, K. (1975). Identification of the vasopressin producing and of the oxytocin producing neurons in the hypothalamic neurosecretory system of the rat. *Cell Tissue Res.* **164,** 153–162.

van Leeuwen, F., and Caffe, R. (1983). Vasopressin-immunoreactive cell bodies in the bed nucleus of the stria terminalis of the rat. *Cell Tissue Res.* **228,** 525–534.

van Vugt, D. A., and Meites, J. (1980). Influence of endogenous opiates on anterior pituitary function. *Fed. Proc., Fed. Am. Soc. Exp. Biol.* **39,** 2533–2538.

Walker, M. D., Henkin, R. I., Harlan, A. B., and Caspar, A. G. T. (1971). Distribution of tritiated cortisol in blood, brain, CSF and other tissues of the cat. *Endocrinology (Baltimore)* **88,** 224–232.

Wang, B. C., Share, L., Crofton, J. T., and Kumura, T. (1982). Effects of intravenous and intracerebroventricular infusion of hypertonic solutions on plasma and cerebrospinal fluid vasopressin concentrations. *Neuroendocrinology* **34,** 212–221.

Wardlaw, S. L., and Frantz, A. G. (1983). Brain β-endorphin during pregnancy, parturition, and the post-partum period. *Endocrinology (Baltimore)* **113,** 1664–1668.

Wardlaw, S. L., Wehrenberg, W. B., Ferin, M., and Frantz, A. G. (1980). Failure of β-endorphin to stimulate prolactin release in the pituitary stalk-sectioned monkey. *Endocrinology (Baltimore)* **107,** 1663–1666.

Westphal, U. (1970). Binding of hormones to serum proteins. *In* "Biochemical Actions of Hormones" (G. Litwack, Ed.), Vol. 1, pp. 209–265. Academic Press, New York.

Wilkes, M. M., Watkins, W. B., Stewart, R. D., and Yen, S. S. C. (1980). Localization and quantitation of β-endorphin in human brain and pituitary. *Neuroendocrinology* **30,** 113–121.

CHAPTER TWO

Mechanisms of Female Reproductive Behavior

S. Schwartz-Giblin, B. S. McEwen, and
D. W. Pfaff

I. Introduction

The reproductive behavior of the female rat, the lordosis posture, occurs in the course of natural events during estrus as a reflex response to appropriate somatosensory stimulation by the male rat (Fig. 1). Lordosis will only occur in response to the male if the female has been exposed to sufficient levels of estradiol and progesterone for some time previously. In fact, the priming effects of estradiol require 18–24 hours of exposure before progesterone will exert its activational influence over a time delay of at least 1 hour. The lordosis behavior is the complex product of the stimulation provided by the male partner operating upon a hormone-primed neural substrate. Neural structures from the levels of the hypothalamus and midbrain to the spinal cord and peripheral somatosensory nerves are involved in the "circuitry" governing lordosis. We now know a great deal about the hormonal requirements for lordosis and about the receptors and brain sites through which estradiol and progesterone influences are produced. We also are beginning to understand the details of neuroanatomy underlying the lordosis "cir-

---

Note: This chapter was submitted in 1984 and revised in 1987.

F I G U R E I. A drawing from a filmed mating encounter showing the lordosis posture of the female rat immediately following the male mount.

cuitry" and the specific role of certain neuropeptides and neurotransmitters within this circuitry. This chapter attempts to provide a current picture of these various facets of the problem, beginning with a discussion of the hormone receptors and hormone actions at the hypothalamic level and then describing, in turn, the neural organization and functional interactions at the level of the brainstem and spinal cord.

II. Receptor-Mediated Functions in Hypothalamic and Preoptic Neurons

A. Estrogen and Progestin Receptors

Recent progress in understanding how and where steroid hormones alter brain function stems in large part from the detection and localization of receptors for the five major classes[1] of steroid hormones in the central nervous system. Estrogen receptors were the first to be identified, characterized, and mapped, whereas receptors for progesterone took much longer before yielding to efforts to study them.

With the introduction of tritiated steroids around 1960, intracellular estrogen receptors were identified and characterized in the uterus and other female reproductive tissues (Jensen and Jacobson, 1962). Studies of the brain soon followed. Like the initial work on the uterus, these studies first established that regions of the brain such as the hypothalamus concentrate and retain systemically administered [^{3}H]estradiol (Eisenfeld and Axelrod, 1965;

---

1 Androgens, estrogens, progestins, glucocorticoids, mineralocorticoids.

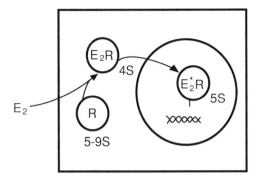

FIGURE 2A. Model of estradiol (E_2) interaction with target cells. Diffusion of E_2 passively across cell membrane. Formation of E_2–receptor complex (E_2R) and translocation to cell nucleus. Activation of E_2R (E_2^*R) and 4S to 5S transformation.

Kato and Villee, 1967; McEwen and Pfaff, 1970). Subsequently, cytosol receptors and cell nuclear retention of the hormone were demonstrated as the basis of the retention (Eisenfeld, 1970; Zigmond and McEwen, 1970). Furthermore, autoradiographic mapping of the cellular sites of uptake and retention of [³H]estradiol played a major role in establishing the brain as an estrogen target organ (Pfaff, 1968; Stumpf, 1968; Pfaff and Keiner, 1973).

What we currently know about the estrogen receptors of the brain is that, from a physicochemical standpoint, they are very similar to those in the pituitary gland and reproductive tract tissues. Moreover, as with the uterus, if [³H]estradiol is applied to estrophilic brain cells, a substantial portion of the radioactivity taken up and retained by the cell can be recovered from the cell nuclei (see Figs. 2A and B). Thus we have every reason to believe that estradiol exerts genomically mediated actions in the brain as well as in the uterus and reproductive tract. Indeed, direct evidence on this point is summarized below.

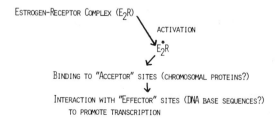

FIGURE 2B. Model of E_2R interaction with postulated acceptor and effector sites. From McEwen (1981) by permission.

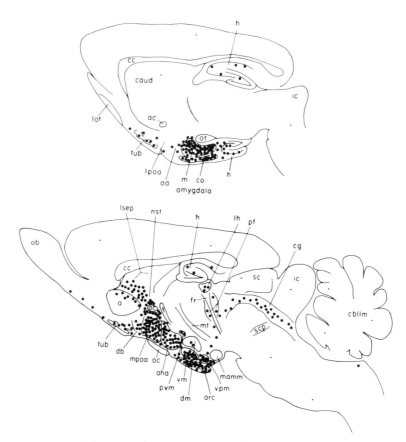

FIGURE 3. Distribution of estrogen-concentrating neurons in the brain of the female rat represented schematically in two sagittal sections. Most labeled neurons could be represented in a medial plane (bottom) based primarily on Fig. L740 in the atlas of Konig and Klippel (1963) and Fig. A35 and A36 in the atlas of Zeman and Innes (1963). Estradiol-concentrating neurons in the amygdala and hippocampus are represented in a more lateral plane (top) based on Fig. L2590 in the atlas of Konig and Klippel (1963). Locations of estradiol-concentrating neurons are represented by black dots. From Pfaff and Keiner (1973), by permission.

What is uniquely interesting about estrogen (as well as other steroid receptors) in the brain is its neuroanatomical distribution (Figs. 3 and 4). [Note that the distributions from autoradiographic localization (Fig. 3) and from cytosol receptor assays (Fig. 4) stand in excellent agreement with each other.] The importance of the neuroanatomical distribution of estrogen receptors will be demonstrated repeatedly below in considering estrogen action

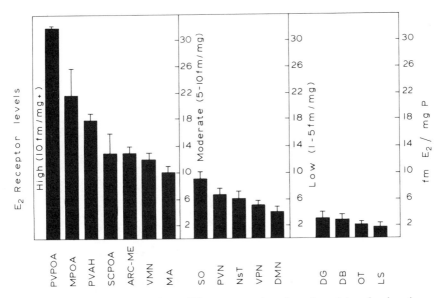

F I G U R E 4. Concentration of E_2 receptors in selected nuclei and subregions. Results are expressed as femtomoles per milligram of E_2 specifically bound per milligram of protein (fm E_2/mg P). The abbreviations used are PVPOA, periventricular preoptic area; MPOA, medial preoptic area; PVAH, periventricular anterior hypothalamus; SCPOA, suprachiasmatic preoptic nucleus; ARC-ME, arcuate nucleus-median eminence; VMN, ventromedial nucleus; MA, medial amygdaloid nucleus; SO, supraoptic nucleus; PVN, paraventricular nucleus; NsT, bed nucleus of the stria terminalis; VPN, ventral premammillary nucleus; DMN, dorsomedial nucleus; DG, dentate gyrus; DB, nucleus of the diagonal band of Broca; OT, olfactory tubercle; LS, lateral septum. From Rainbow *et al.* (1982a) by permission.

on feminine sexual behavior, progesterone receptor induction, and neurochemical effects of estrogens in the brain.

After many unsuccessful attempts using [³H]progesterone, progestin receptors were finally identified using a synthetic progestin, [³H]Ru5020, from Roussel Uclaf in France (McEwen *et al.*, 1982). Like estrogen receptors, neural progestin receptors also appear to be very similar physicochemically to those found in the reproductive tract and pituitary. One important characteristic of these receptors is that in some estrogen-sensitive tissues, such as the uterus, pituitary, and hypothalamus, estrogens induce progestin receptors. Thus, the synergistic interaction between estrogens and progestins on reproductive events including ovulation and sexual behavior may be due in large part to this induction. Figure 5 presents a summary of the areas of the rat brain where progestin receptor induction by estradiol is found. It should

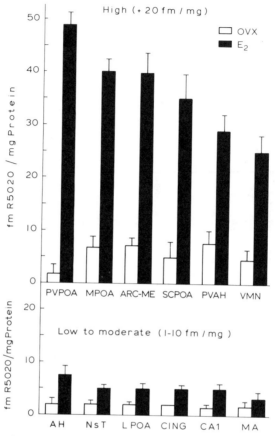

FIGURE 5. Concentration of estrogen-inducible and uninduced progestin receptors in selected nuclei and subregions of the female rat brain. Results are expressed as femtomoles of [³H]R5020 specifically bound per mg of total protein (mean ± SEM). The abbreviations used are PVPOA, periventricular preoptic area; MPOA, medial preoptic area; ARC-ME, arcuate-median eminence; SCPOA, superchiasmatic preoptic area; PVAH, periventricular anterior hypothalamus; VMN, ventromedial nuclei; AH, anterior hypothalamus; NsT, bed nucleus of the stria terminalis; LPOA, lateral preoptic area; CING, cingulate cortex; CA1, subfield of hippocampus; MA, medial amygdala; OVX, ovariectomized rats; E₂, E₂-treated rats. From Parsons *et al.* (1982a), by permission.

be noted that some estrogen-sensitive areas, such as the medial amygdala and bed nucleus, display little if any progestin receptor induction whereas others, such as the ventromedial nucleus and medial preoptic area, display a great deal of induction. Thus progestin receptor induction by estradiol is not a universal feature of estrophilic cells.

B. Spatial and Temporal Aspects of Estradiol and Progesterone Action on Reproductive Behavior

Where in the brain hormones act and how long they take to exert their effects are two of the most fundamental questions to be answered in approaching the mechanisms underlying hormone-dependent sexual behavior. With regard to the spatial aspect, it has been known for 20 years that estradiol exerts its effects on feminine sexual behavior in the rat brain when implanted into the basal hypothalamus. Recently, more precise measurements with much smaller quantities of implanted hormone have established that both estradiol and progesterone activate feminine sexual behavior in rats when applied to the region of the ventromedial nuclei of the hypothalamus, though apparently they do not have this effect when applied anywhere else (Davis *et al.*, 1979, 1982; Rubin and Barfield, 1980, 1983). This seminal finding is complemented by the demonstration that locally applied protein synthesis inhibitor in the ventromedial hypothalamus will block the effects of systemically applied estradiol and progesterone (Rainbow *et al.*, 1982b). Thus it appears that estrogen and progestin action in the vicinity of the ventromedial nuclei, involving protein synthetic steps, are both necessary and sufficient for the activation of lordosis behavior and proceptivity in female rats. This conclusion is not intended to exclude the participation of the other estrogen and progestin receptor-containing hypothalamic and limbic brain areas in the activation of feminine sexual behavior. Indeed, the role of these other brain areas in the full expression of sexual behavior in animals primed systemically with hormone remains a subject of active exploration. Moreover, it remains to be established whether other estrogen/progestin-sensitive brain areas may play a more pivotal role in the sexual behavior which may become reestablished to some degree after extensive damage to the ventromedial hypothalamic area.

The most striking feature of estradiol action to facilitate sexual receptivity is that it takes nearly 24 hours from the initial application of hormone before the behavior can be elicited by a sexually active male or by manual stimulation by the experimenter. The minimally sufficient estrogenic stimulus can actually be reduced in magnitude and time: i.e., estrogen application for the first 6 hours is sufficient to elicit lordosis when progesterone is given at 24 hours, and the estrogenic stimulus can be further reduced to two 1-hour applications of hormone which are at least 4 hours apart but not more than 12–14 hours apart (Fig. 6). Application of protein synthesis inhibitor before the first or second 1-hour estradiol treatment or midway between the first and second treatments blocks the activation of behavior; but protein syn-

	Lordosis Quotient	Lordosis Score	% Proceptive
A	0	0	0
B	•63 ± 12	•0.97 ± .22	75
C	•60 ± 10	•0.83 ± .18	80
D	0	0	0
E	0	0	0
F	0	0	0
G	•40 ± 9	•0.60 + .17	66
H	12 ± 8	0.18 ± .14	0
I	•58 ± 15	•0.88 ± .29	80
J	0	0	0
K	•62 ± 10	•1.04 ± .19	100
L	•65 ± 7	•0.95 ± .13	100
M	3 ± 3	0.03 ± .03	0
N	15 ± 15	0.15 ± 15	25

• Significant Change from Control (p < 0

0 1 2 3 4 5 6 7 8 9 10 11 12 13 14 15 16 17 18   24

Hour of Exposure to E_2

FIGURE 6. Effects of E_2 on the lordosis quotient, lordosis score, and proceptivity (percent proceptive). Females were tested with an experienced male rat 24 hours after the initiation of E_2 treatment. Four hours before testing, each animal received 500 μg progesterone. Results are expressed as the mean ± SE. The temporal paradigms for these experiments are described by the black bars [group A is a control (no E_2)]. Values were compared among groups using a one-way analysis of variance. A significant treatment effect was seen for both variables ($p<0.01$). Newman-Keul's tests revealed that for both lordosis quotients and lordosis scores, groups B, C, G, I, K, and L were different from the control but not from each other ($p<0.01$). From Parsons *et al.* (1982b), by permission.

thesis inhibitor applied after the end of the second 1-hour estradiol treatment is generally without effect (Fig. 7). These results suggest that estradiol initiates a "cascade" of protein synthetic steps culminating in production of gene products sufficient for the activation of the neural pathway for lordosis, in which the early proteins synthesized may prepare the way for the later estrogen action. This conclusion is supported by findings for the actions of the insect steroid hormone ecdysone on chromosomal gene expression in giant salivary gland chromosomes of fruit flies (Clever, 1964; Ashburner, 1974; Ashburner and Berendes, 1978).

Another striking feature of estrogen action on feminine sexual behavior in the female rat is that, under physiological levels of estrogen priming,

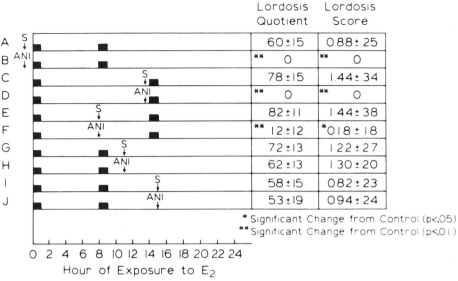

	Lordosis Quotient	Lordosis Score
A	60 ± 15	$0.88 \pm .25$
B	** 0	** 0
C	78 ± 15	$1.44 \pm .34$
D	** 0	** 0
E	82 ± 11	$1.44 \pm .38$
F	** 12 ± 12	* $0.18 \pm .18$
G	72 ± 13	$1.22 \pm .27$
H	62 ± 13	$1.30 \pm .20$
I	58 ± 15	$0.82 \pm .23$
J	53 ± 19	$0.94 \pm .24$

* Significant Change from Control ($p < .05$)
** Significant Change from Control ($p < .01$)

0 2 4 6 8 10 12 14 16 18 20 22 24
Hour of Exposure to E_2

F I G U R E 7. Effects of anisomycin (ANI) on the lordosis quotients and lordosis scores after sufficient treatment. For behavior, animals were assigned to separate mating groups of five to seven animals each and were tested with an experienced male rat 24 hours after the initiation of E_2 treatment. Four hours before testing, each animal received 500 μg progesterone. Experimental groups B, D, F, H, and J received subcutaneous (s.c.) injections of ANI (100 mg/kg) at the times indicated. Control groups A, C, E, G, and I received s.c. injections of saline (S) at the times indicated Behavioral results are expressed as the mean ± SE. The temporal paradigms for these experiments were as indicated in the figure. Lordosis quotients and lordosis scores were compared using Student's *t*-test (each experimental group was compared with its respective control). From Parsons *et al.* (1982c), by permission.

progesterone is required for expression of the behavior. As noted above, the induction of progestin receptors by estradiol is one of the key actions which leads to the expression of sexual behavior. In fact, the conditions depicted in Figs. 6 and 7 for estrogen action on lordosis and protein synthesis blockade apply equally for the induction of progestin receptors (see Fig. 8).

Temporally, progesterone facilitation of lordosis is much more rapid than the priming effects of estradiol, having an onset latency of 1 hour or less under optimal conditions (Fig. 9). Yet progesterone effects are blocked by protein synthesis inhibitors applied systemically or in the ventromedial hypothalamus (Rainbow *et al.*, 1982b). This suggests that rapid, genomically mediated actions of the hormone are required for sexual behavior. However, nongenomic actions of progesterone may also be involved (see below).

	fm [³H]R5020/ mg Protein	%Maximal Induction of Inducible CPRS
A	8.1 ± 0.3	0
B	•12.7± 0.7	•34
C	•11.8± 1.2	•28
D	10.3± 0.9	17
E	9.9± 0.8	14
F	•12.3± 1.1	•32
G	10.0±0.6	15

*Significant Change from Control (p< .05)

0 1 2 3 4 5 6 7 8 9 10 11 12 13 14 15 16 17 18   24

Hour of Exposure to E₂

FIGURE 8. Effects of E_2 cytosol progestin receptors (CPRs) in the mediobasal hypothalamus (MBH)-preoptic area. The number of samples for each observation was four to seven. Each sample consisted of MBH-POA regions from two animals. CPRs (expressed as the mean ± SE and percentage of maximal induction) were measured 24 hours after the insertion of E_2 capsules in the absence of exogenous progesterone. For the latter calculation, CPR levels were determined in animals not exposed to E_2 [represents binding not induced by E_2] and in animals which received s.c. injections of E_2 benzoate (15 μg daily for 3 days) to induce CPRs to maximal levels. The percentage of maximal induction of CPRs was calculated as: ([³H]R5020 binding in experimental group) − ([³H]R5020 binding in control groups)/ ([³H]R5020 binding in E_2 benzoate group) − ([³H]R5020 binding in control group). From Parsons et al. (1982b), by permission.

Under conditions of prolonged, supraphysiological estrogen priming the expression of feminine sexual behavior does not require progesterone treatment. This effect has been shown not to be due to adrenal mediation (i.e., involving the release of adrenal progestins). Rather, a somewhat simpler explanation may be in order, namely, that high levels of estradiol interact with the cytoplasmic progestin receptor and translocate receptors to the cell nuclei, whereupon the effects resemble those of progesterone itself (Parsons et al., 1984). A synthetic estrogen, moxestrol, which does not bind to progestin receptors, is incapable of having these behavior-activating effects.

C. Regulatory Effects of Estradiol and Progesterone on Brain Neurochemistry

Although estradiol and progesterone are capable of rapidly and directly altering electrical activities of some neurons, the time required for hormonal activation of sexual behavior and the ability of protein synthesis inhibitors to block activation indicate that there is much more to the underlying mecha-

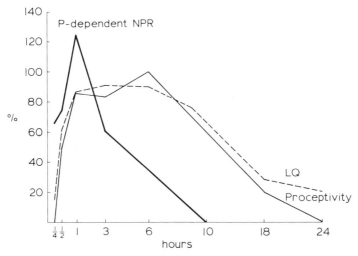

F I G U R E 9. Temporal relationship between proceptivity and lordosis quotient (LQ) scores and progesterone (P)-dependent nuclear [³H]R5020 binding (NPR) expressed as percent elevation above control. From McGinnis *et al.* (1981), by permission.

nism than a hormonal triggering of electrical activity. We must therefore begin to understand the actions of estradiol and progesterone in terms of how they modulate the neurochemical features of neurons which are responsible for electrical activity and neurotransmission. Through genomic actions mediated by intracellular receptors such as those described above, estradiol and progesterone may increase or decrease the levels of gene products involved in biosynthesis, release, reuptake, inactivation, or reception of neurotransmitters or other neuroactive substances (e.g., neuropeptides) (see Fig. 10). In addition, there are instances where steroids may influence these same processes through a more direct interaction not involving genomic action and RNA and protein synthesis (Fig. 10). The following discussion will consider both types of neurochemical events in the ventromedial hypothalamus (VM) as they pertain to the VM itself and then, in the next section, as they pertain to the VM's influence on the midbrain central gray.

Ultrastructural observations of nerve cells in the ventrolateral portion of the ventromedial nucleus of the hypothalamus following estrogen treatment are consistent with the hormone effect on rates of synthesis for some proteins, mediated particularly by increased rates of ribosomal RNA synthesis. Estrogen treatment is followed by the appearance of some neurons

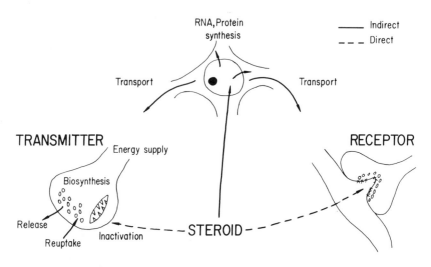

F I G U R E 10.   Genomic and nongenomic effects of steroid hormones on pre- and postsynaptic events. Nongenomic effects (broken line) may involve the action of the hormone on the pre- or postsynaptic membrane to alter permeability to neurotransmitters or their precursors and/or functioning of neurotransmitter receptors. Genomic action of the steroid (solid line) leads to altered synthesis of proteins which, after axonal or dendritic transport, may participate in pre- or postsynaptic events. From McEwen *et al.* (1978), by permission.

with large amounts of rough endoplasmic reticulum, in stacked arrays, together with an increased number of dense cored vesicles in the vicinity of the Golgi apparatus (Cohen and Pfaff, 1981; Carrer *et al.*, 1983). Under the same endocrine conditions, protuberances are seen on the surface of the nucleolus with an increased frequency (R. S. Cohen *et al.*, 1984). Electron microscopic examination of polyethylene glycol-embedded material subjected to enzyme digestion with DNase or RNase indicates that the electron-dense material adhering to the nucleolar surface in ventromedial hypothalamic cells is nucleolus-associated chromatin (Chung *et al.*, 1983, 1984). Thus, the hormone may lead, at least, to increased rates of ribosomal RNA synthesis in particular ventromedial hypothalamic cells, which in turn supports an increased rate of synthesis for particular peptides in those cells.

Within the estrogen- and progestin-sensitive region of the VM, hormonal stimulation appears to influence a number of neurochemical properties which may have the effect of increasing the efficacy of facilitative neural inputs and decreasing the efficacy of inhibitory ones. Two neurotransmitter systems which are being studied along these lines are serotonin and acetylcholine.

From a number of earlier studies with systemically applied drugs (Meyerson *et al.*, 1974; Kow *et al.*, 1974), serotonin was inferred to have an inhibitory influence on feminine sexual behavior. More recent work by Luine and colleagues has focused attention on serotonergic action in the VM. Pargyline implants in this region inhibit feminine sexual behavior and result in elevations of hypothalamic serotonin (Luine and Fischette, 1982; Luine and Paden, 1982), whereas discrete hypothalamic 5,7-dihydroxytryptamine (5,7-DHT) lesions promote increased sexual behavior and specifically reduce the content of VM serotonergic nerve endings (Luine *et al.*, 1983). Thus serotonin input to the VM region appears to be responsible, at least in part, for the inhibitory control of feminine sexual behavior. Regarding hormonal influences on serotonin turnover, it has been suggested (e.g., Kow *et al.*, 1974) that progesterone interferes with serotonin inhibition and thus facilitates sexual behavior. Evidence on this point is incomplete but strongly suggestive of an interaction between estradiol and progesterone in the control of serotonin metabolism. Estradiol treatment decreases type A monoamine oxidase (MAO) in the VM and arcuate nuclei (Luine and Rhodes, 1983). Because serotonin (5HT) is a major substrate for type A MAO, the effect of estradiol might tend to increase 5HT levels. However, progesterone administration increases type A MAO activity in estrogen-primed (but not in unprimed) female rats, which would have the opposite effect on serotonin levels and might promote the postulated disinhibition of sexual behavior (Luine and Rhodes, 1983). Serotonin turnover studies, performed after MAO is inhibited, indicate further influences of estrogen priming, namely, to increase accumulation of serotonin and disappearance of 5-hydroxyindoleacetic acid (5HIAA, a metabolite of 5HT) (Johnson and Crowley, 1983). There are no indications yet of the effects of progesterone on these parameters in estrogen-primed and unprimed rats.

The cholinergic system of the basal forebrain and hypothalamus of the female rat is influenced in several ways by estradiol. In the horizontal nucleus of the diagonal band of Broca, estrogen treatment increases the activity of choline acetyltransferase and acetylcholinesterase (Luine and McEwen, 1983). Estrogen treatment also increases the numbers of muscarinic cholinergic receptor sites in the ventromedial nuclei and in a number of other estrogen-sensitive areas of the preoptic area and hypothalamus (Rainbow *et al.*, 1980; McEwen *et al.*, 1983). Stimulation of the diencephalon of estrogen-primed female rats with a cholinergic agonist facilitates feminine sexual be-havior (Clemens et al., 1980). Thus the separate or combined actions of estradiol on cholinergic properties described above may constitute a mecha-nism for increasing the effectiveness of the cholinergic system on the activa-tion of sexual behavior.

Besides serotonin and acetylcholine, one must also consider the possi-ble involvement and estrogen/progestin sensitivity of other transmitter sys-tems at the hypothalamic level in the control of feminine sexual behavior, e.g., γ-aminobutyric acid (GABA) (Wallis and Lüttge, 1980; Lamberts et al., 1983) and dopamine (Foreman and Moss, 1979). Moreover, there are impor-tant influences of the VM on the midbrain central gray which also involve hormone effects and which influence feminine sexual behavior. These will be considered in the next section.

III. Brainstem Mechanisms and Circuitry

A. Hypothalamic Output

1. ANATOMY

Axons exit in a posterior direction from the basal forebrain, preoptic area, and hypothalamus in a more orderly fashion than the neuroanatomy of the 1950s and 1960s would have supposed. The relative anterior–posterior level of the neuronal cell bodies of origin, their medial–lateral position, and their dorsal–ventral position all determine the routes their axons take de-scending in and around the medial forebrain bundle (Fig. 11). Although by no means do they form a lemniscal system, the orderliness of these axons heading toward the midbrain through and around the medial forebrain bun-dle resembles "laminar flow" (Pfaff and Conrad, 1978). The more anterior the neuronal cell bodies of origin, the further lateral axons are forced to run. Axons from more dorsally placed cell groups tend to run more dorsally. Importantly, very medial cell groups, notably those in periventricular posi-tions, tend to give more strongly medial, periventricular, axonal projections than more laterally placed cell groups.

Since the ventromedial nucleus of the hypothalamus importantly facili-tates female rodent reproductive behavior, its axonal trajectories are especially interesting. Axons descending from the ventromedial nucleus travel in two groups (Krieger et al., 1979). A medial-going group travels a periventricular route to the midbrain central gray (Figs. 12A–C). A lateral-going group of descending axons cuts across the medial forebrain bundle and curves dorsally

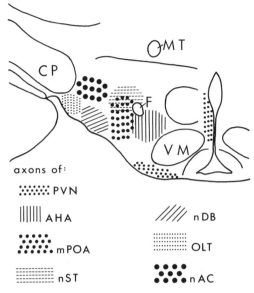

F I G U R E II.   Relative positions of axons descending from the paraventricular nucleus (PVN), anterior hypothalamic area (AHA), medial preoptic area (mPOA), bed nucleus of the stria terminalis (nST), nuclei of the diagonal band of Broca (nDB), olfactory tubercle (OLT), nucleus accumbens (nAC), and neurons in the hypothalamus and medial forebrain bundle at the level of the ventromedial nucleus. CP, cerebral peduncle; F, fornix; MT, mammillothalamic tract; VM, ventromedial nucleus. From Pfaff and Conrad (1978), by permission.

and posteriorly as it runs laterally. These axons continue to curve laterally and dorsally as they descend and then their arc swings medially again and they terminate in the dorsal midbrain reticular formation, in and lateral to the midbrain central gray. While the medial-running group makes some contributions to lordosis behavior, the lateral-running group makes an especially strong quantitative contribution (Manogue *et al.*, 1980).

2. CHEMISTRY

While the electrical activation of midbrain central gray neurons forms a necessary part of the hypothalamic output controlling female reproductive behavior (Pfaff, 1983), the temporal properties of hypothalamic effects suggest that rapid electrical effects are not the whole story. Electrical stimulation of the VM at low frequencies can increase the electrical excitability of midbrain neuronal cell bodies immediately (Sakuma and Pfaff, 1980b), while

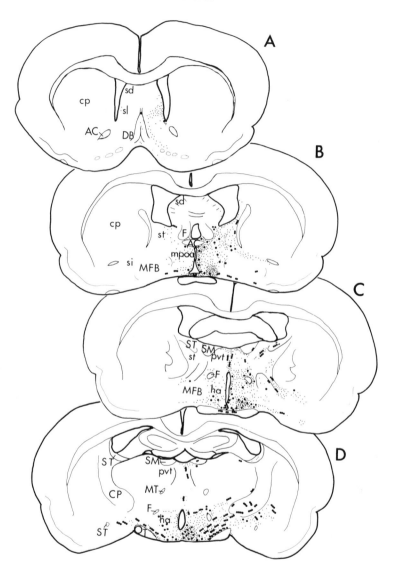

F I G U R E 12.   A representative series of charts from sections of a rat brain (rostral to caudal, A–L) showing the distribution of labeled projections from an injection site in the ventromedial nucleus of the hypothalamus. Large black dots, labeled fibers; small black dots, fields of individual grains; solid black area, region of labeled cell bodies (injection site, section E). Abbreviations: AC, anterior commissure; ac, central nucleus of the amygdala; aco, cortical nucleus of the amygdala; al, lateral nucleus of the amygdala; am, medial nucleus of the amygdala; arc, arcuate nucleus of the hypothalamus; BC, brachium conjunctivum; cg, central gray; Cp, cerebral peduncle; cp,

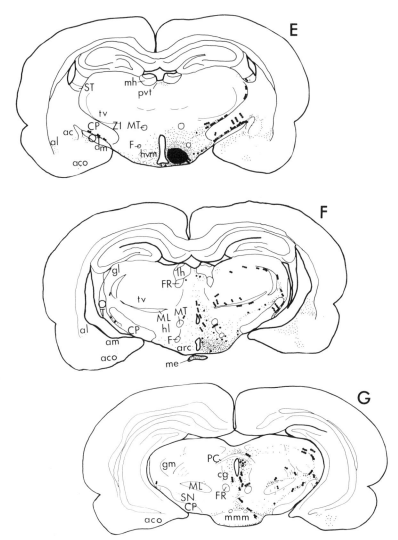

caudoputamen; DB, diagonal hand of Broca; dpm, dorsal premammillary nucleus; dr, dorsal raphe nucleus; dt, dorsal tegmental nucleus of gudden; F, fornix; FR, fasciculus retroflexus; gl, lateral geniculate nucleus; gm, medial geniculate nucleus; ha, anterior hypothalamus; hi, hippocampus; hl, lateral hypothalamus; hvm, ventromedial nucleus of the hypothalamus; hvma, anterior ventromedial nucleus of the hypothalamus; ic, inferior colliculus; ip, interpeduncular nucleus; lc, nucleus locus oneruleus; lh, lateral habenula; LL, lateral lemniscus; lv, lateral ventricle; me, median eminence; MFB, medial forebrain bundle; mh, medial habenula; ML, medial lemniscus; ml, lateral division of the medial mammillary nucleus; mmm, medial division of the medial mammillary nucleus; mp, posterior division of the medial mammillary nucleus; mpoa,

(continued)

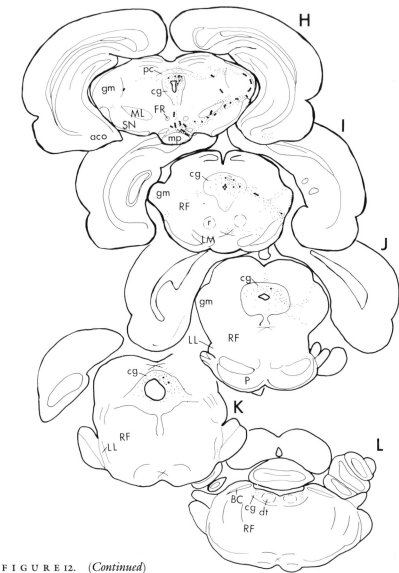

F I G U R E 12. (*Continued*)
medial preoptic area; MT, mammillothalamic tract; OT, optic tract; P, pyramidal tract; PC, posterior commissure; posc, preoptic area pars suprachiasmaticus; pvt, periventricular nucleus of the thalamus; r, red nucleus; RF, reticular formation; sc, superior colliculus; ad, dorsal septal nucleus; si, substantia innominata; sl, lateral septal nucleus; SM, stria medullaris; SN, substantia nigra; ST, stria terminalis; SUM, supramammillary nucleus; tv, ventral nuclei of the thalamus; vpm, ventral premammillary nucleus; ZI, zona incerta. From Krieger *et al.* (1979), by permission.

hypothalamic stimulation with the same parameters will not have its maximal effect on reproductive behavior for 25 to 50 minutes or longer (Pfaff and Sakuma, 1979a). If this time required for a mechanism which supplements the electrical action potentials is construed as the time necessary for the release, diffusion, and action of peptides synthesized in the hypothalamus, then several facts would fall into line. Previously we reviewed ultrastructural evidence (R. S. Cohen and Pfaff, 1981; R. S. Cohen et al., 1984) that estrogen administration can increase the rate of ribosomal RNA synthesis in certain ventromedial hypothalamic cells. If these cells are synthesizing proteins which are transported to endings in the midbrain and which are required for the effect of the hormone on reproductive behavior, interrupting axoplasmic transport would interrupt the effect of the hormone on behavior. This is just what happens; administration of colchicine in and around the VM, to interrupt axoplasmic transport, disrupts estrogen-stimulated lordosis behavior (Harlan et al., 1982). Among the peptides synthesized by hypothalamic or preoptic neurons are three for which evidence has already been gathered as being involved in these behavioral mechanisms. These are luteinizing hormone-releasing hormone (LHRH, the decapeptide required for the ovulatory surge of luteinizing hormone), prolactin, and the opioid peptides.

Given systemically, LHRH can foster female reproductive behavior (Pfaff, 1973; Moss and McCann, 1973). This is an elegant mechanism, for the same protein which stimulates ovulation fosters one of the behaviors required for fertilization. Following the anatomical work referred to above, it was clear that one candidate for a neural site of action would be the midbrain central gray. Indeed, administration of LHRH to the midbrain central gray in ovariectomized female rats given low doses of estrogen can foster female reproductive behavior (Sakuma and Pfaff, 1980d; Riskind and Moss, 1979). A striking demonstration of the LHRH effect indicated not only that it acts in midbrain central gray (Fig. 13), but also that its action is coordinated with opioid peptides (Sirinathsinghji et al., 1983). In these experiments the stimulating effects of naloxone on lordosis behavior could be blocked by an LHRH antiserum. Conversely, the suppressive effects of β-endorphin on lordosis behavior could be overcome by administration of LHRH to the midbrain central gray in female rats given low doses of estrogen (Sirinathsinghji et al., 1983). This is not to say that the only place LHRH may facilitate reproductive behavior is in the midbrain. Infusion into the hypothalamus can increase lordosis behavior (Moss and Foreman, 1976), and there may even be a spinal mechanism. Sirinathsinghji (1983) recently

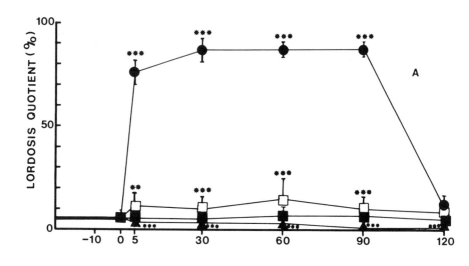

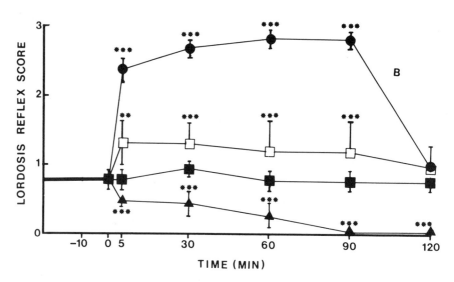

FIGURE 13. The effect of saline (■——■), naloxone (●——●), naloxone plus anti-LHRH antiserum (□——□), and naloxone plus LHRH antagonist [D-Phe², Pro³,D-Phe⁶]LHRH (▲——▲) infusions into the mesencephalic central gray on lordosis behavior in estrogen-primed rats assessed in terms of lordosis quotient (A) and lordosis reflex score (B). Each time point denotes the mean ± s.e.m. of 12 rats. Asterisks (**, $p < 0.01$; ***, $p < 0.001$, Student's t-test) denote statistically significant differences between groups (naloxone versus saline; naloxone versus naloxone plus anti-LHRH antiserum; naloxone versus naloxone plus LHRH antagonist analoge). From Sirinathsinghji *et al.* (1983), by permission.

showed that infusion of LHRH into the spinal subarachnoid space increased lordosis behavior within 5 minutes. It is very unlikely that this effect was due to leakage to another part of the body. Beyond questions about the full range of sites of action, LHRH effects on reproductive behavior pose a special conundrum. The greatest number of cell bodies immunocytochemically recognized as producing LHRH are in and around the preoptic area and diagonal bands (Witkin *et al.*, 1982; Shivers *et al.*, 1983a). Yet lesions of this tissue (Powers and Valenstein, 1972) actually increase lordosis behavior, while electrical stimulation of cell bodies in this region (Pfaff and Sakuma, 1979a; Malsbury *et al.*, 1980) decreases lordosis. How can activation of the tissue which produces a facilitating peptide (LHRH) decrease the behavior as a whole? Is it due to cellular heterogeneity in these preoptic cell groups? Perhaps an important population of LHRH producing cell bodies in the medial basal hypothalamus (as would be predicted from the effects of this decapeptide on reproductive behavior) does not show up immunocytochemically because it is used so quickly that storage levels in the cell body do not reach the threshold for immunocytochemical demonstration. Studies on the possible synthesis of LHRH by medial hypothalamic neurons will answer this question.

Another substance participating in the chemical coding of hypothalamic output is, surprisingly, prolactin. Microinjected locally in the central gray of the midbrain, prolactin increases lordosis behavior in ovariectomized rats given low doses of estrogen priming (Harlan *et al.*, 1983). This experiment had been provoked by immunocytochemical demonstrations (Tubeau *et al.*, 1979; Fuxe *et al.*, 1977; Harlan *et al.*, 1983; Shivers *et al.*, 1983b) that an immunoreactive prolactinlike substance is contained in nerve cell bodies in the basal medial hypothalamus whose axons follow trajectories known to be important for lordosis behavior, including some endings in central gray. Notably, other prolactin-immunoreactive axons descend all the way to the spinal cord (Shivers *et al.*, 1983b). This prolactinlike protein is not simply accumulated in hypothalamic cells from high levels of prolactin secretion by lactotrophs in the pituitary, having traveled up the portal circulation; our immunocytochemical demonstrations include good results from hypophysectomized animals. Also, two lines of evidence suggest that prolactin is actually being produced by cells in which it is detected immunocytochemically. Hybridization experiments with the complementary DNA for prolactin reveal that basal medial hypothalamic cells are expressing the RNA for prolactin (Schacter *et al.*, 1983). Also, following incubation of hypothalamic tissue with ^{35}S-labeled amino acids, a labeled protein can be immunoprecipitated with a prolactin antiserum and has a molecular weight like that of pituitary-produced prolactin (R. E. Harlan *et al.*, unpublished results).

It would make sense that the opiate peptides have some role in governing reproductive behavior. Besides the general consideration that animals threatened or in pain might not have proper circumstances to raise young, the preeminent cutaneous stimulation for evoking lordosis behavior allows several ways in which pain control could be important. Putting together several lines of evidence about female reproductive behavior as an example of motivation, the female rat about to reproduce must be motorically aroused but analgetic (Pfaff, 1982). Modern neuroanatomical work, with chemically specific techniques, suggests, as did Sirinathsinghji *et al.*'s (1983) behavioral work, that β-endorphin could be effective by interacting with steroid-dependent systems of reproductively important peptides. Combining estrogen autoradiography with immunocytochemistry, Morrell *et al.* (1983) found cells in the arcuate nucleus of the hypothalamus which accumulate radioactive estrogen in the cell nucleus and contain immunoreactive β-endorphin in the cytoplasm. Immunocytochemically, the manner in which the periventricular systems for the opiate peptides and LHRH intertwine in the midbrain suggests axo-axonal interactions. Indeed, under the proper conditions, naloxone can increase female rat reproductive behavior (Allen *et al.* 1985). An important site of action for opiate peptides and their antagonists is probably in the midbrain central gray. Lordosis behavior is enhanced by microinjection of naloxone or an antiserum to β-endorphin in the central gray and, conversely, β-endorphin infused in the midbrain inhibits lordosis (Sirinathsinghji *et al.*, 1983). The interactions of opiate peptides with reproductive behavior systems are likely to depend on the precise environmental state and pharmacologic condition of the animal at the beginning of the experiment, as suggested by the work of Moss on the pharmacologic conditions for the naloxone effect (Wiesner and Moss, 1984).

3. SYSTEMS INTERPRETATIONS OF HYPOTHALAMIC OUTPUT

Even with endocrine tools for the study of specific nerve cells as precise as those involved in female reproductive behavior and a circuit (Pfaff, 1980) as manageable, it would be overly simple to assume that the only physiological role of axons exiting the ventromedial hypothalamic region is with reproduction. In the long run, understanding the physiological and chemical details of hypothalamic output will require properly conceptualizing the *modes of action* by which steroid-dependent hypothalamic neurons control behaviors such as lordosis. The broader behavioral contexts for reproduction include chains of communicative behaviors by which reproductively competent conspecifics get together (Pfaff et al., 1972) and emphasize that rodent reproductive behavior is not "merely reflexive" but actually fits the abstract definition of a motivated behavior (Pfaff, 1982).

Along these lines, what the female rat chooses to do under conditions known to influence lordosis behavior has been the subject of research by Emery (1984). Clearly, lesions of the ventromedial nucleus of the hypothalamus can lead to large deficits in lordosis behavior (Mathews and Edwards, 1977; Pfaff and Sakuma, 1979b; Matthews et al., 1983a). While small lesions can lead to reproductive behavior deficits which are only temporary (Okada et al., 1980), the more typical finding is an apparently permanent deficit in lordosis (Mathews et al., 1983b). Emery's experiments showed (Emery and Moss, 1984) that when female rats with small ventromedial hypothalamic lesions control the timing of somatosensory contact with the male rat, they can do lordosis behavior when they allow the stud male to mount. The hypothalamic lesion effect shows up in the females spending much more time in locations where the stud male could not contact them. Contact from the male had been rendered aversive by the hypothalamic lesion. In fact, several experimental manipulations that reduce the noxious aspects of somatosensory stimulation can improve lordosis behavior performance (reviewed by Pfaff, 1982). Thus, an important mode of action of endocrine influences on the output from hypothalamic neurons is that hypothalamic output which fosters female reproductive behavior *changes somatosensory input that might have been noxious into a stimulation which is not aversive*. This formulation will not only be important for interpreting the physiological role of midbrain neurons in reproduction, but also will be useful in understanding steroid hormone effects on aggressive responses to somatosensory input (Malsbury et al., 1980).

Another system with which hypothalamic outputs important for reproduction may interact is the autonomic nervous system. The vigorous forms of locomotor behavior exhibited by female rats soon to ovulate would at least require sympathetic nervous system preparation for energetic muscular action (Pfaff, 1982). In fact, decreasing the sympathetic outflow in female rats by guanethidine sulfate led to deficits in precopulatory behavior (D. E. Emery, unpublished results, 1981). Insofar as the results of precopulatory behavior foster lordosis through its effects on the male as well as direct vestibular effects on the female (Pfaff et al., 1972; Pfaff, 1980), these changes would also lead to decreases in lordosis behavior under circumstances where somatosensory input is not forced upon the female. Given the prominent role of hypothalamic tissues posterior to the preoptic area in stimulating the sympathetic nervous system, it is attractive to put this potential aspect of hypothalamic output together with the considerations above: the female rat optimally prepared for reproductive behavior must be autonomically and motorically aroused, but analgetic in that the somatosensory input appropriate for reproduction will not be treated as aversive.

B. Midbrain Neurons

Lesions of the midbrain central gray lead to decreases in the performance of lordosis behavior (Sakuma and Pfaff, 1979b; Edwards and Pfeifle, 1981; Pfeifle and D. A. Edwards, unpublished data, 1982). In female rats, electrical stimulation of the dorsal portions of the central gray can dramatically facilitate lordosis (Sakuma and Pfaff, 1979a). In female hamsters the midbrain lesion effect takes the following form: central gray destruction leads to an increase in the number of animals that will never do lordosis behavior, but if they start the lordosis posture, the duration of the behavior will not be outside the normal range (Floody and O'Donohue, 1980). The lesion effects in these experiments are not always permanent, and it is not certain that behaviorally important fibers entering and exiting the central gray follow simple, well-defined routes. Especially since neurons whose responses to cutaneous stimulation qualify them for possible participation in this behavior lie both in the central gray and outside the central gray, deep to the superior colliculus and in the intercolliculus region, the field of nerve cells capable of helping to control lordosis behavior probably extends from the central gray through the reticular neurons deep to the superior colliculus (Malsbury et al., 1972; Rose, 1982; Sakuma and Pfaff, 1980c).

At least some of the hypothalamic fibers important for controlling midbrain neurons in the service of reproductive behavior follow a sweeping lateral route as they descend (Manogue et al., 1980). Thus the lesions made by Carrer (1978) might have been effective in reducing reproductive behavior because they demolished hypothalamic descending fibers, somatosensory fibers ascending to the midbrain, long hypothalamic descending fibers, or important cell bodies. The notion that one aspect of midbrain nerve cell participation in the control of reproduction includes responsiveness to somatosensory input comes from a great deal of electrophysiological work showing that some nerve cells respond especially well to behaviorally adequate stimulation (Malsbury et al., 1972; Rose, 1982).

Some estrogen-concentrating neurons in and around the ventromedial nucleus of the hypothalamus send axons to the vicinity of the midbrain central gray, as shown by a combination of retrograde neuroanatomical techniques and steroid autoradiography in the same tissue (Morrell and Pfaff, 1982). Although the medial hypothalamus to periventricular midbrain link is the most obvious, powerful route for neuroendocrine control of reproductive behavior, only 25 to 35% of estrogen-concentrating neurons send their axons in this manner (Morrell and Pfaff, 1982). Thus, it may not be the only logical way that hypothalamic steroid-dependent neurons operate. One possibility is that a substantial action of steroids on hypothalamic neurons is through cells whose axons ramify within their own nuclear groups. Indeed, 65 to 70% of synapses on neurons in the ventromedial nucleus of the hypothalamus are from intrinsic connections (Nishizuka and Pfaff, 1983), and "local neurons" (Rakic, 1976) might operate with mechanisms quite different from our traditional concept of a nerve cell. At the other extreme, a small number of long projecting axons from hypothalamic neurons might have powerful physiological effects at lower levels of the neuraxis. Certainly, paraventricular nucleus neurons can control autonomic function by projections all the way to the spinal cord (Swanson and Kuypers, 1980; Conrad and Pfaff, 1976), and application of sensitive retrograde neuroanatomical methods has revealed large numbers of cells in the lateral hypothalamus, the zona incerta, and some other hypothalamic regions that project all the way to the lumbar spinal cord (Schwanzel-Fukuda *et al.*, 1983, 1984). Behavioral results following axon transections in female hamsters are consistent with the notion, for example, that important hypothalamic synapses could be on neurons in the zona incerta, whose projections would then carry information relevant for behavior over a long distance (Scouten and Malsbury, 1983). In sum, the most sure, physiologically important route for hypothalamic control of female reproductive behavior is to the midbrain central gray, and adjacent reticular neurons, but the possibility of additional, more complex, ladderlike organizations of nerve cells has not been ruled out.

Application of tritiated amino acids to the cell bodies of central gray neurons reveals axons descending as far as the reticular formation of the medulla (Eberhart, J. I. Morrell, M. S. Krieger and D. W. Pfaff, unpublished observations). Electrophysiologically, central gray neurons can be antidromically identified from stimulation points in the nucleus gigantocellularis (NGc) of the medullary reticular formation, and it is these neurons with descending axons whose electrical excitability is increased by ventromedial hypothalamic stimulation (Sakuma and Pfaff, 1980a,b).

C. *Vestibulospinal Involvement*

Through a process of elimination (Pfaff, 1980, pp. 171ff.) the tracts descending from brainstem to spinal cord required for lordosis behavior were shown to be the lateral vestibulospinal tract and the lateral reticulospinal tract. Damage to the source of the lateral vestibulospinal tract, the lateral vestibular nucleus, causes significant decreases in lordosis behavior. The greater the loss in numbers of Deiters cells in the lateral vestibular nucleus (LVN), the greater the behavioral loss. Conversely, electrical stimulation of the LVN is followed by behavioral increases. Since the LVN does not receive inputs from the hypothalamus or from midbrain neurons which in turn receive from the hypothalamus, vestibulospinal neurons must participate in the behavior not by carrying a hormonally dependent facilitation but by helping to manage the posture itself.

D. *Reticulospinal Participation*

Lesions of NGc in the medullary reticular formation led to decrements in lordosis behavior; postoperative lordosis quotients were negatively correlated with the percent loss of giant cells (Pfaff, 1980). Even in those experiments, however, long postoperative recovery periods permitted significant recovery of function. In a subsequent experiment (Modianos and Pfaff, 1979) female rats were tested 4 hours after lesions were made when the animals were lightly anesthetized with halothane. Under these sensitive experimental procedures, NGc appeared to make a larger contribution to the control of the behavior. The biggest behavioral deficits followed lesions which destroyed both NGc and nucleus reticularis magnocellularis (Zemlan *et al.*, 1983). Postural changes in these animals also led to our first ideas about how, in neuromuscular terms, medullary reticulospinal cells might be involved. During the first postoperative days, the proximal limb musculature was hyperextended and there was a marked ventroflexion in the axial musculature (Zemlan *et al.*, 1983). Recovery from these postural changes during the second and third weeks postoperatively was correlated with smaller losses in lordosis behavior.

The exact manner in which reticulospinal cells influence lumbar spinal interneurons or motoneurons to govern female reproductive behavior is unknown. In some way, reticulospinal axons cooperate with lateral vestibulospinal influences and, since vestibulospinal cells could not carry hormone-dependent information, reticulospinal cells must. Electrical recording from single units in the medullary reticular formation revealed cells which responded to a specific convergence between cutaneous stimuli adequate for eliciting lordosis behavior and electrical stimulation of the midbrain central gray. These cells could be part of a "gate" which requires the combination of adequate somatosensory and hormonal inputs (Kow and Pfaff, 1982). Indeed, the ratio of excitatory to inhibitory responses to behavior-eliciting cutaneous stimulation by these medullary nerve cells was significantly increased by estrogen treatment (Kow and Pfaff, 1982).

The manner in which reticulospinal neurons govern reproductive behavior is almost certainly different from the way in which lateral vestibulospinal neurons operate. Very large lesions are required to show an effect of reticulospinal damage. Also, while focal electrical stimulation of lateral vestibulospinal neurons can increase lordosis behavior, similar effects are much harder to demonstrate for reticulospinal axons, and activation of rather widespread reticular cell groups may be required (Cottingham *et al.*, 1987). Electrical stimulation of reticulospinal and vestibulospinal neurons through combinations of sites and parameters while recording lordosis behavior or electrical activity on relevant motor nerves should indicate how reticular and vestibular neurons cooperate to form a complete behavioral mechanism.

IV. Spinal Mechanisms and Circuitry

A. The Nature of the Problem

The lordosis posture can be considered a somatosensory reflex under suprasegmental control but can be distinguished from other such reflexes, e.g., contact placing, by its hormone dependency. Using dye-coated males the receptive fields of the reflex have been delineated; they are the female's flanks, rump, tailbase, and perineum (Pfaff *et al.*, 1977). Loss of the behavior following denervation of the skin of these regions shows that cutaneous receptors are necessary to elicit the lordosis response (Kow and Pfaff, 1976). The appropriate cutaneous stimuli are only adequate when estrogen has been concentrated in the cells of the ventromedial nuclei for about 24 hours prior to stimulation.

Elements in the neuraxis from hypothalamus to spinal cord are required. The suprasegmental input to the final common pathway of the reflex, the motoneuron pools of the lumbar deep back muscles which execute the behavior, descends in the anterolateral tracts of the spinal cord which contain reticulospinal and vestibulospinal fibers (Kow *et al.*, 1977).

Therefore, although a small number of estrogen-concentrating cells have been found in the spinal cord (Morrell *et al.*, 1982), our working hypothesis is that the essential information regarding the hormone state of the animal is carried *to* the spinal cord in the obligatory descending drive from suprasegmental nuclei. Hence, in a female with an intact neuraxis, ovariectomy without estrogen replacement will prevent the reflex from being elicited; on the other hand, an estrogen-treated ovariectomized female will exhibit the reflex upon adequate stimulation but not after selective cord section of the anterolateral columns. Our experiments to date have been designed to answer the question of how this hormone-related obligatory descending drive to the spinal cord affects the input–output relationship between the reflex-eliciting cutaneous afferents and the effector moto-neurons.

Now, for the first time, a role for the intrinsic estrogen-concentrating cells of the spinal cord is also postulated. While few in number their distribution overlaps that of the GABAergic neurons in the dorsal horn responsible for primary afferent depolarization (PAD) leading to presynaptic inhibition. Consequently a small population of estrogen-concentrating neurons could have a powerful effect in gating the sensory input of the reflex if they in fact turned off GABA-mediated PAD. Experiments are now in progress to test this hypothesis.

B. The Form of the Behavior

The lordosis behavior is bilaterally symmetrical and spans the length of the vertebral column: the thoracolumbar spine is dorsiflexed, the rump and tail base are elevated, and the limbs and head are extended (Fig. 1). The posture has a ballistic onset; it is initiated from a tense crouched position following the abrupt halt of precopulatory hopping and darting behavior. Single-frame film analyses of mounting have shown that the rump begins to elevate 161 ms following the first flank contact by the male (Pfaff and Lewis, 1974). Maintenance of the lordosis posture, in particular rump elevation, requires large tensions to be exerted both against the force of gravity and against the constant load of the male. It does not require precise feedback control, however, because the motor performance has no target and feedback control under constant load conditions would be counterproductive. Given the form of the behavior, the lateral longissimus (LL) muscle and transver-sospinalis (TS), whose fibers take origin from the lumbosacral aponeurosis and attach onto the ilium, are appropriately connected to affect rump eleva-tion and would be ideal candidates for participation in the reflex response. In fact, bilateral stimulation of these muscles evokes dorsiflexion centered at the L6 iliac crest level and includes rump elevation (Brink and Pfaff, 1980).

Although the LL and TS are both extensor muscles and cutaneous afferents have been classically described as flexor reflex afferents, Hagbarth

(1952) showed that stimulation of the sural nerve which innervates the skin of the ankle and lateral foot will evoke contraction of the ankle extensors. Similarly, Carlson and Lindquist (1976) reported contraction of cat lumbar back extensor muscles when the skin overlying those muscles was pinched. The LL would also be a likely candidate for participation in the reflex because it has been shown by histochemistry to be predominantly a fast muscle (Schwartz-Giblin et al., 1983). Fast motor units would be ideally suited for recruitment through cutaneous palpation and thrusting pressure by the male during mounting since a reverse recruitment order has been postulated for cutaneous afferents as opposed to spindle afferents. For example, Kanda and co-workers (1977) have shown that sural afferents preferentially excite fast motor units and inhibit the tonically active slow motor units. They suggest that "reverse" recruitment would be functionally advantageous during forceful ballistic movements that require rapid and synchronous activation of an entire motoneuron pool. These would be the precise conditions required for lordosis, and LL being a fast muscle by histochemical criteria could take advantage of reverse recruitment. Furthermore, fast motor units would provide the large tensions required for maintenance of the posture.

The slow fibers in LL are segregated in the superficial one-third of the muscle from L2 through L6, where they comprise 11–18% of the population, and in an oxidative compartment in the medial deep region of the L5 myotome, where they are most concentrated and make up 62% of the muscle fibers. Spindles are most concentrated in this mediocaudal region as well (Schwartz-Giblin et al., 1983).

The anatomical compartmentalization of LL may serve a functional role. The largest portion, which contains predominantly fast fibers, fewer spindles, and yields a weak monosynaptic reflex, would be ideally suited for performance of ballistic-type movements such as lordosis; the medial deep region, on the other hand, containing oxidative slow fibers and more spindles in proximity to the lumbosacral vertebral junction, would be ideally suited for postural control. The slow tonic fibers could strengthen and stabilize the lumbosacral "joint" during movements of the pelvis with respect to the spine during such postures as rearing and turning in which TS and LL participate (Schwartz-Giblin et al., 1984b).

C. Recordings of Axial Electromyography and Axial Movement during Lordosis in Behaving Rats

Using an implanted chip and fine wire bipolar electrodes we have been able to obtain multiunit electromyographic (EMG) recordings with good signal/noise ratios simultaneously from two axial muscles, TS and LL, during

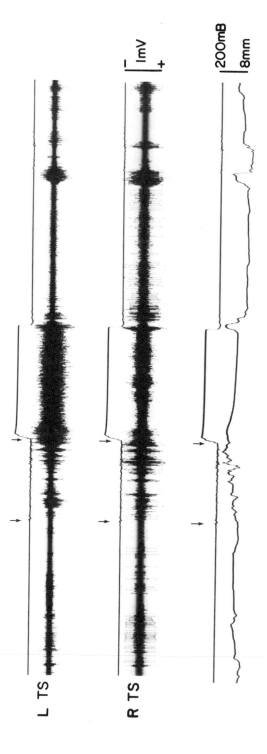

F I G U R E 14. Axial EMG and axial movement evoked by lordosis-eliciting manual stimulation. The upper trace in each record is the stimulus trace; it records the pressure output of a water bulb (WB) on the terminal phalanx of the experimenter's middle finger. Small repetitive fluctuations in the WB record beginning at the first arrow indicate bilateral palpation of the animal's flanks with the thumb and middle finger; the second arrow indicates the onset of perineal pressure applied by the index and middle finger forked around the tail base. The lower trace in the upper and middle records is the EMG of left TS and right TS, respectively. The lower trace in the lower record shows changes in length gauge (LG) resistance, which is linearly related to changes in gauge length. LG shortening is indicated by an *upward* deflection in this figure *only* and reflects rump elevation. The WB is calibrated in millibars (mB): 1 mB = 0.75 mm Hg. The length gauge is calibrated in mm of length change.

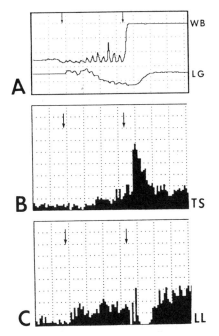

FIGURE 15. Data comparable to that shown in Fig. 14, but after digital processing, showing one pattern of EMG response to lordosis-eliciting stimulation which differentiates TS from LL. The upper channel of the digital scope display in (A) shows the stimulus trace (WB). Bilateral flank stimulation occurs between the arrows and is indicated by repetitive fluctuations in the WB record; it is followed by sustained perineal pressure. The lower channel records changes in length gauge (LG) resistance. In this figure and in Fig. 16 decreased LG resistance, i.e., length gauge shortening, is indicated by a *downward* deflection and reflects rump elevation. Calibrations in this figure and in Fig. 16 are given for one vertical division of the grid and one horizontal division of the grid. Vertical: WB = 127 mB/div, LG = 3.2 mm/div. Horizontal: 1.5 sec/div. (B) and (C) are multiunit firing-rate histograms of EMG activity in TS and LL, respectively. The post-stimulus time (PST) histograms are triggered from the onset of the sweep in (A). Calibrations in this figure and in Fig. 18 are given for one vertical division of the grid and one horizontal division of the grid. Vertical: (B) 8 spikes/div, (C) 8 spikes/div. Horizontal: 1.5 sec/div.

lordosis behavior (Fig. 14) and concomitantly record an independent measure of the axial movement from a length gauge spanning the lumbosacral junction (Schwartz-Giblin *et al.*, 1984b).

Such recordings have demonstrated that manual stimulation of the receptive fields of the lordosis reflex in female rats, as defined either by the areas of skin contacted by the male rat during mounting or by the regions adequate for manually eliciting the reflex, evokes strong EMG responses in both axial muscles, TS and LL. During bilateral palpation of the flanks the EMG activation is accompanied by continuous and cumulative length gauge shortening, reflecting progressive rump elevation (Fig. 15). The earliest length gauge response we recorded had an onset latency of 10 ms.

Characteristically, perineal pressure evokes large amplitude motor units to fire at high frequency. In highly responsive animals, EMG units were sometimes evoked by mere contact to the perineum before the onset of pressure. From concomitant recordings of TS and LL our data suggest (1) that TS is more responsive to the dynamic phase of the pressure than is LL, (2) that LL is more likely to sustain firing during maintained pressure, and (3) that the latency of perineal pressure-evoked TS EMG responses (24/35 within 50 ms) tends to be briefer than those of LL. Figure 15 shows a representative example of a recruitment pattern seen in about half of simultaneous recordings from TS and LL during strong lordoses in one animal. Above threshold pressures, the firing rate of TS motor units increases rapidly with increasing pressure and then falls off during sustained pressure. By contrast, LL motor units reach maximum firing rates after TS and show sustained activity during the plateau phase of pressure. Figure 16 shows an example of the recruitment pattern recorded during the other half of strong lordoses in the same animal; the EMG activity falls off markedly during sustained pressure in both muscles, but the persistent length gauge deflection confirms that lordosis persists throughout and following the pressure stimulus. Given the length–tension and force–velocity relationships of striated muscle originally described by Hill (e.g., 1938), this would in fact be predicted from the characteristics of the behavior. First, during shortening, i.e., during the increased elevation of the rump elicited by perineal pressure, high rates of motor unit recruitment would be required to obtain the force of contraction since a muscle never exerts as much force while shortening as it does while contracting isometrically. Following this initial response, less recruitment would be required to sustain rump elevation particularly in light of the slight lengthening of the muscle as revealed by the small increase in length gauge resistance. A marked decrease in EMG motor unit firing rate has also been observed during sustained maximal voluntary contractions (Bigland-Ritchie et al., 1983). These authors have suggested that during voluntary ballistic contractions high instantaneous firing rates increase the rate of initial force generation but would not be functionally useful thereafter if the force can be maintained at lower discharge rates. The constant load of the male rat may provide a constant stretch on the female's back muscles which would facilitate force production according to the length–tension curve.

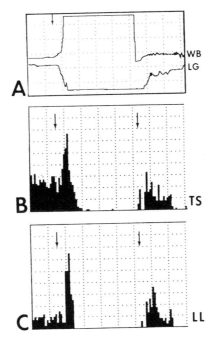

FIGURE 16. Brief axial EMG activation during sustained axial dorsiflexion. (A) The WB and LG traces show that sustained perineal pressure (onset at first arrow) evoked sustained rump elevation. Vertical: 164 mB/div, 1.6 mm/div. Horizontal: 2 sec/div. (B) and (C) The multiunit EMG firing-rate histograms triggered from the onset of the sweep in (A) show that EMG activity in both TS and LL is markedly reduced after the first 2 sec of the pressure application. EMG activity resumes following the pressure offset (second arrow) corresponding in time to the descent of the rump (LG stretch). Vertical: (B) 8 spikes/div, (C) 4 spikes/div. Horizontal: 2 sec/div.

An alternative or additional explanation for the marked decline in back muscle EMG activity during a sustained lordosis is as follows. Once the rump has been elevated by the axial muscles the extended hind limbs, which are an integral part of the lordosis posture, may act like a ratchet device to sustain elevation. This would not only conserve female axial muscle energy under constant load, but prevent the male from "losing ground" during repetitive, intermittent thrusting and intromission, which are characteristic of male rat reproductive behavior. In this regard it is worth noting that the pattern of purely dynamic EMG response in both TS and LL as illustrated in Fig. 16 is not characteristic of lordoses elicited with the limbs off the ground (one of the reflex-eliciting methods used in behavioral testing of the lordosis response).

EMG recordings from the segregated slow-twitch compartment of the muscle showed a different response from that of the rest of the LL muscle. While flank stimulation initially recruited motor units, activity did not persist and subsequent perineal pressure did not elicit characteristic high-frequency firing. Such a pattern of response might be predicted from the reverse recruitment theory of Kanda and co-workers (1977), who found that cutaneous stimuli inhibit tonically firing slow motor units and preferentially excite large motor units. Our data show decreased EMG activity during lordosis performance but not during ineffective lordosis-eliciting stimulation. It appears rather that the EMG in this part of the muscle decreases as the muscle shortens. This could result from reciprocal inhibition evoked by stretch of the antagonist ventral axial muscles during rump elevation; in fact we have evidence of inhibition of the dorsal axial muscles during passive shortening of the spine in anesthetized animals. However, since the monosynaptic reflex is very weak in both rat and cat axial muscle (Brink and Pfaff, 1981; Carlson, 1978b) and the stretch reflex in cat axial muscle is recorded at polysynaptic latency and only from a restricted portion of the muscle, we postulate that the decreased responsiveness of the slow-twitch compartment of LL during axial dorsiflexion results from decreased spindle drive, in this compartment where spindles are selectively concentrated. Accordingly, slow-twitch motor units initially recruited by flank stimulation in the presence of Ia tonic drive may be "turned off" if local spindles are unloaded during the rump elevation, which is activated in large part by contraction of fast-twitch muscle fibers in adjacent parts of the muscle. This has been described for some spindle afferents in medial gastrocnemius during contraction of individual adjacent or distant motor units (Binder and Stuart, 1980; Cameron *et al.*, 1981).

D. Sensory Inflow

1. PRIMARY SENSORY AFFERENTS

Recordings from single units in the L6 dorsal root ganglion show that approximately 30% of cutaneous mechanoreceptors give sustained responses to a lordosis-triggering type of pressure stimulation applied, through a small pressure sensor, to the perineum, tail base, or rump (Kow and Pfaff, 1979). These pressure-sensitive units show no spontaneous activity. The type I (Tylotrich hair-haarscheibe) receptors fire at an initially high rate either to contact or to low-threshold pressure; the discharge rate decreases to a much lower sustained level with maintained contact or with steadily increasing pressure. By contrast, cutaneous pressure-sensitive units do not respond to contact and their firing rate is correlated with stimulus intensity. Estrogen treatment had no effect on the receptive field size or threshold of either of these two types of receptors in ovariectomized rats during recordings from individual neurons.

2. SECOND-ORDER AFFERENTS AND INTERNEURONS

Lumbosacral spinal units recorded by Kow et al. (1980) which respond to the same lordosis-triggering type of pressure stimulation show much larger receptive fields and more complicated responses to the stimulus, and approximately half show resting activity. All of these findings suggest a convergence of multiple primary sensory afferents onto spinal units. The threshold pressure of spinal units is greater than that of primary sensory neurons but is precisely in the range of pressure effective for triggering lordosis in behaving animals. Type I spinal units show maximal responses not only at the onset of pressure but also at the offset; they maintain a lower discharge frequency during sustained contact or during steadily increasing pressures. Input from such units onto the TS motoneuron pool could evoke a response such as is seen in the firing-rate histogram illustrated in Fig. 15 or the response pattern of both TS and LL illustrated in Fig. 16. Of the purely pressure-sensitive units, the majority show discharge frequencies which either are directly correlated with stimulus intensity after a threshold level of pressure is attained or increase rapidly to maximum firing levels at threshold pressures and sustain that level of discharge throughout the increasing stimulus period. Such units could account for the LL firing-rate histogram in Fig. 15. Type I units were located exclusively in the dorsal horn and often in lamina IV; units responding only to pressure were found primarily in the intermediate gray. All pressure-sensitive units were located ipsilateral to their receptive fields. By contrast, contralateral muscle responses have been recorded in both behaving and anesthetized animals (to be described below), implying that crossing occurs at the input to the motoneuron.

Estrogen treatment has no effect on the receptive-field size, threshold, or resting activity of individual lumbosacral spinal neurons (Kow et al., 1980). In behaving animals, however, estrogen has been shown by these same investigators to increase the effectiveness of lordosis-eliciting cutaneous pressure in a dose-related fashion: e.g., in response to 300 mB of pressure to the perineum, 80% of rats pretreated with 10 μg estradiol benzoate for 4 days exhibited lordosis compared with 40% of females pretreated with a single 10-μg injection. Given the lack of evidence for estrogen effects on individual primary sensory and spinal interneurons and the small effect that has been reported for receptive-field expansion of the pudendal nerve

(Komisaruk *et al.*, 1972; Kow *et al.*, 1973), it is more likely that the important facilitation of estrogen on the behavior is by way of facilitating the throughput between the low-threshold cutaneous pressure receptors and the lumbar motoneurons which are the final common pathway of the reflex.

The next section will describe the characteristics of multiunit responses recorded from the axons of the effector lumbar motoneurons evoked by electrical stimulation of lordosis-relevant cutaneous nerves in anesthetized rats. These experiments were conducted to elucidate the temporal characteristics of the response to precisely timed maximal synchronous afferent volleys in order to evaluate, by comparison, the possible integrative processes occurring in the spinal cord during performance of the behavior.

E. Motoneuron Output

The motoneurons innervating LL and TS have been demonstrated by retrograde labeling following HRP injections into the muscle. They form a continuous string of cells situated in the medial aspect of the ventral horn from approximately T_{11} through $L6$ (Brink *et al.*, 1979). The low density of epaxial motoneurons visualized by this technique and identified by anti-dromic invasion from muscle nerves (Brink and Pfaff, 1981) predicted that an attempt to investigate the properties of the cutaneous reflex underlying the lordosis response by a microelectrode study of single units would not be fruitful. Therefore, we chose instead to record from the segmental motor nerves which innervate the axial muscles while stimulating segmental cutaneous afferents from the skin of the flanks and rump in the receptive field of the lordosis reflex (Schwartz-Giblin *et al.*, 1984a). In ovariectomized rats with or without estrogen replacement, brief trains of shocks to the cutaneous afferents evoked a polysynaptic multiunit response with a mean latency of 9.5 ms in about 50% of trials (Fig. 17). This mean latency is in good agreement with the shortest latency length gauge response we recorded in behaving animals following flank stimulation. The nerve response could be reciprocally evoked between neighboring segmental nerves without any apparent rostrocaudal asymmetry. As classically described for cutaneous reflexes by Sher-

---

F I G U R E 17. A cutaneous-evoked polysynaptic nerve response recorded in an estrogen-treated rat. The bottom four rows are continuous records from 0–103 sec recorded from the muscle branch of the L_4 nerve during 2/sec stimulus trains of 25 μA to the distal cutaneous branch of the L_3 nerve. The upper row of records (0–26 sec) shows the early response at a faster sweep speed. All records contain five superimposed traces. From Schwartz-Giblin *et al.* (1984a), by permission.

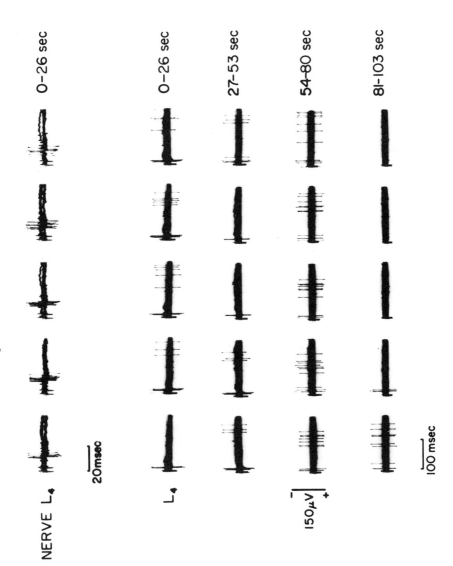

L₃ STIMULATION

NERVE L₄   0–26 sec

L₄   0–26 sec

27–53 sec

54–80 sec

81–103 sec

20msec

150μV

100 msec

rington (1906), the response habituated to repeated stimulation at 2/sec within 0.5–1.5 minutes but could be evoked again after several minutes' rest. Stimulation at 10/sec, which would more closely mimic the flank palpation rate of the male rat, enhanced the cutaneous-evoked response, but it too habituated. The most behaviorally significant ($p = .05$) result of these nerve-recording experiments was that in rats pretreated with estradiol, 60% of the 9.5-ms responses were followed by late discharges at latencies greater than 40 ms and as long as 200 ms as compared to only 24% in untreated control animals (Fig. 17). Given that 40 ms is the approximate duration of postfiring feedback inhibition (Renshaw IPSPs) onto motoneurons, these late discharges may represent the effect of a motoneuron postinhibitory rebound (Wilson et al., 1960) on either a tonic subthreshold excitatory drive or a suprasegmental reflex loop. Alternatively, they may manifest a truncation of PAD, which is evoked strongly in cutaneous afferents by other cutaneous afferents and which is the mechanism of presynaptic inhibition. Since late discharges are facilitated by estrogen, estrogen could act (1) to increase the tonic excitatory drive on motoneurons, (2) on the suprasegmental loop, or (3) to depress PAD. Evidence from the literature bearing on these hypotheses will be presented later in this chapter. During mating behavior, asynchronous late discharges may be required to produce fused tetanic tensions in axial muscles since in cat LL (Carlson, 1978a) the tetanus/twitch ratio does not exceed one until nerve frequencies of 15 Hz; maximum fusion occurs at 50 Hz. Therefore, even with male flank palpation rates as high as 12–14/sec and thrusting rates of 14–16/sec (Pfaff and Lewis, 1974), prolonged excitation would increase the tension output of female axial dorsiflexors.

Stimulation of the pudendal nerves which innervate the perineal region, a major receptive field of the lordosis reflex, evokes LL motor nerve activity. The LL nerve responses can be ipsilaterally or contralaterally evoked; bilateral stimulation facilitates the response. Such data provide evidence for the statement above that sensory information crosses at the level of the input to the motoneuron. PST histograms of LL motor nerve activity following pudendal nerve stimulation show in nearly all cases an early synchronous burst of activity which peaks at 11 ms and a late period of increased activity at about 100 ms. The nerve discharge rate is reduced below baseline from approximately 34 to 50 ms after the stimulus (Cohen et al., 1985). While no difference in response pattern between males and females has been recorded, females (with or without estrogen treatment) respond earlier to pudendal nerve stimulation: onset latency 5.8 versus 8.4 ms.

A descending facilitatory influence on the pudendal nerve-evoked response has been revealed (Sullivan *et al.*, 1986; Cohen *et al.*, 1987a). This descending facilitation may reflect an estrogen influence; it will be discussed in more detail later in the chapter.

F. Theories of Descending and Hormonal Control

I. DESCENDING CONTROL

Chronic spinal rats do not perform lordosis; selective cord sections show the anterolateral columns to be the necessary and sufficient pathway for the supraspinal influence on the behavior (Kow *et al.*, 1977). Zemlan and co-workers (1979) confirmed for the rat that the reticulospinal and vestibulo-spinal tracts descend in the anterolateral columns, because following micro-transections of these columns at T10 and local application of HRP, labeled cells were found mostly in the ventral portion of the NGc, ipsilaterally at the level of the caudal half of the nucleus of the VII cranial nerve, and were also found in the dorsal portion of the LVN. Fink-Heimer silver impregnation below the transection showed extensive preterminal degeneration; at the L3 level, degenerating terminals were most concentrated in lamina VII, pre-dominantly ipsilaterally but contralaterally as well, and were also seen ip-silaterally in lamina IX.

These anterolateral tract fibers could exert a tonic direct facilitatory effect on either motoneurons or interneurons such that they decreased the threshold for excitation. Thus they would be functionally effective in com-bination with afferent inflow, depolarizing the membrane potential to threshold for firing. This could occur, for example, at the latencies of early excitation evoked by cutaneous afferents from flank and perineum or by decreasing recurrent Renshaw inhibition (Haase and Van Der Meulen, 1961).

Alternatively, the descending control could be recruited as part of a suprasegemental loop initiated by the segmental reflex afferents. Under such conditions the descending effect could contribute to prolonging the early cutaneous-evoked polysynaptic response or might be responsible solely for the late discharges seen at much longer latencies (>40 ms) following cutane-ous nerve stimulation from both flank and perineum (Sullivan *et al.*, 1986; Cohen *et al.*, 1987a,b).

The preceding theories of descending control do not preclude an addi-tional effect on the afferent limb of the reflex; preterminal degeneration after anterolateral column transection was also seen in laminae I–IV at spinal level L1 (Zemlan *et al.*, 1979). A direct or indirect inhibitory influence at this site on the GABAergic neurons responsible for PAD would increase the synaptic efficacy of cutaneous reflex afferents (Lundberg and Vyklicky, 1966). This would be predicted to be a powerful way of enhancing lordosis since cutaneous afferents receive the strongest negative feedback of all primary somatosensory afferents and the action is exerted by the cutaneous afferents themselves, indirectly through the GABA pathway.

2. HORMONAL CONTROL

The estrogen-concentrating cells in the medial hypothalamus must exert the main hormonal control on the reflex since local implants of estradiol can stimulate high levels of lordosis behavior with proven absence of leakage (Davis *et al.*, 1979). As described in Section III of this chapter, a polysynaptic pathway from these cells to NGc of the medulla has been demonstrated. Therefore, the main hormonal control of the behavior may ultimately be exerted on the final common pathway of the reflex by at least part of the identical conventional neurophysiological mechanisms postulated for the descending control.

The estrogen-concentrating cells in the cord are few in number and not sufficient to stimulate the behavior, at least not in the absence of the descending input to the cord; the descending input we believe to be identical, at least in part, with the necessary hormonal influence. The intrinsic spinal E-concentrating cells are located in the dorsal portion of the gray matter (Morrell *et al.*, 1982). On that basis estrogen would not be predicted to directly modulate the effector motoneurons as has been postulated for the androgen-concentrating neurons in the sexually dimorphic motor nucleus in male rats involved in penile reflexes (Breedlove and Arnold, 1980) and in the androgen-concentrating motoneurons which innervate the clasping muscles in frog (Erulkar *et al.*, 1981).

The distribution of E-concentrating cells in laminae II–IV and lamina X overlaps the distribution of dorsal horn GABAergic terminals (McLaughlin *et al.*, 1975); GABAergic cells in the substantia gelatinosa (SG) make axo-axonic synaptic contacts with primary afferent fibers and could thus be responsible for evoking PAD (Barber *et al.*, 1978). If spinal GABA neurons contain estrogen, as they do in the preoptic anterior hypothalamic area (Flügge *et al.*, 1986), then estrogen could exert a genomic effect on the synthesis or release of GABA to depress PAD. Mendell and Wall (1964) reported examples of presynaptic disinhibition at the segmental level. By depressing PAD, a small population of spinal estrogen neurons could reduce the negative feedback effect that cutaneous afferents exert upon themselves. The increased synaptic efficacy of the cutaneous afferents along with the estrogenic increased descending drive would enhance lordosis reflex strength. On the other hand, a consequence of opening a gate to cutaneous input (i.e., reducing "cutaneous negative feedback") may be the irritability that one often sees in estrogen-primed rats as compared to rats treated with estrogen plus progesterone (Emery, 1984).

G. Evidence Bearing on the Theories

1. DESCENDING FACILITATION OF THE LUMBAR DEEP BACK MUSCLES

In acute, urethane-anesthetized rats we (Femano et al., 1984a,b) recorded ipsilateral activation of lateral longissimus, medial longissimus, and transversospinalis EMG during low-intensity (10–40μA) trains of high-frequency pulses (200–250 Hz) to the brainstem reticular formation. Stimulation sites which produced activation were ipsilateral and included NGc, the caudal tip of the medial portion of the caudal pontine nucleus (PC), and the lateral reticular formation. Stimulation of contralateral medial reticular formation with comparable currents would generally inhibit activation evoked from corresponding sites, ipsilaterally.

The latency of EMG activation from the first pulse in the train of stimuli to NGc showed marked facilitation with successive stimulus trains. Increasing the train repetition rate from 2 trains/sec to 4 trains/sec increased the facilitation. Multiunit interspike interval distributions of the EMG responses within a train are multimodal (Fig. 18); the modes are multiples of the stimulus interpulse interval. Only a fixed latency response with low probability would generate such a family of distributions. These results imply that the probability of any individual pulse at 200 Hz evoking an EMG response is small but that the connection between NGc and the axial motor unit is a tight one.

The mean onset latency from NGc to axial EMG is 6.0 ms (range: 4–7 ms). This latency is compatible with a monosynaptic connection between reticulospinal (RS) fibers and lumbar axial motoneurons in rat. This assessment is based on an efferent conduction velocity of 30 m/sec for the fastest RS fibers (Kow and Pfaff, 1982), a conduction distance of 90 mm from the medullary reticular formation to the L3 spinal cord, 1.5 ms for the shortest central delay and efferent conduction time of a monosynaptic reflex recorded on an LL muscle nerve (Brink and Pfaff, 1981), 0.9 ms for neuromuscular delay in LL, and an additional 1 ms for muscle conduction in LL at 7 m/sec (Schwartz-Giblin et al., 1984a); the estimated latency based on this calcula-

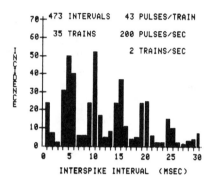

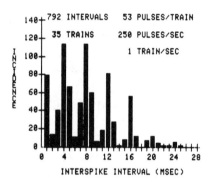

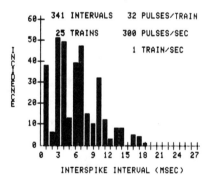

FIGURE 18. Computer analysis of EMG interspike intervals recorded from medial longissimus during repetitive train stimulation within NGc. Distributions were obtained with a stimulus intensity of 40 μA and with stimulus pulse rates of 200, 250, and 300 Hz. The first mode in each histogram reflects the multiunit nature of the EMG records. From Femano *et al.* (1984b), by permission.

tion equals 6.4 ms. Peterson and his colleagues (1979) have recorded mono-synaptic excitatory postsynaptic potentials (EPSPs) in neck and thoracic back muscle motoneurons in cat following stimulation of the ventral caudal reticular formation, but not in limb motoneurons.

Cohen *et al.* (1987a) have found that from several hours up to 1 day following bilateral spinal cord transection in the midthoracic region, the probability of LL nerve response to pudendal nerve stimulation is markedly reduced. When such responses were recorded they peaked at a fixed latency of 17 ms as compared to 11- and 22-ms peaks in control animals. No long latency responses at 100 ms have been recorded after cord transection, although they had a probability of 0.9 in controls (Cohen *et al.*, 1985). These data suggest that the descending input *facilitates* the early cutaneous-evoked polysynaptic response and is *necessary* for evoking the delayed discharges. The data do not allow discrimination of a tonic descending drive from a long suprasegmental reflex loop.

Evidence for a spino–bulbo–spinal (SBS) reflex which involves both the medial and the lateral reticular formation comes from the work of Shimamura and Livingston (1963) in cat. The characteristics of the SBS reflex that are consistent with cutaneous-evoked responses we record from rat lumbar axial motor nerves are that it is elicited by activation of cutaneous and high-threshold muscle afferents and that it habituates to repetitive stimulation. The SBS reflex was recorded in limb flexor motoneurons but not in limb extensor motoneurons; axial motoneurons were not examined.

2. INTERACTIONS OF BRAINSTEM RETICULAR NEURONS WITH CUTANEOUS AFFERENTS

In anesthetized and ovariectomized rats, Kow and Pfaff (1982) have shown that 28% of reticulospinal and 17% of unidentified single neurons in the medial medullary reticular formation respond to lordosis-eliciting "fork" perineal pressure. Estrogen treatment had no effect on the total number of responsive units but did increase the ratio of units excited to that inhibited (E/I ratio) by specific lordosis-relevant stimulation. Estrogen had no effect on the E/I ratio of units responsive to other forms of somatosensory stimulation such as pinching or brushing hairs. From these experiments it was concluded that estrogen had a specific effect of increasing the net reticular formation excitatory output drive from sensory input relevant for eliciting lordosis.

The ventromedial caudal medulla has been shown under various experimental conditions to have similar effects on transmission in spinal reflex pathways as do cutaneous afferents. For example, stimulation of cutaneous afferents and reticulospinal fibers inhibits Renshaw cells (Wilson *et al.*, 1964; Haase and Van Der Meulen, 1961), thus decreasing negative feedback on the motoneuron and thereby increasing reflex gain. They are also both involved in the regulation of primary afferent synaptic transmission. The efficacy of synaptic transmission in primary afferent fibers can be reduced by PAD (presynaptic inhibition) and increased by primary afferent hyperpolarization (PAH) (presynaptic disinhibition). Lundberg and Vyklicky (1966) have reported that from some ventromedial regions of the caudal brainstem it is possible to evoke tonic PAD; this can be temporarily interrupted by cutaneous afferent volleys. Stimulation of other ventromedial sites produces PAH and depression of presynaptic inhibition in cutaneous afferent fibers.

Rudomin and co-workers (1983) have reported that stimulation of the cells of origin of the reticulospinal fibers traveling in the anterolateral columns and stimulation of cutaneous afferents have the same effect on PAD evoked in Ia and Ib muscle afferents by the antagonistic muscle afferent, viz., to *depress* Ia and to *increase* Ib PAD. This is also relevant from the point of view of lordosis since both specific brainstem descending pathways and cutaneous afferents are necessary for the performance of lordosis and both excite axial motoneurons. Furthermore, to the extent that stretch of back muscles contributes to the overall force of contraction during lordosis, both descending reticulospinal fibers and cutaneous afferents would decrease PAD in Ia afferents, thereby increasing the efficacy of the stretch reflex, and produce PAD in Ib afferents, thereby reducing the effect of autogenic inhibition. Picrotoxin (1 mg/kg) occluded the inhibitory effect of reticulospinal fibers and cutaneous afferents on Ia PAD. This finding supports the hypothesis that the reticulospinal and cutaneous afferent fibers are having their PAD effect via the same pathway whose last-order neuron is GABAergic. The picrotoxin action began 5 minutes after i.v. administration with only partial recovery by 1–2 hours.

Furthermore, stimulation of reticulospinal fibers and cutaneous afferents also has the same effect on PAD in cutaneous afferents, viz., to evoke PAD. The PAD in cutaneous afferents can be evoked by a single cutaneous volley; it peaks from 7–20 ms and has a duration of 100–200 ms (Eccles and Krnjevic, 1959). Therefore, repetitive cutaneous volleys evoked by male thrusting at a rate of 14–16/sec (Pfaff and Lewis, 1974) would fall well within the effectiveness of cutaneous PAD. High-frequency NGc stimulation evokes a PAD in large- and small-diameter cutaneous fibers which peaks at 40 ms and has a duration up to 300 ms (Martin *et al.*, 1979). One way the estrogen influence could act at the brainstem level would be to suppress those reticulospinal pathways that evoke PAD in cutaneous fibers. This would enhance and prolong transmission in the polysynaptic reflex evoked by lumbar cutaneous and pudendal nerves and thereby facilitate occurrence of late discharges which are, respectively, estrogen dependent and dependent on descending drive.

3. SPINAL PAD PATHWAYS

Jankowska et al. (1981) reported that PAD of cutaneous afferents evoked by intraspinal stimulation had the shortest latency and lowest threshold when stimuli were restricted to laminae III and IV. This suggests that the last-order neuron in the PAD pathway is in lamina III or IV. There are large pyramidal-shaped cells situated on the border between laminae III and IV which have dendrites and axon terminals in the SG (Rethelyi and Szentágothai, 1969); the dendrites are postsynaptic to terminals of descending suprasegmental fibers and the axon terminals are presynaptic to primary afferent terminals. The SG is also reported to have the highest density of glutamic acid decarboxylase (GAD) immunoreactivity in the spinal cord (McLaughlin et al., 1975), including axo-axonic synapses between GAD terminals of intrinsic SG cells and primary afferent terminals of cutaneous origin (Barber et al., 1978). Both the GAD terminals and the primary afferent terminals are presynaptic to dendrites presumed to derive from the large lamina III–IV pyramidal cells. The SG is also the site of highest density of E-concentrating cells in the spinal cord of rat (Morrell et al., 1982).

Another pathway has been postulated to be responsible for presynaptic inhibition of small-diameter (Aδ and C) dorsal root fiber terminals (Jessel and Iversen, 1977). This pathway involves enkephalin-containing neurons in the SG (Elde et al., 1978), opiate receptors on dorsal root fiber terminals (LaMotte et al., 1976), and substance P, the presumptive neurotransmitter of small-diameter dorsal root fibers (Hokfelt et al., 1975).

Naloxone has been shown to facilitate lordosis behavior when administered into the spinal subarachnoid space and to prevent the inhibition of sexual receptivity by intrathecal morphine (Wiesenfeld-Hallin and Sodersten, 1984). The same has been shown for direct applications into the midbrain central gray (Sirinathsinghji et al., 1983). In the midbrain central gray, pretreatment with an LHRH antagonist prevented the naloxone-induced facilitation of lordosis, whereas LHRH reversed endorphin inhibition of lordosis.

How might the spinal cord action of naloxone occur? LHRH has also been reported to facilitate lordosis behavior when administered into the spinal subarachnoid space (Sirinathsinghji, 1983). While immunoreactive LHRH terminals have not been localized in the spinal cord, it is conceivable that the endogenous opiates may be acting to regulate some aspect of an LHRH effect; regulation of LHRH release by opiates has been shown in the basomedial hypothalamus (Rotsztejn *et al.,* 1978) and postulated for the midbrain central gray.

Alternatively, morphine has been reported to decrease the stimulated release of substance P from small nerve terminals (Jessel and Iversen, 1977) and to decrease the excitability of "wide dynamic range" spinal neurons in lamina V (Zieglgansberger and Bayerl, 1976; Le Bars *et al.,* 1976). Since lordosis-relevant cutaneous responses are also recorded from lamina V, these might be attenuated by morphine. Naloxone by itself does not affect the spontaneous or evoked release of substance P although it has a pronounced behavioral effect: it increases lordosis, decreases the threshold for vocalization to shock, and decreases the latency to withdrawal of tail from hot water; it does not alter the response to innocuous stimuli. Therefore, during reproductive behavior under the influence of spinal naloxone, lordosis-relevant cutaneous afferents may be unaffected presynaptically but their postsynaptic action could be enhanced by increased excitability of shared interneurons in nociceptive cutaneous reflex pathways. Furthermore, Renshaw feedback inhibition on axial motoneurons could be decreased for the same reason.

Enkephalin has also been reported to *depress* presynaptic inhibition in the spinal cord in a naloxone-reversible manner (Nicoll *et al.,* 1980; Suzue and Jessell, 1980). Theoretically this should increase the synaptic efficacy of lordosis-relevant afferents and hence reflex strength. We have reported that the most striking effect of blockade of spinal GABA transmission by intrathecal bicuculline is hyperalgesia to light tactile stimulation and increased contact avoidance between females and males in mating encounters (Schwartz-Giblin and Pfaff, 1985). It may be the balance between these two effects which is important for reproductive behavior (see, e.g., Sawynok, 1987; Stein *et al.,* 1987). Wiesenfeld-Hallin and Sodersten also state that their intrathecal morphine antagonism of reproductive receptivity suggests that enkephalin in the spinal cord may participate in the coding of a broader range of sensory information than pain perception (see Iadorola *et al.,* 1988).

4. BEHAVIORAL EVIDENCE THAT SUPPORTS AN INHIBITORY ROLE FOR GABA IN THE LORDOSIS REFLEX

In preliminary behavioral studies, S. Schwartz-Giblin, P. A. Femano, and D. W. Pfaff (unpublished observations) have found that following a single dose of diazepam (0.5–1.0 mg/kg), a GABA enhancer, in estrogen-primed ovariectomized rats, the mean lordosis scores (rated 0–4) of combined manual tests at 5 and 10 minutes, were statistically lower than in the 10 minutes preceding drug treatment ($p < .02$). By contrast, following a single dose of picrotoxin (100–300 µg/kg, an average of one-tenth the seizure dose), a GABA antagonist, the mean lordosis scores were greater than in the control period ($p < .05$) (cf. Fernandez-Guasti et al., 1986). Saline injection did not affect lordosis behavior. Although the mean percentage change in lordosis scores is small, the direction of change for individuals in a group is largely consistent, and the two antagonistic drugs affect lordosis in opposite ways. A striking consequence of systemic picrotoxin, in some animals, was a prolongation of lordosis following the offset of manual stimulation. Prolonged lordoses are rarely seen following manual stimulation in control rats, even when well primed with estrogen.

Given that these behavioral data are based on systemic injections, we cannot know the site of action of the GABA-related drug effects. In the hypothalamus, Fernandez-Guasti and co-workers (1986) have reported that picrotoxin injections *facilitate* lordosis in estrogen-treated ovariectomized rats, whereas Margaret McCarthy (personal communication) has demonstrated that direct applications of bicuculline into the VM *suppress* lordosis performance in ovariectomized rats treated with estrogen plus progesterone. A possible explanation for this apparent discrepancy is what Wallis and Lüttge (1980) reported, that estrogen and progesterone individually decrease GAD activity in the arcuate and anterior hypothalamus but combined estrogen plus progestrone treatment does not.

The effects of systemic picrotoxin and diazepam on the lordosis reflex are significant within 10 minutes; this is the time frame for effects of the same systemic doses of picrotoxin and diazepam on spinal cord potentials related to PAD (Schmidt, 1971; Levy, 1977; Eccles et al., 1963). The behavioral effect seen after systemic picrotoxin, however, is not the same effect observed after intrathecal injection of GABA antagonists. For example, (1) intrathecal picrotoxin produced hyperalgesia but systemic picrotoxin did not (Schwartz-Giblin and Pfaff, 1985); and (2) intrathecal bicuculline produced hyperalgesia and increased avoidance behavior without affecting the lordosis quotient (the lordosis score was not rated in the intrathecal tests).

Another possible site of action for a facilitatory effect of picrotoxin on lordosis is the periacqueductal gray (PAG). GABA antagonists increase the output of the PAG (Moreau and Fields, 1986) and electrical stimulation of the PAG facilitates lordosis and axial EMG responses to lordosis-relevant reticulospinal inputs (Cottingham et al., 1987). However, according to Moreau and Fields (1986) GABA antagonists injected into the PAG also facilitate morphine, but injections of endorphin into the PAG suppress lordosis behavior (Sirinathsinghji *et al.*, 1983). The electrical stimulus and the chemical stimulus must be activating two different cell groups in the PAG; we are currently investigating whether GABA antagonists facilitate the cell group whose output facilitates axial muscles.

Our preliminary data are consistent with the hypothesis that one action of estrogen may be to *turn off* GABA PAD pathways in the spinal cord. A decrease in PAD in small-diameter high-threshold cutaneous fibers could account for the avoidance behaviors seen in animals primed with estrogen alone as contrasted with rats treated with estrogen plus progesterone (Emery, 1984), whereas a decrease in PAD of large-diameter low-threshold cutaneous afferents could account for the increased reflex strength (i.e., increased lordosis score to manual testing) seen in our studies after systemic picrotoxin. Progesterone, on the other hand, decreases avoidance behaviors and decreases the effectiveness of opiates to suppress lordosis behavior (Sirinathsinghji *et al.*, 1983); a naturally occurring metabolite of progesterone enhances GABA-evoked inhibition (Majewska *et al.*, 1986).

5. EFFECTS OF OTHER STEROIDS ON SPINAL CORD TRANSMISSION

Erulkar and co-workers (1981) studied reflex transmission in segments of isolated frog spinal cord containing the motoneuron pool of the sternoradiallis muscle, which effects the male clasping reproductive reflex. The cord preparations were from male frogs castrated 40–50 days before the experiment. Reflex responses were recorded from the sternoradialis muscle nerve and compared before and after administration of dihydrotestosterone to the bathing fluid. They found that with paired shocks, steroid facilitated the response to the second stimulus; it was more synchronous and greater in amplitude at all interpulse intervals out to 800 ms. Intracellular recordings from motoneurons in control preparations showed that single shocks to dorsal root characteristically evoked a single spike preceded by a small synaptic potential. In a steroid bathing medium motoneuron responses to single dorsal root shocks consisted of doublet discharges riding on large EPSPs. Doublet discharges were also recorded from steroid-bathed motoneurons following antidromic stimulation. Motoneurons of the sternoradialis muscle concentrate dihydrotestosterone. Therefore, changes in reflex excitability which are related to increased postsynaptic excitability might be predicted. However, the long latency facilitation out to 800 ms would be more likely related to changes in synaptic transmission or interneuronal pathways.

Hall and collaborators (1978) tested the effect of chronic daily triamcinolone treatment (8 mg/kg) in cats on the properties of the monosynaptic reflex elicited by supramaximal stimulation to the dorsal roots. With single shocks they found no difference in the initial monosynaptic component of the reflex but observed a significant increase in the amplitude of late polysynaptic potentials in treated animals. The responses to paired dorsal root shocks revealed facilitation at long interpulse intervals comparable to the effect seen in isolated frog spinal cord after dihydrotestosterone administration to the bath: i.e., triamcinolone increased the synaptic recovery of the response to the second of paired shocks at interpulse intervals of 100–400 ms. They found no steroid-related effect on postsynaptic inhibition but found a steroid-related increase in presynaptic inhibition. Glucocorticoid treatment also produced an increased effect of posttetanic potentiation on a test monosynaptic reflex. Hall and his collaborators concluded from their data that there is strong support for the hypothesis that glucocorticoids enhance excitatory synaptic transmission at least in part through a direct action on primary afferent endings.

H. A Model for Descending Estrogen-Related Control of the Lordosis Reflex

Most of the above data from our laboratory and from the literature are consistent with the model of the ventromedial caudal brainstem and spinal cord shown in the circuit diagram in Fig. 19. The model reflects data which show that descending pathways from the basomedial hypothalamus and the PAG facilitate lordosis behavior and specifically that electrical stimulation of the PAG facilitates lordosis and axial EMG responses evoked by reticulospinal stimulation. The effects of electrical stimulation of the PAG with respect to analgesia and nociceptive reflexes are mimicked by injections of opiates into the PAG, implying that the opiates increase the output of the PAG. By contrast, β-endorphin injections into the PAG depress lordosis while injections of naloxone facilitate lordosis. Consequently, a single model cannot satisfy both the effects of electrical and opiate stimulation of the PAG on lordosis behavior.

It does not escape our attention that the motoneurons in the model innervate axial extensor muscles and thus must be distinguished from PAG control over various flexion reflexes; in fact, electrical stimulation of PAG facilitates axial extensor EMG but decreases nociceptive reflexes. Furthermore, the antinociceptive effects produced by the PAG are blocked by lesions in the dorsolateral funiculus, whereas lordosis behavior depends upon the integrity of the anterolateral tracts.

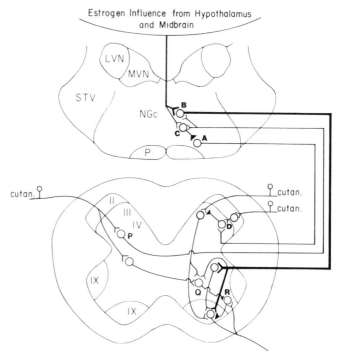

F I G U R E 19. A diagram of neuronal connections consistent with the electrophysiological data discussed in the chapter. Black triangles are inhibitory synapses; all other synaptic connections are excitatory. Cells A, B, and C are brainstem reticular neurons; cell D is the last-order neuron in the PAD pathway; cell P is a spinoreticular projection neuron; cell Q is an inhibitory interneuron that acts on Renshaw cells (cell R) and thus reduces negative feedback on axial motoneurons. See text for complete explanation.

In Fig. 19, electrical stimulation of reticulospinal cell B evokes EMG activity in axial muscles at monosynaptic or disynaptic latencies. Input to cell B from cutaneous afferents in the pudendal nerve facilitates descending drive to axial motoneurons at both brief and long latency. Estrogen-related inputs to cell B from more rostral regions of the brain would fire or increase their excitability and thereby enhance the effect of cutaneous afferents from flank and rump. Pudendal afferents also make connections with interneurons in all laminae of the dorsal horn; these in turn excite both ipsilateral and contralateral axial motoneurons.

The negative feedback effect of Renshaw (R) cells on axial motoneurons is inhibited both by cutaneous reflex afferents (Schwartz-Giblin and

Pfaff, 1988) and by descending fibers from the ventromedial caudal brainstem. Electrical stimulation of the ventromedial caudal brainstem in the region of cell A would evoke PAD in Ib (and cutaneous afferents). The action of cell A is inhibited by cutaneous afferent input via cell C. While Lundberg and Vyklicky's data are consistent with cutaneous inhibition of brainstem-evoked PAD occurring at the spinal level, they do not rule out the possibility that it may occur suprasegmentally. Electrical stimulation of the ventromedial caudal brainstem in the region of cell C would produce PAH by the mechanism of presynaptic disinhibition.

Recording from B cells or from unidentified (nonprojecting) C cells, one would see excitatory effects from stimulation of lordosis-relevant afferents; on the other hand, reticulospinal A cells would show inhibitory effects from stimulation of lordosis-relevant afferents.

If the descending estrogen-related projection to the caudal brainstem synapsed on both B and C cells and was weak due to tonic inhibition at the midbrain central gray (Moreau and Fields, 1986), and if the cutaneous input onto B and C cells was relatively weak, then during estrogen treatment the enhanced descending drive would increase the number of B and C cells excited by the lordosis-relevant cutaneous input. More A cells would be inhibited by the descending drive, thereby occluding the potential inhibitory effect of lordosis-relevant afferents on the A cells. The overall effect of estrogen treatment would be to increase the E/I ratio of reticular cells activated by the lordosis-eliciting stimulus without increasing the total number of reticular cells activated by the stimulus.

According to this model the main estrogenic control of the lordosis reflex would be exerted via descending systems in the anterolateral tracts which would act (1) directly or indirectly to excite the motoneurons of the final common pathway, and (2) to suppress PAD in the lordosis-relevant cutaneous fibers. We know that lesions of the anterolateral columns of the spinal cord decrease lordosis behavior. This is consistent with our model since in estrogen-primed animals, interrupting the subset of reticulospinal fibers derived from B cells of origin would decrease excitatory input to the axial motoneuron pool, while interrupting reticulospinal fibers of A cells of origin would have minimal effect since they would have been suppressed by estrogen at the brainstem level.

V. Summary

Intracellular estrogen and progestin receptors in the ventromedial nuclei of the hypothalamus are necessary and sufficient for the expression of hormone control over lordosis and feminine proceptive reproductive behavior in the rat. The hormonal control is mediated, at least in part, by genomic action since local applications of protein synthesis inhibitors block the action of systemic hormone administration. The effectiveness of discontinuous estrogen exposure over a 24-hour period followed by the rapid priming effect of progesterone suggests that estrogen initiates a "cascade" of protein synthetic steps, the products of which are necessary for later estrogen action, and that these estrogen actions prepare the way for the more rapid protein synthetic actions of progesterone. It is also evident that the ovarian steroids have effects on the processes involved in synaptic transmission and that these effects may be genomic as well as nongenomic. Evidence has been presented of a role for the serotonergic and cholinergic systems acting in the hypothalamus and basal forebrain on lordosis behavior and of a neurochemical role for estradiol and progesterone in the metabolism of these neurotransmitters.

A medial and lateral group of axons travel from the ventromedial hypothalamus to the midbrain central gray; the lateral group is more important in the regulation of lordosis. From considerations of the difference in time course of effects resulting from stimulation and ablation studies in hypothalamus as compared to midbrain, the mechanisms by which these two important brain sites regulate feminine sexual behavior can be evaluated. The long latency of response to manipulations of neurons in the ventromedial hypothalamus indicate that more than conduction of action potentials in anatomical pathways connecting these two sites is involved. Estrogen can increase the rate of ribosomal RNA synthesis in the ventromedial nucleus, and peptides such as LHRH and prolactin which are synthesized in diencephalic structures have brief latency facilitatory effects when injected locally into the midbrain central gray. The data strongly suggest a role for axoplasmic transport of peptides between these two sites. The more rapid effects on lordosis from manipulations of midbrain neurons and the more conventional response of midbrain neurons to both antidromic electrical stimulation from medullary gigantocellular neurons and somatosensory stimulation suggest that the facilitatory effects from central gray are mediated by more classical electrophysiological conduction and synaptic transmission. Facilitation of lordosis by spinal subarachnoid injections of LHRH and naloxone and the presence of immunoreactive prolactin fibers in the spinal cord implicate the spinal cord itself as an additional regulatory site for reproductive behavior. Furthermore, the action of the opiate antagonist, naloxone, and its blockade by LHRH antiserum was discussed in relation to the role of pain in reproductive behavior.

The integrative importance of brainstem reticular formation neurons was emphasized since they receive hormone-related inputs from midbrain central gray and lordosis-relevant cutaneous inputs and have a tight mono- or disynaptic connection with the axial motor units of the final common pathway. Therefore, the obligatory descending pathway to the spinal cord may be identical with the main hormonal control of the reflex; data show that descending fibers and estrogen are necessary for the late discharges, which prolong motor outflow, to be evoked from cutaneous afferents from perineum and from flank and rump, respectively. In addition, spinal estrogen-concentrating neurons may function to depress the PAD pathway and thereby potentiate the synaptic efficacy of these lordosis-relevant cutaneous afferents.

Data from electromyography and length gauge recordings during lordosis behavior indicate that the axial muscles lateral longissimus and transversospinalis are both activated and that their response patterns are predictable both from classical descriptions of the physiological properties of skeletal muscle and from the characteristics of spinal dorsal horn neurons responding to lordosis-relevant stimulation. High-frequency electromyographic activation coincides with the ballistic phase of the lordosis reflex, suggesting that the role of asynchronous cutaneous-evoked late discharges, i.e., high instantaneous firing rates, is to produce the large tetanic tensions required from the axial muscles during the performance of feminine reproductive behavior. Cutaneous afferents would also amplify reflex gain by inhibiting Renshaw-mediated negative feedback onto the motoneuron pool.

Taken together, these various observations as well as additional supporting data (see Pfaff and Schwartz-Giblin, 1988) are providing a picture for the first time of the anatomical, neurophysiological, and neurochemical organization of a hormone-dependent behavior in a higher vertebrate species.

References

Allen, D. L., Renner, K. J., and Luine, V. N. (1985). Naltrexone facilitation of sexual receptivity in the rat. *Horm. Behav.* **19,** 98–103.

Ashburner, M. (1974). Sequential gene activation by ecdysone in polytene chromosomes of *Drosophila melanogaster. Dev. Biol.* **39,** 141–157.

Ashburner, M., and Berendes, H. D. (1978). Puffing of polytene chromosomes. *In* "The Genetics and Biology of *Drosophila*" (M. Ashburner, and T. R. F. Wright, eds.), pp. 315–395. Academic Press, London.

Barber, R. P., Vaughn, J. E., Saito, K., McLaughlin, B. J., and Roberts, E. (1978). GABAergic terminals are presynaptic to primary afferent terminals in the substantia gelatinosa of the rat spinal cord. *Brain Res.* **141,** 35–55.

Barber, R. P., Vaughn, J. E., and Roberts, E. (1982). The cytoarchitecture of GABAergic neurons in rat spinal cord. *Brain Res.* **238,** 305–328.

Bigland-Ritchie, B., Johansson, R., Lippold, O. C. J., and Woods, J. J. (1983). Contractile speed and EMG changes during fatigue of sustained maximal voluntary contractions. *J. Neurophysiol.* **50,** 313–324.

Binder, M. D., and Stuart, D. G. (1980). Responses of Ia and spindle group II afferents to single motor-unit contractions. *J. Neurophysiol.* **43,** 621–629.

Breedlove, S. M., and Arnold, A. P. (1980). Hormone accumulation in a sexually dimorphic motor nucleus of the rat spinal cord. *Science* **210,** 564–566.

Brink, E. E., and Pfaff, D. W. (1980). Vertebral muscles of the back and tail of the albino rat *(Rattus norvegicus albinus). Brain, Behav. Evol.* **17,** 1–47.

Brink, E. E., and Pfaff, D. W. (1981). Supraspinal and segmental input to lumbar epaxial motoneurons in the rat. *Brain Res.* **226,** 43–60.

Brink, E. E., Morrell, J. I., and Pfaff, D. W. (1979). Localization of lumbar epaxial motoneurons in the rat. *Brain Res.* **170,** 23–41.

Cameron, W. E., Binder, M. D., Botterman, B. R., Reinking, R. M., and Stuart, D. G. (1981). "Sensory partitioning" of cat medial gastrocnemius muscle by its muscle spindles and tendon organs. *J. Neurophysiol.* **46,** 32–47.

Carlson, H. (1978a). Morphology and contraction properties of cat lumbar back muscles. *Acta Physiol. Scand.* **103,** 180–197.

Carlson, H. (1978b). Observations on stretch reflexes in lumbar back muscles of the cat. *Acta Physiol. Scand.* **103,** 437–445.

Carlson, H., and Lindquist, C. (1976). Exteroceptive influence on the lumbar back muscle tone and reflexes in the cat. *Acta Physiol. Scand.* **97,** 332–342.

Carrer, H. F. (1978). Mesencephalic participation in the control of sexual behavior in the female rat. *J. Comp. Physiol. Psychol.* **92,** 877–887.

Carrer, H. F., and Aoki, A. (1982). Ultrastructural changes in the hypothalamic ventromedial nucleus of ovariectomized rats after estrogen treatment. *Brain Res.* **240,** 221–233.

Chung, S. K., Cohen, R. S., and Pfaff, D. W. (1983). Ultrastructure of ventromedial hypothalamic nucleoli and nuclei revealed by enzyme digestions in de-embedded thin sections. *Soc. Neurosci. Abstr.*

Chung, S. K., Cohen, R. S., and Pfaff, D. W. (1984). Ultrastructure and enzyme digestion of nucleoli and associated structures in hypothalamic nerve cells viewed in resinless sections. *Biol. Cell.* **51,** 23–34.

Clemens, L. G., Humphrys, R. R., and Dohanich, G. P. (1980). Cholinergic brain mechanisms and the hormonal regulation of female sexual behavior in the rat. *Pharmacol. Biochem. Behav.* **13,** 81–88.

Clever, U. (1964). Actinomycin and puromycin: Effects on sequential gene activation by acdysone. *Science* **146,** 794–795.

Cohen, M. S., Schwartz-Giblin, S., and Pfaff, D. W. (1985). The pudendal nerve-evoked response in axial muscle. *Exp. Brain Res.* **61,** 175–185.

Cohen, M. S., Schwartz-Giblin, S., and Pfaff, D. W. (1987a). Effects of total and partial spinal transections on the pudendal nerve-evoked response in rat lumbar axial muscle. *Brain Res.* **40,** 103–112.

Cohen, M. S., Schwartz-Giblin, S., and Pfaff, D. W. (1987b). Brainstem reticular stimulation facilitates back muscle motoneuronal responses to pudendal nerve input. *Brain Res.* **405,** 155–158.

Cohen, R. S., and Pfaff, D. W. (1981). Ultrastructure of neurons in the ventromedial nucleus of the hypothalamus in ovariectomized rats with or without estrogen treatment. *Cell Tissue Res.* **217,** 451–470.

Cohen, R. S., Chung, S. K., and Pfaff, D. W. (1984). Alteration by estrogen of the nucleoli in nerve cells of the rat hypothalamus. *Cell Tissue Res.* **235,** 485–489.

Conrad, L. C. A., and Pfaff, D. W. (1976). Efferents from medial basal forebrain and hypothalamus in the rat. II. An autoradiographic study of the anterior hypothalamus. *J. Comp. Neurol.* **169,** 221–262.

Cottingham, S. L., Femano, P. A., and Pfaff, D. W. (1987). Electrical stimulation of the midbrain central gray facilitates reticulospinal activation of axial muscle EMG. *Exp. Neurol.* **97,** 704–724.

Davis, P. G., McEwen, B. S., and Pfaff, D. W. (1979). Localized behavioral effects of tritiated estradiol implants in the ventromedial hypothalamus of female rats. *Endocrinology (Baltimore)* **104,** 898–903.

Davis, P. G., Krieger, M. S., Barfield, R. J., McEwen, B. S., and Pfaff, D. W. (1982). The site of action of intrahypothalamic estrogen implants in feminine sexual behavior: An autoradiographic analysis. *Endocrinology (Baltimore)* **111,** 1581–1586.

Eccles, J. C., and Krnjevic, K. (1959). Potential changes recorded inside primary afferent fibres within the spinal cord. *J. Physiol. (London)* **149,** 250–273.

Eccles, J. C., Schmidt, R. F., and Willis, W. D. (1963). Pharmacological studies on presynaptic inhibition. *J. Physiol. (London)* **168,** 500–530.

Edwards, D. A., and Pfeifle, (1981). Hypothalamic and midbrain control of sexual receptivity in female rats. *Physiol. Behav.* **26,** 1061–1067.

Eisenfeld, A. J. (1970). ^{3}H-Estradiol: *In vivo* binding to macromolecules from the rat hypothalamus, anterior pituitary and uterus. *Endocrinology (Baltimore)* **86,** 1313–1318.

Eisenfeld, A. J., and Axelrod, J. (1965). Selectivity of estrogen distribution in tissues. *J. Pharmacol. Exp. Ther.* **150,** 469–475.

Elde, R., Hokfelt, T., Johansson, O., Ljungdahl, A., Nilsson, G., and Jeffcoate, S. L. (1978). Immunohistochemical localization of peptides in the nervous system. *In* "Centrally Acting Peptides" (J. Hughes, ed.), pp. 17–35. Macmillan, London.

Emery, D. E. (1984). Effects of endocrine state on sociosexual behavior of female rats tested in a complex environment. *Behav. Neurosci.* **100,** 71–78.

Emery, D. E., and Moss, R. L. (1984). Lesions confined to the ventromedial hypothalamus alter motivation for copulation in female rats. *Horm. Behav.* **18,** 313–329.

Erulkar, S. D., Kelley, D. B., Jurman, M. E., Zemlan, F. P., Schneider, G. T., and Krieger, N. R. (1981). Modulation of the neural control of the clasp reflex in male Xenopus laevis by androgens: A multidisciplinary study. *Proc. Natl. Acad. Sci. U.S.A.* **78,** 5876–5880.

Femano, P. A., Schwartz-Giblin, S., and Pfaff, D. W. (1984a). Brainstem reticular influences on lumbar axial muscle. I. Effective sites. *Am. J. Physiol.* **246,** R389–R395.

Femano, P. A., Schwartz-Giblin, S., and Pfaff, D. W. (1984b). Brainstem reticular influences on lumbar axial muscle. II. Temporal aspects. *Am. J. Physiol.* **246,** R3936–R401.

Fernandez-Guasti, A., Larsson, K., and Beyer, C. (1986). Lordosis behavior and GABAergic neurotransmission. *Pharmacol., Biochem. Behav.* **24,** 673–676.

Floody, O. R., and O'Donohue, T. L. (1980). Lesions of the mesencephalic central gray depress ultrasound production and lordosis by female hamsters. *Physiol. Behav.* **24,** 79–85.

Flügge, G., Oertel., W. H., and Wuttke, W. (1986). Evidence for estrogen-receptive GABAergic neurons in the preoptic/anterior hypothalamic area of the rat brain. *Neuroendocrinology* **43,** 1–5.

Foreman, M. M., and Moss, R. L. (1979). Role of hypothalamic dopaminergic receptors in the control of lordosis behavior in the female rat. *Physiol. Behav.* **22,** 283–289.

Fuxe, K., Hokfelt, T., Eneroth, P. Gustafsson, J.-A., and Skett, P. (1977). Prolactin-like immunoreactivity:localization in nerve terminals of rat hypothalamus. *Science* **196,** 899–900.

Haase, J., and Van Der Meulen, J. P. (1961). Effects of supraspinal stimulation on Renshaw cells belonging to extensor motoneurones. *J. Neurophysiol.* **24,** 510–520.

Hagbarth, K.-E. (1952). Excitatory and inhibitory skin areas for flexor and extensor motoneurons. *Acta Physiol. Scand.* **26,** Suppl. 94, 1–58.

Hall, E. D., Baker, T., and Riker, W. F. (1978). Glucocorticoid effects on spinal cord function. *J. Pharmacol. Exp. Ther.* **206,** 361–370.

Harlan, R. E., Shivers, B. D., Kow, L.-M., and Pfaff, D. W. (1982). Intrahypothalamic colchicine infusions disrupt lordotic responsiveness in estrogen-treated female rats. *Brain Res.* **238,** 153–167.

Harlan, R. E., Shivers, B. D., and Pfaff, D. W. (1983). Midbrain microinfusions of prolactin increase the estrogen-dependent behavior, lordosis. *Science* **219,** 1451–1453.

Hill, A. V. (1938). The heat of shortening and the dynamic constants of muscle. *Proc. R. Soc. London, Ser. B* **126,** 136.

Hokfelt, T., Kellerth, J. O., Nilsson, G., and Pernow, B. (1975). Experimental immunohistochemical studies on the localization and distribution of substance P in cat primary sensory neurons. Loss of substance P after dorsal rhizotomy. *Brain Res.* **100,** 235–252.

Iadarola, M. J., Douglass, J., Civelli, O., and Naranjo, J. R. (1988). Differential activation of spinal cord dynorphin and enkephalin neurons during hyperalgesia: Evidence using cDNA hybridization. *Brain Res.* **455,** 205–212.

Jankowska, E., McCrea, D., Rudomin, P., and Sykova, E. (1981). Observations on neuronal pathways subserving primary afferent depolarization. *J. Neurophysiol.* **46,** 506–516.

Jensen, E. V., and Jacobson, H. I. (1962). Basic guides to the mechanism of estrogen action. *Recent Prog. Horm. Res.* **18,** 387–408.

Jessel, T. M., and Iversen, L. L. (1977). Opiate analgesics inhibit substance P release from rat trigeminal nucleus. *Nature, (London)* **268,** 549–551.

Johnson, M. D., and Crowley, W. R. (1983). Acute effects of estradiol on circulating luteinizing hormone and prolactin concentrations and on serotonin turnover in individual brain nuclei. *Endocrinology (Baltimore)* **113,** 1935–1941.

Kanda, K., Burke, R. E., and Walmsley, B. (1977). Differential control of fast and slow twitch motor units in the decerebrate czt. *Exp. Brain Res.* **29,** 57–74.

Kato, J., and Villee, C. A. (1967). Preferential uptake of estradiol by the anterior hypothalamus of the rat. *Endocrinology (Baltimore)* **80,** 567–575.

Komisaruk, B. R., Adler, N. T., and Hutchison, J. (1972). Genital sensory field: enlargement by estrogen treatment in female rats. *Science* **178,** 1295–1298.

Kow, L.-M., and Pfaff, D. W. (1973). Effects of estrogen treatment on the size of receptive field and response threshold of pudendal nerve in the female rat. *Neuroendocrinology* **13,** 299–313.

Kow, L.-M., and Pfaff, D. W. (1976). Sensory requirements for the lordosis reflex in female rats. *Brain Res.* **101,** 47–66.

Kow, L.-M., and Pfaff, D. W. (1979). Responses of single units in sixth lumbar dorsal root ganglion of female rats to mechanostimulation relevant for lordosis reflex. *J. Neurophysiol.* **42,** 203–213.

Kow, L.-M., and Pfaff, D. W. (1982). Responses of medullary reticulospinal and other reticular neurons to somatosensory and brainstem stimulation in anesthetized or freely-moving ovariectomized rats with or without estrogen treatment. *Exp. Brain Res.* **47,** 191–202.

Kow, L.-M., Malsbury, C. W., and Pfaff, D. W. (1974). Effects of progesterone on female reproductive behavior in rats: Possible modes of action and role in behavioral sex differences. *In* "Reproductive Behavior" (W. Montagna and W. A. Sadler, eds.), pp. 179–210. Plenum, New York.

Kow, L.-M., Montgomery, M., and Pfaff, D. W. (1977). Effects of spinal cord transections on lordosis reflex in female rats. *Brain Res.* **123,** 75–88.

Kow, L.-M., Zemlan, F. P., and Pfaff, D. W. (1980). Responses of lumbosacral spinal units to mechanical stimuli related to analysis of the lordosis reflex in female rats. *J. Neurophysiol.* **43,** 27–45.

Krieger, M. S., Conrad, L. C. A., and Pfaff, D. W. (1979). An autoradiographic study of the efferent connections of the ventromedial nucleus of the hypothalamus. *J. Comp. Neurol.* **183**, 785–816.

Lamberts, R., Vijayan, E., Graf, M., Mansky, T., and Wüttke, W. (1983). Involvement of preoptic-anterior hypothalamic GABA neurons in the regulation of pituitary LH and prolactin release. *Exp. Brain Res.* **52**, 356–362.

LaMotte, C., Pert, C. B., and Snyder, S. H. (1976). Opiate receptor binding in primate spinal cord:distribution and changes after dorsal root section. *Brain Res.* **112**, 407–412.

Le Bars, D., Guilbaud, G., Jurna, I., and Besson, J. M. (1976). Differential effects of morphine on responses of dorsal horn lamina V type cells elicited by A and C fibre stimulation in the spinal cat. *Brain Res.* **115**, 518–524.

Levy, R. A. (1977). The role of GABA in primary afferent depolarization. *Prog. Neurobiol.* **9**, 211–267.

Luine, V. N., and Fischette, C. T. (1982). Inhibition of lordosis behavior by intrahypothalamic implants of pargyline. *Neuroendocrinology* **34**, 237–244.

Luine, V. N., and McEwen, B. S. (1983). Sex differences in cholinergic enzymes of diagonal band nuclei in the preoptic area. *Neuroendocrinology* **36**, 475–482.

Luine, V. N., and Paden, C. M. (1982). Effects of monoamine exidase inhibition on female sexual behavior, serotonin levels and type A and B monoamine oxidase activity. *Neuroendocrinology* **34**, 245–251.

Luine, V. N., and Rhodes, J. (1983). Gonadal hormone regulation of MAO and other enzymes in hypothalamic areas. *Neuroendocrinology* **36**, 235–241.

Luine, V. N., Frankfurt, M., Rainbow, T. C., Biegon, A., and Azmitia, E. (1983). Intrahypothalamic 5,7-dihydroxytryptamine facilitates feminine sexual behavior and decreases [³H]imprimine binding and 5-HT uptake. *Brain Res.* **264**, 344–348.

Lundberg, A., and Vyklicky, L. (1966). Inhibition of transmission to primary afferents by electrical stimulation of the brain stem. *Arch. Ital. Biol.* **104**, 86–97.

Majewska, M. D., Harrison, N. L., Schwartz, R. D., Barker, J. L., and Paul, S. M. (1986). Steroid hormone metabolites are barbiturate-like modulators of the GABA receptor. *Science* **232**, 1004–1007.

Malsbury, C. W., Kelley, D. B., and Pfaff, D. W. (1972). Responses of single units in the dorsal midbrain to somatosensory stimulation in female rats. *Int. Congr. Ser.—Excerpta Med.* **273**, 205–209.

Malsbury, C. W., Pfaff, D. W., and Malsbury, A. M. (1980). Suppression of sexual receptivity in the female hamster: Neuroanatomical projections from preoptic and anterior hypothalamic electrode sites. *Brain Res.* **181**, 267–284.

Manogue, K., Kow, L.-M., and Pfaff, D. W. (1980). Selective brain stem transections affecting reproductive behavior of female rats: The role of hypothalamic output to the midbrain. *Horm. Behav.* **14**, 277–302.

Martin, R. F., Haber, L. H., and Willis, W. D. (1979). Primary afferent depolarization of identified cutaneous fibers following stimulation in medial brain stem. *J. Neurophysiol.* **42**, 779–790.

Mathews, D., and Edwards, D. A. (1977). Involvement of the ventromedial and anterior hypothalamic nuclei in the hormonal induction of receptivity in the female rat. *Physiol. Behav.* **19**, 319–326.

Mathews, D., Donovan, K. M., Hollingsworth, E. M., Hutson, V. B., and Overstreet, C. T. (1983a). Permanent deficits in lordosis behavior in female rats with lesions of the ventromedial nucleus of the hypothalamus. *Exp. Neurol.* **79**, 714–719.

Mathews, D., Greene, S. B., and Hollingsworth, E. M. (1983b). VMN lesion deficits in lordosis: Partial reversal with pergolide mesylate. *Physiol. Behav.* **31**, 745–748.

McEwen, B. S. (1981). Neural estrogen receptors in the life cycle of the albino rat. *Exp. Brain Res., Suppl.* **3**, 61–79.

McEwen, B. S., and Pfaff, D. W. (1970). Factors influencing sex hormone uptake by rat brain regions. I. Effects of neonatal treatment, hypophysectomy, and competing steroid on estradiol uptake. *Brain Res.* **21**, 1–16.

McEwen, B. S., Krey, L. C., and Luine, V. N. (1978). Steroid hormone action in the neuroendocrine system: When is the genome involved? *In* "The Hypothalamus" (S. Reichlin, R. J. Baldessarini, and J. B. Martin, eds.), pp. 255–268. Raven Press, New York.

McEwen, B. S., Davis, P. G., Gerlach, J. L., Krey, L. C., MacLusky, N. J., McGinnis, M. Y., Parsons, B., and Rainbow, T. C. (1982). Progestin receptors in the brain and pituitary gland. *In* "Progesterone and Progestins" (C. W. Bardin, E. Milgrom, and P. Mauvais-Jarvis, eds.), pp. 59–76. Raven Press, New York.

McEwen, B. S., Snyder, L., and Rainbow, T. C. (1983). Correlation of nuscarinic receptor induction in the ventromedial nucleus with the activation of feminine sexual behavior by estradiol. *Abstr. Soc. Neurosci., 13th Annu. Meet.* Vol. 9, No. 314.4, p. 1077.

McGinnis, M. Y., Parsons, B., Rainbow, T. C., Krey, L. C., and McEwen, B. S. (1981). Temporal relationship between cell nuclear progestin receptor levels and sexual receptivity following intravenous progesterone administration. *Brain Res.* **218**, 365–371.

McLaughlin, B. J., Barber, R., Saito, K., Roberts, E., and Wu, J. Y. (1975). Immunocytochemical localization of glutamate decarboxylase in rat spinal cord. *J. Comp. Neurol.* **164**, 305–321.

Mendell, L. M., and Wall, P. D. (1964). Presynaptic hyperpolarization: A role for fine afferent fibres. *J. Physiol. (London)* **172**, 274–294.

Meyerson, B. J., Carrer, H., and Eliasson, M. (1974). 5-Hydroxytryptamine and sexual behavior in the female rat. *Adv. Biochem. Psychopharmacol.* **11**.

Modianos, D., and Pfaff, D. W. (1979). Medullary reticular formation lesions and lordosis reflex in female rats. *Brain Res.* **171**, 334–338.

Moreau, J.-L., and Fields, H. L. (1986). Evidence for GABA involvement in midbrain control of medullary neurons that modulate nociceptive transmission. *Brain Res.* **397**, 37–46.

Morrell, J. I., and Pfaff, D. W. (1982). Characterization of estrogen-concentrating hypothalamic neurons by their axonal projections. *Science* **217**, 1273–1276.

Morrell, J. I., Wolinsky, T. D., Krieger, M. S., and Pfaff, D. W. (1982). Autoradiographic identification of estradiol-concentrating cells in the spinal cord of the female rat. *Exp. Brain Res.* **45**, 144–150.

Morrell, J. I., McGinty, J., and Pfaff, D. W. (1983). Some steroid hormone concentrating cells in the medial basal hypothalamus (MBH) and anterior pituitary contain B-endorphin or dynorphin. *Soc. Neurosci. Abstr.* **9** (Part 1), 90.

Moss, R. L., and Foreman, M. M. (1976). Potentiation of lordosis behavior by intrahypothalamic infusion of synthetic LHRH. *Neuroendocrinology* **20**, 176–181.

Moss, R. L., and McCann, S. M. (1973). Induction of mating behavior in rats by luteinizing hormone releasing factor. *Science* **181**, 177–179.

Nicoll, R. A., Alger, B. E., and Jahr, C. E. (1980). Enkephalin blocks inhibitory pathways in the vertebrate CNS. *Nature (London)* **287**, 22–25.

Nishizuka, M., and Pfaff, D. W. (1983). Pattern of synapses on ventromedial hypothalamic neurons in the female rat: An electron microscopic study. *Soc. Neurosc. Abstr.* **9** (Part 1), 317.

Okada, R., Watanabe, H., Yamanouchi, K., and Arai, Y. (1980). Recovery of sexual receptivity in female rats with lesions of the ventromedial hypothalamus. *Exp. Neurol.* **68**, 595–600.

Parsons, B., Rainbow, T. C., MacLusky, N. J., and McEwen, B. S. (1982a). Progestin receptor levels in rat hypothalamic and limbic nuclei. *J. Neurosci.* **2**, 1446–1452.

Parsons, B., McEwen, B. S., and Pfaff, D. W. (1982b). A discontinuous schedule of estradiol treatment is sufficient to activate progesterone-facilitated feminine sexual behavior and to increase cytosol progestin receptors in the hypothalamus of the rat. *Endocrinology (Baltimore)* **110**, 613–619.

Parsons, B., Rainbow, T. C., Pfaff, D. W., and McEwen, B. S. (1982c). Hypothalamic protein synthesis essential for the activation of the lordosis reflex in the female rat. *Endocrinology (Baltimore)* **110**, 620–624.

Parsons, B., Rainbow, T. C., Snyder, L., and McEwen, B. S. (1984). Progesterone-like effects of estradiol on reproductive behavior and hypothalamic progestin receptors in the female rat. *Neuroendocrinology* **38**, 25–30.

Peterson, B. W., Pitts, N. G., and Fukushima, K. (1979). Reticulospinal connections with limb and axial motoneurons. *Exp. Brain Res.* **36**, 1–20.

Pfaff, D. W. (1968). Uptake of estradiol-17B-3H in the female rat brain: An autoradiographic study. *Endocrinology (Baltimore)* **82**, 1149–1155.

Pfaff, D. W. (1973). Luteinizing hormone releasing factor (LRF) potentiates lordosis behavior in hypophysectomized ovariectomized female rats. *Science* **182**, 1148–1149.

Pfaff, D. W. (1980). "Estrogens and Brain Function: Neural Analysis of a Hormone-Controlled Mammalian Reproductive Behavior." Springer-Verlag, New York.

Pfaff, D. W., ed. (1982). "The Physiological Mechanisms of Motivation." Springer-Verlag, New York.

Pfaff, D. W. (1983). Impact of steroid hormones on hypothalamic neurons: Chemical, electrical and ultrastructural studies. *Recent Prog. Horm. Res.* **39,** 127–179.

Pfaff, D. W., and Conrad, L. C. A. (1978). Hypothalamic neuroanatomy: Steroid hormone binding and patterns of axonal projections. *Int. Rev. Cytol.* **54,** 245–265.

Pfaff, D. W., and Keiner, M. (1973). Atlas of estradiol-concentrating cells in the central nervous system of the female rat. *J. Comp. Neurol.* **151,** 121–158.

Pfaff, D. W., and Lewis, C. (1974). Film analyses of lordosis in female rats. *Horm. Behav.* **5,** 317–335.

Pfaff, D. W., and Sakuma, Y. (1979a). Facilitation of the lordosis reflex of female rats from the ventromedial nucleus of the hypothalamus. *J. Physiol. (London)* **288,** 189–202.

Pfaff, D. W., and Sakuma, Y. (1979b). Deficit in the lordosis reflex of female rats caused by lesions in the ventromedial nucleus of the hypothalamus. *J. Physiol. (London)* **288,** 203–210.

Pfaff, D. W., and Schwartz-Giblin, S. (1988). Cellular mechanisms of female reproductive behaviors. *In* "The Physiology of Reproduction" (E. Knobil and J. D. Neill, eds.), Chapter 35, pp. 1487–1568. Raven Press, New York.

Pfaff, D. W., Lewis, C., Diakow, C., and Keiner, M. (1972). Neurophysiological analysis of mating behavior responses as hormone-sensitive reflexes. *Prog. Physiol. Psychol.* **5,** 253–297.

Pfaff, D. W., Montgomery, M., and Lewis, C. (1977). Somatosensory determinants of lordosis in female rats: Behavioral definition of the estrogen effect. *J. Comp. Physiol. Psychol.* **91,** 134–145.

Rainbow, T. C., DeGroff, V., Luine, V. N., and McEwen, B. S. (1980). Estradiol 17β increases the number of muscarinic receptors in hypothalamic nuclei. *Brain Res.* **198,** 239–243.

Rainbow, T. C., Parsons, B., MacLusky, N. J., and McEwen, B. S. (1982a). Estradiol receptor levels in rat hypothalamic and limbic nuclei. *J. Neurosci.* **2,** 1439–1445.

Rainbow, T. C., McGinnis, M. Y., Davis, P. G., and McEwen, B. S. (1982b). Application of anisomycin to the lateral ventromedial nucleus inhibits the activation of sexual behavior by estradiol and progesterone. *Brain Res.* **233,** 417–423.

Rakic, P. (1976). "Local Circuit Neurons." MIT Press, Cambridge.

Rethelyi, M., and Szentagothai, J. (1969). The large synaptic complexes of the substantia gelatinosa. *Exp. Brain Res.* **7,** 258–274.

Riskind, P., and Moss, R. L. (1979). Midbrain central gray: an extrahypothalamic site for LHRH potentiation of lordosis behavior in female rats. *Brain Research Bull.* **4,** 203–208.

Rose, J. D. (1982). Midbrain distribution of neurons with strong, sustained responses to lordosis trigger stimuli in the female golden hamster. *Brain Research* **240,** 364–367.

Rotsztejn, W. H., Drouva, S. V., Pattou, E., and Kordon, C. (1978). Met-enkephalin inhibits *in vitro* dopamine-induced LHRH release from mediobasal hypothalamus of male rats. *Nature (London)* **274,** 281–282.

Rubin, B. S., and Barfield, R. J. (1980). Priming of estrous responsiveness by implants of 17B-estradiol in the ventromedial hypothalamic nucleus of female rats. *Endocrinology* **106,** 504–509.

Rubin, B. S., and Barfield, R. J. (1983). Progesterone in the ventromedial hypothalamus facilitates estrous behavior in ovariectomized estrogen-primed rats. *Endocrinology* **113,** 797–804.

Rudomin, P., Jimenez, I., Solodkin, M., and Duenas, S. (1983). Sites of action of segmental and descending control of transmission on pathways mediating PAD of Ia- and Ib-afferent fibers in cat spinal cord. *J. Neurophysiol.* **50,** 743–769.

Sakuma, Y., and Pfaff, D. W. (1979a). Facilitation of female reproductive behavior from mesencephalic central gray in the rat. *Am. J. Physiol.* **237,** R278–R284.

Sakuma, Y., and Pfaff, D. W. (1979b), Mesencephalic mechanisms for integration of female reproductive behavior in the rat. *Am. J. Physiol.* **237,** R285–R290.

Sakuma, Y., and Pfaff, D. W. (1980a). Cells of origin of medullary projections in central gray of rat mesencephalon. *J. Neurophysiol.* **44,** 1002–1011.

Sakuma, Y., and Pfaff, D. W. (1980b). Excitability of female rat central gray cells with medullary projections: Changes produced by hypothalamic stimulation and estrogen treatment. *J. Neurophysiol.* **44,** 1012–1023.

Sakuma, Y., and Pfaff, D. W. (1980c). Convergent effects of lordosis-relevant somatosensory and hypothalamic influences on central gray cells in the rat mesencephalon. *Exp. Neurol.* **70,** 269–281.

Sakuma, Y., and Pfaff, D. W. (1980d). LH-RH in the mesencephalic central gray can potentiate lordosis reflex of female rats. *Nature (London)* **283,** 566–567.

Sawynok, J. (1987). GABAergic mechanisms of analgesia: An update. *Pharmacol., Biochem. Behav.* **26,** 463–474.

Schacter, B., Shivers, B., Harlan, R., and Pfaff, D. W. (1983). Evidence for prolactin messenger RNA in the rat brain. *Endocr. Soc. Abstr.* **346,** 167.

Schmidt, R. F. (1971). Presynaptic inhibition in the vertebrate central nervous system. *Ergebn. Physiol., Biol. Chem. Exp. Pharmakol.* **63,** 20–101.

Schwanzel-Fukuda, M., Morrell, J. I., and Pfaff, D. W. (1983). Polyacrylamide gel provides slow release delivery of wheat germ agglutinin (WGA) for retrograde neuroanatomical tracing. *J. Histochem. Cytochem.* **31,** 831–836.

Schwanzel-Fukuda, M., Morrell, J. I., and Pfaff, D. W. (1984). Localization of forebrain neurons which project directly to the medulla and spinal cord of the rat by retrograde tracing with wheat germ agglutinin. *J. Comp. Neurol.* **226,** 1–20.

Schwartz-Giblin, S., and Pfaff, D. W. (1985). Intrathecal bicuculline in rats evokes hyperalgesia and exaggerates avoidance of contact by other rats. *Soc. Neurosci. Abstr.* **83,** 5.

Schwartz-Giblin, S., and Pfaff, D. W. (1988). Control of amplitude in a cutaneous reflex subserving lordosis: A role for a progesterone metabolite. *Soc. Neurosci. Abstr.* **14,** No. 79.14, 184.

Schwartz-Giblin, S., Rosello, L., and Pfaff, D. W. (1983). A histochemical study of lateral longissimus muscle in rat. *Exp. Neurol.* **79,** 497–518.

Schwartz-Giblin, S., Halpern, M., and Pfaff, D. W. (1984a). Segmental organization of rat lateral longissimus, a muscle involved in lordosis behavior: EMG and muscle nerve recordings. *Brain Res.* **299,** 247–257.

Schwartz-Giblin, S., Femano, P. A., and Pfaff, D. W. (1984b). Axial electromyogram and intervertebral length gauge responses during lordosis behavior in rats. *Exp. Neurol.* **85,** 297–315.

Scouten, C. W., and Malsbury, C. W. (1983). Labeling knife cuts used to locate the cells of origin of axons necessary for lordosis behavior. *Soc. Neurosci. Abstr.* **9** (Part 1), No. 150.21, 519.

Sherrington, C. (1906). "The Integrative Action of the Nervous System." Yale Univ. Press, New Haven, Connecticut.

Shimamura, M., and Livingston, R. B. (1963). Longitudinal conduction systems serving spinal and brain stem coordination. *J. Neurophysiol.* **26,** 258–272.

Shivers, B. D., Harlan, R., Morrell, J. I., and Pfaff, D. W. (1983a). Immunocytochemical localization of luteinizing hormone-releasing hormone in male and female rat brains. *Neuroendocrinology* **36,** 1–12.

Shivers, B. D., Harlan, R., and Pfaff, D. W. (1983b). Immunocytochemical mapping of immunoreactive prolactin in female rat brain. *Soc. Neurosci. Abstr.* **9** (Part 2), No. 294.18, 1018.

Sirinathsinghji, D. J. S. (1983). GnRH in the spinal subarachnoid space potentiates lordosis behavior in the female rat. *Physiol. Behav.* **31,** 717–723.

Sirinathsinghji, D. J. S., Whittington, P. E., Audsley, A. R., and Fraser, H. M. (1983). B-endorphin regulates lordosis in female rats by modulating LH-RH release. *Nature (London)* **301,** 62–64.

Stein, C., Morgan, M. M., and Liebeskind, J. C. (1987). Barbiturate-induced inhibition of a spinal nociceptive reflex: Role of GABA mechanisms and descending modulation. *Brain Res.* **407,** 307–311.

Stumpf, W. E. (1968). Estradiol-concentrating neurons: Topography in the hypothalamus by dry mount autoradiography. *Science* **162,** 1001–1003.

Sullivan, J. M., Schwartz-Giblin, S., and Pfaff, D. W. (1986). Correlations between EEG state and spontaneous and evoked axial muscle EMG. *Brain Res.* **368,** 197–200.

Suzue, T., and Jessell, T. (1980). Opiate analgesics and endorphins inhibit rat dorsal root potential in vitro. *Neurosci. Lett.* **16,** 161–166.

Swanson, L. W., and Kuypers, H. G. J. M. (1980). A direct projection from the ventromedial nucleus and retrochiasmatic area of the hypothalamus to the medulla and spinal cord of the rat. *Neurosci. Lett.* **17,** 307–312.

Tubeau, G., Desclin, J., Parmentier, M., and Pasteels, J. L. (1979). Cellular localization of a prolactin-like antigen in the rat brain. *J. Endocrinol.* **83,** 261–266.

Wallis, C. J., and Lüttge, W. G. (1980). Influence of estrogen and progesterone on glutamic acid decarboxylase activity in discrete regions of rat brain. *J. Neurochem.* **34,** 609–613.

Wiesenfeld-Hallin, Z., and Sodersten, P. (1984). Spinal opiates affect sexual behaviour in rats. *Nature (London)* **309,** 257–258.

Wiesner, J. B., and Moss, R. L. (1984). Beta-endorphin suppression of lordosis behavior in female rats; lack of effect of peripherally-administered naloxone. *Life Sciences* **34**(2), 1455–1462.

Wilson, V. J., Talbot, W. H., and Diecke, F. P. J. (1960). Distribution of recurrent facilitation and inhibition in cat spinal cord. *J. Neurophysiol.* **23,** 144–153.

Wilson, V. J., Talbot, W. H., and Kato, M. (1964). Inhibitory convergence upon Renshaw cells. *J. Neurophysiol.* **27,** 1063–1079.

Witkin, J. W., Paden, C. M., and Silverman, A.-J. (1982). The luteinizing hormone-releasing hormone (LHRH) systems in the rat brain. *Neuroendocrinology* **35,** 429–438.

Zemlan, F. P., Kow, L.-M., Morrell, J., and Pfaff, D. W. (1979). Descending tracts of the lateral columns of the rat spinal cord: A study using the horseradish peroxidase and silver impregnation techniques. *J. Anat.* **128,** 489–512.

Zemlan, F. P., Kow, L.-M., and Pfaff, D. W. (1983). Effect of interruption of bulbospinal pathways on lordosis, posture, and locomotion. *Exp. Neurol.* **81,** 177–194.

Zieglgansberger, W., and Bayerl, H. (1976). The mechanism of inhibition of neuronal activity by opiates in the spinal cord of cat. *Brain Res.* **115,** 111–128.

Zigmond, R. E., and McEwen, B. S. (1970). Selective retention of oestradiol by cell nuclei in specific brain regions of the ovariectomized rat. *J. Neurochem.* **17,** 889–899.

CHAPTER THREE

Maternal Behavior: Sensory, Hormonal, and Neural Determinants

Judith M. Stern

I. Introduction

A. *Nursing as Defining Behavior of Class Mammalia*

A mammal is defined in Webster's New World Dictionary as "any of a large class (Mammalia) of warm-blooded, usually hairy vertebrates whose offspring are fed with milk secreted by the female mammary glands." Therefore, nursing behavior, including both the mother's behavior and that of her young, is the defining behavioral category of the class Mammalia. The young come equipped with a set of behaviors which enable them to find and ingest, usually by suckling, mother's milk. In turn, mothers at the least are tolerant of young while they are locating and obtaining their milk, but often engage in behaviors as well which enhance the likelihood of the young being successful. The degree to which the mother is actively involved in ensuring the young's access to her milk spans the class, from monotremes, in which the fetal-like young make their way unaided from the cloaca to the teat, to our own species, in which the young must be held and brought to the breast.

A comprehensive neuroethological analysis of mammalian maternal behavior, grounded in an understanding of the sensory, neural, and motor mechanisms, has been impeded in large part by the complexity of maternal behavior, which includes all behaviors engaged in by the mother related to

Psychoendocrinology, copyright © 1989 by Academic Press, Inc. All rights of reproduction in any form reserved.

her young, behaviors very different in appearance and function from nursing. These other behaviors provide the young with heat, shelter, cleaning, defense, and guidance, in addition to nourishment. To make matters even more confusing, males of certain species regularly participate in the care of their young, with behaviors that may overlap with those of the mother; these behaviors are termed paternal. However, by focusing on nursing behavior and its immediate antecedents, I believe that great strides will be made in our understanding of the neurobiological organization of maternal behavior, which is of such key importance to the survival of all mammals.

B. Norway Rats (Rattus norvegicus)

Mammalian species vary widely, of course, in the details and extent of their nurturance of offspring (Eisenberg, 1981; Rosenblatt et al., 1985). At one extreme are marsupials, some species of which gestate for a shorter period than the length of the estrous cycle and therefore appear to have no endocrine recognition of pregnancy (Hunsaker and Shupe, 1977, pp. 290–300). Indeed, aside from an increase in pouch cleaning (by licking) prior to and after the emergence of the young from the cloaca, no active nurturant behaviors are displayed until the joey is old enough to detach from the teat and emerge from the pouch. At the other extreme of infraprimate mammals, Norway rats (rats, heretofore) have a rich repertoire of maternal behaviors which they engage in frequently. Laboratory rodents, rats in particular, have received a great deal of attention from students of the psychobiology of maternal behavior and rats are the subject of my own research. Therefore, research on rats will be the main focus of this chapter, but with a comparative perspective interspersed throughout and at the finale.

Patterns of nurturance, including behavioral and physiological adaptations, are affected by ecology (e.g., terrestrial or aquatic; degree of sociability). Rats are a burrowing species and have a maternal burrow which is well defended from both conspecific and allospecific intruders, a characteristic consistent with the birth of *altricial* young (Ewer, 1968). Rodents which live above ground, such as guinea pigs and spiny mice, have *precocial* young. The burrow isolation also means that rats do not require an olfactory mechanism for "imprinting to," and exclusive preference for, their young, which is true for herd (e.g., sheep) and colony (e.g., bats) species. Despite intensive nurturance for a short period of time, rats are *polytocous;* they have 12 teats, and at least in the benevolent atmosphere of the laboratory, litters of 12 or more are common. In nature, male Norway rats are not known to engage in paternal behavior of any sort.

Maternal behavior is usually studied in a natural model, i.e., as an outcome of pregnancy and parturition, with the postpartum female in a state of lactation, and the rat is no exception. However, several other models are appropriate for rats, including postpartum maternal behavior in nonlactating dams, sensitized maternal behavior which is induced by extensive cohabitation with pups, and hormone-induced maternal behavior. The results from these different models have provided valuable insights into the organization of maternal behavior and its hormonal control. There have been very exciting developments in the last few years in elucidating which hormones activate the display of maternal behavior, especially with respect to prolactin and intracerebral oxytocin. The next step is to understand how the various hormones work, for which a detailed understanding of sensory, neural, and motor mechanisms is required.

Interest in maternal behavior in recent years has culminated in several edited books (e.g., Bell and Smotherman, 1980; Gubernick and Klopfer, 1981; Elwood, 1983) and chapters (e.g., Hart, 1985; Numan, 1985; Rosenblatt *et al.*, 1985). Clearly, there are many aspects of maternal behavior that are beyond the scope of this chapter, such as maternal influences on development (Bell and Smotherman, 1980; Hofer, 1981). As to the older literature on rat maternal behavior, Wiesner and Sheard's remarkable monograph (1933) provided a generation of researchers with valuable leads, as did the early papers of Beach (1937, 1938; Beach and Jaynes, 1956a,b,c) and the analysis of Rosenblatt and Lehrman (1963). Lehrman's masterful 1961 chapter not only reviewed the early literature on all nonhuman mammals and birds, but provided heuristic and still-current conceptualizations of the psychobiology of maternal behavior.

The aim of this chapter is to begin a neuroethological analysis of maternal behavior, with emphasis on nursing behavior and its immediate antecedents. It is based on all the relevant information published to date and as yet unpublished work from my laboratory on Norway rats. Unfortunately, the literature reveals that (1) all too often the dramatic and easily quantifiable behavior of retrieval was the sole or principal measure of rat maternal behavior; (2) nursing behavior was not assessed adequately; and (3) virtually no attention has been paid as yet to actual motor mechanisms. Therefore, there are many gaps in our knowledge at this writing, but a neuroethological analysis provides the basis for testable hypotheses. The approach I am taking, in this chapter and in my current research, was inspired directly by the work of Pfaff and co-workers on feminine sexual behavior (Pfaff and Modianos, 1985) and of Zeigler and others (1985) on ingestion.

II. Description of Components, Chronology, and Functional Analysis

A. Maternal Aggression

Some form of maternal aggressiveness is widespread among mammals and is usually the most intense and reliable type of aggression among females (Svare, 1981). Shortly prior to term a number of behavioral changes take place in rats. The late pregnant female frequently becomes more aggressive and "adopts a special part of the burrow system as a breeding nest from which she now excludes" conspecifics (and allospecifics) (Ewer, 1968); these are aspects of *maternal aggression,* i.e., the defense of the nest site and the young by fighting off intruders (Svare, 1981). With an enclosed nest having a small opening, the female may be able to ward off the intruder simply by guarding the entrance (Lore, 1987). As studied in the typical laboratory setting, the female leaves the nest site to investigate the intruder, with sniffing concentrated around the face. This often leads to aggressive grooming (nosing and pawing at intruder's fur, mostly anterior to neck, especially snout), usually in the upright posture. Biting (isolated, at a variety of locations) and attacking (sudden lunge, grabbing with paws, biting, rolling, and pinning) follows swiftly. The female may also display hip throws (thrusts to the male with the female's rear leg) and lateral postures (sidewise posture in which female presents lateral aspect to the male). A typical response of the intruder is the upright subordinate posture, in which the intruder sits back on his haunches, forepaws raised off the floor, with back markedly arched. Sometimes highly aggressive intruders, usually older males, will take the offensive and kill pups in the nest. Maternal aggression develops in late term (Mayer and Rosenblatt, 1984) and reaches a peak between days 1 and 5 postpartum (except during postpartum estrus) (Mayer *et al.,* 1987), although it may persist at high levels for about another week (Erskine *et al.,* 1978b).

B. Nest Building

The nest provides both shelter and warmth for the vulnerable, largely immobile, and hairless pups, which are not able to maintain body temperature. In the laboratory, nests are constructed from a variety of presented materials, including straw, paper strips, and string, which are carried to the nest site and manipulated with both mouth and paws. There is a sudden rise in anticipatory nest building activity starting the day prior to parturition and usually on the darkened side of the home cage (Beach, 1937). The maternal nest does not reach its maximum size and quality until after the litter is born.

Whereas the preparturient nest tends to be flat, the postparturient nest may have walls and cover the pups completely (Rosenblatt and Lehrman, 1963). Alternatively, the nest, along with food pellets and other available objects, forms a wall between the pups and the world. In the absence of nest materials (or sufficient pups), the exuberant dam may retrieve her tail. Nests are reconstructed at a new nest site in response to the disturbance or destruction of the original nest (Beach, 1937; Calhoun, 1962; Brewster and Leon, 1980a) and sometimes for no apparent reason (Wiesner and Sheard, 1933). Nest repair or reconstruction declines beginning about 13 days postpartum (Rosenblatt and Lehrman, 1963), a time when the now-furred pups move about the home cage. There are several problems with the use of nest building as an accurate measure of maternal responsiveness: (1) because of the many parameters involved, nest building is difficult to quantify and so is usually rated; (2) there are large individual differences among normal rats in nest-building propensity (Wiesner and Sheard, 1933); (3) no correlations were found between nesting activities and other components of postpartum maternal behavior (Slotnick, 1967a); and finally, (4) nest building appears and is modified in other contexts.

C. Retrieval and Grouping

Simultaneous with the prepartum changes described above is an increased propensity to *retrieve* alien young placed opposite the nest. Retrieval is normally accomplished by picking the pup up gently by the dorsal skin with the incisors, after briefly sniffing it, and carrying the pup to the nest site. Repeated retrievals in this way result in the *grouping* of the pups in the nest. Once the pups are all gathered in the nest, the maternal female will make one or more excursions from the nest, apparently in search of additional pups. Little or no rapid retrieval (i.e., within 6 minutes of pup presentation) occurs in cycling rats or during most of pregnancy. About 50% of Wistar and Long-Evans rats retrieve and group rapidly when tested within 3 hours prior to the onset of parturition (Slotnick *et al.*, 1973; Stern, 1987a), whereas over 90% of a colony of Sprague-Dawley rats (with a low retrieval threshold) do so at this stage (Mayer and Rosenblatt, 1984).

During parturition, retrieval typically takes place at a short distance, with the female stretching from the nest to bring the newborn closer to her. Later, dams retrieve pups that have wandered away from the nest or are brought out of the nest by the dam while still attached to a nipple. Being dramatic and easily quantifiable, retrieval (usually of no more than six pups) has been a favorite measure to assess the onset and maintenance of maternal responsiveness ever since the comprehensive monograph of Wiesner and Sheard (1933). In these tests, any young are removed from the cage and then the dam's own young or alien young are introduced, either in a cluster diagonally opposite the nest site or spread about the cage. Retrieval of this sort declines beginning about 14 days postpartum (Rosenblatt and Lehrman, 1963), even when standard-aged pups (5–10 days old) are used. The carrying of larger pups, in the second week of their life, is facilitated by the pup's assumption of a limp "transport response" posture; this posture gradually develops during the first 10 days of life and occurs in response to cutaneous stimulation on the dorsal surface (Brewster and Leon, 1980b).

D. Licking

Except for nursing, licking is the most common mammalian pup-directed maternal behavior in most species (Lehrman, 1961, p. 1311). Furthermore, in rats and many other mammals, licking is normally the *first pup-oriented maternal behavior* to occur. Licking of the dam's own anogenital region occurs before and during the emergence of the fetus. During parturition licking serves to remove the pup's fetal membranes, thereby stimulating respiration and freeing limbs for movement. Postpartum, extended bouts of licking often occur while the pup is held in the forepaws. The licking movements proceed typically from the head to the trunk to the anogenital region, the latter being particularly attractive to the dam. Perineal stimulation of the pups elicits a response of limb extension and postural immobility (Moore and Chadwick-Dias, 1986), permitting maternal licking to continue long enough to eventually stimulate the pup's urination and defecation (Rosenblatt and Lehrman, 1963). The amount of water obtained from the ingestion of the pups' urine contributes significantly to the dam's own increasing water needs as lactation progresses (Friedman *et al.*, 1981). Once maternal behavior is established, pup licking occurs most often when the dam returns to the nest, prior to the initiation of a nursing bout. Maternal licking is maintained at high levels into the third postpartum week and then declines gradually, in large part due to age-related changes in the pups' behavior (Moore and Chadwick-Dias, 1986).

E. Nursing Behavior

I. DESCRIPTION

The most characteristic nursing behavior of rats and other altricial rodents is the *active crouching posture,* in which the dam stands over the young with a pronounced dorsal arch and with legs splayed to accommodate all or most of the pups. The posture may be described as a *ventriflexion,* involving all or most of the spinal cord, bilateral symmetry, and inhibition of locomotion. In the absence of the high arch, which may occur when pups manage to crawl under a nonmaternal female, the pups are squashed, a rare occurrence with a maternal dam (Wiesner and Sheard, 1933). Pups remain attached to nipples for up to an hour while the female is in this posture. The quiescence of the dam is required not only to enable the pups to become and remain attached, but also for milk ejection to occur in response to the suckling stimuli (see Section IV,A,3). Although the dam rearranges pups and retrieves them from the outskirts of the nest prior to assuming the crouching posture, once nursing is initiated and the dam is quiescent she becomes oblivious to the fate of pups that are not under her. The first nursing usually occurs within an hour after the last pup is born (Rosenblatt and Lehrman, 1963; Holloway *et al.,* 1980). The dam's licking and self-grooming, including both of her nipple lines, may spread amniotic fluid on her ventrum, which may be an olfactory cue for the pups' first suckling response (Teicher and Blass, 1976). The active crouching posture is predominant in the first week postpartum and then begins to decline in frequency as the *passive nursing posture* becomes more common (Stern and Levine, 1972). In the latter, the mother lies on her side while the pups suckle (catlike); this posture may occur when a dam falls to her side from an active crouching posture or when a nursing bout is initiated by pups from a recumbent dam.

While the young are small and immobile, *dam-initiated suckling* occurs by the return of the mother to the nest (Rosenblatt and Lehrman, 1963). Several minutes are spent sniffing, nuzzling, licking, handling, and rearranging the pups and self-grooming. This serves both to activate the pups (P. E. Pedersen *et al.,* 1982) and to stimulate the dam's continued maternal attentions (Stern and Johnson, 1989). The thigmotactic pups (Sturman-Hulbe and Stone, 1929) crawl under the dam's ventrum, search for and attach to nipples and, by this activity, stimulate the dam's active crouching behavior (Johnson and Stern, 1987; Stern, 1989; see Sections III,D,3 and III,D,4). The onset of *pup-initiated suckling* occurs at about day 12 postpartum, when the pups are able to locomote and seek out their mother away from the nest (Bolles and Woods, 1964; Rosenblatt and Lehrman, 1963). In this situation, the pups may provoke a nursing posture (passive or active) by burrowing under their dam's ventrum, without the preliminaries of maternal oral attentions. The *weaning phase* begins when the pups begin to eat solid food at day 16 (Bolles and Woods, 1964). At this time the mother increasingly rejects and avoids the attempts of her older offspring to suckle (Reisbick *et al.*, 1975). Weaning is often completed by days 24–28 (e.g., Bolles and Woods, 1964; Babicky, *et al.*, 1970), but often the offspring continue to be successful in at least some of their attempts to suckle up to day 40 when in the presence of a younger litter, the one conceived during their dam's postpartum estrus (Stern and Rogers, 1988). Apparently, when the mother is nursing her preweaning-aged litter, she is tolerant of the suckling of much older rats (Pfister *et al.*, 1986).

2. REGULATION AND TIMING OF NURSING BOUTS

The rat mother nurses her young at frequent intervals throughout the day (Bolles and Woods, 1964; Rosenblatt and Lehrman, 1963). By monitoring the activity of rats in a dual-chambered cage, one chamber containing the litter and nest and the other containing food and water, Grota and Ader (1969, 1974) found that immediately after delivery dams spent 85% of their time in the litter chamber, most of it nursing; there was a progressive decline in this measure so that by day 17 postpartum the mothers spent only 30% of their time in the litter chamber. On a 24-hour basis, mothers in midlactation spent almost twice as much time in the litter chamber when the lights were on than when the lights were off. This time distribution is related to a diurnal pattern of milk intake (i.e., weight gain) (Levin and Stern, 1975). However, by restricting the dam's food and water intake to the 12-hour light period, Stern and Levin (1976) showed that the circadian patterns of maternal behavior and litter weight gain are an inverse of the dam's food and water intake and presumably general activity and temperature (see below) rhythms.

Due to the work of Leon and his co-workers, much is known about the termination of nursing bouts and, therefore, the decline in the total duration of nursing time (e.g., Croskerry *et al.*, 1978; Leon *et al.*, 1978, 1985; Adels and Leon, 1986). In brief, the chronically heightened heat production of lactating dams renders them vulnerable to the acute temperature rise that often occurs during nursing, and it is the temperature rise which contributes to the termination of the nest bout. Various maternal physiological factors (see Section VI,C), as well as the growing size of the pup mass, contribute to the decline in nest bout and total nesting duration seen with elapsed time postpartum. The thermal regulation of mother–young contact helps explain many aspects of rat maternal behavior, such as the previously puzzling finding of Seitz (1954) that mothers spend more time with small litters than with large ones. However, Kittrell and Satinoff (1988) found that body temperature rises considerably more when the dam is active than it does during a nursing bout, thereby casting doubt on the belief that nursing bouts are terminated because dams cannot tolerate higher temperatures. Also, not all nest-bout terminations can be explained by a rise in maternal temperature (e.g., mothers with chilled pups), and mechanisms underlying the initiation of pup contact by mothers are just beginning to be worked out (Section III).

F. Parturition: The Normal Emergence of Maternal Behavior

1. IS PARTURITION NECESSARY?

The experimental results on the *prepartum onset of maternal behavior* in Norway rats described above are analogous to anecdotal naturalistic observations of late pregnant animals, such as sheep and goats, that steal away the newborn of another female (Ewer, 1968). These observations indicate that the hormonal events leading up to late pregnancy have prepared the female to behave nurturantly toward young and they are consonant with the well-known principle that systems develop and are functional before they are actually needed (cf. Gottlieb, 1971, on embryonic development of sensory systems). However, it would be wrong to conclude from these results, as well as from others on Cesarean delivery of offspring (Wiesner and Sheard, 1933), that the experience of parturition per se is irrelevant to the complete emergence of maternal behavior and physiology.

First, not all rats behave maternally either prepartum or following near-term pregnancy termination, and there are strain differences in the percentage of females that are maternally responsive in these circumstances. Following pregnancy termination on day 21 or 22, a high proportion of females in albino strains rapidly become maternal [94% of Wistar dams, within 30 minutes (Moltz *et al.*, 1966); Sprague-Dawley (Mayer and Rosenblatt, 1979)], but the proportion of pigmented Long-Evans dams is not as high [(Peters and Kristal, 1983); ~60% in 24 hours (Stern, 1983, 1985)]. Surgery stress was not responsible for the low rate and long latencies (Stern, 1985; see Fig. 1). In contrast, almost all intact, healthy, undisturbed female rats, pigmented or not, do become maternal at term.

Second, parturition experience may be required for the complete *suppression of infanticide* toward live young exhibited by nonpregnant and primigravid dams. The cannibalism rate is 22% among primiparous Cesarean-delivered Long-Evans dams compared to 0% among dams that experience parturition (Stern, 1985). Although some of this reduction is already evident in late-pregnant Long-Evans rats (Peters and Kristal, 1983), the experience of parturition may well be required for complete infanticide suppression. In response to Cesarean-delivered pups that were "lively, uncleaned of fetal fluids, and presented without placentas, nearly all virgins killed and

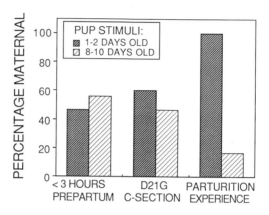

MATERNAL CONDITION

F I G U R E I. Percentage of primiparous Long-Evans female rats which became maternal toward 1- to 2- or 8- to 10-day-old rat pups in three different peripartum situations. On the left, dams were tested for retrieval for 6 minutes every 3 hours prior to the onset of parturition (Stern, 1987a). In the middle, pups were presented the day following Cesarean delivery on gestation day 21, and on the right, pups were presented 24 hours after removal of the dam's own young, one by one, during parturition (Stern, 1985). In the last two categories, the percentage maternal refers to full maternal behavior (retrieval, grouping, and crouching) up to 24 hours after pup presentation. (Adapted from Stern, 1985, 1987a.)

nearly all mothers did not" (Peters and Kristal, 1983). Further, this suppression of infanticide is maintained for the first 2 weeks postpartum, an effect not due to continued exposure to young per se (Peters, 1980).

Third, the mother's normal behavior during parturition is necessary for optimal survival of her young. Deficiencies at this time, e.g., due to anosmia (Benuck and Rowe, 1975; Holloway et al., 1980) or infraorbital denervation (Stern et al., 1987), have both short- and long-term consequences for pup viability.

Fourth, the parturition experience alone, including heightened arousal and many novel stimuli, "organizes" long-term retention of short-latency maternal responsiveness and influences pup preferences 24 hours later (see Section II,G).

Last, but certainly not least, it is during parturition that maternal behavior normally emerges in most female mammals. An understanding of how this complex behavior becomes organized, so that the various components occur in a coherent sequence, may be enhanced by a careful analysis of the parturition process. Indeed, manipulations that disrupt the onset of maternal behavior when it is dissociated from parturition are far less effective when parturition is experienced (Numan, 1974, vs. Noonan and Kristal, 1979; Ahdieh *et al.*, 1987).

2. EMERGENCE OF MATERNAL BEHAVIOR

The rapid and rather reliable appearance of nurturance at birth (for placental mammals) is a marvel of nature. A careful examination of this phenomenon, however, with romantic notions of "mother love" cast aside, reveals that the various components of maternal behavior emerge gradually and that the female is dependent upon specific stimuli, from her own body and the young, to bring about the transition to the maternal state. Parturition in the rat has been described qualitatively by Rosenblatt and Lehrman (1963) and quantitatively by Holloway *et al.* (1980).

For several days prior to the onset of labor, the female is lethargic and remains largely in one corner of her cage (Rosenblatt and Lehrman, 1963). Prior to the birth of the first, and to varying extents each subsequent pup, there is a series of *contractions,* during which the female typically lies in a stretched posture. As the contractions become more forceful and occur at shorter intervals, the fetus emerges into the birth canal, heralding the onset of *delivery.* By adjustments in her posture, the female aids in the expulsion of the fetus; these adjustments include "stretching her body full length, arching her back downward and pushing her hind legs against the floor, while contracting the lower part of her body" (Rosenblatt and Lehrman, 1963).

In response to the first fetus emerging at the vaginal orifice, the female assumes a "head between the legs posture," while sitting on her lower back, with her legs spread, thereby stretching the vaginal opening. At this point, the female's most critical and *active participation in delivery* occurs. With tongue and teeth, she vigorously licks the vaginal opening area and pulls and bites at tufts of hair and the protruding fetal membranes, which are often broken in this way. The resulting birth fluids are spread about the floor, the nesting materials, and the female's own body, and are profusely licked. The actual emergence of the fetus is often aided by the pulling of its mother, along with a final series of forceful contractions (Rosenblatt and Lehrman, 1963). The pup is then a major focus of its mother's licking (Section II,D). This sequence is repeated for the delivery of other fetuses, but not all. In some cases, the contractions, stretching movements, previous widening of the vaginal opening, and perhaps also the push of other fetuses, result in the delivery of a fetus, or even two almost simultaneously, without the dam's active participation. After the second pup is born, licking movements are distributed among the pups, the afterbirths, and the female herself (i.e., self-grooming).

Birch (1956) suggested that the female's own anogenital licking during late pregnancy is functionally related to her transition to birth fluids and pup odors at the time of parturition. However, prevention of self-licking with collars until shortly prior to parturition did not interfere with the onset and establishment of maternal behavior (Christopherson and Wagman, 1965; Kirby and Horvath, 1968). Thus, the stimuli eliciting licking during parturition are too insistent to be diminished by prior self-licking deprivation. Indeed, the intense licking by the dam of her own anogenital region in response to the emergence of a fetus is stimulus bound, reminiscent of licking that occurs in response to tail-pinch-induced arousal (e.g., Szechtman *et al.*, 1977); perhaps similar somatosensory and nigrostriatal dopaminergic mechanisms (Antelman *et al.*, 1975) are activated in response to both of these stimuli.

The placenta is expelled within a few minutes after a pup is born. The dam typically picks this up with her forepaws and ingests it, followed by the umbilical cord. The proximate and ultimate causes of *placentophagia* have been reviewed by Kristal (1980). Apparently, placentas smell and taste good to rats, especially parturient rats, and their ingestion provides a short-lasting enhancement of analgesia stemming from the endogenous release of opioids during parturition (Kristal *et al.*, 1986a,b). Other behaviors occurring between births are *sniffing* of pups and *self-grooming*, especially of the head region, which occurs often in conjunction with ingestion of birth products (Holloway *et al.*, 1980). Perhaps fortunately for the dam, which has so many new things to pay attention to, the majority of the pups are born in the second half of parturition (Holloway *et al.*, 1980). In addition, nest repair and additional *nest building* occur at intervals during and following parturition, as well as the initial *retrieval* of the young. The pups are born in fairly close proximity to each other because the dam engages in very little general activity, i.e., locomoting (Holloway *et al.*, 1980). Repeated retrievals, completed during or after the last pup's birth, result in the grouping of the pups near the dam, thus setting the stage for the first nursing.

G. Maternal Memory

The display of maternal behavior seems to have an indelible impact on a rat, much like the long-term retention of sexual behavior in sexually experienced males of many species (Bermant and Davidson, 1974). On a short-term basis, mothers learn to prefer the odor of their own pups versus alien pups (Beach and Jaynes, 1956a) and their retrieval behavior becomes quicker and more efficient with practice (Beach and Jaynes, 1956b). And 24 hours after removal of pups during parturition a pup selectivity based on size is apparent which is not present in the absence of the parturition experience (Stern, 1985; see Fig. 1). On a long-term basis, virgins that have been sensitized previously show quicker reinduction of maternal behavior, and females that have experienced parturition, or even Cesarean delivery followed by brief maternal experience, will become maternal again quickly even several weeks later (Bridges, 1975, 1977; Cohen and Bridges, 1981; Fleming and Rosenblatt, 1974c; Mayer and Rosenblatt, 1975). Developmentally, interaction with alien pups or younger siblings as weanlings increases the likelihood of maternal responsiveness to alien pups later in life (Gray and Chesley, 1984; Stern and Rogers, 1988). Further, maternally experienced rats are less vulnerable than naive ones to the maternal-behavior-suppressing effects of various endocrine (Moltz and Weiner, 1966; Moltz et al., 1966, 1969a) and sensory/neural (Fleming and Rosenblatt, 1974a; Schlein et al., 1972; Schwartz and Rowe, 1976; Franz et al., 1986) manipulations. How and where these experiential effects are encoded has received no attention, but the influence of the cerebral cortex on maternal behavior suggests that it plays a role (see Section V,A).

III. Sensory Regulation

A. The Concept of Multisensory Control: A Revised View

In invertebrates and nonmammalian vertebrates, favorite subjects of ethologists and neuroethologists, it is commonplace to find that a specific stimulus activates a specific response. Thus, the red color of a male robin's breast, a herring-gull parent's bill tip, and a female stickleback's belly are "sign stimuli" for particular responses of conspecifics (Hinde, 1970). In contrast, the sensory regulation of many mammalian species-specific behaviors appears to be more complex. Maternal behavior in rats has been considered to be under multisensory control since the work of Beach and Jaynes (1956c). By manipulating both pup characteristics and the sensory status of lactating dams with some selective denervations, and assessing maternal behavior by the retrieval response in experienced retrievers, these workers concluded that

". . . the lactating female rat's retrieving behavior normally involves a multisensory pattern of stimulation, including visual, chemical, tactile, thermal, and possibly other cues. At the same time it is evident that no one type of stimulation is essential. Any single modality can be dispensed with as long as alternative routes of sensory input remain. Elimination of two different sensory avenues does not abolish the maternal reaction but may decrease the efficiency with which it is performed. The additional elimination of a third exteroceptive pathway creates even greater interference but still does not totally eliminate the retrieving response." (pp. 120–121)

Beach and Jaynes, in this seminal 1956 paper, went on to compare the female rat's maternal behavior to that of maze learning and masculine sexual behavior in rats, also thought to be under multisensory control, and to place these three types of behaviors into a context of neocortical control. I will return to the latter issue; at this point I wish to reexamine the meaning of multisensory control and offer a revised view.

A major problem with the conclusion of Beach and Jaynes, which has been widely accepted until this day (e.g., Numan, 1985), is that it is based entirely on studies of experienced retrievers, with retrieving behavior as the only endpoint. No information was given on the number of retrieving tests each dam had received prior to a given test or of the day postpartum when the various tests and manipulations were administered. Therefore, even if one or two sensory modalities are critical for the *onset* of maternal behavior, rapid conditioning of pup cues from the other sensory modalities probably takes place, perhaps during the course of parturition or within the first day postpartum. By analogy, whereas olfactory bulbectomy has little or no effect on the sexual behavior of sexually experienced male rats (Heimer and Larsson, 1967), it greatly impairs the copulatory ability of naive male rats (Larsson, 1975). Note also that in mice, the sense of smell may be essential for masculine sexual behavior (Murphy and Schneider, 1970) and maternal behavior (see below), so the ability of rats to rely upon multisensory control, at least once relevant experience is gained, is not necessarily generalizable to other mammals. Another important point is that the sensory regulation of each separate component of maternal behavior needs to be considered as well (cf. Herrenkohl and Sachs, 1972).

Thus, a modified conclusion from the results of Beach and Jaynes (1956c) is that *the retrieving behavior of experienced postpartum lactating rats appears to be under multisensory control, probably in large part due to conditioning*. I say "appears to be" instead of "is" because of other limitations of the report of Beach and Jaynes (1956c): (1) There is no indication of the interval between surgical desensitization and testing. In particular in the case of somatosensory input from the snout region, the operation–testing interval is critical (Kenyon *et al.*, 1983). (2) The full array of possibly relevant sensory inputs was not tested, e.g., sound, taste, and other tactile afferents, both trigeminal and ventral somatosensory. And I say "probably in large part due to conditioning" because a thorough examination of the stimulus inputs required for the *onset* or the *induction* (i.e., sensitization) of maternal behavior has not been completed. However, on the basis of relevant experiments published since 1956, as well as preliminary evidence, I have tentatively concluded that *somatosensory inputs from pups are most critical in a primary sense, whereas visual, auditory, and chemosensory inputs, though utilized and subject to conditioning, are not essential for the onset or induction of pup-oriented maternal behavior in rats, or the maintenance of its major components.*

The evidence for this contention will now be reviewed. There is information on the responsiveness to normal and altered pup cues during the performance of already established maternal behavior. Studies of the onset of maternal behavior following late-pregnancy Cesarean delivery or during parturition and on pup-induced maternal behavior have utilized both pup cues and sensory deprivations, although a large proportion of these studies are on olfaction. The reliable induction of rapid-onset maternal behavior with hormones (see Section IV,C) is so recent that no one has yet utilized such a procedure to examine sensory regulation.

B. Sensory Induction of Maternal Behavior (Sensitization)

I. DESCRIPTION AND STRAIN DIFFERENCES

Various forms of nurturant behavior toward young by juvenile, adult female (nonpregnant, nulliparous) and male conspecifics occur naturally in a number of mammalian species (Spencer-Booth, 1970). In contrast to the behavior of the postpartum rat, most adult, maternally naive rats avoid pups after some brief exploratory sniffing; a few either retrieve pups spontaneously or attack them. However, in many laboratory strains of Norway rats cohabitation with small young for several days (*concaveation*) provokes a change in behavior toward the alien pups from avoidance or indifference to maternal caretaking behavior (*sensitization*), which includes retrieval, grouping at the nest site, licking, and crouching over pups, and eventually maternal-type nest building (Wiesner and Sheard, 1933; Cosnier and Couturier, 1966; Rosenblatt, 1967). The occurrence of sensitized maternal behavior is itself strong evidence that sensory stimuli from pups, if effective, can activate the underlying neural circuits. Analogously, with sufficiently strong and effective stimulation of flanks, rump, and cervix, the lordosis posture can be elicited in ovariectomized rats (Komisaruk and Diakow, 1973).

Sensitization occurs after a variable latency, typically a mean of 4–7 days, but there are striking strain differences in the initial responsiveness to pups, in the propensity to become sensitized, and in responsiveness to distal pup cues alone (cf. Stern, 1983). For example, Jakubowski and Terkel (1985) found that Sprague-Dawley females (stock originally from Charles River, MA) were highly maternal (35% spontaneously; 92% of the remaining virgins became sensitized), whereas Wistar females (stock originally from The Weitzmann Institute, Rehovot, Israel) were not (10% spontaneously; only 29% became sensitized during 15 days of concaveation). (Often "spontaneously maternal" refers to pup retrieval only, but Stern and MacKinnon (1976) showed that while spontaneous retrievers did become fully maternal more rapidly than initially neutral responders, i.e., median latencies of 1 versus 7 days, only 25% were fully maternal on the first day.) Furthermore, as Festing (1981) cautioned, colonies of rats bearing the same strain name, worldwide, are likely to be genetically different. Whereas the Wistar population of Jakubowski and Terkel (1986) required *contact* with pups over a 48-hour period postpartum to establish maternal behavior and its subsequent maintenance (similar to Long-Evans females, Stern, 1983), the Wistar population (Woodlyn, Ontario) of Orpen *et al.* (1987) retained maternal responsiveness for 7 days after Cesarean delivery, with no pup contact in between; and one 24-hour period of *distal* stimulation from very young pups, 1 day after pregnancy termination, increased the proportion of females displaying rapid-onset maternal behavior 10 days after surgery. If a low threshold of maternal responsiveness among adult rats is an anomaly of domestication, as it appears to be in mice (Jakubowski and Terkel, 1982), then strains with a high threshold should be a more meaningful model of the "real" Norway rat (cf. Jakubowski and Terkel, 1985). Clearly, sensory, neural, experiential, and hormonal mechanisms interact with genotype in the expression of maternal behavior.

2. ONTOGENY AND FUNCTIONAL SIGNIFICANCE OF SENSITIZATION

There is an age-related increase in sensitization latency in both Wistar and Long-Evans female and male rats, such that weanlings, e.g., 24-day-olds, tend to become maternal in 0–2 days, and adult latencies of 4–6 days are reached peripubertally, e.g., at 42 days (Bridges *et al.*, 1974a; Stern, 1987b; Stern and Rogers, 1988). In brief, *periweaning expression is followed by pubertal repression* (Stern, 1986). Some investigators have argued that while juvenile rats are capable of showing components of maternal behavior, they are lacking in the mature organization, integration, and consistency of this behavior, resulting in its characterization as *social rather than maternal* (Mayer and Rosenblatt, 1979; Brunelli *et al.*, 1985; Moretto *et al.*, 1986). However, the elicitation of full maternal behavior *is* demonstrable in weanlings when standard procedures for sensitization of adults are followed, i.e., weaning at a normal age and housing separately for 2 days in the test cage and room (cf. Stern and Rogers, 1988).

Hormone manipulations that advance puberty (Bridges *et al.*, 1974a; Mayer and Rosenblatt, 1979) or day-21 gonadectomy or hypophysectomy (Stern, 1987b) did not influence the age-related decline in maternal responsiveness. Based on these and other data, Stern (1987b) concluded that pituitary-independent maturation of inhibitory neural pathways is responsible for this decline. A related behavioral correlate of maturing inhibitory pathways is fearfulness and associated defensive reactions (e.g., Bronstein and Hirsch, 1976), the development of which are correlated with an increase in pup avoidance (Mayer and Rosenblatt, 1979).

Maternal behavior in mammals is most relevant to the well-being of young when it occurs postpartum. What then is the adaptive significance, if any, of maternal behavior emerging in nonlactating females, or even males, animals incapable of fully nurturing young? Rosenblatt's suggestion (1970)— that sensitization occurs because maternal behavior in rats is normally *maintained* postpartum by pup stimulation rather than by hormones—does not explain the possible functional significance for *induction* of maternal behavior that is independent of reproductive state or gender, or for *ontogenetic changes in latency*. The most likely situation in which Norway rats, other than parturient dams, would have extensive interaction with newborn pups is that in which juveniles remain in the maternal nest after the arrival of a new litter, an apparently common occurrence in nature (Davis and Hall, 1951; Calhoun, 1962). Weanlings display components of maternal behavior toward their younger littermates (Gilbert *et al.*, 1983; Stern and Rogers, 1988). Furthermore, 50% of rats ("responders") which lived with their younger siblings for either 4 or 11–18 days after their birth became maternal rapidly when tested for sensitization at day 42 (Stern and Rogers, 1988). These data provide an ethological explanation for sensitization in rats, as well as its ontogeny. A prediction from this line of reasoning (Stern, 1986) is that short-latency onset of maternal behavior among weanlings would not occur in rodent species with no postpartum estrus, which seems to be true for golden hamsters (H. I. Siegel and J. S. Rosenblatt, unpublished observations). Another prediction, as yet untested, is that females responsive to prior experience with younger siblings may be more successful mothers with their own first litter under challenging conditions. Primiparous mothers display less efficient maternal behavior than multiparous rats (Gilbert *et al.*, 1984) and are known to be more vulnerable to a variety of insults (see Section II,G).

C. Distal Stimuli: Sight, Sound, and Smell of Pups

Once maternal behavior is fully established following parturition, the retrieving drive may be rather indiscriminate: highly maternal lactating rats have reared mouse pups successfully (Denenberg *et al.*, 1966) and, most dramatically, some have retrieved (or rather dragged!) to the nest kittens, baby rabbits, and resistant chicks (Wiesner and Sheard, 1933). In contrast to this picture of stimulus nonspecificity, other data do indicate that specific aspects of the pup stimulus complex are discriminable and play a role in rat maternal behavior. However, while rats respond to the sight, sound, and smell of pups, each of these senses can be dispensed with for both postpartum and sensitized maternal behavior.

I. SIGHT AND SOUND OF PUPS

Given the darkness of rat burrows, it is not surprising that the parturition behavior of blinded dams was excellent (Beach, 1937). Subsequently, during tests in which visual ability undoubtedly facilitated performance, blinded rats were slightly impaired in their efficiency compared to sighted and otherwise intact controls. Blinding had no effect on sensitization latencies in Long-Evans virgin females (J. M. Stern, unpublished observations, 1987).

Lactating rats can orient to a pup on the basis of sight alone (i.e., pup in an airtight bottle), although additional exteroceptive cues enhance this response (i.e., pup in a double wire-mesh restraining cage) (Beach and Jaynes, 1956c). Newborn pups are more effective than older pups at eliciting retrieval postpartum (Wiesner and Sheard, 1933; Peters and Kristal, 1983) and at hastening the onset of sensitization (Stern and MacKinnon, 1978). When pups were removed from primiparous Long-Evans dams during parturition and the dams were presented with varying pup stimuli 24 hours later, newborn pups and 6- to 8-day-old hamsters elicited a high percentage of maternal behavior, whereas newborn hamsters and 8- to 10-day-old rat pups did not; clearly, this discrimination cannot be based on species-specific visual, auditory, or olfactory characteristics. In contrast, neither Cesarean-delivered rats

(Stern, 1985) nor pregnant rats tested for maternal behavior onset prepartum (Stern, 1987a) responded differently toward newborn or older rat pups (Fig. 1). Apparently, dams learn to prefer pups the *size* of newborn rats during parturition; although vision is the most obvious mediator of this preference, it is possible that a somatosensory map is utilized.

Rat pups emit ultrasonic "distress" calls in response to isolation, low temperatures, and unfamiliar tactile and olfactory stimuli (Okon, 1972; Oswalt and Meier, 1975). These calls elict orientation by lactating dams (Allin and Banks, 1972) and increase the likelihood of the entire litter being transported elsewhere when the nest site is endangered (Brewster and Leon, 1980a). In the presence of olfactory cues, ultrasonic calls from pups provide directional cues for retrieval (Smotherman *et al.,* 1974). Freshly killed or anesthetized pups, which are of course silent and motionless, are retrieved readily, though slightly more slowly than awake pups (Beach and Jaynes, 1956c; Kenyon *et al.,* 1983; Stern, 1985; Johnson and Stern, 1987). Although Terkel *et al.* (1979a) reported stimulation of prolactin release in lactating dams by rat pup ultrasounds, this finding has not been replicated (Stern *et al.,* 1984; Voloschin and Tramezzani, 1984).

The parturition behavior and litter-rearing ability of deafened rat dams has not been studied, but maternal behavior onset following late-pregnancy Cesarean delivery appeared to be normal, except for slight delays in beginning to sniff and retrieve pups (Herrenkohl and Rosenberg, 1972). Although deafening had no effect on sensitization latencies, deafened rats displayed a significantly higher incidence of injuring or killing a pup on at least one occasion (J. M. Stern, unpublished observations, 1987). Further, once maternal, we noticed that deaf rats were more likely to inadvertently injure pups, apparently because they could not hear the pups' cries in response to being stepped on. Genetically deaf male mice show a higher than normal incidence of infanticide (D'Amato, 1987). Together, these results suggest that pup vocalizations may contribute to the suppression of infanticide.

2. SMELL OF PUPS

Olfaction has been the most studied sense to date with regard to maternal behavior in rats. A problem in assessing this literature is that there simply is no ideal way to induce anosmia without causing nonspecific side effects. The only sure way to induce complete anosmia is the removal of the olfactory bulbs, but this often results in "nonsensory" ramifications, including irritability and heightened aggressiveness (Alberts and Friedman, 1972), which may be due largely to secondary telencephalic damage (Cain, 1974). An alternative approach is the temporary destruction of the olfactory epithelium by intranasal administration of 5% $ZnSO_4$ (Alberts and Galef, 1971). Because this treatment is thought to be specific to the main olfactory system, it has an advantage over olfactory bulbectomy, which destroys the accessory, as well as the main, olfactory system. For long-term experiments, the treatment is usually administered at 2- to 3-day intervals (Slotnick and Gutman, 1977). The major disadvantages of this peripherally induced anosmia are the difficulty of determining whether the anosmia is complete and the possibility of non-specific illness (Slotnick and Gutman, 1977; Mayer and Rosenblatt, 1975), although improvements in the administration procedure can ameliorate the ill effects (Mayer and Rosenblatt, 1977).

The effects of anosmia and pup odors on maternal behavior in rats are summarized in Table I (including references). Olfactory bulbectomy has been associated with an increased incidence of *infanticide,* both postpartum and during sensitization in virgins, but it is not inevitable. This effect is diminished or eliminated with maternal or hormonal (i.e., pregnancy) experience and it does not occur with peripheral anosmia or accessory olfactory system damage alone, but the combination has not been assessed. Indeed, $ZnSO_4$-induced anosmia eliminated cannibalism in a strain with a high incidence of this behavior (Wamboldt and Insel, 1987). Thus, *secondary telencephalic damage* (Cain, 1974), *rather than anosmia per se,* may be responsible for the effect on infanticide, and appropriate experiences can overcome the irritability or hyperresponsiveness that such damage causes. This interpretation is supported by a study on hamsters that systematically explored the roles of the main and accessory olfactory systems (Marques, 1979). In contrast to rats, virgin female hamsters are either maternal or cannibalistic upon initial contact with pups. Olfactory bulbectomy reduced both of these responses or converted maternal females into killers; vomeronasal nerve cuts (VNCC) converted most killers into carriers (i.e., maternal); intranasal $ZnSO_4$ treatment had little effect on its own, but when combined with VNCC, it converted into carriers those females which continued to kill after VNCC. Note that this is the only report to date to take full advantage of intranasal anosmia by combining it with vomeronasal-system damage (although vomeronasal organ removal is preferable to VNCC because it avoids the possibility of olfactory-bulb damage). Because hamsters typically cannibalize a portion of their litter early postpartum (Day and Galef, 1977), perhaps anosmia would reduce the frequency of this response. In maternally naive mice, anosmia, via bulbectomy or intranasal $ZnSO_4$ treatment, replaced the onset of spontaneous maternal behavior at term and in virgins with cannibalism, although prior

TABLE I   Effects of Odors or Anosmia on Maternal Behavior in Rats

Behavior	References[a]
A. Postpartum Maternal Behavior	
1. Suppression of infanticide during or after parturition?	
a. Prepartum bulbectomy[b] resulted in presumed or observed cannibalism during or after parturition in primiparas	A,B
b. Prepartum bulbectomy resulted in little or no cannibalism in primiparas or multiparas	B,C,D,E
c. Prepartum peripheral anosmia or vomeronasal nerve section resulted in no cannibalism in primiparas	E,F
2. Parturition behavior deficits?	
a. Prepartum bulbectomy resulted in parturition deficits and increased pup mortality, first birth	B,E,G
b. Parturition behavior deficits were not observed after bulbectomy before first or second delivery	D,B
c. Prepartum peripheral anosmia or vomeronasal nerve section resulted in no such deficits in primiparas	E,F
3. Retrieval and pup-growth deficits?	
a. Prepartum bulbectomy resulted in long-term retrieval and pup-growth deficits of primiparas, but not multiparas	E B
b. Peripheral anosmia resulted in some retrieval impairment in primiparas	E
c. Prepartum vomeronasal-system destruction in primiparas did cause retrieval impairment, or did not cause retrieval impairment	H F
d. Prepartum bulbectomy, plus blinding or deafening, resulted in a high incidence of litter mortality	I
e. Postpartum bulbectomy alone had no significant effect on retrieval ability	J
f. Postpartum bulbectomy, combined with blinding or snout desensitization, resulted in retrieval impairment	J
4. Preference for odors associated with rat pups?	
a. Preference for own versus alien pups abolished by postpartum bulbectomy	K
b. Unfamiliar odor (lavender) placed on pups reduced retrieval and increased infanticide	J

(continued)

TABLE I (*Continued*)

Behavior	References[a]
c. Unfamiliar odor (lavendar) placed on pups reduced maternal behavior onset and increased infanticide, but peripartum familiarization with odor eliminated deleterious effects	L
d. Maternal behavior toward allospecifics; thus, species-specific odor is not necessary	M,N,O
e. Maternal licking is greater toward male pups due to androgen-dependent preputial gland odor	P,Q
f. Maternal licking of pups is decreased by anosmia or odor masking	D,P
g. Attraction to own bedding from late gestation to second week postpartum	R
5. Maternal aggression toward strange intruder	
a. Odor of inaccessible pups maintained maternal aggression for at least 4 hours	S
b. Postpartum surgical anosmia reduced incidence and intensity of maternal aggression	T
B. Sensitized Maternal Behavior	
1. Anosmia and infanticide?	
a. Presensitization bulbectomy increased incidence of infanticide in maternally naive virgins, but not in females experienced with maternal behavior or pregnancy hormones	C,D
b. Presensitization peripheral anosmia did not increase infanticide	B,C
c. Pups newly delivered and uncleaned of birth products elicit a high rate of infanticide	U,V,W,X
2. Anosmia induces short-latency onset of maternal behavior	Y
a. Rapid retrieval with presensitization bulbectomy	D
b. Rapid retrieval with presensitization peripheral anosmia	U,V,W
c. Synergy between partial olfactory bulb damage and vomeronasal nerve section	Z

[a](A) Herrenkohl and Rosenberg, 1972; (B) Schwartz and Rowe, 1976; (C) Schlein *et al.*, 1972; (D) Fleming and Rosenblatt, 1974a; (E) Benuck and Rowe, 1975; (F) Jirik-Babb *et al.*, 1984; (G) Holloway *et al.*, 1980; (H) Saito, 1986; (I) Mena and Grosvenor, 1971; (J) Beach and Jaynes, 1956c; (K) Beach and Jaynes, 1956b; (L) Stern, 1987a; (M) Wiesner and Sheard, 1933; (N) Denenberg *et al.*, 1966; (O) Stern, 1985; (P) Moore, 1981; (Q) Moore and Samonte, 1986; (R) Bauer, 1983; (S) Ferreira and Hansen, 1986; (T) Ferreira *et al.*, 1987; (U) Fleming and Rosenblatt, 1974b; (V) Mayer and Rosenblatt, 1975; (W) Mayer and Rosenblatt, 1977; (X) Wambolt and Insel, 1987; (Y) Peters and Kristal, 1983; (Z) Fleming *et al.*, 1979. See text for further description, interpretation, and references.

[b]Bulbectomy refers to bilateral olfactory bulbectomy.

maternal experience counteracted these effects (Gandelman *et al.*, 1971; Seegal and Denenberg, 1974).

Several investigators observed *parturitional maternal-behavior deficits* during the first delivery of rats subjected to prepartum bulbectomy that were severe enough to increase pup mortality. According to Pollak and Sachs (1975), there are no deleterious effects of neonatal bulbectomy on the subsequent postpartum maternal behavior exhibited by these rats as adults, but Risser and Slotnick (1987) demonstrated recently that complete neonatal bulbectomy is inconsistent with survival and incomplete neonatal bulbectomy is consistent with both survival and the ability to smell. In the positive reports, the impairment included maternal deficiencies in licking the pups clean of their fetal membranes and of completing placentophagia. The most detailed observations of parturition also revealed a higher incidence of non-pup-oriented behaviors, including general activity and ingestion, as well as a lower incidence of behaviors that facilitate pup survival (e.g., pulling with the teeth at anogenital area) and the early establishment of nursing (Holloway *et al.*, 1980). Because of the poor start that these pups had during parturition, their survival rate by weaning age was decreased, even with an intact foster dam (Holloway *et al.*, 1980). Dams bulbectomized prior to their first mating that displayed parturition behavior abnormalities also had *long-term retrieval deficits* and their *pups gained less weight,* deficits not seen in multiparous dams. These problems are reminiscent of those seen in dams with extensive neocortical damage (Beach, 1937; Section V,A). It is not clear to what extent these impairments are due to olfactory loss per se or some other effect of bulbectomy.

Olfactory bulbectomy abolishes the preference for own versus alien pups, but the latter are still accepted readily by anosmic laboratory rats. However, the killing of alien pups by wild rat mothers is probably elicited by their odor (King, 1939). Postpartum bulbectomy produced in experienced retrievers has no effect on retrieval ability unless it is combined with another sensory loss, i.e., blinding or snout desensitization (Beach and Jaynes, 1956c). Anosmia combined with blinding results in high litter losses and failure of exteroceptively induced release of prolactin (PRL) (Mena and Grosvenor, 1971), perhaps related to inadequate PRL secretion in general (Blask and Reiter, 1975). Olfaction is the prime mediator for the exteroceptive release of both PRL (Mena and Grosvenor, 1971) and corticosterone (Zarrow *et al.*, 1972) in response to pup cues. The reasons for the high litter losses among the anosmic–blind and anosmic–deaf dams (60% versus 0% for deaf–blind dams) are not clear from Mena and Grosvenor's (1971) report, but inadequate parturition behavior and milk secretion may be among them.

A number of studies suggest that rat odor from the pups is not a requirement for a pup to elicit retrieval from a rat mother and that *preferences for odors associated with pups are learned,* including initially aversive ones, such as the artificial odor of lavender or the natural odor of newly delivered, uncleaned pups. Mothers probably learn pup-related odors during parturition or they may learn them from their own secretions, possibly starting in late pregnancy. Bauer (1983) found that "from just before birth until the second week postpartum, rats are attracted to their own bedding but cannot or do not distinguish between their own and that of other maternal animals," a finding which suggests a hormonal basis both for the odor and for a possible alteration in its perception. However, we have been unable to replicate Bauer's results in two experiments (J. M. Stern, unpublished observations, 1987). Ewes learn, and become attracted to, the odor of their own lamb via extensive licking of it within hours after its birth, and olfaction is essential for the specificity of the resulting bond; i.e., anosmia does not abolish maternal responsiveness, only its selectivity (Poindron and Le Neindre, 1980). A preference for *male pup odor,* based on an androgen-dependent preputial gland substance secreted in the urine (Moore, 1982), may be innate, since maternally naive, adult female rats are also attracted to this odor (Moore, 1985). The increased genital stimulation from maternal licking which males receive because of this odor contributes, along with their own later self-grooming, to masculine sexual behavior development (Moore, 1984; Moore and Rogers, 1984).

Pup odors appear to be vital to the display of *maternal aggression.* The odor of pups (that are not emitting distress cries), passing through a nylon mesh bag, maintained aggression toward a strange intruder for at least 4 hours, whereas total separation from pups or the sight of pups through a glass flask did not (Ferreira and Hansen, 1986). Furthermore, lactating rats rendered anosmic on postpartum day 2 or 3 by surgical destruction of the olfactory epithelium and mucosa are much less aggressive toward a strange female intruder on the fourth postoperative day than are sham-operated controls (Ferreira et al., 1987). It would be of interest to know the effects of previous maternal aggression experience and of longer-term anosmia on maternal aggression. In mice, vomeronasal organ removal inhibits maternal aggression but not other aspects of maternal behavior (Wysocki et al., 1986); however, in postpartum rats maternal aggression does not seem to be inhibited by this loss (J. M. Kolunie and J. M. Stern, unpublished observations, 1988).

Finally, central or peripheral anosmia *hastens the onset of maternal behav-*

ior in virgins, and this has been observed under both experimental and "clinical" (i.e., colonies with respiratory infections) conditions. Fleming and co-workers (Fleming and Rosenblatt, 1974a,b; Fleming *et al.,* 1979; Fleming and Luebke, 1981; Fleming, 1986) believe that to a maternally naive virgin the odor of pups is fearful or aversive, and this inhibitory input directly affects sites known to stimulate maternal behavior (see Section V,B). Once the subject habituates to the pup odor so that it is no longer fearful, maternal behavior emerges. This is an appealing hypothesis, but continuous exposure to pup odors has no effect on subsequent sensitization or postdelivery maintenance of maternal responsiveness in some rat populations (Stern, 1983; Jakubowski and Terkel, 1985). Further, this hypothesis does not address the question of what are the positive sensory stimuli from pups which elicit nurturance. A more limited (but not mutually exclusive) hypothesis to explain the anosmia results is that retrieval is stimulated, perhaps nonspecifically. According to Welker (1964), anosmic rats are rather indiscriminant in their tendency to retrieve strange objects and bring them back to their nesting site. Once a pup so retrieved is at the nesting site, the likelihood of further contact with it is enhanced, which culminates in a more rapid onset of full maternal behavior (Stern and MacKinnon, 1976). Indeed, anosmic rats tend to show retrieval alone before exhibiting full maternal behavior (Mayer and Rosenblatt, 1977).

D. Contact Stimuli: Taste and Feel of Pups

Gustatory stimuli from pups affects duration of pup licking, but probably not other aspects of maternal behavior. Perioral and ventral somatosensory stimulation from pups are probably the primary and critical determinants of maternal responsiveness. The specific details and mediating neural pathways of this stimulation have not yet been worked out.

I. TASTE OF PUPS

As we have seen, pup licking is a major component of rat maternal behavior from parturition until weaning. Kristal and co-workers have demonstrated the attractiveness of placental fluids, and placenta per se, in some instances (Kristal, 1980; Kristal *et al.,* 1981; Steuer *et al.,* 1987). Lactating rats are known to be responsive to pup tastes: they preferred water flavored with pup anogenital secretions to water alone (Charton *et al.,* 1971), although salinity was not controlled. Lactation enhances the salt preference and appetite of female rats; providing them with a 0.15 M NaCl solution decreases their anogenital licking of pups (Gubernick and Alberts, 1983). Some of the findings discussed above on odor cues from pups are probably germane to taste sensations as well. For example, the odorant in male urine which heightens the dam's pup-anogenital licking probably provides a taste cue as well (Moore 1981, 1985; Moore and Samonte, 1986). The elicitation of infanticide toward pups coated with oil of lavender under some circumstances is probably related to the bitter taste of this intense odorant.

Does the taste of pups contribute to maternal motivation? I suspect that, as in the case of ingestion (Jacquin, 1983), taste has only a modulating role, and discerning what that role may be would be a formidable task because of the extreme complexity of the gustatory system, both centrally and peripherally. One simple approach, which reduces both gustatory and somatosensory sensations from the tongue, is to anesthetize the tongue with a short-acting, local anesthetic such as lidocaine. Lactating dams treated in this way 30 minutes prior to reunion with their pups showed no deficit in retrieval latency or duration, and the dams then went on to lick their pups, though for a reduced duration (Stern and Johnson, 1989). Deprivation of licking per se, by suturing the dam's mouth shut temporarily, does not prevent pup-licking and self-licking movements or crouching; in fact, crouching begins sooner, probably due to the lack of afferent feedback from licking (Stern and Johnson, 1989; Section III,D,3).

2. TRIGEMINAL AFFERENTS

The rat's snout, probably as sensitive as the human finger tip (MacIntosh, 1975), and the first point of contact with pups (and other objects prior to mouth opening; Zeigler et al., 1984), is an obvious choice for reception of both motivating (cf. MacDonell and Flynn, 1966; Zeigler et al., 1985) and guiding cues from pups. Brief or prolonged sniffing precedes mouth opening, during which snout and vibrissae contact with the object occur repeatedly (Welker, 1964). Clearly, discrimination of pups occurs at this level, even prior to mouth opening. Indeed, the most drastic inhibition of retrieval by intact lactating dams during simultaneous choice tests occurs in response to altered thermal and tactile pup cues (Beach and Jaynes, 1956c). And a rubber toy the size of a newborn rat is not retrieved readily by either postpartum mothers (Plume et al., 1968) or sensitized maternal virgins (Rosenblatt, 1975; Stern and MacKinnon, 1976), although visual and olfactory as well as perioral somatosensory cues may have contributed to this discrimination.

Given the importance for retrieval of thermal and tactile cues associated with the skin of pups, the further finding of Beach and Jaynes (1956a), that *anaptia* (i.e., desensitization of snout and lip region by partial trigeminal

TABLE II Effects of Trigeminal Denervations on Behaviors Executed with the Mouth[a]

Behavior/denervation	Deficit	References[b]
Ingestion		
1. Sphenopalatine, superior alveolar (SA), +inferior alveolar (IA)	Prolonged and severe aphagia and adipsia; inhibition of mouth opening in response to food	A,B
2. Infraorbital (IO)	Slight, brief aphagia and adipsia	A,B
Suckling in neonates		
1. IO	Nipple attachment severely impaired; no recovery in 7-day-olds	C
Attack behavior		
1. IA + IO	Severe inhibition of biting attack (cats)	D
2. IO	Moderate disruption of mouse killing	E,F
Self-grooming		
1. IA, SA, IO, +lingual	Intact; slight changes in form only	G
2. IO	No change in percentage responding or duration	H

[a] See Table III for maternal behavior; all studies on rats, except D. Denervation key: Sphenopalatine: soft and hard palate, cheeks, nasal mucosa; superior alveolar: upper incisors, molars, gums; inferior alveolar: lower lip, lower incisors, molars, gums, floor of mouth, chin, chin hairs; infraorbital: upper lip, rhinarium, vibrissae, facial pads; lingual: anterior two-thirds of tongue, gums, mouth mucosa, sublingual glands.

[b] (A) Jacquin and Zeigler, 1983; (B) Zeigler et al., 1985; (C) Hofer et al., 1981 (D) MacDonnell and Flynn, 1966; (E) Grégoire and Smith, 1975; (F) Welle and Coover, 1978; (G) Berridge and Fentress, 1986; (H) J.M. Stern, unpublished observations, 1987.

deafferentation) had little effect on retrieval performance, seems paradoxical. Indeed, "their performance was entirely normal except for slight difficulties in grasping the young." This outcome was interpreted in terms of the *multisensory control of retrieval*, i.e., in the absence of snout sensation, distal cues can be used instead by dams well-practiced in retrieving. However, recent research on the effects of trigeminal sensations on behavior in rats (Table II), including maternal behavior, suggests the need for alternative interpretations.

Trigeminal deafferentations have been found to cause major disruptions in biting attack in cats, ingestion in adult and suckling in neonatal rats and, most recently, maternal aggression toward a strange intruder and pup-oriented maternal behavior in rats (Tables II and III). The maternal-behavior effects have also been demonstrated by anesthetizing the mystacial (whisker) pads with lidocaine; this treatment blocks impulses in the infraorbital nerve, an enormous branch of the trigeminal maxillary division which innervates the

TABLE III Effects of Perioral Desensitization on Oral Components of Maternal Behavior in Rats[a]

A. Retrieval
  1. Lidocaine injected s.c. in the mystacial pads (LMP) 30 minutes before testing usually eliminated retrieval.
  2. Retrieval occurred after LMP with appropriate prior experience; when present, interretrieval intervals were prolonged.
  3. Following infraorbital denervation (IO-X), retrieval was eliminated up to 12 hours, but it largely recovered by 24 hours.
  4. Recovery of retrieval at 24 hours after IO-X was diminished by separation from pups between operation and testing.
  5. Similar pattern of deficit and recovery after IO + mental denervations.
  6. After prepartum IO-X, retrieval delayed substantially, with long-term deficits in completeness and speed.
  7. Postpartum vibrissae removal, masseter muscle anesthesia, or facial nerve cuts did not affect retrieval.
B. Nest Building
  1. After prepartum IO-X, little or no nest building, before, during, or after parturition.
  2. After postpartum IO-X, decline and recovery of nest building paralleled that of pup retrieval.
C. Pup Licking
  1. With LMP, during retrieval or crouching tests pup licking was virtually eliminated.
  2. With postpartum IO-X, recovery of pup licking paralleled recovery of retrieval.
  3. With prepartum IO-X, pup licking began during parturition but dams were not persistent, so many pups were only partially cleaned. Also, deficiency in self anogenital licking and pulling with teeth to facilitate delivery of emerging fetuses.
D. Maternal Aggression
  1. With LMP, there was little or no aggression toward a strange male intruder, without or with prior fighting experience.
  2. Dams which did fight under LMP did so with increased latency and decreased vigor.
  3. With postpartum IO or IO + mental denervation, aggression initially eliminated, with partial recovery by 24 hours.
  4. Postpartum vibrissae removal did not decrease maternal aggression.

[a]References: Postpartum LMP: Kenyon et al., 1981, 1983; Stern and Kolunie, 1989; Kolunie and Stern, 1987. Postpartum IO-X: Kenyon et al., 1983; J.M. Stern, unpublished observations, 1987, 1988; J.M. Stern and J.M. Kolunie, unpublished observations, 1987. Prepartum IO-X: Stern et al., 1987.

skin of the nose, muzzle, upper lip, and mouth mucosal membrane (Greene, 1968). The retrieval deficit during perioral anesthesia is profound, but not absolute: with repeated testing under lidocaine (Kenyon et al., 1981) or with prior experience retrieving in a strange apparatus (Stern and Kolunie, 1989), retrieval occurred. For a short time (~12 hours) following infraorbital denervation, retrieval is abolished, but recovery of this behavior (and other oral maternal behaviors) is evident by 24 hours after surgery (Kenyon et al., 1983; J. M. Stern, unpublished observations, 1988). Since the operation–testing interval which Beach and Jaynes (1956c) used (unspecified) was probably at least 24 hours, their failure to find an effect of anaptia can now be understood. But what can account for the abrupt change from severe deficit to rapid recovery? And is the trigeminal role in maternal behavior important if recovery from deafferentation is so rapid?

Recovery of function after peripheral denervation can be explained in several ways (Goldberger, 1980). Of these, (1) *neural regeneration,* is not relevant to the retrieval recovery because the latter is too fast. (2) *Behavioral substitution* occurs to an extent, but cannot entirely explain the recovery of retrieval. In the situation in which retrieval occurs during perioral anesthesia, dams often use their snout, chin, and paws to aid in retrieval; however, retrieval by the usual means occurs as well, though more slowly (Stern and Kolunie, 1989). Interaction with pups during the interval between postpartum infraorbital denervation and testing 24 hours later contributes to the recovery, but it is not essential (J. M. Stern, unpublished observations).

The most likely explanation for the rapid recovery of retrieval following infraorbital denervation is (3) *neural reorganization* of the somatotopic projections (e.g., Wall and Egger, 1971; Merzenich et al., 1983). Following deafferentation, the regions normally served by the infraorbital are then served by the remaining trigeminal branches. Rapid behavioral recovery is probably due to the "unmasking" of synapses that existed before deafferentation but which previously were nonfunctional. Thus, infraorbital denervation results in only a partial sensory loss, which would not be acceptable in any other sensory modality. The experiments summarized in Table III suggest that deafferentation of the alveolar nerves, which innervate the teeth, in addition to the infraorbital, will produce a more long-lasting, and perhaps permanent, deficit in maternal oral activities; these nerves were not included in the deafferentation procedure of Beach and Jaynes (1956c). Preliminary observations from my lab support this suggestion but more extensive work is needed to assess the effects of various combinations of trigeminal deafferentations on maternal behavior, as well as to maintain the body weight of the deafferentated subjects with a palatable diet (such as pablum; Jacquin and

Zeigler, 1983) and, if necessary, orogastric intubations (Berridge and Fentress, 1986).

During the time when perioral desensitization inhibits retrieval, it also reduces or eliminates all other maternal behavior activities executed with the mouth (Table III), such as pup licking, even though desensitized dams are interested in being with their pups upon reunion with them, as evidenced by sniffing, pushing their snout under the pile of pups, and attempts to retrieve. Ability to open the mouth and use the tongue for licking are not impaired because grooming, i.e., self-licking, occurs (Stern and Kolunie, 1987, 1989). Indeed, only subtle aspects of grooming are changed by extensive orosensory deafferentation, demonstrating the predominant central programing of this behavior (Berridge and Fentress, 1986). What is different about retrieval, pup licking, ingestion, suckling, and biting attack is that they are dependent upon proximate cutaneous stimuli. Ingestion in adults is the most highly practiced oral activity, yet after trigeminal orosensory deafferentation the jaw-opening reflexes cannot be elicited by distal stimuli from food. According to Jacquin and Zeigler (1983), oral ingestive responses "are organized in chain-reflex fashion; trigeminal orosensation appears not only to elicit those reflexes but to provide a flow of sensory input that links these elementary movement patterns into adaptive sequences. . ." (p. 93).

In the case of maternal behavior, since a persistent deficit has not yet been demonstrated after trigeminal deafferentation, perhaps the primary role of perioral somatosensory afferents would be more evident at the *onset of maternal behavior* in dams delivering for the first time, before presumed conditioning to distal stimuli can have taken place. By carrying out infraorbital denervations shortly prior to term (4–43 hours) we found severe disruptions in parturition behavior and continued impairments in retrieval (Stern *et al.*, 1987; Table III). Although initial pup licking and placentophagia appear normal, the females seem to "lose steam" as parturition proceeds: they walk around, leaving pups strewn about, incompletely cleaned, and often with cord and placenta still attached. We observed astonishing instances in which a pup, half out of its mother's vagina, was dragged about for 20 minutes or more. Clearly, the normal active participation in the expulsion and cleaning of the pups with tongue and teeth is often highly deficient in dams with infraorbital denervation. Little or no retrieval occurs, nest building is absent both before and during parturition, and nursing is delayed. Given these deficits, it is not surprising that pup mortality is much higher in the denervated than in the sham-operated group, both by the end of parturition and subsequently. Pup retrieval continued to be incomplete, awkward, and slow on days 1–3 postpartum. However, there was no cannibalism of living pups, and only one dam lost her entire litter due to a failure to crouch. Thus, infraorbital denervation carried out prepartum is more disruptive to maternal behavior than the same operation carried out postpartum. I suspect that more extensive trigeminal denervation prepartum will thoroughly and persistently disrupt maternal behavior.

The most unexpected (and exciting) outcome of our studies on perioral somatosensation to date is that it is critical to nursing behavior, which is not carried out with the mouth! (See Table IV,A.) In the initial experiments on perioral anesthesia (Kenyon *et al.*, 1981; Stern and Kolunie, 1989), failure to crouch may have occurred because the pups were not in the nest, the dam having failed to retrieve and group them there. Therefore, we placed the pups in the nest and additionally increased the dam's motivation to nurse with a 4- to 6-hour separation from her litter prior to testing (Stern and Johnson, 1989). Using either perioral anesthesia or a muzzle to prevent snout contact with pups, we found an almost complete disruption of crouching behavior during early lactation. An interesting difference between these procedures is that lidocaine-treated dams became inactive, away from the pups, after their initial exploration of them, while muzzled dams continually attempted to make contact with the pups for 2 hours, resulting in the scattering of the pups and nest material. Whereas contact without adequate feedback leads to a decline in maternal interest, distal stimulation from pups in the absence of perioral contact maintains interest in making such contact but is not sufficient to permit crouching to occur. Whether any distal stimulus is effective in maintaining the muzzled dam's interest in her pups or whether one stimulus is critical (e.g., olfactory) remains to be determined. In late lactation (or with older pups), when pups ordinarily are able to initiate nursing, in or out of the nest, perioral-stimulus deprivation can be dispensed with in some cases. For example, on day 14 postpartum, pups in five of eight litters were able to provoke the crouching posture from their lidocaine-treated dams, lying away from the nest, by burrowing under the dam's ventrum.

Which aspect of perioral contact with young pups is necessary to induce crouching? The longest-duration activity prior to the onset of crouching is licking which, as we have seen, serves to stimulate the pups' urination and activates them (Section II,D), and which is eliminated by perioral anesthesia of the dam. Deprivation of the afferent feedback from licking (by anesthetizing the tongue with lidocaine) or of actual licking (by suturing the mouth closed), however, did not affect crouching behavior or subsequent litter weight gains. Although sutured dams could not open their mouths, they engaged in movements that simulated both pup licking and self-grooming. Apparently, 2-day-old pups are sufficiently activated by tactile stimulation without tongue contact to attach to nipples and suckle effectively. Thus, during early lactation, the antecedents to nursing that involve snout but not tongue contact with pups are requisites for the mother's crouching posture. The mechanism mediating this effect has yet to be elucidated, but the projection of the trigeminal complex to the afferent milk-ejection pathway in the mesencephalic lateral tegmentum (Dubois-Dauphin *et al.*, 1985a) may be relevant.

Somatosensory contact with pups, especially perioral contact initially, probably mediates the induction of maternal behavior during sensitization. This tentative conclusion is supported by a detailed assessment of subject–pup interactions during sensitization, using 24-hour time-lapse videotape recordings (J. M. Stern, unpublished observations). When the subject's behavior is sampled only for a short time after the introduction of fresh pups each day, the onset of maternal behavior appears to be sudden (Fleming and Rosenblatt, 1974c), although preceded by the cessation of pup avoidance (Fleming and Luebke, 1981). Our findings indicate that during sensitization, subjects that were initially neutral toward pups sniffed pups hundreds of times, initially very briefly, licked them briefly and then extensively, retrieved them at short distances, and were increasingly tolerant of physical proximity with them prior to the onset of full maternal behavior. A similar pattern occurs in subjects that retrieve pups spontaneously, though at a faster pace. Many of the pup contacts were facilitated by the pups (1–4 days of age) moving about the cage, making it difficult for the subject to continue avoiding proximity with them. Confining pups to one quadrant of the cage, thereby reducing subject–pup interaction, may have been the cause of the low incidence of sensitization in an otherwise highly maternal strain (Gray and Chesley, 1984). Taken together with the ineffectiveness of blinding or deafening on sensitization latencies and the facilitating effect of anosmia (see previous discussion), these results suggest that somatosensory contacts with pups constitute the pup stimulation necessary for sensitization. Verification of this hypothesis with appropriate trigeminal deafferentations will be worthwhile.

3. VENTRAL SOMATOSENSORY AFFERENTS

Is perioral stimulation from pups (see previous section) sufficient to cause the rat dam to adopt an active crouching posture over the litter? Alternatively, is ventral somatosensory stimulation from the pups, during their search for a nipple, necessary to elicit the active crouching posture? The

TABLE IV Perioral and Ventral Somatosensory Determinants of Crouching Behavior in Postpartum Rats[a]

Somatosensory manipulations	Day	Results
A. Perioral		
1. Lidocaine, dam's mystacial pads	2	Only 1 of 8 dams crouched (versus 8 of 8 controls); little weight gain
2. Lidocaine, dam's mystacial pads	14	Delayed crouching of only 5 of 8 dams (versus 8 of 8 controls); little weight gain
3. Lidocaine, dam's tongue	1	Normal crouching and litter weight gains
4. Dam's mouth sutured	2	Normal crouching and litter weight gains
5. Dam muzzled	2	No crouching; dams persisted in trying to make perioral contact with pups
6. Dam muzzled	14	Reduced percentage of crouching in 30 minutes with delayed onset
7. Infraorbital denervation (IO-X)	3	Little or no crouching prior to recovery of retrieval
8. Prepartum IO-X	0–3	Delayed onset of crouching; failure to crouch in one dam
B. Ventral		
1. Dead pups	2	No crouching; licking, handling, and hovering over pups
2. Dead pups	10	No crouching; licking, handling, and hovering over pups
3. Lidocaine, pups' mystacial pads	7	No crouching; licking, handling, and hovering over, and reretrievals
4. Lidocaine, pups' mystacial pads	14	No crouching; licking, handling, and hovering over, and reretrievals
5. Chilled pups	2	No crouching until pups warmed up to near nest temperature by dam
6. Test with 1, 2, 4, or 8 pups	4	No crouching with 1–2 pups, but weight gain possible; delayed crouching in some with 4 pups; normal crouching with 8 pups.

[a]Dam and litter are separated for 4–6 hours prior to crouching test. Litters were kept at nest temperature, except in condition B.5, in which litter was chilled between 10 and 30 minutes prior to test. Lidocaine was injected, under Metofane anesthesia, 30 minutes prior to start of test. Mouth suturing and placement of the muzzle (which permitted seeing, hearing, and smelling the pups) were carried out under Metofane anesthesia 90 minutes prior to testing; control sutures were on the lower lip only and the control muzzle permitted snout contact. In the pup number experiment (B,6), litter size was 8 except during the test. References: (A) Stern and Johnson, 1989; (B) Johnson and Stern, 1988; Stern and Johnson, in preparation.

answer seems to be that perioral stimulation is necessary for the dam to hover over the pups, but such stimulation is not sufficient to elicit the crouching posture, which requires the additional stimulation of the ventrum by the pups, at all stages of lactation (Table IV,B). This was demonstrated conclusively by providing dams with pups that were freshly killed and still warm, rendered incapable of rooting effectively due to anesthetization of their own mystacial pads, inactive due to chilling, or inadequate in number. In all of these cases, the pups were retrieved readily (though dead and chilled pups were retrieved later) and dams hovered over them for a long time while licking intermittently. However, the crouching posture, characterized by a pronounced dorsal arch and splayed legs, was never adopted. Crouching in response to chilled pups occurred only after the pups were warmed, by the dam's ventrum and tongue, to a temperature close to normal. This finding suggests that the chronic nest chilling (22.5°C) carried out by Leon *et al.* (1978) should have reduced or prevented crouching and suckling (which were not assessed); therefore, the prolonged nest bouts that this procedure provoked were probably due to nursing never having been initiated rather than to a lowering of the dam's temperature. Indeed, since maternal core and ventrum temperatures were *lower* at the end than at the beginning of the prolonged nest bout with cooled pups, factors other than a rise in temperature must come into play to terminate a nesting bout, and failure of pups to stimulate nursing behavior must be one of them.

Support for the above conclusions comes from a reinterpretation of data collected by Leigh and Hofer (1973). These workers reduced litters to one pup on day 12 postpartum and compared the mother–pup interactions with those of litters of eight. Mothers spent more time in contact with the singleton than with the full litter, and of that contact time, the mothers of singletons spent more time pup licking (27 versus 12%) and less time in a nursing position (46 versus 79%), while the singleton pups spent more time pushing the dam than did the pups in the full litter (25 versus 8%). Upon returning to the nest, the single pup is the sole object of the mother's licking; without other pups to move to her ventrum, thereby stimulating her crouching response, the licking continues longer with the singleton than it does normally. The ventral stimulation provided by the single pup is only barely sufficient, or insufficient, to provoke the crouching response. Indeed, according to Leigh and Hofer (1973), the stimulation provided by a single pup is insufficient to maintain lactation prior to day 12 postpartum, despite extensive nursing.

In sum, whereas distal stimulation from pups maintains contact-seek-

SENSORY REGULATION OF RAT
MATERNAL BEHAVIOR

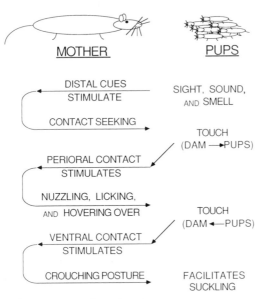

F I G U R E 2. The sequence of mother–litter interactions which culminates in a nursing bout is depicted schematically in this figure and described in detail in the text. No one has yet tested the maternal behavior of rats deprived of vision, hearing, and olfaction, but severe deficits are likely. Once the mother and her young litter are in proximity, however, the dam must receive feedback from her snout and vibrissae contact with the pups, which stimulates more of this contact, i.e., nuzzling, licking, and rearranging pups in the nest, while she hovers over them. The hovering position allows the thigmotactic pups to orient to the dam's ventral surface and begin rooting for a nipple. The ventral stimulation which the dam receives from the pups stimulates her crouching posture; this posture, and concomitant immobility, facilitates nipple attachment and suckling. Pups which are incapable of rooting effectively, e.g., due to mystacial-pad anesthesia or chilling-induced inactivity, do not stimulate crouching. Older pups are capable of stimulating crouching even from a recumbent dam, with little or no prior perioral contact from her, by burrowing under her ventrum.

ing, perioral contact with them is necessary in early lactation to cause hovering over them, and ventral somatosensory stimulation from pups is necessary at all stages of lactation to elicit the crouching posture (Fig. 2).

Although there is probably overlap in the afferent paths that stimulate crouching and milk ejection, the two responses are clearly dissociable. Ventral somatosensory afferents involved in stimulating crouching are probably more diffuse than those that elicit the milk-ejection reflex. When it occurs, the upright crouching posture necessarily precedes milk ejection; in rats, at least several minutes of suckling are required before a milk ejection occurs, and conscious mothers must become calm and inactive since the state of electroencephalographic (EEG) slow-wave sleep accompanies oxytocin release (see Section IV,C,4). In addition, the crouching response occurs in rats not capable of a milk ejection due to the removal of teats (Wiesner and Sheard, 1933), or the removal of mammary glands and teats (Moltz et al, 1967), or because they are maternal virgins not treated with hormones (Wiesner and Sheard, 1933; Rosenblatt, 1967). Recordings from the abdominal segmental nerves of lactating rabbits revealed not only a "profusion of teat receptors," but also widespread cutaneous receptors from the skin overlying the mammary glands (Findlay, 1966). In rats, thelectomized (nipple-removed) postpartum mothers not only crouch, but also secrete prolactin in response to the pups (Moltz et al, 1969b; Stern, 1977; Stern and Siegel, 1978), and the maintenance of the prolactin-induced lactational diestrus is dependent on tactile stimulation of the ventrum (Zarrow et al., 1973). Of course, milk ejections occur in the absence of the upright crouching posture, in anesthetized dams when the dam lies on her side (catlike, most common during late lactation) and when there is only one pup (Drewett and Trew, 1978).

Lehrman (1961) suggested that when lactating mammary glands are present, their engorgement may provide a motivation to nurse, but not in all species or all circumstances. The relationship between mammary engorgement and nursing behavior has been worked out best in rabbits. Although mammary distension can play a role in the initiation of a nursing bout (Cross, 1952; Findlay, 1969), such distension is neither necessary nor sufficient for its initiation (Findlay, 1969; Findlay and Roth, 1970) nor for determining its duration (Lincoln, 1974). Instead, the conscious experience of nursing is a more critical factor. Reduced suckling stimulation during conscious nursing, e.g., by teat anesthesia, increases the duration of the nursing bout and greatly increases the likelihood that the mother will nurse a second time that day (Findlay and Tallal, 1971). Thus, teat sensations are not necessary to induce lactating does to nurse, but rather they influence the termination of nursing. Given the marked differences in nursing pattern between rabbits (which nurse once per day) and rats (which nurse continually), differences in the mechanisms mediating nursing behavior would not be surprising.

In mice, suckling stimulation is considered to be essential for the onset and establishment of postpartum maternal aggression (Svare, 1977). In contrast, thelectomized rats are as aggressive as controls, both prepartum and postpartum. This species difference may be due to greater sensitivity of rats to ventrum afferents other than those from the teats (Mayer and Rosenblatt, 1987). It is unlikely, however, that ventrum afferents are not involved in maternal aggression in rats (see Section V,D).

E. Pelvic Afferents

Stimulation of uterine, cervical, and/or vaginal afferents by fetuses and pups may affect maternal responsiveness by a neural or neuroendocrine route. Uterine distension of day-10 or day-11 pseudopregnant rats resulted in a shorter latency to onset of maternal behavior than the same procedure applied to cycling rats or than occurred in pseudopregnant controls without the uterine manipulation (Graber and Kristal, 1977). Vaginal-cervical stimulation has been found to stimulate maternal behavior in multiparous estrogen-primed sheep (Keverne et al., 1983) and rats (Yeo and Keverne, 1986). Oxytocin secretion may mediate these effects (Kendrick et al., 1986). Alternatively or additionally, there is evidence that vaginal-cervical stimulation increases activity in the noradrenergic innervation of the olfactory system (Rosser and Keverne, 1985), lesions of which impair the display of maternal selectivity in sheep by preventing olfactory recognition (Pissonnier et al., 1985).

IV. Hormonal Regulation: One Hormone, Two Hormones, Three Hormones, . . . More?

It is a long and widely held belief that hormones are largely responsible for the essentially immediate appearance of maternal behavior during or after birth in placental mammals. Certainly it makes good sense that the hormonal changes that make pregnancy, parturition, and lactation possible should also play a role in the behavior that makes all that physiological effort worthwhile; by analogy, the ovarian hormonal pattern preceding ovulation in rats also induces sexual receptivity. However, mostly because the endocrine events of pregnancy and associated with maternal behavior in placental mammals are enormously complex and incompletely deciphered, the hormonal induction of maternal behavior, in rats or any other mammal, has not yet become a routine and widely replicable procedure, as is the hormonal induction of sexual receptivity in an extensive variety of female mammals.

There are two principal reasons for wanting to know the hormonal regimen necessary and sufficient to induce rapid-onset maternal behavior: (1) scientists would like to understand thoroughly the type, sequence, and level of hormones which normally induce nurturance; and (2) a reliable, and preferably fairly easy, method to induce maternal behavior rapidly can be used as a tool to study other influential factors, e.g., necessary sensory input or neurochemical mediation. This tool need not reflect a thorough knowledge of all the hormones that contribute to the behavior in question. The estradiol (E) + progesterone (P) induction of sexual receptivity is such a tool and had been long before it was discovered that gonadotropin releasing hormone contributes to the expression of sexual receptivity in hypophysectomized, ovariectomized rats (Pfaff, 1973).

With respect to the current status of the hormonal induction of maternal behavior in rats, reasonable models, based on E and high levels of P followed by its decline, are now available for two widely used strains. But old and heated controversies about the roles of prolactin (PRL) and oxytocin (OT) have been flamed by recent data, some of which even question whether E has a direct action on the neural substrate for maternal behavior or whether its action is indirect, e.g., via the stimulation of PRL. Thus, a fascinating but incomplete story is available on the hormonal induction of maternal behavior in Norway rats, by far the best studied mammalian species in this respect.

Three cautionary notes are in order, (1) *Redundancy:* It is likely that maternal behavior at term is overdetermined with respect to both hormonal and nonhormonal mechanisms. Redundancy of hormonal mechanisms may help to explain inconsistencies in results. (2) *Strain differences:* There are strain differences in maternal responsiveness with respect to both hormonal and nonhormonal induction of maternal behavior, including population differences within the "same" strain (see Section III,B,1). Therefore, generalizations about "the rat" are imprudent until results are replicated using various strains. (3) *Nature of effect:* (a) The most typical endpoint is the latency to onset of maternal behavior. However, experimenters have differed in their *criteria* for this endpoint. Typically, retrieval and grouping of all presented pups within a specified time interval (15, 60, or 180 minutes) on two consecutive days (the latency being the first of these days) is a minimal criterion. Sometimes other components, including the fundamentally important nursing behavior, are included in the criteria as well, but not always. (b) Another problem is that while significant group differences in latency are evidence of an effective hormone treatment, in most cases the percentage of rats displaying maternal behavior with *rapidity*, i.e., within 3 hours of initial pup presentation, is low. (c) Finally, it is rare for studies on hormonal regulation to include measures other than the onset latency, i.e., to consider the *extent and intensity* of the induced behavior. Therefore, studies which have done so will be treated separately (see Section IV,D).

A. Neuroendocrine Events of Pregnancy, Parturition, and Lactation

1. PREGNANCY

There are enormous endocrine changes during pregnancy among eutherian mammals which involve the maternal ovary and pituitary, the maternal–fetal placental unit, and the fetus. I will focus on those hormones that have been most implicated in the hormonal regulation of maternal behavior.

Estrogen and progesterone prepare the endometrium of the uterus for implantation of fertilized ova, maintain pregnancy (Casey *et al.*, 1985), and stimulate mammary gland development, i.e., duct and lobuloalveolar growth (Lyons, 1958). After the first week of pregnancy in rats there is a shift from pituitary to placental stimulation of E and P secretion and a shift in dominant source from corpora lutea to placenta. Prolactin maintains P secretion in rats in the early part of pregnancy (McLean and Nikotovich-Winer, 1973) and stimulates mammary gland development and lactogenesis (Lyons, 1958).

Serum levels of *estradiol* and *progesterone* have been measured in the same laboratory during pregnancy in rats. According to Bridges (1984), E is maintained at moderately high levels of ~30 pg/ml in the first half of pregnancy (compare to proestrus levels of 35.0–52.0 pg/ml; Feder, 1985) and then reaches even higher levels of close to 80 pg/ml from day 15 on. Progesterone rises from low estrus values (~5 ng/ml) on day 1 to a first peak of ~70 ng/ml between days 4 and 10, then rises markedly to a higher peak in the second half of pregnancy with mean values of ~100–150 ng/ml, and then declines sharply after day 20. Thus, the E/P ratio becomes very high in the days before term, and because P inhibits certain actions of E in rats, this ratio is particularly valuable (Soloff *et al.*, 1979); it shows that the effectiveness of E (for parturition and possibly for maternal behavior) in late pregnancy when P declines is much greater than the change in its secretion would indicate.

In response to the cervical stimulation of mating, there are semicircadian surges of *prolactin* (Butcher *et al.*, 1972), comparable in magnitude to the proestrus peak of ~70 ng/ml (Smith *et al*, 1975); the diurnal and nocturnal surges last until days 8 and 10, respectively (Smith and Neill, 1976). Rat *placental lactogen*, which has prolactinlike properties, becomes detectable on day 8 and reaches a peak on day 11 (e.g., Linkie and Niswender, 1973; Kelly *et al.*, 1975), when the pituitary is no longer required to maintain the corpora lutea. Pituitary PRL rises again at the end of pregnancy (Morishige *et al.*, 1973) in response to the periovulatory E surge.

In summary, pregnant rats have high levels of E, high and then very high levels of P followed by its precipitous decline, a sharp rise in the E/P ratio around the time of parturition, and high levels of either PRL or placental lactogen.

2. PARTURITION

For obvious reasons, theories and data on the endocrine control of parturition have played a heuristic role in research on the endocrine control of the onset of maternal behavior. Based on the opposite roles of E and P on uterine contractility—stimulation versus suppression, respectively—a simple theory emerged based on the late-pregnancy fall in P to explain the endocrine trigger of parturition, thought to be generalizable to many eutherian mammals (Csapo, 1969). Radioimmunoassay (RIA) of maternal steroids in various species (see Rosenblatt and Siegel, 1981, for illustrations) has not supported this theory (Fuchs, 1985). There is a sharp decline in P prior to labor in many species (rat, mouse, rabbit, sheep, goat, cow, pig, dog, cat), but in others there is no decrease (hamster, guinea pig, primates), and in some there is an increase (mare). Estrogen does rise in many species, but parturition occurs during greatly decreased levels in some species (mare, cow). "Moreover, progesterone does not prolong gestation in all species, and does not inhibit myometrial contractions or response to oxytocin in a number of species" (Fuchs, 1985, p. 208).

Oxytocin, secreted during the expulsive stage of labor in response to cervical-vaginal distension, and possibly earlier, is a powerful uterotonic agent which plays a major endocrine role in parturition (Fuchs, 1985). In rats, uterine OT receptors rise markedly at term, reaching maximal levels during labor; apparently, these receptors are stimulated by estrogens, inhibited by P (Soloff *et al.*, 1979), and potentiated by uterine distension (Fuchs, 1985). Plasma OT levels rise somewhat during the first stage of labor and then dramatically at the beginning of fetal expulsion (Fuchs, 1985). Synchronized myometrial activity, of critical importance in polytocous species, develops shortly before parturition in rats (Fuchs, 1985). When a sufficient number of uterine OT receptors is present, endogenous or exogenous OT elicits laborlike contractions. It may not be possible to determine the effect of a total elimination of OT secretion on parturition because (1) the fetus secretes OT; (2) hypophysectomy does not completely eliminate OT since the severed stalk secretes OT and a miniature neurohypophysis forms rapidly from the reorganized severed axons (Bintarningsih *et al.*, 1958; Dogterom *et al.*, 1977; Moll and de Wied, 1962); (3) lesions of the paraventricular and supraoptic nuclei, which synthesize OT, are also incompletely effective because magno-cellular OT-secreting islands exist outside these nuclei; and (4) OT antiserum may not be effective against the massive release of OT during delivery (Fuchs, 1985). Nonetheless, the conclusion that OT is necessary for parturition to take place is supported by ample data.

The recent evidence implicating OT in the onset of maternal behavior in rats and sheep (Section IV,C,4) raises the question of whether the events of parturition also increase levels of intracerebral OT (see Section VI,D).

3. LACTATION

a. Hormonal Background. In rats, the mammary gland develops under the influence of various hormones. Duct growth requires the combined ac-tion of estrogen, glucocorticoids, and growth hormone, while lobuloalveolar development additionally requires P and PRL (Meites, 1966). Lactogenesis is principally under the control of PRL, but is facilitated by insulin and glucocorticoids. Milk let-down, or release, is due to OT. The suckling stim-ulus is the afferent limb for neuroendocrine reflexes culminating in the release of hormones necessary for milk synthesis and release, i.e., PRL (Amenomori *et al.*, 1970), ACTH (Voogt *et al.*, 1969), and OT (Fuchs *et al.*, 1984). In addition, the suckling stimulus inhibits gonadotropin secretion, resulting in anovulation, both directly and via hyperprolactinemia (Smith, 1978); this function contributes to long birth spacing in human populations which do not practice contraception (cf. Stern *et al.*, 1986).

b. Afferent Pathways of Suckling-Induced Oxytocin and Prolactin. When attached to nipples, rat pups suck every few seconds, and they do so asynchronously except when there is a milk ejection. There is no specificity regarding which of the 12 nipples are suckled in order to elicit a milk ejection, although lactation may not be supported by only one pup prior to day 12 postpartum (Leigh and Hofer, 1973), and a larger number of pups is needed to elicit a milk-ejection response from anesthetized than from conscious dams (Drewett and Trew, 1978; Lincoln and Wakerley, 1973). Recent neu-roanatomical and neurophysiological evidence demonstrated that a given dorsal root ganglion innervates 2–3 different mammary glands (Tasker *et al.*, 1986) and individual spinal neurons show convergence and summation of the afferent input (Poulain and Wakerley, 1986). Furthermore, the effective path-ways may be more diffuse than is revealed in experiments with anesthetized dams. Indeed, efforts to block milk ejection in conscious dams by brain transections reveal a need to make very large cuts (e.g., Voloschin and Dottaviano, 1976). Most of the studies tracing the afferent pathways of the suckling-induced neuroendocrine secretions have been done on anesthetized rats, rabbits, guinea pigs, and goats and have utilized transections, discrete electrolytic lesions, electrical stimulation, and neuroanatomical tracing to locate the pathways. The dependent variables were litter weight gain, intramammary pressure, and/or OT RIAs to assess milk ejection.

Spinal nerves carry the suckling stimuli into the lateral funiculus ipsilaterally (although there is a minor contralateral component in rats) and from there to the contralateral midbrain lateral tegmentum (Tindal et al., 1967b, 1969; Tindal and Knaggs, 1975; Juss and Wakerley, 1981). Recent evidence suggests that before the decussation, there is a synapse in the lateral cervical nucleus (LCN) (Dubois-Dauphin et al., 1985a,b). The LCN is located in the dorsal horn of the uppermost three segments of the spinal cord, lacks somatotopic organization, receives inputs from a large part of the body, and therefore is not involved with fine discrimination of somesthetic information (Geisler et al., 1979). In short, the LCN provides a basis for the further integration of diffuse inputs based on pattern rather than precise location; this is consistent with the characteristics of the rat litter's suckling. The integrity of the lateral tegmentum is essential for OT release, whereas the ventral tegmentum, while not essential, "may contribute to the timing of the intermittent milk ejection responses" (Juss and Wakerley, 1981).

The pathways from the midbrain to the diencephalon, which eventually must reach the magnocelluar OT-synthesizing nuclei of the posterior hypothalamus, are widely dispersed into dorsal (via central gray and periventricular structures) and ventral (via subthalamus) paths. Although these are not yet fully delineated, they may include the nucleus accumbens (Smith and Holland, 1975) and limbic input via the bed nucleus of the stria terminalis (Sawchenko and Swanson, 1983). Alerting stimuli activate a brainstem-forebrain mechanism that exerts central inhibition of OT secretion (Taleisnik and Deis, 1964), which runs through the hippocampus and subiculum, and goes from there to the hypothalamus (Tindal and Blake, 1984). Both the habituation to distracting stimuli, which interferes with milk let-down, and the positive conditioning to suckling-related stimuli, which occurs in a variety of mammals [e.g., rats (Deis, 1968); ewes (Fuchs et al., 1987)], probably occurs readily at the hippocampal level. Slow-wave sleep is a correlate of milk ejection in rats (Voloschin and Tramezzani, 1979; Lincoln et al., 1980) but not in rabbits (Paisley and Summerlee, 1983).

Effective stimuli for PRL release may be even more diffuse than those for OT release because some PRL secretion occurs in thelectomized rats (see below). Also, in rats PRL release is conditionable to distal stimuli, principally olfactory (Grosvenor, 1965; Mena and Grosvenor, 1971). The pathway for release of PRL initially follows that for OT but in the diencephalon it departs from the OT pathway by ascending laterally and anteriorly in the medial forebrain bundle, finally fanning out in the preoptic region (Tindal and Knaggs, 1977). The amygdala is also involved in the regulation of PRL secretion: marked increases in PRL were observed following lesions of (Moore *et al.*, 1965) or E implants in the amygdala (Tindal *et al.*, 1967a).

4. RELEVANT CNS HORMONES

The past decade has seen a burgeoning of information on peptides secreted by the CNS which have an *in situ* action or which are carried by axonal transport to distant CNS receptor sites where they have a neurotransmitter or neuromodulatory role. Of relevance to the present topic are the molecules that are the same as or highly similar to hormones synthesized by the hypothalamus or anterior pituitary. Elucidation of the secretory control and function of these peptides is just beginning, but the evidence to date suggests that there is a functional relationship between the physiological and behavioral actions, e.g., cholecystokinin and feeding, vasopressin and drinking, gonadotropin releasing hormone and sexual behavior. Clearly, circulating hormone levels no longer provide sufficient information from which to draw conclusions about the hormonal regulation of a given behavior, including maternal behavior.

In rats a prolactinlike substance has been identified in the brain (Shivers *et al.*, 1983), and PRL-synthesis capacities have been traced to the hypothalamus (Schachter *et al.*, 1984). Prolactinlike immunoreactivity (PLI) has been quantified and characterized in both hypothalamic and extrahypothalamic sites (Emanuele *et al.*, 1986, 1987) and shown to be independent of hypophysectomy and stress, suggesting that it is manufactured in the brain rather than in the pituitary. However, cross-reactivity of PLI with proopiomelanocortin remains a plausible, but as yet untested, possibility. Although more work on the activity, specificity, and secretory control of PLI is needed, the possibility of its involvement with maternal behavior cannot be dismissed (see Section VI,C).

Oxytocin is measurable in the cerebrospinal fluid (CSF) of a number of species, including rats (Dogterom *et al.*, 1977). Based on neuroanatomical, electrophysiological, and lesion studies, the paraventricular nucleus (PVN) is considered to be the major source of brain and CSF oxytocin. From there, OT is transported to various parts of the brain and spinal cord (e.g., Swanson and Sawchenko, 1980; Buijs *et al.*, 1985). The sites of OT concentration which may be playing a role in maternal behavior will be considered later (Section VI,D).

B. Sensitized Maternal Behavior

I. IS SENSITIZED MATERNAL BEHAVIOR NONHORMONAL?

Because maternal behavior can be induced by pup stimulation in hypophysectomized females, and with similar latencies as controls (Rosenblatt, 1967; Bridges *et al.*, 1985; Stern, 1987b), Rosenblatt (1967) concluded that sensitization is nonhormonal in nature. Further, Rosenblatt (1970) suggested that because postpartum maternal behavior is maintained by pup stimulation (Rosenblatt and Lehrman, 1963) rather than hormones, it is possible to induce maternal behavior in this way; consequently the similarities, rather than the differences, between postpartum (or hormone-induced) and pup-induced maternal behavior have been emphasized (Fleming and Rosenblatt, 1974c; Rosenblatt *et al.*, 1985, p. 258).

In light of the evidence on centrally elaborated neuroregulatory peptides, as well as the continuation of OT secretion from the magnocellular hypothalamic nuclei following hypophysectomy (Section IV,A), I suggest that pup-induced maternal behavior not be termed "nonhormonal," but rather "pituitary independent." The evidence for a role of OT in postpartum maternal behavior will be reviewed later (Section IV,C,4); it is at least conceivable that changes in intracerebral OT play a role in sensitization of maternal behavior as well. Further, tonic ovarian secretions contribute to the quality of induced maternal behavior in female rats. Le Roy and Krehbiel (1978) demonstrated that intact virgin females were superior to ovariectomized virgin females in retrieval latency, time spent crouching over and licking pups, and nest building; Mayer and Rosenblatt (1979) also found nest building to be enhanced by ovarian influences. And the fact that sex differences in various aspects of responsiveness to pups have been reported (see next section) further indicates that hormones secreted neonatally and/or in adulthood may influence the expression of pup-induced maternal behavior.

2. SEX DIFFERENCES

Male Norway rats have no paternal behavior because they play no role in nature in the care of young. In fact, given the infanticidal propensity of

TABLE V   Sex Differences in Responsiveness toward Pups in Rats

Topic	References[a]
A. Sensitization When Intact	
1. Weanlings: shorter latency in males than females	A,B,C,D
2. 36 or 42 days of age: no sex difference in latency	A,E
3. Adulthood: no sex difference in latency,	F,G
or shorter in females than males	H,I,J,K
4. Induced behavior maintained less reliably by males	I
5. Sex difference in latency abolished by olfactory bulbectomy	K
but not by peripheral anosmia	I
B. Effects of Adult Gonadectomy on Sensitization	
1. No effect in females or males,	F,G,L,M,N,O
with one exception	H
2. Robustness of induced maternal behavior similar in intact males and	
castrated males and females but inferior to that of intact females	L
C. Effects of Perinatal Androgen on Sensitization	
1. Neonatal androgen: percentage of males responding decreased,	M,N
or not decreased	P
2. Prenatal androgen: percentage of females responding decreased	Q
3. Neonatal androgen: percentage of males and females retrieving pups	
from a T-maze decreased	P
D. Infanticide Rate	
1. Where there is a sex difference, males higher than females	e.g., R
2. Castration in infancy, but not adulthood, reduces high male rate	S,T,U
3. Testosterone administration to intact males in adulthood increased	
the response	V
4. Response reduced by rearing in testing laboratory or cohabitation	
with female and pups	V

[a](A) Bridges et al., 1973; (B) Gray and Chesley, 1984; (C) Stern and Rogers, 1988; (D) Kinsley and Bridges, 1988; (E) Stern, 1987b; (F) Quadagno et al., 1973; (G) Rosenblatt, 1967; (H) Mayer and Rosenblatt, 1977; (I) Mayer and Rosenblatt, 1979; (J) McCullough et al., 1974; (K) Fleisher et al., 1981; (L) Le Roy and Krehbiel, 1978; (M) Quadagno and Rockwell, 1972; (N) Rosenberg and Herrenkohl, 1976; (O) Siegel and Rosenblatt, 1975b; (P) Bridges et al., 1972; (Q) Ichikawa and Fujii, 1982; (R) Jakubowski and Terkel, 1985; (S) Rosenberg et al., 1971; (T) Rosenberg, 1974; (U) Rosenberg and Sherman. 1975; (V) Brown, 1986a.

many male rats, it is wise for a mother to continue agonistic actions toward even the father of her current litter (Brown, 1986b). Therefore, maternal behavior might be sexually differentiated in rats, subject to the same perinatal hormonal events that influence sexual behavior and physiology. But weaning and adult male rats can be sensitized, which indicates that maternal behavior "circuits" exist in male rats and are not completely eliminated by either the neonatal or the adult hormonal milieus. Table V summarizes the relevant data on possible sex differences in responsiveness to pups. (Note that disparate findings have emerged from the same investigators, using the same strain; in the case of Quadagno's work with Long-Evans rats, the sex difference was dependent upon the use of a large cage.) While weanling-aged males show maternal behavior more quickly than female age-mates, an age-related PRL effect (Kinsley and Bridges, 1988), there is either (1) no sex difference in sensitization latency or percentage responding at later ages, or (2) an advantage for females. Similarly, when there is a sex difference in infanticide rate, females display it less often. Both of these sex differences are probably related to the lower amount of androgen normally present perinatally in females than in males.

Because perinatal treatment of female rats with androgen is usually inconsistent with fertility (Gorski, 1971), only one study has focused on postpartum maternal behavior as a function of such treatment (Stern and Strait, 1983). Golden hamsters treated with a high dose of testosterone propionate (300 µg) on postnatal days 6, 8, or 10 had treatment-day-dependent deficiencies in reproductive success which fell short of sterility in most females. However, the maternal behavior of these dams, as measured by retrieval and time spent crouching, appeared to be normal. Perhaps any perinatal androgen treatment that permits pregnancy and parturition also allows for normal postpartum maternal behavior.

C. Hormones and the Onset of Maternal Behavior

I. HIGHLIGHTS OF EARLY STUDIES AND VARIOUS MODELS

a. Peripartum Events In an early attempt to demonstrate that humoral factors incite maternal responsiveness in rats, Stone (1925) joined a maternal and a nonmaternal rat via parabiosis, but this noble attempt failed. Subsequent efforts with hormone injections up until about 20 years ago also were unsuccessful (see Rosenblatt and Siegel, 1981). In retrospect, these efforts were not enlightening, in part because of a failure (1) to take sensitization (Section III,B) into account (e.g., Riddle et al. 1942), (2) to use more than one hormone per group (Riddle et al., 1942; Lott, 1962; Lott and Fuchs, 1962), or (3) to use P and its withdrawal along with E + PRL (Beach and Wilson, 1963). In contrast, the hormonal induction of nest building in mice [with E + P (Lisk et al., 1969; Lisk, 1971)], hamsters [E + P (Richards, 1969)], and rabbits [E + P + PRL (Zarrow et al., 1962)] was successful.

At that point, Terkel and Rosenblatt (1968) found that maternal behavior could be induced more rapidly in intact virgins that received an injection of plasma (~3.5 cc) from a recently (within 48 hours) parturient dam than in those receiving plasma from various control donors (2.5 versus 4–7 days); this was the first successful demonstration of an endogenous humoral factor underlying maternal responsiveness toward pups. Later Terkel (1972) perfected a technique for continuous cross-transfusion in freely behaving rats, resulting in 50% mixing of the two blood pools after 3 hours. With this technique, Terkel and Rosenblatt (1972) showed that the blood of a rat that had given birth within 24 hours, but not just before or after, effectively promoted retrieving and crouching in 88% of virgins with a mean latency of 14.5 hours. As expected, transfusion from virgin females that were maternal "nonhormonally" (i.e., spontaneous retriever or sensitized) was not successful in inducing maternal behavior in a virgin recipient (Terkel and Rosenblatt, 1971).

Although this work provided an elegant demonstration of humoral, presumably hormonal, factors underlying maternal responsiveness at parturition, it remains a mystery as to what those factors may be. It did provide the impetus for an approach used extensively by Rosenblatt and his co-workers which I will call *the pregnancy termination model*. The basis of this model is the belief that the hormonal events immediately around the time of parturition are at least as critical to the stimulation of rapid-onset maternal behavior as is the long preceding period of pregnancy. The role of the latter was shown by the finding that the latency (in days) to become maternal decreases as pregnancy advances from day 13 on (the "pregnancy effect;" Rosenblatt and Siegel, 1975). However, short of the period within a day of parturition when retrieval occurs in some rats immediately (Section II,C), much more dramatic effects occur with pregnancy termination by hysterectomy. Following this procedure, the ovarian secretion of P declines and gonadotropin release is stimulated, culminating in ovulation and a surge of E, thus simulating the events of late pregnancy. The later in pregnancy the hysterectomy is done, the more rapid is the induction of ovulation and maternal behavior (Rosenblatt and Siegel, 1975). After hysterectomy on day 16, with pup presentation 48 or 72 hours later, short-latency-onset maternal behavior declines if the ovary is removed as well, but it is restored with an injection of estradiol benzoate (EB) (5 μg/kg) at the time of surgery (Siegel and Rosenblatt, 1975a).

The above findings of Siegel and Rosenblatt reinforced the belief of these workers, still held (see Rosenblatt *et al.*, 1985), that the *E surge* at the end of pregnancy in rats is critical. Whatever its role may be, it is important to point out that not all data are consistent with this belief. In the same paper by Siegel and Rosenblatt (1975a) both E- and oil-treated groups were highly responsive when tested 24 hours after surgery, with 40% becoming maternal on the day of pup presentation. MacKinnon and Stern (1977) compared the latency to become maternal of Sprague-Dawley rats hysterectomized and ovariectomized on days 16 and 21 of pregnancy, i.e., before and after the onset of the late-pregnancy E surge (or increased E/P ratio); pups were presented 18 hours later. There was no difference between these groups in maternal behavior onset, with median latencies of 0 and 1 day, respectively. Taken together, these data indicate that, *following late-pregnancy termination*, the *recent presence of E* may be sufficient to induce rapid-onset maternal behavior, even when the current levels are negligible and when the peak E surge or E/P ratio was eliminated.

Bridges *et al.* (1978) demonstrated that P begins to decline sooner when the ovaries and uterus are removed than when the uterus alone is removed, and that heightened responsiveness to foster young begins about 24 hours after the onset of the P decline. The critical feature of *progesterone withdrawal* is underscored by these results and buttressed by numerous reports that the maintenance of high P levels seriously interferes with the onset of maternal behavior in primiparous rats (Bridges and Feder, 1978; Bridges *et al.*, 1978; Bridges and Russell, 1981; Moltz *et al.*, 1969a; Numan, 1978; Siegel and Rosenblatt, 1975c).

At the natural termination of pseudopregnancy, a state hormonally similar to pregnancy in terms of suppression of estrous cyclicity (elevated P, diurnal and noctural PRL surges), mice, rabbits, cats, lions, and dogs may display maternal responses (see Steuer *et al.*, 1987, for review). Recently, Steuer *et al.* (1987) found that there is a 50% reduction in latency to onset of maternal behavior in Long-Evans rats after natural pseudopregnancy termination. Further, a more dramatic facilitation of maternal responsiveness occurs at this time when subjects are proferred placenta-smeared, rather than clean, 3- to 8-day-old pups. In this case, five of eight rats had a median latency of maternal behavior onset of 2 hours; however, the remaining three rats were initially infanticidal and required ≥2.5 days to become maternal. These results suggest that in rats (1) a treatment which lasts as long as pseudopregnancy (i.e., 2 weeks) is sufficient to induce rapid-onset maternal behavior in many females if the pup stimuli are sufficiently attractive; and (2) the hormonal events of pseudopregnancy and its termination are not sufficient to suppress infanticide completely. Perhaps a higher, or longer-lasting, P peak is required for the latter.

The use of the pregnancy-terminated (or pseudopregnancy-terminated) rat can be considered a lengthy treatment because of the endogenous endocrine changes which precede experimental manipulations. When Siegel and Rosenblatt (1975b,c) applied their findings on pregnancy-terminated rats to *nonpregnant, nulliparous, ovariectomized females*, they found that a *single injec-*

tion of EB also was effective in inducing short-latency onset of maternal behavior (e.g., 29% maternal on day 0; median onset of 0.5 day, with pups presented at 48 hours after manipulations), but only with 20 times the dose (*100 μg/kg*) and only *in conjunction with hysterectomy*. The maximal effectiveness occurred when pups were presented 48–72 hours after surgery + EB, as in the pregnancy-terminated rat. Following hysterectomy/ovariectomy +100 μg/kg EB and pup presentation 24 hours later, Siegel and Rosenblatt (1975b) found a median maternal behavior latency of 1.0 day (versus 4.5 days for oil-treated controls), with 100% maternal by day 3. Using the same treatment, however, Stern and MacKinnon (1976) found a median maternal behavior latency of 3 days (versus 6 days for intact, untreated virgins), with only 57% maternal by day 3. In these two experiments, the strain was the same (Sprague-Dawley) but the supplier different (Charles River for Rosenblatt's lab; Blue Spruce Farms for my lab), and rats were tested on a reversed light schedule in Rosenblatt's, but not in my lab. Therefore, the effectiveness of this treatment varies greatly in different laboratories. Nonetheless, because of its ease, a large dose of EB has been used to prime females in experiments on the role of intracerebral OT.

Obvious questions concerning the Siegel and Rosenblatt procedure for inducing maternal behavior rapidly in virgins concern the role of hysterectomy and the unphysiological dose of estrogen used. It was later demonstrated that the removal of the uterus enhances the effectiveness of exogenous estradiol with respect to lordosis (Siegel *et al.*, 1978), possibly because this organ concentrates E to a great extent. As to the dose of E, 5 μg/kg was later found to be effective in nulliparous rats if preceded by P, either in the form of an implant for 15 days or a single 5-mg injection 1 day before EB injection (Doerr *et al.*, 1981). Thus, P pretreatment, even for a short time, enhances behavioral responsiveness to E, perhaps by affecting E uptake and retention in neural tissues (Lisk and Reuter, 1977). However, the role of strain differences in E-stimulated maternal responsiveness must be considered; whereas E alone enhanced maternal responsiveness in Sprague-Dawley rats (e.g., Doerr *et al.*, 1981; Bridges, 1984; Giordano, 1987), even lengthy treatment with E alone was ineffective in my laboratory with Long-Evans rats (Stern and McDonald, 1989).

b. Pregnancy Simulation Models The other major approach to elucidating the hormonal basis of maternal behavior has been *the pregnancy simulation model,* in which the endocrine events necessary to maintain pregnancy and to stimulate mammary gland development (Section IV,A) are simulated to varying extents with exogenous hormone administration to ovariectomized subjects. Moltz *et al.* (1970), working with Wistar rats, used injections of EB (12 μg/day, days 0–11), P (6 mg/day, days 6–9), and PRL (2.5 mg/day, day 9, P.M., and day 10, A.M.), with presentation of newborn pups on day 10, P.M. All their rats became maternal 34–40 hours after pup presentation compared with latencies of 7 days for vehicle-injected controls concaveated with pups. Later, Siegel (1974) was unable to replicate these results with Sprague-Dawley rats. Zarrow *et al.* (1971) used an even lengthier hormone procedure (presumably with Wistars) designed to stimulate lactation: EB (2 μg) and P (4 mg), days 1–20, followed by 20 IU (1 mg) PRL, with or without cortisol acetate (2 mg), until the termination of testing; three 2- to 8-day-old pups were presented 24 hours after the fourth PRL injection. In this study, the mean retrieval latency was about the same as that found by Moltz *et al.* (1970), but 41% of their animals displayed maternal behavior on the first day and 82% within 24 hours, compared to 4–5 days for controls. Using a slight modification (3 μg EB, days 1–20; 4 mg P, days 3–17; 1 mg PRL, day 21, A.M.; pups presented day 21, P.M.), Bridges *et al.* (1973) also obtained reduced retrieval latencies, but Krehbiel and Le Roy (1979), who followed the Bridges *et al.* (1973) procedure with Holtzmann rats, did not.

In sum, these studies enhanced our understanding of the hormonal pattern necessary to induce rapid-onset maternal behavior, but additional fine-tuning of doses and sequences was called for. In retrospect, the P treatment was too short in the Moltz *et al.* procedure and the interval between P withdrawal and testing was probably too long in the Zarrow *et al.* and Bridges *et al.* protocols. A major problem concerns replicability in another laboratory, perhaps under slightly different testing conditions, and especially with a different strain of rats. (Replicability remains a problem, in large part because the population of maternal behavior researchers is small, and a hormone-induction study that simulates pregnancy and includes several groups can take many months to complete.)

2. RECENT ESTROGEN PLUS PROGESTERONE-BASED MODELS

In an effort to simulate the ovarian hormone profile of pregnancy as closely as possible, Bridges (1984) evaluated the serum hormone levels and behavioral effects of various combinations of E and P, each administered as subcutaneously implanted silastic capsules, the secretion from which is rather steady. Capsules of E that are 1–2 mm in length approximated levels of serum E during the first 10 days of pregnancy (~35–40 pg/ml); the 5-mm-length capsule approximated levels between gestation days 15 and 22 (~75–85 pg/ml); the 10-mm capsule produced levels (~100 pg/ml) higher than Bridges observed during pregnancy. For P, 90 mm (3 × 30 mm) of silastic capsules yields a mean serum level of ~32 nm/ml, which does not come close to the pregnancy levels of ~80 ng/nl between days 4 and 10 and ≥120 ng/ml between days 12 and 20; thus, the relatively low P levels are the main drawback of this pregnancy simulation model. The main behavioral effects are as follows: (1) The most effective treatments include a 1- or 2-mm E capsule from day 1, with 90 mm P between days 3 and 13 and pup presentation on day 14. Mean retrieval and grouping latencies of 1–2 days occur (versus 5–7 days for ovariectomized, blank-implant controls), with 20–40% showing this behavior rapidly, i.e., in ≤1 hour. (2) The results are similar whether or not the E capsule is removed on day 13. (3) The synergism between E and P is not apparent with 5- or 10-mm capsules of E because these doses of E resulted in retrieval latencies (mean of 2 days) not significantly different from the E + P groups. (4) Maximal effectiveness occurs with P treatment for 10 days. (5) A similarly effective treatment occurs with P pretreatment for 10 days + 2 mm E at time of P removal, and testing 1 day later (2-day mean latency). Therefore, for the synergistic effect of E + P on maternal behavior, "elevated blood levels of E need not be present both during and after P exposure" (Bridges, 1984, p. 938).

Recently we followed the essential features of Bridges' (1984) protocol using Long-Evans rats (Stern and McDonald, 1989) (Fig. 3). Because this strain is not as maternally responsive in a variety of circumstances as the Sprague-Dawley strain, and because pilot work suggested that the 5-mm dose was not as effective as the 10-mm capsule, the latter was used. In support of Bridges' results, rats given E + P, with the E left in when the P was removed, become fully maternal significantly faster than controls given blank capsules (1.8 versus 5.7 days). In striking contrast to Bridges' results, if the E capsule is removed along with the P capsules on day 13, or if treatment consists of 10 mm E only, latencies are no different from blank controls (5.4 and 5.2 days). Thus, continued E treatment following P withdrawal is necessary, and 2 weeks of high-dose E is not sufficient to hasten maternal onset in Long-Evans rats. Because only 20 and 60% became maternal on days 0 and 1 of testing, I tested another recent model as well.

Giordano (1987) reported onset of maternal behavior in Sprague-Dawley females within 3 hours in 93% of rats treated with his laboratory's most effective steroid regimen, designed to reach higher P levels than that achieved by Bridges' model. Using a modification of this model, Long-Evans rats were given a 10-mm E silastic capsule on day 1, P by injection (4 mg/day, days 3–16), and tested for maternal behavior beginning on day 18. Females

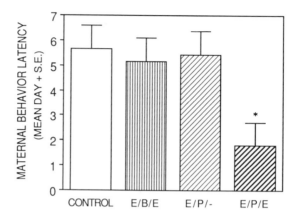

FIGURE 3. Maternal behavior latency in ovariectomized, virgin, maternally naive Long-Evans rats as a function of estradiol (E) alone (vertical stripes) or in conjunction with progesterone (P) (diagonal stripes). X/Y/Z: X represents 10 mm silastic capsules of E, s.c., or blank (B) capsule on day 1; Y represents 3 × 30 mm silastic capsules of P, s.c., in the sides, or blank capsules, days 3–13; Z represents E capsule which, if present, either remains in during testing (E) or is removed (−) on day 13; half of the no-hormone controls (white bar) had their blank neck capsule removed on this day. Pups were presented on day 14 and daily thereafter for 10 days or until maternal behavior occurred for 2 days consecutively. * represents significant difference from other groups. (From Stern and McDonald, 1989.)

receiving this treatment (Stern and McDonald, 1989; Fig. 4), with or without an additional injection of 5 μg EB on treatment day 16 to simulate the parturitional ovulatory surge, show substantially reduced latencies to become maternal compared to nonsteroid-treated females or those treated with P or E alone. In the most effective groups, the percentage maternal was 22–36% on test day 0 and 82–100% by day 1. As Bridges also found, short-term treatment with E following P withdrawal is somewhat effective. Finally, in the third experiment of this series, neither hysterectomy (part of the Giordano procedure, following the work of Siegel and Rosenblatt described above) nor duration of P treatment (13 versus 10 days) affected maternal behavior latencies. The major difference between these results and those of Giordano has to do with the *different baseline levels of responding for the two strains*. Thus, Giordano (1987) found that the test-day-0 percentage maternal is ~48 and 10% for E treatment alone and nonsteroid-treated controls, respectively, as compared to 0% for both of these groups in my lab. Nonetheless, it is fair to say that reasonably replicable procedures are now available, which are lengthy

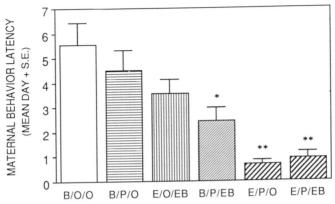

F I G U R E 4. Maternal behavior latency in ovariectomized, virgin, maternally naive Long-Evans rats as a function of P alone (horizontal stripes), E alone (vertical stripes), or various combinations of E and P (diagonal stripes). X/Y/Z: X represents 10 mm silastic capsule of E, s.c., or blank (B) capsule on day 1, which remained in throughout testing; Y represents 4 mg/day P, s.c., days 3–15, or oil vehicle alone (O); Z represents a single injection of estradiol benzoate (EB) (5 μg/rat, s.c.) on day 16, or oil vehicle alone. Pups were presented on day 18 and daily thereafter for 10 days or until maternal behavior occurred for 2 days consecutively. * represents significant difference from no-hormone group; ** represents significant difference from no-hormone (white bar), P-only, and E-only groups. (From Stern and McDonald, 1989.)

but easy to administer, and which can be used as tools for studying mechanisms of hormonal action or other influences on the expression of maternal behavior.

3. PROLACTIN

The recent E + P models described above do not include PRL because it had become clear that the E treatment is sufficient to stimulate endogenous PRL secretion (Amenomori *et al.*, 1970; MacLeod, 1975). Also, a number of lines of evidence questioned whether PRL is a vital ingredient in the hormonal induction of maternal behavior in rats. Riddle *et al.* (1942) was the first to propose PRL as the maternal hormone, but his evidence was flawed and his findings were not subsequently replicated, either with injections of PRL (Lott and Fuchs, 1962; Beach and Wilson, 1963) or with PRL-secreting pituitary grafts (Baum, 1978). In addition, inhibition of the late-pregnancy PRL surge did not interfere with the subsequent onset of maternal behavior, as demonstrated by midpregnancy hypophysectomy (Obias, 1957), late pregnancy ovariectomy (Bridges *et al.*, 1974b), and by administration of drugs that inhibit PRL secretion starting either preterm (Stern, 1977), or after surgical termination of pregnancy (Numan *et al.*, 1977; Rodriguez-Sierra and Rosenblatt, 1977). Still, other intriguing evidence continued to suggest a possible role for PRL in maternal behavior. In mice, PRL implanted in the hypothalamus enhances pup-oriented maternal behaviors (Voci and Carlson, 1973). In golden hamsters, PRL inhibition with ergocornine reduces postpartum maternal behavior and aggression (Wise and Pryor, 1977). In rabbits, Anderson *et al.* (1971) demonstrated that ovarian steroid-induced nest building in rabbits is dependent upon the presence of the pituitary, and Zarrow *et al.* (1971) identified PRL as the critical pituitary factor.

Bridges *et al.* (1985) demonstrated a *critical role for prolactin* by showing that the E + P stimulation of maternal behavior was ineffective in hypophysectomized rats, but effectiveness was restored by replacement of PRL, in the form of either PRL-secreting ectopic pituitary grafts or injections of PRL (twice daily for 13 days, with pup presentation on day 12). How can this striking finding be reconciled with the lack of effects of PRL in rats described above? First, PRL does not act alone, but in synergy with ovarian steroids. Second, elimination of the late-pregnancy PRL surge (or its simulated equivalent) is clearly not sufficient to rule out a role for PRL in maternal behavior stimulation. As described earlier (Section IV,A,1), there are semicircadian surges of PRL in rats during the first 8–10 days of pregnancy, followed by a rise in placental lactogen. Subsequently Loundes and Bridges (1986) showed that prolonged PRL treatment of hypophysectomized rats (0.5 mg twice daily, from day 1 to end of testing), along with E + P, stimulates the rapid onset (latencies of <1 day) of all aspects of maternal behavior, whereas acute PRL treatment (from day 11) stimulates pup-carrying only. Therefore, prolonged exposure to PRL, or PRL-like molecules, like placental lactogen, apparently influences the rapid onset of maternal behavior at parturition. There has been no direct test as yet of the possible behavioral effectiveness of rat placental lactogen. However, MacKinnon and Stern (1977) showed that, following pregnancy termination on day 16, maternal behavior latency is not affected, but subsequent T-maze retrieval of pups is impaired in Sprague-Dawley rats carrying ≤6 fetuses compared to controls carrying ≥7 fetuses; this finding may reflect a "dose-dependent" effect of placental lactogen.

Finally, *P pretreatment + PRL, in the absence of estradiol*, facilitated the onset of maternal behavior in rats (Bridges and Dunckel, 1987). This remarkable finding was demonstrated in ovariectomized, hypophysectomized, and nonhypophysectomized rats treated with P implants (6 × 30 mm, resulting in serum P levels of 65–70 ng/ml; removed 1 day before testing) and PRL-secreting ectopic pituitary grafts. The reduction in maternal-behavior latencies compared to nonhormone-treated controls was more substantial in the experiment with pituitary-intact rats (from 6 to 1.5 days, with 7 of 12 hormone-treated females retrieving within the first test hour) than in the experiment with hypophysectomized rats (from 6 to 3 days, with no immediate responses reported). This is possibly because PRL serum levels are substantially higher in the former condition than in the latter (e.g., 48.7 versus 14.0 ng/ml on the day before testing). It is not known yet if the effects would be more dramatic with even higher PRL levels. To date, PRL alone has not been shown to facilitate maternal responsiveness (Baum, 1978; Bridges and Dunckel, 1987). Also, *growth hormone*, secreted by the grafts (Bridges *et al.*, 1987), may be involved in the stimulation of maternal behavior in these experiments.

Given the emphasis on the role of estradiol in mediating rat maternal behavior by Rosenblatt and his associates, the suggestion of Bridges and Dunckel (1987) that E may not have direct actions at all, but only an indirect one via stimulation of PRL (e.g., Amenomori *et al.*, 1970), is notable. Even the evidence for central sites of E action on maternal behavior (Section VI,A) and E priming of OT-stimulated rapid-onset maternal behavior (Section IV,C,4) must now be reevaluated in hypophysectomized rats. However, it is important to determine if estrogen priming, long term or even short term, enhances the effectiveness of P + PRL in hypophysectomized rats.

4. OXYTOCIN

Klopfer (1971) first hypothesized that oxytocin, essential for parturition, may be responsible for the rapidity of the onset of "mother love" at term. In 1979, Pedersen and Prange provided evidence for this hypothesis with intracerebroventricular (icv) infusions of OT (400 ng) in maternally naive, ovariectomized, Sprague-Dawley (Zivic Miller) rats primed with EB (100 μg/kg) 48 hours prior to infusions and pup presentations. Whereas only 2 of 11 saline-infused rats responded maternally within 1 hour of pup presentation, 11 of 13 OT-infused rats displayed full maternal behavior in this time. This OT effect is dependent on E priming (Pedersen and Prange, 1979) and is dose dependent and peptide specific (C. A. Pedersen *et al.*, 1982).

The robustness of these findings was soon questioned when two laboratories reported their failure to replicate the effect with Sprague-Dawley (Charles River) (Rubin *et al.*, 1983) or Wistar (Bolwerk and Swanson, 1984) rats. However, by following the procedure of Pedersen and Prange (1979) more exactly, including their source of rats, Farhrbach et al. (1984, 1985) were successful in replicating the effect. Other critical conditions include the source of the OT and testing in a novel cage after 2 hours of habituation (Farhbach et al., 1986). The use of Zivic Miller rats became suspect upon reports of a chronic low-grade respiratory infection, possibly resulting in hyposmia. Because anosmia (Fleming and Rosenblatt, 1974a,b) and hyposmia (Mayer and Rosenblatt, 1975) alone hasten the onset of maternal behavior in virgins, Wamboldt and Insel (1987) investigated the interaction between icv OT and decreased olfaction in a healthy strain of Sprague-Dawley rats (from Taconic Farms). Using the successful procedures described above, they found that OT resulted in rapid-onset maternal behavior only in rats rendered hyposmic with intranasal $ZnSO_4$. As during parturition, these rats engaged in pup licking prior to retrieval and grouping. Thus, the significance of centrally acting OT to maternal behavior might be dependent upon reduced or altered olfaction peripartum, a combination that has not yet been examined.

Another line of evidence concerning the role of OT in rat maternal behavior involves the use of antisera to OT or OT antagonists to block E-stimulated maternal behavior in virgins and in 16-day pregnancy-terminated rats (Fahrbach et al., 1985; Pedersen et al., 1985). Even more convincing evidence is the demonstration of a delay in the onset of postpartum maternal behavior (in Wistar rats) by icv injection of the OT analog antagonist, D-(CH_2)5,8-ornithine-vasotocin (van Leengoed et al., 1987). Because this treatment, administered during parturition, had no effect on the course of parturition, the effect of the antagonist was presumed to be limited to centrally acting OT (Buijs et al., 1985). Furthermore, prepartum lesions of the paraventricular nucleus, which supplies most of the intracerebral OT, disrupts onset (E. B. Keverne, unpublished observations) but not maintenance (Numan and Corodimas, 1985) of maternal behavior in rats. Taken together, these results suggest that centrally released OT, in conjunction with other neuroendocrine events of parturition, plays a modulatory role in the onset of maternal behavior in rats at term (Fahrbach et al., 1985). Whether the interaction is with E directly, as has been assumed, or with PRL, or both, remains to be determined. A link between E, OT, and olfaction in sheep has been demonstrated by Keverne et al. (1983). Vaginal distension, a powerful stimulator of neurohypophyseal OT secretion into the peripheral circulation (Section IV,A,2), promoted acceptance of lambs in E-primed virgin ewes and of alien lambs in recently parturient ewes already imprinted to the odor of their own young.

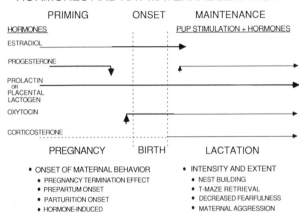

HORMONES AND RAT MATERNAL BEHAVIOR

FIGURE 5. It is widely acknowledged that hormones prime maternal responsiveness, such that nurturance emerges rapidly after presentation of pups. The hormones that have been implicated in hormone-induced maternal behavior in rats are estradiol (E), progesterone (P), and prolactin (PRL) (and possibly placental lactogen when pituitary PRL is low during the middle of gestation); oxytocin, perhaps of intracerebral origin, also may contribute during labor. The duration of action of P and PRL is ≥10 days, and P must be withdrawn ~2 days before pups appear. The timing and duration of action of E are not clear. Although pup stimulation alone, which is capable of inducing maternal behavior in virgins (i.e., sensitization), albeit slowly, maintains the essential features of maternal behavior postpartum, it is not capable of duplicating the intensity and extent of maternal characteristics present in rats which have been primed with hormones (endogenous or exogenous). In addition, the hormones which are increased by suckling and contribute to milk secretion, including PRL, P, and adrenal corticoids, also influence thermoregulation, which affects the timing of nest bouts as well as food intake and metabolism. Therefore, hormones contribute substantially to the maintenance, as well as to the onset, of maternal behavior in rats.

Most recently, icv OT stimulated maternal behavior in ovariectomized, E-treated multiparous ewes (Kendrick *et al.*, 1987).

In sum, estradiol, progesterone, prolactin, and possibly placental lactogen, as well as oxytocin have been implicated in the priming of maternal responsiveness. Figure 5 provides a schematic illustration of these factors and their approximate times of action.

D. Hormones and the Maintenance of Maternal Behavior

Beginning with the finding of Bridges *et al.* (1972) that lactating rats are much more likely to retrieve pups from a T-maze extension of their home cage than are sensitized females or sensitized males, there have been many reports on the differences in the expression of maternal behavior as a function of how the behavior was induced. Further work with the T-maze revealed the following: (1) Postpartum mothers that cannot suckle due to prior nipple removal (thelectomy) retrieve as well as intact mothers (Stern and MacKinnon, 1976) even when prolactin secretion of the thelectomized mothers (Stern and Siegel, 1978) is inhibited with ergocryptine (Stern, 1977). (2) Subjects that become maternal rapidly following hormone stimulation retrieve as readily as postpartum females (Bridges *et al*, 1973; Stern and MacKinnon, 1976; Bridges, 1984). (3) The longer the pregnancy prior to termination by hysterectomy/ovariectomy, the greater the propensity to retrieve pups from the T-maze (MacKinnon and Stern, 1977). (4) A low level of placental hormones, indicated by carrying ≤6 fetuses, does not affect latency to onset of maternal behavior but is associated with a low likelihood of T-maze retrieval. (5) The presence of androgen neonatally in males and females does not reduce latency to become maternal in response to hormonal stimulation in adulthood, but it does reduce T-maze retrieval markedly (Bridges *et al.*, 1973). (6) Sensitized females can be induced to retrieve pups from the T-maze if tested under less frightening conditions in the dark, but ovariectomized females are much less likely to do so than intact virgins (Mayer and Rosenblatt, 1979). Note that in two of these situations [(4) and (5)], the T-maze test is more discriminating than is maternal-behavior-onset latency. Females that are maternal due to endogenous or exogenous hormones are more likely to explore the T-maze fully (Stern and MacKinnon, 1976). Because spontaneous activity of lactating rats compared to cycling females is either depressed [running wheel (Richter, 1956)] or not different [open-field ambulation scores (Stern *et al.*, 1973)], the T-maze may be a more meaningful measure of the postpartum mother's willingness to engage in goal-directed activities related to the care of her young than are other activity indices. Lactating rats are less fearful in other situations as well (Hard and Hansen, 1985).

Whereas sensitized virgins display little or no aggression toward a strange intruder, postpartum rats are very aggressive (Erskine *et al.*, 1978a,b); this difference is largely attributable to hormonal priming (Krehbiel and Le Roy, 1979; Mayer and Rosenblatt, 1987) and not to postpartum pituitary changes (Erskine *et al.*, 1980a,b). Nest-building scores are also higher in hormone-treated than in sensitized virgins (Krehbiel and LeRoy, 1979; Mayer and Rosenblatt, 1979).

The normal decline in nursing duration in the second postpartum week is influenced by the dam's levels of adrenal corticoids and P (Woodside and Leon, 1980; Adels and Leon, 1986) via the effect of these steroids on the dam's thermoregulatory control of nest-bout duration (Leon *et al.*, 1978). Because the levels of these hormones are directly related to suckling-induced changes (Section IV,A,3; ovarian P stimulated by PRL), it would be interesting to know how nest-bout duration is regulated in nonlactating rats, both postpartum (thelectomized) and nulliparous (sensitized). Since nonlactating rats have a lower thermal "setpoint" (Leon *et al*, 1978, 1985; Adels and Leon, 1986) and the pups with them are not heated by warm milk, the thermoregulatory model would predict *longer* nest-bout durations when these mothers are kept with progressively older pups than when lactating rats are kept with similarly aged young. In contrast, during the decline phase of maternal behavior, sensitized virgins showed *decreases* in maternal care and increases in withdrawal behaviors sooner than lactating dams (Reisbick *et al.*, 1975); hormone-induced virgins have not been tested in this way. Finally, in the last part of lactation, Wistar rats secrete a prolactin-dependent maternal pheromone (Leon, 1974) which is attractive to the young; however, the presence of such a pheromone in Long-Evans rats is questionable (Galef and Muskus, 1979).

Taken together, these results suggest that hormones that facilitate the *onset* of maternal behavior also influence the *intensity* and *extent* of the maternal behavior repertoire (Fig. 5). In the case of the T-maze retrieval results, the enhancing effects of hormonal stimulation persist even when these hormones are no longer present. Also, the *patterning of nursing behavior*, at least in lactating dams, is influenced by levels of hormones that are affected by suckling-induced hormonal changes. Perhaps it would be better to consider the establishment and maintenance of complete postpartum maternal behavior to be the result of a complex interaction of hormonal priming, contact with pups, and lactational endocrine physiology (Fig. 5).

V. Brain Lesions and Maternal Behavior

Given the complexity of maternal behavior, in both number of components and their sequencing, it is not surprising that many parts of the brain are involved in its expression. Although these parts are intricately connected with each other, at this writing there are few specific details concerning

TABLE VI Brain Lesions and Disruptions of Maternal Behavior in Rats[a]

Lesion[b]	Status[c]	Nest building	Parturition[d]	Retrieval	Nursing	References[e]
A. Cortex, Limbic Forebrain System, and Hypothalamus						
Neocortex (large)	V, premating	Little or none	Disrupted	Slow, incomplete	Decreased	A,B,C
Median cortex	M, premating	Little or none	Disrupted	Erratic, incomplete	Decreased	D
Cingulate cortex	M, premating	Poor	Normal	Erratic, incomplete	Decreased	E
Hippocampus	V, premating	Poor	?	Somewhat inferior	Decreased	F
Fornix transection	First G, D 5–8	None	?	Little or none	Absent	G
Septum	V, premating	Poor	Disrupted	Highly erratic	Decreased	H
PVN	P, D 4 PP	Inferior	—	Normal	Normal	I
PVN	First G, D 12–16	?	Normal	5 of 8 –	Absent	J
LH	First G, late G	None	Cannibalism	None	Absent	K,L
B. Medial Preoptic Area, Its Connections, and Midbrain Projections						
MPOA (large)	P, D 5 PP	Little or none	—	Little or none	Almost none	M
MPOA (large)	V, sensitized	Reduced	—	None	Absent	N
MPOA	First G, D 21	Little or none	Disrupted: 3 of 7	Normal	Normal	O
MPOA-LH lateral cuts	P, D 5PP	Little or none	—	Little or none	Almost none	M (+I)
MPOA-LH lateral cuts	P, D 5–10 PP	12 of 23 –	—	12 of 23 –	Present	P
MPOA-LH lateral cuts	First G, D 20	Little or none	Some disruption	Little or none	Delayed	Q
MPOA-LH lateral cuts	B, G, D 20	Little	Normal	Little or none	Almost normal	Q
MPOA-LH lateral cuts	D 16 G, HO + EB	Reduced	—	Reduced	~50% +	R
Dorsal MPOA	P, D 3 PP	Reduced	—	Reduced	Present	S
Dorsal MPOA cuts	D 16 G, HO + EB	Normal	—	Normal	Reduced	R

Dorsal LH cuts	P, D 4 PP	None	—	Little	Very little	T
Rostral MPOA	V, Sensitized	Reduced	—	Reduced	Reduced	U
Rostral MPOA cuts	D 16 G, HO + EB	Reduced	—	Reduced	~50% +	R
MFB cuts (posterior)	First G, D 20	Little or none	Some disruption	Delayed onset	Delayed onset	Q
MFB cuts (posterior)	Second G, D 20	Little	Some disruption	Normal	Normal	Q
U-MPOA + C-MFB cuts	First G, D 20	Delayed	Some disruption	Delayed onset	Delayed onset	Q
U-MPOA + C-MFB cuts	Second G, D 20	Delayed	Normal	Normal	Normal	Q
MFB: NL or FL cuts	V, sensitized	Reduced	—	None	None	V
MFB: NL or FL cuts	P, after birth	Reduced	—	Little or none	None	V
SN	P, D 4 PP	Delayed	—	Brief reduction	Brief reduction	W
VTA	P, D 4 PP	Little or none	—	Little or none	Reduced	X
VTA	M, premating	None	Cannibalism	None	None	Y

[a] Lesions summarized in this table were aspiration, radiofrequency, electrolytic, or via knife cuts.

[b] PVN, paraventricular nucleus; LH, lateral hypothalamus; MPOA, medial preoptic area; MFB, medial forebrain bundle; U, unilateral; C, contralateral; NL, near lateral; FL, far lateral (spares MPOA efferents); SN, substantia nigra; VTA, ventral tegmental area.

[c] V, virgin; P, primiparous; B, biparous; M, multiparous; D, day; G, gestation; PP, postpartum; HO + EB, pregnancy termination by hysterectomy/ovariectomy + one injection of estradiol benzoate (20 μg/kg).

[a2] ?, not possible to assess; "some disruption" refers to placentophagia.

[e] (A) Beach, 1937; (B) Stone, 1938; (C) Davis, 1939; (D) Stamm, 1955; (E) Slotnick, 1967b; (F) Kimble et al., 1967; (G) Steele et al., 1979; (H) Fleischer and Slotnick, 1978; (I) Numan and Corodimas, 1985; (J) E.B. Keverne, unpublished observations, 1987; (K) Avar and Monos, 1967; (L) Avar and Monos, 1969a; (M) Numan, 1974; (N) Numan et al., 1977; (O) Noonan and Kristal, 1979; (P) Terkel et al., 1979b; (Q) Franz et al., 1986; (R) Numan and Callahan, 1980; (S) Jacobson et al., 1980; (T) Numan et al., 1985; (U) Gray and Brooks, 1984; (V) Miceli et al., 1985; (W) Numan and Nagle, 1983; (X) Numan and Smith, 1984; (Y) Gaffori and Le Moal, 1979. See text for further references and descriptions.

which connections are vital to maternal behavior and whether identified connections are ascending or descending, or both. I have retained a traditional organization for reviewing the literature on neural mechanisms of maternal behavior (cf. Slotnick, 1975; Numan, 1985), i.e., a "top-down" analysis, which notably also happens to be roughly chronological (Table VI). Eventually, an analysis in terms of sensory inputs and motor outputs will be more valuable. Note that postpartum lesions have the advantage of not being able to interfere with the hormonal onset of maternal behavior, but the disadvantage of not assessing the effect on parturition and, therefore, the normal onset of maternal behavior.

A. Neocortex

Beach (1937) provided us with the first systematic study of the neural mechanisms of maternal behavior in rats, or for any mammal[1]. He tested the prevailing view that innate behavior is organized at a subcortical level but, in support of Lashley's alternative view (1933), he found that lesions of the neocortex are detrimental to performance of maternal behavior, the degree of impairment being related to the size of the lesion rather than to its specific site. However, lesions made in infancy (Beach, 1938) are not as disruptive to maternal behavior as those of equal size made in adulthood (cf. Murphy et al., 1981, on hamsters), consistent with current knowledge of recovery of function and brain reorganization after early neural insults.

Specifically, Beach (1937) found that anticipatory nest building declined in proportion to lesion size, in terms of quantity, quality, and time of onset. At the first discovery of the litter, 97% of control pups were alive, while only 57% of the pups of dams with the largest lesions (≥40%) were alive. This was due, not to cannibalism, but to a "failure . . . to free their pups of the fetal membranes." Also, in the groups with smaller lesions there was a substantial increase in the number of partially cleaned pups, characterized as "freeing the pup of the membraneous sac, but failing to sever the cord, to devour the placenta, or to lick the pup clean." The most deficient

---

[1] This was the first publication of the late Frank Beach, one of the most influential comparative psychologists of our time. It was based on his doctoral dissertation work at Harvard, one of his mentors being Karl Lashley. In his forward to "Sex and Behavior: Status and Prospectus" (McGill et al., 1978), a "festschrift" for Beach, Robert Hinde reported that Beach "used to sleep in the laboratory so that he could watch mother rats giving birth." As my own work on the sensory regulation of maternal behavior in rats makes clear, Beach's early work on maternal behavior remains a splendid well from which to draw.

females were still able to carry a pup about the cage, but they did not "retrieve the scattered litter to a single corner and hover over the young in a nursing posture." Failures to retrieve and gather the young completely or transport the nest and litter during the subsequent daily disruption tests (involving nest illumination, blowing with a fan, and heating) surely led directly to the further pup mortality from exposure and starvation. These tests were so challenging that only about 60% of the *control* pups survived to postpartum day 5, whereas none of the offspring of the most deficient mothers, those with 40–57% of their cortical tissue removed, survived. With less rigorous testing (Kimble *et al.*, 1967), as well as replacement of weak or dead pups (Stamm, 1955), mothers with large neocortical lesions fared better than did Beach's mothers.

The cortical deficit is not principally sensory in nature because cortically blind rats were inferior in most tests of maternal behavior to rats rendered blind peripherally. Also, there did not seem to be a motor impairment, demonstrated by the presence of pup carrying. Therefore, Beach concluded that the cortical deficit is integrative in nature, in Lashley's (1933) mass-action sense. Partial or total decortication of rats also disturbs or abolishes masculine sexual behavior (Beach, 1940), grooming, food hoarding, and social behavior (Vanderwolf *et al.*, 1978). In Vanderwolf's terminology (1985), spontaneous behavior in rats can be divided into type 1: voluntary, appetitive, instrumental movements such as walking, rearing, manipulating objects with the forelimbs, and postural shifts, and type 2: automatic, reflexive, consummatory, respondent actions such as licking, chewing, grooming, pelvic thrusting, and startle, as well as alert immobility in any posture. The nonspecific afferents ascending from the brainstem to the cortex may be the basis of Lashley's general cortical function by playing a role in the initiation and regulation of type 1 behaviors. Maternal behavior clearly consists of elements of both type 1 and type 2 behaviors. Thus, during the licking of a newly emerged pup, the licking element is present in the decorticated rat, but the postural and forelimb movements necessary to continue licking the pup to completion, including severing and eating the umbilical cord and attached placenta, are not activated. Similarly, the elements of picking up the pup and carrying it are present, but depositing it in the right place and continuing to retrieve the remaining pups in sequence are not. Decorticate rats are not lacking in "arousal," but in fact may exhibit hyperreactivity to novel, intense stimuli (Beach, 1937; Vanderwolf *et al.*, 1978), which further disrupts their behavior. In sum, decortication results in only minor effects on response topography, but drastic effects on response probability (Vanderwolf, 1985).

B. Limbic Forebrain System

1. CINGULATE CORTEX

By the time the next major report on brain mechanisms and maternal behavior appeared in 1955, much work had implicated the limbic system, an interconnected group of cortical and subcortical structures (MacLean, 1952), with the regulation of emotional and species-specific behaviors. Lesions restricted to the allocortex, including cingulate cortex, and involving <20% of total cortex, produced a similar but not identical maternal-behavior-deficiency syndrome as that described by Beach following large neocortical lesions (Stamm, 1955; Slotnick, 1967b). The deficit produced by these lesions did not begin until the pup-gathering phase during parturition, and it was persistent. Dead or dying pups were replaced with foster pups, which managed to induce some maternal responsiveness in the deficient dams over several days; however, testing in a strange, brightly lit cage disrupted their behavior again (Slotnick, 1967b). A hormonal explanation for the deficits is unlikely because a similar disruption occurred in the sensitized maternal behavior of castrated male rats (Slotnick, 1967b). Stamm concluded that the median cortex has a "unique function in the integration of complex unlearned behavior patterns." It is unclear why the retrieval sequence is disrupted after this lesion, but not licking and placentophagia during parturition. Unlike rats, mice with cingulate-cortex lesions show only a moderate disturbance of maternal behavior (Carlson and Thomas, 1968; Slotnick and Nigrosh, 1975), possibly due to a species difference in degree of corticalization of function (Slotnick and Nigrosh, 1975).

2. HIPPOCAMPUS

Damage to other limbic-system structures also disrupts maternal behavior. Kimble et al. (1967) found that premating lesions produced in the dorsal hippocampus (which usually included the fimbria) resulted in postpartum deficits, including inferior nest building, "somewhat inferior" retrieval, and an insufficient amount of time spent nursing (<20% versus 50% for controls), even though time spent licking and handling the pups was normal. There was a high rate of pup mortality, which the authors attributed to maternal cannibalism, probably due to pups being eaten after dying of neglect and starvation. Evaluation of this neglect is difficult because parturition was not observed, subjects were not disturbed until postpartum day 6, and there was no replacement of dead pups. Destruction of only the fimbria, which interconnects the hippocampus and the septum, allows for rearing of normal litters to weaning (Brown-Grant and Raisman, 1972), although there is a tendency to retrieve pups and nesting material to more than one part of the cage (Terlecki and Sainsbury, 1978), suggesting a spatial orientation deficit. Although the hippocampus receives nonspecific sensory afferents, lesions of this structure produce relatively mild deficits in the array of behaviors severely disrupted by extensive neocortical destruction (Vanderwolf, 1985).

A syndrome of complete postpartum maternal neglect in rats was observed postpartum following fornix transections on days 5–8 of pregnancy (Steele *et al.*, 1979). The fornix is the main efferent pathway of the hippocampus, and close to 50% of these fibers terminate in the mammillary bodies of the posterior hypothalamus (MacLean, 1985). In this context, it is of interest that premating lesions of the mammillary bodies produced a severe disturbance in nursing behavior and a deficiency in nest building; pup licking, retrieval, and grouping were intact (Slotnick, 1969, 1975). However, termini of other fornix efferents include the nucleus accumbens, anterior thalamus, medial preoptic area, medial basal hypothalamus, and midbrain central gray, and the fornix receives afferents from the septum (MacLean, 1985). Therefore, although Steele *et al.* (1979) interpreted the disruptive effects of fornix transections on postpartum maternal behavior in terms of an observed 25% reduction in hippocampal norepinephrine (Steele *et al.*, 1979), additional explanations for their intriguing results are likely.

3. SEPTUM

Premating lesions of the *septum* cause deficits in all aspects of maternal behavior, resulting in a high rate of pup mortality (Slotnick, 1969, 1975; Fleischer and Slotnick, 1978). This key structure is interconnected with the hippocampus, preoptic area, hypothalamus, and mammillary body region. Rats with septal lesions "delivered pups in different parts of the cage [and] were seen carrying pups about the cage between births, even while other pups were half-emerged from the vaginal canal" (Fleischer and Slotnick, 1978). The condition of the mammary glands shortly after parturition was normal, suggesting that a deficiency in circulating hormones was not involved. Retrieval was not absent, but rather impaired, being characterized by erratic carrying of pups, often out of the nest ("removals"), although the total time spent in maternal activities during retrieval tests was not significantly different from controls. That the septal lesion does not diminish "maternal motivation" was further demonstrated by showing that virgins with these lesions developed pup-induced maternal behavior, but also with erratic retrieval. Postpartum lesions produced the typical hyperresponsive septal syndrome and a massive disruption of maternal behavior, including infanticide. Premating septal lesions also severely disrupt maternal behavior in mice (Carlson and Thomas, 1968; Slotnick and Nigrosh, 1975) and almost completely abolish it in rabbits (Cruz and Bayer, 1972).

4. AMYGDALA AND CENTRAL OLFACTORY SYSTEM

Given the role of odors in pup-oriented maternal behavior and maternal aggression toward intruders, the integrity of central processing sites for odors may be important to the expression of these behaviors. The *amygdala* is one such site, portions of which receive direct and major afferents from both the main and the accessory olfactory systems, and which projects to the hypothalamic medial preoptic area, important in maternal behavior (see below). However, large lesions of the amygdala, carried out premating in virgins, have little or no effect on postpartum pup-oriented maternal behavior of rats (Slotnick, 1969, 1975) or mice (Slotnick and Nigrosh, 1975). Also, lesions of the stria terminalis, the major efferent pathway of the amygdala, have no apparent effect on the onset (Brown-Grant and Raisman, 1972) or maintenance (Numan, 1974) of maternal behavior in rats. Rather, lesions of the amygdala, especially the corticomedial nuclei, in virgin rats hasten the onset of sensitized maternal behavior (Fleming *et al*, 1980), apparently due to reduced "fearfulness." These results buttress Fleming's earlier hypothesis that pup odors are initially aversive to virgins (Fleming and Rosenblatt, 1974a,b); in the absence of the odors themselves (anosmia) or neural centers that appraise their relevance (e.g., amygdala), neophobia toward pups is reduced. Alternatively, because lesions of the amygdala are critical in coding sensory stimuli of emotional significance (LeDoux, 1986), the reduced fear of pups after amygdala lesions may involve all sensory stimuli from pups, not just their odor.

Ferreira *et al*. (1987) investigated the effects on maternal aggression of lesions to two parts of the olfactory system: the *mediodorsal thalamic nucleus*, which receives afferents from the olfactory bulb, and one of its projection sites, the *prefrontal insular cortex*. These lesions cause impaired odor discriminations and altered odor preferences rather than anosmia. Consonant with these effects, bilateral electrolytic lesions in either of these sites markedly reduce offensive aggression toward a male intruder, but they have no effect on food intake, freezing behavior, or pup-oriented maternal behavior. As the authors pointed out, nonolfactory interpretations for these results ought to be considered as well.

C. Hypothalamus and Medial Preoptic Area (MPOA)

1. PARAVENTRICULAR NUCLEUS (PVN)

This nucleus is of interest because (1) the MPOA, important to maternal behavior (Section V,C,3), projects to the PVN, in part via the MPOA's lateral efferents; and (2) the PVN is the source of intracerebral OT (e.g., Buijs et al., 1985), which is implicated in maternal behavior (Section IV,C,4). In postpartum rats, neither radiofrequency lesions of the PVN nor cuts that severed its lateral connections with the MPOA disrupted maternal behavior, except for producing inferior nest building and litter weight losses due to milk-ejection failure (Numan and Corodimas, 1985). However, in a preliminary study, similar lesions carried out during pregnancy disrupted the postpartum onset of maternal behavior, apparently without affecting parturition behavior or inducing infanticide (E. B. Keverne, unpublished observations, 1987). These results, which require confirmation with neurotoxic lesions, suggest that intracerebral OT, stimulated perhaps by the events of parturition, is critical to the onset of maternal behavior; it is apparently not critical for the suppression of infanticide. In contrast, when Herrenkohl and Rosenberg (1974) deafferented the medial basal portion of the hypothalamus of late-pregnant primigravid rats, they found apparently normal maternal behavior in spite of milk-ejection failure; perhaps PVN collaterals to extrahypothalamic sites were intact after deafferentation.

2. LATERAL HYPOTHALAMUS (LH)

The first indication of a hypothalamic lesion that disrupts maternal behavior in rats directly, apparently independently of the neuroendocrine control of pregnancy, parturition, or lactation, came from the work of Avar and Monos (1967, 1969a,b). When rats were subjected to lesions of the *midlateral parafornical hypothalamus,* which substantially damaged the *medial forebrain bundle* (MFB), on day 16 or 17 of their first pregnancy, their maternal responsiveness was severely depressed; this included (1) the absence of prepartum nest building, (2) cannibalism of pups during and after parturition, and (3) failure to retrieve, group, and nurse the young. During parturition, lesioned dams not only licked young and ate placentas, but also "nibbled" the pups. Although lateral hypothalamic lesions cause a well-known syndrome of hypophagia and hypodipsia, there was little or no evidence of this syndrome at the time of parturition and no histology was reported. Further, 25% of the pair-fed controls also displayed some cannibalism, but otherwise their maternal behavior was normal. General locomotor activity and body temperature did not seem to be affected in the lesioned rats. However, far-lateral hypothalamic lesions produced a more severe general syndrome of persistent aphagia and adipsia with no recovery, as well as reduced locomotion and body temperature; in these cases, fetuses died *in utero* (Avar and Monos, 1969b).

Lateral hypothalamic lesions are well known to produce a variety of motivational deficits (Epstein, 1971), including "sensory neglect" (Marshall *et al.*, 1971). Therefore, a problem in interpreting the rather dramatic findings of Avar and Monos is whether the maternal behavior deficit can be said to be at all specific. The answer to this problem came from the subsequent work on the preoptic area and its lateral connections.

3. MEDIAL PREOPTIC AREA (MPOA)

By the time that Numan began to study the role of the preoptic area (POA) in the regulation of maternal behavior (1974), it was known that the POA–anterior hypothalamic region concentrates both estradiol and testosterone (Pfaff, 1968) and that it regulates autonomic thermoregulation (Lipton, 1968), osmosensory regulation of thirst (Blass and Epstein, 1971), and various aspects of reproduction. Its reproductive functions include regulation of PRL and ovulatory LH secretion (Raisman and Field, 1971), inhibition of feminine sexual behavior (Powers and Valenstein, 1972), activation of masculine sexual behavior (Heimer and Larsson, 1966–1967), sociosexual behavior in a variety of male mammals (Hart, 1974), and maternal behavior (Numan, 1974, 1985). The POA, consisting primarily of medial, lateral, and periventricular nuclei, possesses elaborate reciprocal connections with other hypothalamic regions, limbic system, and midbrain (Conrad and Pfaff, 1976).

Large lesions of the MPOA, or the severing of its (lateral) connections with the LH but not major connections with the amygdala or hippocampus, produce a syndrome of severe maternal neglect, without cannibalism or the more general motivational deficits of the LH syndrome. The retrieval deficit is not due to a general oral motor impairment since successful candy-hoarding occurs in females with a pup-retrieval deficit due to cuts of the lateral MPOA (Numan and Corodimas, 1985). Nursing behavior deficiencies resulting from similar cuts do not seem to be due to altered thermoregulation (Miceli *et al.*, 1983). Lesions which disrupt maternal behavior are consistent with normal (or enhanced) female sexual receptivity (Numan, 1974). By using lesions produced by a neurotoxin that spares axons, N-methyl-DL-aspartic acid (NMA), Numan *et al.* (1988) showed that the disruptive effects of lesions on maternal behavior are due to destruction of cell bodies in the MPOA rather than to fibers of passage in the LH.

The question of whether the MPOA regulates the *onset* of maternal behavior, as well as its *maintenance*, has been addressed (Table VI). Both the *induction* of maternal behavior in virgins (Numan *et al.*, 1977) and the *short-latency onset* of maternal behavior following pregnancy termination and EB treatment were disrupted by MPOA lesions or mediolateral or anterior cuts (Numan and Callahan, 1980), respectively. Nevertheless, 57% of the rats with the effective cuts displayed nursing behavior. Noonan and Kristal (1979) examined the *spontaneous onset of maternal behavior at parturition* in primiparous Long-Evans rats following lesions of the MPOA on pregnancy day 21; the lesions were made through permanently indwelling electrodes so that the lesioning procedure itself was not debilitating. There were severe disruptions in parturition behavior, including long delays to complete placentophagia, pup retrieval, and manipulation of nest material, as well as birthing all over the cage in 3 of 7 of the lesioned dams, and all showed deficits in nest building. But in striking contrast to the results of Numan and co-workers, all dams became maternal and maintained healthy litters. There are a number of procedural differences which might account for the discrepancy in these results, including a different strain and the lesion location, which was more medial in Noonan and Kristal's (1979) than in Numan's (1974) study. However, Franz *et al.* (1986) found that lateral knife cuts of the MPOA in late pregnancy of primiparas resulted in severe deficiencies in the oral components of maternal behavior and delayed, but did not prevent, nursing behavior. In contrast, biparous dams, also deficient in nest building and retrieval, showed normal placentophagia and almost normal nursing behavior. Taken together, these results suggest that in rats (1) MPOA insults are less effective in completely eliminating maternal behavior at term than in other situations; (2) oral components are disrupted more than is nursing; and (3) prior maternal experience attenuates the lesion effects.

Several studies have addressed the issues of which site within the POA and which of its connections are most important for the expression of maternal behavior (Table VI). Both small lesions of the MPOA and restricted lateral knife cuts indicate that the integrity of the dorsal part is more important to maternal behavior than is that of the ventral part. In all of these studies, nursing behavior was much less affected than nest building and retrieval and, in one study, 11 of 23 dams continued to show the full repertoire of maternal behavior. On the other hand, dorsal MPOA cuts only reduced nursing behavior, which Numan and Callahan (1980) attributed to hippocampal connections. Rostral MPOA lesions and cuts were more effective in preventing maternal behavior onset than caudal MPOA lesions and cuts, again with nursing behavior less affected; the anterior lesion effect may be related to septal connections. By using testosterone-treated sensitized females, Gray and Brooks (1984) were able to reveal dissociations in the control of maternal and masculine sexual behaviors within the MPOA–anterior hypothalamic (AH) continuum; whereas AH lesions had little or no effect on maternal behavior (cf. Jacobson *et al.*, 1980), rostral AH lesions were the most effective in reducing male-like mounting behavior. Some MPOA efferents descend through the MFB; however, MFB transections contribute to, but do not completely account for, maternal behavior deficits resulting from cuts to the lateral MPOA (Table VI).

Whereas maternal behavior disruption in rats following MPOA lesions or cuts is a robust phenomenon, no other investigators have replicated as complete a disruption as that found by Numan and his co-workers. It is likely that some of this disparity is due to the size of the lesion or knife cut produced, while some, as we have seen, is due to the subject's reproductive state and previous maternal experience. Also, some of the disparity may be related to details of lesion production, behavioral criteria, and testing procedures; these details should be evaluated in future research. For example, if nursing behavior occurs in the "absence" of retrieval and grouping at a nest site, it behooves the investigator to find out how this happens. Specific crouching tests, in which the pups are placed together at the rat's preferred location (see Section III,D), would be helpful, especially if the investigator previously scattered the pups and destroyed the nest of rats with a retrieval and nest-building deficit! Hovering over pups, even if some are attached to nipples, and nursing while lying down should be distinguished from the active crouching posture. Furthermore, because infanticide has not been reported in any of the studies on MPOA insults in rats, suppression of infanticide is probably mediated elsewhere. If the MPOA provides "the motivation" and integration for all components of maternal behavior, then adequate lesions should reliably eliminate all components. Alternatively, there may not be such a single "center," or if there is a single "center," the MPOA is not it.

Unfortunately, almost nothing is known about the maternal-behavior role of the MPOA in animals other than rats. Miceli and Malsbury (1982) performed lateral cuts that either severed (near-lateral, NL) or spared (far-lateral, FL) connections between the MPOA and the MFB and tested female hamsters for responsiveness to pups as virgins or soon after parturition. The major findings were that both NL and FL cuts interfered with suppression of infanticide (cf. Day and Galef, 1977), rather than maternal activation, and that a situational variable (i.e., food-hoarding opportunity) profoundly reduced infanticide in both intact and lesioned hamsters. In fact, lesioned hamsters allowed to hoard raised larger litters than sham-lesioned hamsters not allowed to hoard. The two types of knife cuts had similar effects except that the NL hamsters had a nest-building deficit suggestive of their being hyperthermic. In postpartum cats, large MPOA–AH lesions, which abolish masculine sexual behavior in males, caused a moderate reduction in maternal licking of kittens and in retrieval or effort to join a separated kitten as compared to sham-lesioned mothers, but did allow for the successful rearing of robust litters (Voith, 1982). Therefore, the MPOA does not appear to be critical to maternal motivation in either hamsters or cats.

D. Midbrain

I. PERIPEDUNCULAR NUCLEUS (PPN)

Lesions on postpartum day 8 of the PPN in the center of the *lateral midbrain tegmentum,* which interrupt the lactation-dependent afferent path of the suckling stimulus (Hansen and Gummesson, 1982; Hansen and Kohler, 1984), also reduced lactational hyperphagia and blocked maternal aggression (Hansen and Ferreira, 1986a), apparently without disrupting maternal caretaking behaviors. Because premating nipple removal did not block maternal aggression (Mayer et al., 1987a), the interrupted pathway presumably carries more diffuse ventral somatosensory inputs than that provided by the teats alone. Unfortunately, specific tests for maternal behavior toward pups were not carried out and it is not clear whether hormone secretion was affected. PPN lesions suppressed both the appetitive (courting) and consummatory (lordosis) components of feminine sexual receptivity in rats (Carrer, 1978). Thus, the PPN appears to be an important site for the receipt, and possibly integration, of behaviorally relevant somatosensory stimulation.

2. SUBSTANTIA NIGRA (SN) AND VENTRAL TEGMENTAL AREA (VTA)

In recent years, Numan and his co-workers have focused on the connections between the MPOA and mesencephalic sites that might be responsible for maternal behavior in rats. The sites studied, the SN and the VTA, were chosen because they receive efferents from the LPOA and LH (Conrad and Pfaff, 1976; Numan et al., 1985) and are the source of dopaminergic projections in the nigrostriatal and mesolimbic-cortical pathways (Section VI,F,3). These sites have many functions, including sensorimotor integration and extrapyramidal motor-system regulation. Bilateral electrolytic lesions of the SN carried out on postpartum day 4 resulted in temporary maternal neglect (but not infanticide) as well as the induction of stereotyped motor activity indicative of dopamine depletion; the severity and duration of these various effects were related to the size of the lesion. A similar syndrome was seen after unilateral mediolateral POA cuts combined with either ipsilateral or contralateral SN lesions, which was more pronounced in the contralateral group (Numan and Nagle, 1983).

Using a similar strategy, Numan and Smith (1984) investigated the VTA; the indirect projections to this region from the dorsal MPOA via the lateral preoptic area (LPOA) have been shown to be more important to maternal behavior than the direct projections from the ventral MPOA (Numan et al., 1985). Postpartum bilateral electrolytic lesions of the VTA resulted in lesion-size-related, severe, long-lasting deficits in maternal behavior, most pronounced for retrieval. Pup growth was also affected, even when nursing occurred, but lactation was not assessed. The combination of a unilateral mediolateral POA cut with a contralateral VTA lesion produced a similar disruption of maternal behavior, but with some recovery. Reductions in dams' body weight were not correlated with degree of maternal-behavior deficit. As with SN and MPOA lesions, the deficits were characterized by neglect, not infanticide. Taken together, Numan and Smith concluded that (1) maternal behavior deficits following VTA lesions are more severe than those following SN lesions; (2) both regions seem to be more involved with the oral components of maternal behavior than with nursing; and (3) the assymetrical-lesion study supports a role for the descending MPOA–LPOA–VTA pathway in mediating the display of maternal behavior.

A complete elimination of maternal behavior was found by Gaffori and Le Moal (1979) following premating bilateral VTA radiofrequency lesions. Multiparous dams with these lesions did not build nests, before or after parturition; 100% killed and ate their pups either during parturition or shortly thereafter and displayed no maternal behavior. These dramatic results suggest that the integrity of this part of the mesencephalon is critical for the onset of maternal behavior and for the suppression of infanticide, even in maternally experienced dams.

The VTA lesions may disrupt maternal behavior by interrupting a *motor activation* pathway. One of the projection sites of the A10 dopaminergic cells in the VTA is the *nucleus accumbens,* which in turn projects to midbrain sites involved in the initiation of motor activity via a descending, multisynaptic route (see Swanson et al., 1987). Premating lesions of nucleus accumbens resulted in severely impaired lactation and maternal behavior, including neglect and a high incidence of cannibalism (Smith and Holland, 1975). The authors attributed the lactation deficit to interference with OT secretion and the behavioral disturbance to hyperemotionality because it was eliminated by posterior hypothalamic lesions. Using both anatomical and electrophysiological techniques, a direct pathway (which is reciprocal) (Simerly and Swanson, 1986), from both lateral POA (Swanson *et al.*, 1984) and dorsal MPOA (Swanson *et al.*, 1987) to both subthalamic (zona incerta) and mesencephalic [VTA and *pedunculopontine nucleus* (PPN_g)] locomotor regions, was convincingly demonstrated. *Nondopaminergic* fibers from throughout the SN and VTA innervate the PPN_g, which is also a main target of basal ganglia outflow (Heimer *et al.*, 1985). While the existence of this pathway goes a long way toward explaining the effects of MPOA, SN, and VTA lesions on maternal behavior in terms of motor activation, the complexity of VTA anatomy deserves further analysis.

The VTA is considered an important component of the *limbic-midbrain area* (Nauta, 1958). There are ascending and descending MFB connections (Nauta, 1958), ascending serotonergic and NE fibers (Dahlström and Fuxe, 1964), as well as ascending dopaminergic mesolimbic-cortical projections. Also, there is the possibility that *nonspecific damage to nearby* regions during lesioning affects *sensory pathways* critical to maternal (and other) behaviors. For example, there is a trigeminal sensory pathway in the mesencephalon which runs dorsal to the SN and across the VTA (R. L. Smith, 1973).

The finding of Gaffori and Le Moal (1979) of infanticide during or shortly after parturition following premating VTA lesions does not suggest a motor-activation impairment. Nakamura and Nakamura (1976) studied the behavioral and neurochemical changes in rats following lesioning of the VTA A10 cells by infusion of 6-hydroxydopamine. The behavioral changes included increased spontaneous motor activity, increased muricide, and aggressiveness if handled, while the most widespread neurochemical change was a significant reduction in 5-HIAA, a serotonin metabolite, in all telencephalic, diencephalic, and mesencephalic structures examined. The periparturitional infanticide, therefore, may have been due to irritability induced by serotonin depletion (see Section VI,E,3).

VTA lesions result in a variety of behavioral disruptions that have an *oral* component: food hoarding (Stinus *et al.*, 1978), angiotensin-induced drinking (Swanson *et al.*, 1978), eating, and gnawing in rats (Roberts, 1980; Nadaud *et al.*, 1984), and biting attack in cats (Goldstein and Siegel, 1980, 1981). An elegant analysis of the aphagia and adipsia syndrome following VTA lesions revealed that the most severe deficits result from damage to the intermediate zone between the SN and VTA, involving both dopaminergic and nondopaminergic neurons, the latter deriving from the trigeminal sensory pathway (Nadaud *et al.*, 1984). Indeed, ingestive behavior disorders following trigeminal deafferentation have long been compared to the LH syndrome in rats (Zeigler and Karten, 1974). Naudaud *et al.* (1984) showed that the dopaminergic and trigeminal systems contribute separately, but synergistically, to normal ingestive behavior. Thus, the *sensory regulation* of oral maternal behaviors may be influenced by the VTA and related regions.

Finally, based on studies of addictive drugs and electrical self-stimulation, activation of the VTA-dopamine system has been identified as a physiological substrate for the "rewarding" aspect of "basic appetitive motivation" (Bozarth, 1987). Presumably, it is the receipt of particular sensory stimuli and the performance of appropriate motor responses, with intervening interneuronal activation, which constitutes the basis for such reward (cf. Glickman and Schiff, 1967). Physical contact with young may be as "rewarding" to caregivers as it is to the young (Stern, 1986). The evidence reviewed above suggests that the VTA may be involved in the mediation of these various aspects of maternal behavior.

VI. Neuroendocrine Integration of Maternal Behavior: Possible Sites and Mechanisms of Hormone Action

In his landmark 1948 book, "Hormones and Behavior," Frank Beach (the "father" of a field he named—behavioral endocrinology) listed the multitude of ways that hormones influence behavior, a heuristic list which probably provoked research that bore fruit many years later. His principles include the following: A detailed and accurate behavioral analysis must precede any attempt to understand how a given hormone affects behavior. Hormones act on various sensory, motor, and neural mechanisms, as well as on metabolism and peripheral structures. A single hormone rarely, if ever, acts alone. And perhaps most importantly, hormones never cause, but merely activate, the appropriate neuromotor mechanisms.

Surprisingly little is known about specific sites and mechanisms of action of hormones in activating the onset or regulating the maintenance of maternal behavior in rats, much less in other mammals. There are limited data on the effects of E on the MPOA, but much additional work is needed. Obviously, the complexity of both the endocrine and the neural regulation of maternal behavior, as well as the recency of the persuasive findings on the roles of PRL and OT, have impeded progress on this important issue to date. It is likely that several hormones affect many central neural sites involved in the expression of maternal behavior. Furthermore, hormones may influence components of maternal behavior, such as nest building and the patterning of nursing bouts, via temperature regulation. And the analysis of each major component of maternal behavior in terms of its stimulus determinants expands conceptualizations of where and how hormones affect maternal responsiveness to peripheral structures. These various directions are fulfilling the expectations of Beach (1948), and later Lehrman (1961), that the whole organism, not just a single "motivational center," must be taken into account.

A. Estradiol

Radioactivity following an injection of $[^3H]E$ is found throughout the CNS in a pattern common to a variety of vertebrates, including members of each class in the phylum (Pfaff, 1976). In rats, sites of greater and longer-lasting concentration include the limbic forebrain (cingulate gyrus, olfactory tubercle, hippocampus, amygdala, septum, nucleus of the stria terminalis), hypothalamus (MPOA, anterior hypothalamus, arcuate nucleus, VMH, ventral premammillary nucleus, PVN), and to a lesser extent, the mesencephalic central gray, olfactory bulbs, and spinal cord (Pfaff, 1968; Pfaff and Keiner, 1973). Many of these sites have already been implicated in the regulation of various aspects of maternal behavior in rats. Of course, other functions of E have been identified for some of these sites, notably sexual behavior and pituitary regulation (Pfaff, 1980).

In many species including rats and humans, E is a potent stimulator of both adenohypophyseal PRL secretion (Amenomori et al.,1970; MacLeod, 1975) and neurohypophyseal OT secretion (Robinson et al., 1981). Given the recent evidence on PRL and OT involvement in the stimulation of rapid-onset maternal behavior, these findings are critical for interpreting the role of E. Estradiol acts at the level of the anterior pituitary by increasing the number of lactotropes and the rate of PRL synthesis (MacLeod, 1975). In addition, E induces an afternoon surge of PRL for at least 6 days which is dependent upon neural feedback. Using E implants, Pan and Gala (1985) found that in rats the MPOA was the most sensitive and the VMH the next most sensitive area for inducing the afternoon PRL surges, whereas implants in other areas, including the suprachiasmatic nuclei, corticomedial amygdala, cerebral cortex, and anterior pituitary, were ineffective.

The interaction of E and OT is complex and its functional significance largely unknown. Estrogen increases uterine sensitivity to OT by increasing the number of OT receptors shortly prior to term; mammary gland OT receptors are regulated differently (Soloff et al., 1979), indicating that E does not act uniformly on OT receptors in different tissues. Colocalization studies have revealed that E and OT are concentrated in the same cells of the PVN, predominantly in the posterior subnucleus (Rhodes *et al.*, 1982), which projects to the medulla and spinal cord rather than to the posterior pituitary (Armstrong *et al.*, 1980). This suggests an autonomic role for the posterior subnucleus. Because fewer than 3% of the hypothalamic magnocellular neurons that project to the posterior pituitary concentrate E, Rhodes *et al.* (1982) suggested that E-induced OT secretion occurs by a mechanism other than binding to magnocellular neurons. Similarly, in ovariectomized rats, E administration increased plasma levels of OT but not vasopressin; this facilitative effect was abolished by anterior hypothalamic deafferentation, indicating a site of action distant from the neurohypophysis (Yamaguchi *et al.*, 1979).

From the extensive work on sexual receptivity, it is well known that E increases the activity of brain cells in response to somatosensory stimulation (Pfaff, 1980; Rose and Bieber, 1984). Comparable work on maternal behavior does not yet exist, but there are three studies implicating the MPOA (see Section V,C) as a site of action of E in facilitating maternal responsiveness.

Numan *et al.* (1977) used the 16-day-pregnant, hysterectomized, ovariectomized rat model and found that MPOA implants of E significantly reduced the latency to onset of maternal behavior (0-day median), with 80% becoming maternal on the first day of pup presentation, while E in the VMH, mammillary bodies, or subcutaneously (same amount as in MPOA implant) was not significantly more effective than cholesterol in the MPOA (3-day median). Although other findings with PRL-release inhibitors support the belief that E acts directly on the MPOA to stimulate rapid-onset maternal behavior rather than indirectly via PRL secretion, the effects of the array of endogenous hormones up to day 16 and the presence of the pituitary make the interpretation difficult. Fahrbach and Pfaff (1986) compared the effects of implants of (diluted) E versus cholesterol in the MPOA of ovariectomized, nulliparous Zivic Miller Sprague-Dawley rats. Recall from the discussion on OT induction of maternal behavior in virgins that this rat population has a very low threshold for maternal responsiveness, which might be due to hyposmia. Estradiol resulted in only a slight facilitation of retrieval. In future work, rats with a high threshold of maternal responsiveness should be used and other E-concentrating sites that have been implicated in maternal behavior via lesion studies should be assessed.

The third report implicating a maternal-behavior effect of E by its action on the MPOA also suggests that retrieval is the component predominantly affected. Ahdieh *et al.* (1987) implanted the antiestrogen, tomoxifen, into either the MPOA or the VMH of pregnant rats on day 20 of pregnancy. Controls received cholesterol in the MPOA. MPOA implants of tomoxifen markedly reduced the prepartum onset of retrieval, but maternal behavior during and after parturition was not different from controls, with the exception of 2 of 16 dams which did not become maternal during the 4-hour test. In a group of similarly treated dams that were delivered by Cesarean section on day 22, tomoxifen produced a 30% reduction in the percentage that were maternal the next day compared to cholesterol-implanted controls. VMH, but not MPOA, implants of tomoxifen reduced the percentage displaying postpartum estrus. Since E implants in the VMH are almost as effective as those in the MPOA in inducing a PRL surge (Pan and Gala, 1985), these results further support the belief that E has a direct effect on MPOA neurons in mediating maternal responsiveness. However, the limited nature of the antiestrogen effect, especially if dams were allowed to give birth normally, suggests that the physiological or behavioral aspects of the parturition experience, or both, are sufficient to largely counteract the deficiency of E in the MPOA.

Gonadal steroids, especially estradiol, influence all sensory modalities by affecting both peripheral structures and CNS processing (Gandelman, 1983). Both taste (Zucker, 1969) and odor (Pietras and Moulton, 1974) perception are affected by these steroids in rats. Tactile sensitivity is enhanced by E in various species, including women [touch detection and 2-point discrimination thresholds (Herron, 1933; Henkin, 1974)], canaries [brood patch (Hinde and Steel, 1964)], and rats. The receptive fields of the pudendal nerve innervating genital skin were enlarged by E in rats, an effect that continued after transection of the pudendal nerve central to the recording site (Komisaruk *et al.*, 1972; Kow and Pfaff, 1973–1974). These results suggest a peripheral mechanism of action and a function of enhancing sensitivity to the tactile stimuli that provoke lordosis. However, the finding that E increased the size and sensitivity of facial receptive fields in rats led Bereiter and Barker (1975) to suggest that "estrogenic influence on tactile sensitivity is a general phenomenon, and not restricted to body loci contacted by the male during mating." In further work, it was found that the enlargement of facial fields is specific to rapidly adapting mechanoreceptors and is estrogen specific in that it occurred during natural estrus, was not affected by P, either with or without E, was affected only slightly by testosterone (and probably via conversion to E), and was induced in castrated males treated with E (Bereiter and Barker, 1980). Because the E-induced enlargement of the receptive field persisted after trigeminal root section, peripheral mechanisms are responsible that may include alterations in the viscoelastic properties of the facial skin (E caused an acute and marked epidermal hyperplasia), with catecholamine involvement (Bereiter *et al.*, 1980).

The functional significance of the estrogenic effects on facial tactile sensitivity is not known. I propose that it functions around the time of parturition to enhance sensitivity to stimuli which the female then acts on with her mouth, i.e., nest materials, emerging fetuses, newly born pups, and other birth products such as placentas. It is likely that there is a similar E-induced enhancement of skin sensitivity in the teats and surrounding ventral skin, rendering the dam more likely to respond by crouching to the pups' tactile stimulation of her ventrum. Once maternal behavior is established, other mechanisms, perhaps hormonal in part, would then continue to enhance the mother's sensitivity to oral and ventral somatosensory stimuli.

B. Progesterone

The evidence reviewed above indicates that the facilitation of maternal behavior in rats brought about by either E or PRL treatment is largely ineffective in the absence of first a high level of, then a decline in P. In contrast, treatment with P alone has no effect, while maintenance of high levels of P inhibits maternal-behavior onset [in primiparae (Moltz *et al.*, 1969a)]. This is consonant with the well-known antagonism between P and E. The mechanisms of the facilitative and inhibitory effects of P on maternal behavior are entirely unknown.

Like the other steroid hormones, P is taken up from the circulation by various sites in rat brain (Warrembourg, 1980). Progesterone acts on the same brain site, the VMH, as does E, both to stimulate sexual receptivity in female rats and to bring about sexual refractoriness, although other effective sites have been identified as well (Barfield *et al.*, 1983). Hypothalamic implants of P facilitated nest building in mice (Voci and Carlson, 1973). Midbrain neurons of E-primed hamsters are highly responsive to P, which may be due to genomic and/or rapid membrane effects (Rose, 1986). Bridges and Dunckel (1987) speculated that P might alter blood–brain barrier characteristics to enable PRL to reach central structures more readily. Using the hysterectomized, ovariectomized, E-treated day-16-pregnant rat model, Numan (1978) implanted P (unilaterally) into a variety of sites (MPOA, VMH, midbrain tegmentum, raphe nuclei) and did not obtain any inhibition of onset of maternal behavior. The more plausible reasons offered by Numan for this failure were (1) insufficient P (28-gauge cannulae were used), (2) a central site not yet explored, (3) multiple central sites of action, and (4) a peripheral site of action. Also, bilateral implants might be more effective and, given the antagonism by P of the E-stimulated sensitivity to oxytocin in the uterus, it is possible that a similar effect occurs with respect to brain OT.

Lactating rats have a high heat production, which renders them vulnerable to acute hyperthermia during nursing bouts, thereby limiting the duration of these bouts. High levels of P and corticosterone, secreted following suckling-induced increases in PRL and ACTH, respectively, act together to increase maternal heat load, apparently via an increase in maternal thermal set point by P and a metabolic effect on heat production by corticosterone (Adels and Leon, 1986).

C. Prolactin

Despite a half-century-old belief, first promulgated by Riddle, that PRL is *the* maternal-behavior hormone, convincing evidence that it does play a role in rat maternal behavior is recent (see Section IV,C,3), while information on its mechanism of action in this regard, in any species, remains sparse. In mice, hypothalamic implants of PRL stimulate a full pattern of maternal behavior, including nest building, retrieval, licking, and hovering over pups (Voci and Carlson, 1973). In ring doves, systemically administered PRL maintains incubation behavior (Lehrman and Brody, 1964; Janik and Buntin, 1985); in contrast, icv PRL does not, although it does promote hyperphagia and inhibit gonadotropin secretion without any apparent systemic effect (Buntin and Tesch, 1985). In rats, icv PRL stimulates grooming (Drago *et al.*, 1980), midbrain microinfusions of PRL facilitate E-dependent lordoses (Harlan *et al.* 1983), icv PRL inhibits female sexual receptivity (Dudley *et al.*, 1982), and iontophoretically applied PRL modulates the activity of single neurons (Yamada, 1975). PRL enters the ventricular system (Perlow, 1982) from the peripheral circulation by binding to receptors in epithelial cells of the choroid plexus (Walsh *et al.*, 1978, 1984, 1987). This unique polypeptide access to CSF (Walsh *et al.*, 1987) suggests that periventricular structures may be most responsive to circulating PRL.

To date, there are no known roles for brain-elaborated prolactinlike immunoreactivity (PLI), which has been found in the hypothalamus, cerebellum, thalamus, brainstem, hippocampus, cerebral cortex, and caudate (Emanuele *et al.*, 1986, 1986). However, the fact that the E + P treatment that stimulates rapid-onset maternal behavior is ineffective in hypophysectomized rats (Bridges *et al.*, 1985) suggests that brain PLI is not involved in maternal behavior or that it is not stimulated by estrogen.

D. Oxytocin

Beyond the demonstrations that intracerebral OT plays a role in the onset of maternal behavior in rats, there is no information on a specific site or mechanism of action. However, oxytocinergic pathways, emanating from the PVN, project to several structures implicated in the regulation of maternal behavior by lesion studies, including the dorsal hippocampus, amygdala, septum, SN (Buijs *et al*, 1985), and MPOA (Brooks *et al.*, 1986). On the basis of changing levels of immunoreactivity during pregnancy and postpartum, Caldwell *et al.* (1987) provided evidence that OT synthesized in the anterior commissural nucleus may be transported to extrahypothalamic sites, including the ventral septum, around the time of parturition. Following experience with lactation there is a restructuring of PVN synapses (e.g., Hatton and Ellisman, 1982). Most recently, ultrastructural dendritic changes resulting from pup-induced maternal behavior have been found in the supraoptic nucleus (Modney *et al.*, 1987), a brain region subject to rapid modification from olfactory input (Yang and Hatton, 1987). These exciting results provide an insight into how sensitization and long-term effects of maternal experience may be mediated at the neuronal level.

Masculine sexual behavior in rats is modified by OT (Stoneham *et al.*, 1985; Hughes *et al*, 1987), possibly in part by its action on the lumbar spinal cord, which includes the motoneuron innervation of the bulbocavernosus muscles involved in erections and copulatory behavior (Monaghan and Breedlove, 1987). Could similar mechanisms be operating in female rats, subserving sensory, motor, or autonomic functions involved in motherhood? In ewes vaginocervical stimulation results in increased CSF levels of OT (Kendrick *et al.*, 1986), while in rats, pregnancy and lactation are associated with marked increases in OT levels throughout the spinal cord, with the largest changes being in the lumbosacral region (Miaskowski *et al.*, 1987). An obvious role in parturition is suggested by these data, but given that the largest changes were found during lactation, it is possible that spinal cord OT is involved in the neuroendocrine pathways related either to suckling, or more germane to our interests herein, to nursing behavior, or both.

E. Neurochemical Correlates

I. OPIOIDS

Bridges and his co-workers have several lines of evidence suggesting that *reduced* opioid levels of the preoptic area, which are affected by ovarian steroids, are necessary to the expression of maternal behavior in rats. Systemic administration of morphine disrupted the rate of onset and quality of maternal behavior in pregnancy-terminated rats (Bridges and Grimm, 1982; Grimm and Bridges, 1983) and it greatly reduced the incidence of maternal aggression in lactating rats (Kinsley and Bridges, 1986), effects blocked by simultaneous administration of naloxone. Morphine implants into the MPOA, but not the VMH, blocked established maternal behavior (Rubin and Bridges, 1984). Morphine did not depress open-field activity of lactating rats, and dextrorphan, a relatively inactive opiate stereoisomer, did not disrupt maternal behavior (Grimm and Bridges, 1983). Opioid levels and receptors in the POA increase during pregnancy and decrease during lactation (Hammer and Bridges, 1987); these changes were not seen in the hypothalamus (Bridges and Ronsheim, 1987). Finally, pregnancy-induced changes in POA levels of immunoreactive β-endorphin were mimicked by a 2-week E + P regimen known to stimulate rapid onset of maternal behavior (Bridges and Ronsheim, 1987). Presumably, it is the increased E/P ratio around the time of parturition which decreases the POA opioid levels, thereby allowing maternal behavior to be expressed. The authors suggested that morphine may disrupt maternal behavior by induction of hyperthermia, but interference with release of PRL (Panerai et al., 1978) and OT (Haldar and Sawyer, 1978) is also possible.

Antagonism of endogenous opioids in parturient rats reduced placentophagia and pup cleaning, but not other aspects of maternal behavior and the former was reduced in nonpregnant primiparous rats as well (Mayer et al., 1985). The authors interpreted these effects in terms of hyperalgesia-induced stress or altered taste perception. Apparently, optimal levels of opioids are required for completely normal maternal behavior during parturition.

2. γ-AMINOBUTYRIC ACID (GABA)

Three behavioral characteristics of lactating rats—hyperphagia, aggression toward conspecifics, and decreased fearfulness—are also elicited by benzodiazepines, which probably enhance GABA's inhibitory actions. Hansen and his associates have investigated these links. Three functional benzodiazepine antagonists decrease food intake and maternal aggression, while strengthening freezing in lactating rats (Hansen et al., 1985). The CSF level of GABA in lactating rats is 40-fold above the detectability level when mothers are with their pups and undetectable 6 hours after the removal of the pups (Qureshi et al., 1987). Finally, as demonstrated by unilateral infusions of the GABA receptor blocker, bicuculline, the site of action of GABA may be the

VMH alone for food intake and the VMH and amygdala for aggression; freezing behavior (in response to a loud sound) was not affected by GABA at either site (Hansen and Ferreira, 1986b).

3. MONOAMINES

The complications in attempting to discern the role of a given neurotransmitter in the regulation of a given behavior are legion. Problems in assessing the effects of depletion include neurochemical specificity; partial depletion, leading to supersensitivity and recovery of function; interaction between neurotransmitters; and nonspecific damage. All structures implicated in the regulation of maternal behavior by brain lesions contain one or more of the monoamines.

From the *norepinephrine* (NE) neurons of the *locus coeruleus,* in the pontine tegmentum, there are five major efferent bundles which supply (1) other brainstem nuclei (e.g., principal sensory trigeminal nucleus) and the spinal cord, (2) cerebellar cortex, (3) central gray and periventricular hypothalamus, (4) VTA and MFB in hypothalamic region (*ventral bundle*), and (5) rostral MFB and forebrain, including thalamus, amygdala, septum, hippocampus, medial cortex, olfactory tubercle and bulbs, and neocortex (*dorsal bundle*) (Loughlin and Fallon, 1986). This widespread NE system has been likened to a central autonomic nervous system, "well-suited for setting levels of central neural activity underlying different behavioral states" (Shepherd, 1988). It is also involved in certain forms of sensory recognition, including tactile (Hansen *et al.,* 1981), and olfactory (Rosser and Keverne, 1985), as well as neuroendocrine regulation, via an NE projection from the lateral tegmentum to the hypothalamus.

Depletions of NE during pregnancy in rats resulted in deficits in maternal behavior ranging from mild [minor disruption of nest building (Bridges *et al,* 1982)], to moderate [in nest building and crouching, but not retrieval (Rosenberg *et al.,* 1977)], to severe [in nest building, crouching, and retrieval (Steele *et al.,* 1979)], although pregnancy and parturition were not affected. Litter-growth deficits despite nursing behavior (Bridges *et al.,* 1982; Rosenberg *et al.,* 1977) may have been due to impairments either in PVN synthesis of OT or in sensory information from the suckling stimulus reaching the hypothalamus (Bridges *et al.,* 1982). Degree of depletion or differences in recovery of function probably do not account for the disparity in these results. In the study with the mildest deficits, both the dorsal and ventral NE bundles were transected on days 14–17, resulting in 70% depletion of hypothalamic NE (other areas not assayed), whereas in the study with the most severe deficits, the dorsal bundle only was transected on days 5–8, resulting in

25% depletion in hippocampal NE. However, it is likely that the severity of the effects on maternal behavior in the Steele *et al.* (1979) study is due to consequences of fornix transection other than hippocampal depletion of NE (see Section V,B,2). Postpartum day-4 depletion of NE, via icv administration of the neurotoxin, 6-hydroxydopamine, had no discernible effect on maternal behavior or physiology (Rosenberg *et al.*, 1977). Thus, in rats, depletion of forebrain NE may be disruptive to the onset of maternal behavior and lactation, but not to their maintenance. In sheep, prepartum depletion of olfactory-bulb NE prevents the parturient mother "from forming a selective bond with her own lamb, enabling her to adopt alien lambs" (Pissonnier *et al.*, 1985).

The other relevant catecholamine, *dopamine* (DA), derives principally from midbrain nuclei in the *SN and VTA,* which contribute to the nigrostriatal and mesolimbic-cortical pathways, respectively, although there are contributions from each of these areas to the other DA pathway (Fallon and Loughlin, 1985). The former innervates the striatum, especially the caudate-putamen, and is important in the control of motor activation. The latter innervates many sites, including the nucleus accumbens, olfactory structures, septum, bed nucleus of the stria terminalis, and prefrontal and cingulate cortex, and is important in regulating responsiveness to external stimuli (Caggiula *et al.*, 1979; Fink and Smith, 1980; Marshall *et al.*, 1980; Mogenson and Vim, 1981). There are also several short-axon sites, including the tuberoinfundibular DA system, that regulate hypophyseal secretion. The studies reviewed previously on SN or VTA lesions and their projection sites (Sections V,B and V,D) are the only ones published that concern a possible role for DA in the regulation of maternal behavior. However, by using the DA receptor antagonist, haloperidol, previously shown by Hansen *et al.* (1981) to markedly inhibit the motorically active components of female sexual behavior (proceptivity), but not the lordosis reflex (receptivity), L. A. Taylor and I are now finding a dose-dependent disruption of pup retrieval and licking, while nursing behavior (crouching posture) and physiology (litter weight gain after 30 minutes of suckling) are enhanced.

The *raphe nuclei,* mostly in the hindbrain, are the major source of the indoleamine, *serotonin* [5-hydroxytryptamine (5-HT)]. The projections include preoptic area, hypothalamus, thalamus, caudate-putamen, septum, hippocampus, cerebral cortex, olfactory bulb, brainstem (e.g., VTA and SN), and spinal cord (Tork, 1986). Like the NE system, the widespread 5-HT system influences "arousal, sensory perception, emotion, and higher cognitive functions" (Shepherd, 1988).

Systemic neurotoxin treatment during pregnancy (parachlorophenylalanine, by gavage), resulting in 97% depletion of brain 5-HT, caused 100% infanticide at birth (W. T. Moore and Hampton, 1974), but caused a decrease in maternal and fetal weight gain, an increase in neonatal morbidity, as well as a possible increase in irritability due to massive depletion of brain 5-HT (Koe and Weissman, 1968). For these reasons, the results are difficult to interpret. Postpartum neurotoxic (icv parachloroamphetamine) or electrolytic (dorsal raphe) depletions of 5-HT resulted in pup mortality due to interference with suckling-induced PRL release (Kordon et al., 1973–1974), with no effect on maternal behavior (Rowland et al., 1978; Barofsky et al., 1983b). However, lesions of the median raphe, produced before mating or on postpartum day 1, resulted in a 55% depletion of hypothalamic 5-HT and in litter growth deficiencies due to insufficient nursing behavior rather than to inadequate PRL release (Barofsky et al., 1983a). Other behavioral deficiencies included a high incidence of cannibalism during early lactation and failure to retrieve pups, but these problems affected a portion of the subjects only and were not associated with complete maternal neglect. Thus, hypothalamic 5-HT, dissociated from that which regulates PRL, may be involved in certain aspects of maternal behavior.

In sum, there is limited but suggestive evidence for an involvement of NE, DA, and 5-HT in the expression of maternal behavior in rats. A specific mother–newborn sensory processing function of olfactory bulb NE has been identified in sheep, suggesting that future research will benefit from site-specific depletions that are based on well-formulated hypotheses about maternal behavior components.

VII. Conclusions and Comparative Perspective

A. Sensorimotor Regulation

A major contribution of this chapter is a revision of the concept of the multisensory control of rat maternal behavior. I propose that perioral and ventral somatosensory inputs from pups are critical to the onset and maintenance of maternal behavior (Fig. 2). Olfaction contributes to maternal aggression and to adequate parturition behavior in rats, but its loss causes only subtle effects on other aspects of pup caretaking; however, the loss of both olfactory bulbs may have additional deleterious effects, especially in inexperienced rats (Table I). I propose that conditioning to distal cues emanating from pups occurs during parturition and in the early postpartum period. The sight, sound, and smell of pups are all stimuli to which rat mothers are responsive, which facilitate their localization of pups displaced from the nest and which maintain their interest in the pups until physical contact can be made. Although multiple distal denervations interfere with the performance and efficiency of maternal behavior, no one is essential for the release of pup caretaking activities.

Most maternal activities in rats are carried out with the mouth, i.e., nest building, ingestion of birth products, pup licking and retrieval, and biting an intruder. The evidence concerning a crucial role for perioral somatosensation includes the following: (1) Alterations of the tactile or thermal characteristics of pup skin are more likely to hinder retrieval than other kinds of pup alterations. (2) Postpartum snout desensitization (perioral anesthesia; infraorbital denervation) severely decreases the likelihood of retrieval, licking, and aggression toward a strange intruder. (3) Prevention of snout, but not tongue, contact with pups also prevents crouching in the early postpartum period, when pups are unable to provoke a crouch directly by moving to the dam and burrowing under her ventrum. (4) Prepartum infraorbital denervation severely disrupts nest building and parturition behavior, retrieval, and onset of first nursing. More extensive trigeminal denervations are required to extend these observations because there is substantial recovery of maternal functions within 24 hours of postpartum infraorbital denervation, probably due to neural reorganization. Also, the mechanisms underlying "carrying" and "depositing at the nest site," which follow picking up a pup (or nest material) with the incisors, have yet to be elucidated.

Active ventral stimulation from pups during their search for a nipple is required at all stages of lactation to diminish dams' pup licking and provoke the crouching response. This was demonstrated by providing dams with freshly killed but still warm pups or live pups incapable of rooting effectively due to mystacial pad anesthesia or to chilling; such pups are retrieved readily, licked, and hovered over, but are incapable of provoking the dam's crouching posture, characterized by a pronounced dorsal arch (ventriflexion), splayed legs, and quiescence. Other ventral stimuli play a modulatory role; for example, rises in the dam's ventrum temperature limit the duration of her nursing bouts. Also, lesions of the afferent suckling path in the lateral midbrain tegmentum greatly reduce maternal aggression toward an intruder.

How generalizable to other mammals are the above conclusions? For mammals that give birth in groups (e.g., herd animals, such as sheep, and communal nesters, such as bats, seals, and some mice), the ability to identify the odor of one's own young rapidly is crucial; for these mammals, unlike rats, therefore, odor undoubtedly plays a more important role in the details of maternal responsiveness or selectivity.

I expect that perioral stimulation from offspring is important, if not requisite, to the onset of nursing behavior only in species and at times in the maternal behavior cycle in which maternal oral activities are predominant prior to initiation of nursing. Nest building of a maternal nature is displayed by rabbits and by rodents with altricial young. During parturition, the pattern of self-licking, licking the newborn, ingesting the birth products, and in general facilitating the birth and survival of the neonate which is displayed by Norway rats is also typical of other altricial rodents and such diverse mammals as the domestic cat, elk, goat, sea lion, and rhesus monkey; exceptions to this pattern include marsupials, camelids, and aquatic mammals which give birth at sea (Lehrman, 1961). Licking that precedes and helps to bring about suckling also occurs in the sheep, goat, elk, domestic cat, and dog (Rheingold, 1963). A variety of rodents and cats retrieve their young in a ratlike fashion; dogs use a combination of carrying and licking, while others, such as aquatic mammals, use their snout to guide the young (Lehrman, 1961; Rheingold, 1963). Two interesting predictions arise from these considerations. The first prediction is that house mice are dependent upon perioral somatosensation for both maternal caretaking and aggression, as they are dependent upon ventral somatosensation, i.e., suckling, for maternal aggression (Svare, 1977), despite the belief that pup-oriented maternal behavior in this species is under the unisensory control of olfaction. It is likely that intact olfaction is a necessary, but not a sufficient, condition for these activities, a possibility that has not been tested. The second prediction is that rabbits are not dependent on perioral stimulation for nursing behavior (though possibly for nest building). In this species mothers do not retrieve young, and the pups become active and uncovered prior to their mother's once-daily nursing visit and subsequently urinate on their own, with no licking assistance from their mother (Hudson and Distel, 1982).

The active crouching posture displayed by rats may be unique to altricial rodents and other similarly sized altricial mammals. In other mammals, however, often there are specific postural adjustments that make it easier for the young to gain access to the teats. What probably is most generalizable is that there is an attraction to the young, based largely on learned distal cues, so that physical proximity occurs in conjunction with species-specific nursing intervals, and then tolerance and probably enjoyment of the ventral contact from the young culminates in suckling. The physical contact with the young further serves to keep the mother immobile, or at least relatively immobile (e.g., whales) (Slijper, 1979, pp. 381–382), so that milk can be obtained. Support for the present view comes from the work of Vince (1987) on tactile communication between ewe and lamb. Both postural adjustments and immobility of the dam are elicited by touches on her ventrum by the experimenter that simulate those given by the lamb as it pushes its muzzle up against its mother.

B. Neuroendocrine Regulation

The ability to induce the full repertoire of maternal behavior rapidly by hormonal manipulation in a large proportion of maternally naive subjects from a population with a high threshold of maternal responsiveness currently is possible only for Norway rats. There are now available protocols of estrogen administration, along with high levels of progesterone followed by its decline, which result in a large percentage of ovariectomized rats becoming maternal within 1 day of pup presentation, and these are effective in two widely used strains, Sprague-Dawley and Long-Evans. However, E-stimulated prolactin release is part of the necessary hormonal milieu, because the E + P treatment is ineffective in hypophysectomized rats. The recent demonstration that P pretreatment plus PRL, in the absence of E, stimulated a reduced latency of maternal behavior suggests that E may be acting by a humoral route instead of, or in addition to, a direct neural route. The fourth key maternal hormone is oxytocin, which was demonstrated to facilitate onset of maternal behavior in E-primed rats when infused icv. After a period when this evidence seemed to be restricted to a particular rat strain with a very low threshold of maternal responsiveness (Zivic Miller), there is now support from demonstrations that an OT antagonist or PVN lesions can delay the onset of postpartum maternal behavior.

A maternal-behavior function for OT at the time of parturition is probably generalizable across (placental) mammals; it is both parsimonious and sensible because this hormone is probably involved in the birth of all eutherian mammals. Further, the OT stimulation of maternal behavior in rats has now been extended to sheep. PRL is also probably available at the time of parturition in all or most mammals, as well as during pregnancy, and it has been implicated in various aspects of maternal behavior in mice, hamsters, and rabbits as well as rats. Details of E and P levels (and ratios) during pregnancy vary widely across species, but both probably contribute to priming of maternal responsiveness, as they do to the maintenance of pregnancy and to mammary gland development. However, whether they have a direct neural role or only a humoral role via stimulating secretion of and/or enhancing sensitivity to PRL and OT remains to be determined. No one has yet taken all four of these hormones into account in the same experiment.

Estrogen may help to prime maternal responsiveness in various ways and at various sites. With respect to its action on the MPOA, the direct neural versus indirect humoral actions of E must now be clarified. Other CNS sites that both concentrate E and contribute to the regulation of maternal behavior, notably the septum, have not been explored with E implants. Enhancement of facial and ventral skin sensitivity by E is a predicted peripheral mechanism of action. Progesterone contributes to the thermal regulation of nest-bout duration. Intracerebral OT-bearing axons project to many brain sites implicated in the regulation of maternal behavior, e.g., septum and hippocampus. Prolactin, taken up by periventricular brain structures from the peripheral circulation, in conjunction with placental lactogen and/or a neurochemical with prolactinlike immunoreactivity found in hypothalamic and extrahypothalamic sites, may influence maternal responsiveness. Peripherally, PRL contributes to thermal regulation of nest bouts and to the production of the maternal pheromone in late lactation. Demonstrations of actual CNS sites of action of P, OT, and PRL await future research.

I suggest that sensitized maternal behavior be termed "pituitary independent" rather than "nonhormonal" because OT, now implicated in the expression of maternal behavior, continues to be secreted into the peripheral circulation after hypophysectomy and is elaborated intracerebrally. Further, other subtle neurochemical changes may underlie the sensitization phenomenon.

Finally, the data suggest that hormones contribute not only to the onset of maternal behavior, but also to its intensity and extent (Fig. 5). Rats that are maternal following hormone stimulation (endogenous or exogenous) differ from rats that are induced to be maternal by pup stimulation, without exogenous hormones, in the intensity and extent of their maternal-behavior repertoire. The former display better nest building, decreased fearfulness, and increased maternal aggression when compared to the latter. In addition to this priming role of hormones, hormones necessary for lactation that are stimulated by the suckling of pups may influence maternal behavior indirectly, e.g., thermoendocrine influences on nest-bout duration.

C. Neurobiological Regulation of Maternal-Behavior Components

Table VII represents a composite of the various sensory, neural, or neurotransmitter manipulations, carried out either before or after parturition, that have been found to disrupt aspects of rat maternal behavior. Many brain sites have been implicated, but the ways in which they interact have yet to be elucidated.

The delineation of neural circuits underlying the expression of each

TABLE VII   Components of Rat Maternal Behavior and Their Disruption by Biological Insults[a]

Maternal behavior components in sequence	Sensory, neural, or neurochemical insult resulting in disruption[b]
1. Nest building: anticipatory and postpartum	IO-X; lesions of neocortex (large), cingulate cortex, dorsal hippocampus, fornix, mammillary bodies, septum, LH, MPOA, VTA
Parturition Behavior: Steps 2–8	
2. Oral facilitation of births	IO-X; lesions of neocortex (large), MPOA
3. General activity suppressed	BOB, IO-X; lesions of septum, MPOA
4. (a) Lick pups clean; (b) eat birth products	BOB, IO-X; lesions of neocortex (large), septum; opiate antagonism
5. Infanticide suppressed	BOB?; lesions of dorsal hippocampus?, septum (postpartum), LH, VTA; massive serotonin depletion
6. Persistence: repeat steps 2–5 to end	With above manipulations, behavior tends to worsen as parturition proceeds
7. Retrieve and group pups	IO-X; lesions of neocortex (large), cingulate cortex, septum, LH, MPOA, VTA
8. Hover over pups, pups provoke crouching	Normal onset of first nursing blocked or delayed with deficits in steps 2–7
Postpartum Behavior: Steps 9–11	
9. Repeat oral activities: steps 1, 4a, and 7	Perioral anesthesia and IO-X; lesions of neocortex (large), cingulate cortex, septum, LH, MPOA, SN, VTA; DA antagonism
10. Active crouching posture during nursing	Oral activities deficits (step 9); lesions of dorsal hippocampus, fornix, mammillary bodies, median raphe
11. Protect litter from intruders	Anosmia, perioral anesthesia, IO-X; lesions of prefrontal insular cortex, mediodorsal thalamus, peripeduncular nucleus; GABA infusions into amygdala

[a]See relevant parts of text for details of effects, which differ in severity, duration, and quality.
[b]IO-X, bilateral infraorbital denervation; BOB, bilateral olfactory bulbectomy; LH, lateral hypothalamus; MPOA, medial preoptic area; VTA, ventral tegmentum area; SN, substantia nigra; GABA, γ-aminobutyric acid.

component of maternal behavior requires separate analyses. The control of maternal aggression is similar to that of pup retrieval and licking with respect to perioral somatosensory regulation, but apparently different with respect to the importance of additional olfactory and ventral somatosensory determinants. Nursing behavior, i.e., crouching, is undoubtedly regulated by a ventral somatosensory pathway, with probable projections to hippocampus, fornix, and mammillary bodies. Maternal oral activities are regulated by trigeminal somatosensory afferents, with probable connections with the MPOA, LH, and VTA; aberrancies in these activities, especially retrieval, result from (premating) lesions of cerebral cortex (large), cingulate cortex, or septum. Because perioral stimulation from pups is requisite to nursing in early lactation (or with young pups), deficits that may be specific to retrieval and nest building should be tested more carefully when these oral activities are dispensible because the pups initiate nursing. The regulation of infanticide suppression seems to be different from the regulation of maternal responsiveness. Thus, MPOA lesions in rats result in maternal neglect (or maternal oral activities deficit), but not in infanticide. The most dramatic display of infanticide in rats around the time of parturition resulted from premating VTA lesions; given the complexity of this mesencephalic region, a number of interpretations of this effect are plausible, including irritability due to serotonin depletion.

The trigeminal sensory system may affect maternal behavior in many ways and via many paths. Local circuits probably directly affect oral reflexes such as licking. An ascending path, through the VTA and near the LH, may directly influence both ingestion and maternal oral activities. Trigeminal influences on motor activities may derive from such "nontrigeminal" projections as dorsal horn, gracile and cuneate nuclei, and cerebellum, while influences on other sensory modalities can occur via solitary nucleus and reticular formation projections (Jacquin *et al.*, 1982). A possible influence on the suckling (and nursing behavior?) afferent pathway may occur via a projection to the mesencephalic lateral tegmentum (Dubois-Dauphin *et al.*, 1985a).

Severe parturition-behavior deficits have been seen following large neocortical lesions, recent infraorbital denervation, recent MPOA lesions (in 3 of 7 rats) and, to a lesser extent, following olfactory bulbectomy. The syndrome includes (1) decreased facilitation of births by oral behavior, (2) incomplete pup cleaning and ingestion of birth products, (3) increased general activity, and (except for bulbectomy) (4) little or no nest building and retrieval of pups. What could account for the overlap in effects caused by these very different insults? I suggest that it is decreased perception and/or processing of the powerful sensory stimuli provided by the emerging fetuses, neonates, and birth products which produces this syndrome. Following large neocortical lesions in adulthood, there is a paucity of nonspecific afferents ascending from the brainstem, resulting in an inability to switch easily from respondent to instrumental movements. Following infraorbital and olfactory deafferentation, there is insufficient sensory feedback to keep the female focused appropriately and, after the former, the stimuli to elicit maternal oral activities are greatly diminished (and should be obliterated after more extensive trigeminal denervation). The parturitional as well as postpartum maternal oral activities deficits resulting from MPOA lesions may be due, at least in part, to the loss of trigeminal afferent processing.

Comparative data on the neural regulation of maternal behavior are sparse. There is little or no effect on maternal behavior of large neocortical lesions in infancy in rats and hamsters or of amygdalectomy in rats and mice. Cingulate cortex lesions are much more disruptive to rats than they are to mice. And, perhaps surprisingly given the attention paid to them, MPOA lesions are much more disruptive to rats than they are to hamsters or cats. In contrast, septal lesions cause severe disruptions in the maternal behavior of rats, mice, and rabbits.

Philip Teitelbaum once suggested to me that pup retrieval is a "reflex"; it took me a decade to realize that he was essentially correct (but better late than never!). Of course, the variability and complexity of the actions involved in rat maternal behavior, as well as the formerly prevailing concept of multi-sensory control, made it difficult to see the behaviors in apparently simplistic terms. However, the reflex concept used by Pfaff and Modianos (1985) for their studies on feminine sexual receptivity in rats is complex and applicable as well to maternal behavior in rats: "a well-defined set of stimuli reliably determines a well-defined constellation of muscular responses." The task ahead remains to determine completely what those stimuli and muscular responses are for each component of maternal behavior. As this chapter reveals, a good start has been made in characterizing the stimuli and in describing the responses, but much more precision is needed. At this point, nothing is known about the motor responses in terms of specific muscles and motoneurons except what may be gleaned from other behaviors which utilize similar movements. For example, a great deal is known about the control of jaw-muscle reflexes (Zeigler et al., 1985); this information can be applied to the actions involved in picking up pups and nesting materials as well as biting intruders. I suspect that much less is known about the muscles involved in the crouching, or "ventriflexion," response.

Attempts to understand the neuroendocrine, and more broadly, the neurobiological, mechanisms regulating maternal behavior will be impeded by continuing to think about the MPOA or some other area as a motivational "center" of maternal behavior which is separate from the integration of sensorimotor mechanisms. Such thinking leads to an overemphasis on the brain localization of goal-directed actions occurring in an all-or-none fashion at the expense of studies which analyze the detailed elements of behavior and their control mechanisms at various levels (e.g., Marshall *et al.*, 1971; Teitelbaum *et al.*, 1983).

". . . Given the flexibility that is the hallmark of goal-directed (motivated) behavior and the presumed rigidity of the neural mechanisms underlying movement, it is not surprising that most psychologists placed goals at the top of their intellectual hierarchy and movement at its bottom. The study of motivation in psychology became the search for brain structures mediating central motive states (drive), leaving the study of sensorimotor mechanisms to the physiologist. Yet goals are ultimately achieved by the initiation, direction, and termination of behaviors, and behavior, after all, consists entirely of movements and postural adjustments. So somehow, the twain (movements and motives) must meet." (Zeigler *et al.*, 1985, p. 186)

I suggest that the dominant meeting place, in the case of maternal behavior, is in the somatosensory system, and that there are hormones acting at multiple sites from skin to brain which enhance responsiveness to pup stimuli.

Acknowledgments

The recent research from my lab described herein was supported by NIMH Grant MH-40459 (JMS) and carried out with the help of Jane Kolunie, Susan Johnson, Cheryl McDonald, Gwyn Gronlund, and Lisa Taylor. A Faculty Academic Study Program leave from Rutgers University, 1987–1988, facilitated the writing process enormously. I am grateful to Mark Kristal for his thorough and thoughtful reading of the penultimate version of this chapter. I am endebted to Seymour Levine, not only for his keen editorial judgments, but more importantly for his invaluable professional support since 1970.

References

Adels, L. E., and Leon, M. (1986). Thermal control of mother-young contact in Norway rats: Factors mediating the chronic elevation of maternal temperature. *Physiol. Behav.* **36**, 183–196.

Ahdieh, H. B., Mayer, A. D., and Rosenblatt, J. S. (1987). Effects of brain antiestrogen implants on maternal behavior and on postpartum estrus in pregnant rats. *Neuroendocrinology* **46**, 522–531.

Alberts, J. R., and Friedman, M. I. (1972). Olfactory bulb removal but not anosmia increases emotionality and mouse-killing. *Nature (London)* **238**, 454–455.

Alberts, J. R., and Galef, B. G. (1971). Acute anosmia in the rat: A behavioral test of peripherally induced anosmia. *Physiol. Behav.* **6**, 619–621.

Allin, J. T., and Banks, E. M. (1972). Functional aspects of ultrasound production by infant albino rats. *Anim. Behav.* **20**, 175–185.

Amenomori, Y., Chen, C. L., and Meites, J. (1970). Serum prolactin levels in rats during different reproductive states. *Endocrinology (Baltimore)* **86**, 506–510.

Anderson, C. O., Zarrow, M. X., Fuller, G. B., and Denenberg, V. H. (1971). Pituitary involvement in maternal nest-building in the rabbit. *Horm. Behav.* **2**, 183–189.

Antelman, S. M., Szechtman, H., Chin, P., and Fisher, A. E. (1975). Tail pinch-induced eating, gnawing and licking behavior: Dependence on the nigrostriatal dopamine system. *Brain Res.* **99**, 319–337.

Armstrong, W. E., Warach, S., Hatton, G. I., and McNeill, T. H. (1980). Subnuclei in the rat hypothalamic paraventricular nucleus: A cytoarchitectural, horseradish peroxidase and immunocytochemical analysis. *Neuroscience* **5**, 1931–1958.

Avar, Z., and Monos, E. (1967). Effect of lateral hypothalamic lesions on maternal behavior and foetal vitality in the rat. *Acta Med. Acad. Sci. Hung.* **23**, 255–261.

Avar, Z., and Monos, E. (1969a). Biological role of lateral hypothalamic structures participating in the control of maternal behaviour in the rat. *Acta Physiol. Acad. Sci. Hung.* **35**, 285–294.

Avar, Z., and Monos, E. (1969b). Behavioural changes in pregnant rats following far-lateral hypothalamic lesions. *Acta Physiol. Acad. Sci. Hung.* **35**, 295–303.

Babicky, A., Ostadolova, I., Parizek, J., Kolar, J., and Bibr, B. (1970). Use of radioisotope techniques for determining the weaning period in experimental animals. *Physiol. Bohemosl.* **19**, 457–467.

Barfield, R. J., Rubin, B. S., Glaser, J. H., and Davis, P. G. (1983). Sites of action of ovarian hormones in the regulation of oestrous responsiveness in rats. *In* "Hormones and Behaviour in Higher Vertebrates" (J. Balthazart, E. Prove, and R. Gilles, eds.), pp. 2–17. Springer-Verlag, Berlin.

Barofsky, A. L., Taylor, J., Tizabi, Y., Kumar, R., and Jones-Quartey, K. (1983a). Specific neurotoxin lesion of raphe serotonergic neurons disrupt maternal behavior in the lactating rat. *Endocrinology (Baltimore)* **113**, 1884–1893.

Barofsky, A. L., Taylor, J., and Massari, V. J. (1983b). Dorsal raphe-hypothalamic projections provide the stimulatory serotonergic input to suckling-induced prolactin release. *Endocrinology (Baltimore)* **113**, 1894–1903.

Bauer, J. H. (1983). Effects of maternal state on the responsiveness to nest odors of hooded rats. *Physiol. Behav.* **30**, 229–232.

Baum, M. J. (1978). Failure of pituitary transplants to facilitate the onset of maternal behavior in ovariectomized virgin rats. *Physiol. Behav.* **20**, 87–89.

Beach, F. A. (1937). The neural basis of innate behavior. I. Effects of cortical lesions upon the maternal behavior pattern in the rat. *J. Comp. Psychol.* **24**, 393–436.

Beach, F. A. (1938). The neural basis of innate behavior. II. Relative effects of partial decortication in adulthood and infancy upon the maternal behavior of the primparous rat. *J. Genet. Psychol.* **53**, 109–148.

Beach, F. A. (1940). Effects of cortical lesions upon copulatory behavior in male rats. *J. Comp. Psychol.* **29**, 193–244.

Beach, F. A. (1948). "Hormones and Behavior." Harper (Hoeber), New York.

Beach, F. A., and Jaynes, J. (1956a). Studies of maternal retrieving in rats. I. Recognition of young. *J. Mammal.* **37**, 177–180.

Beach, F. A., and Jaynes, J. (1956b). Studies on maternal retrieving in rats. II. Effects of practice and previous parturitions. *Am. Nat.* **90**, 103–109.

Beach, F. A., and Jaynes, J. (1956c). Studies of maternal retrieving in rats. III. Sensory cues involved in the lactating female's response to her young. *Behaviour* **10**, 104–125.

Beach, F. A., and Wilson, J. (1963). Effects of prolactin, progesterone and estrogen on reactions of nonpregnant rats to foster young. *Psychol. Rep.* **13**, 231–239.

Bell, R. W., and Smotherman, W. P. (1980). "Maternal Influences and Early Behavior." SP Med. Sci. Books, New York.

Benuck, I., and Rowe, F. A. (1975). Centrally and peripherally induced anosmia: Influences on maternal behavior in lactating female rats. *Physiol. Behav.* **14**, 439–447.

Berieter, D. A., and Barker, D. J. (1975). Facial receptive fields of trigeminal neurons: Increased size following estrogen treatment in female rats. *Neuroendocrinology* **18**, 115–124.

Bereiter, D. A., and Barker, D. J. (1980). Hormone-induced enlargement of receptive fields in trigeminal mechanoreceptive neurons. I. Time course, hormone, sex and modality specificity. *Brain Res.* **184**, 395–410.

Bereiter, D. A., Stanford, L. R., and Barker, D. J. (1980). Hormone-induced enlargement of receptive fields in trigeminal mechanoreceptive neurons. II. Possible mechanisms. *Brain Res.* **184**, 411–423.

Bermant, G., and Davidson, J. M. (1974). "Biological Bases of Sexual Behavior." Harper and Row, New York.

Berridge, K. C., and Fentress, J. C. (1986). Contextual control of trigeminal sensorimotor function. *J. Neurosci.* **6**, 325–330.

Bintarningsih, I. W. R., Johnson, R. E., and Li, C. H. (1958). Hormonally induced lactation in hypophysectomized rats. *Endocrinology (Baltimore)* **63**, 540–547.

Birch, H. G. (1956). Sources of order in the maternal behavior of animals. *Am. J. Orthopsychiatry* **26**, 279–284.

Blask, D. E., and Reiter, R. J. (1975). Pituitary and plasma LH and prolactin levels in female rats rendered blind and anosmic: Influence of the pineal gland. *Biol. Reprod.* **12**, 329–334.

Blass, E. M., and Epstein, A. N. (1971). A lateral preoptic osmosensitive zone for thirst in the rat. *J. Comp. Physiol. Psychol.* **76**, 378–394.

Bolles, R. C., and Woods, P. J. (1964). The ontogeny of behavior in the albino rat. *Anim. Behav.* **12**, 427–441.

Bolwerk, E. L. M., and Swanson, H. H. (1984). Does oxytocin play a role in the onset of maternal behaviour in the rat? *J. Endocrinol.* **101**, 353–357.

Bozarth, M. A. (1987). Ventral tegmental reward system. *In* "Brain Reward Systems and Abuse" (J. Engel and L. Oreland, eds.), pp. 1–17. Raven Press, New York.

Brewster, J. A., and Leon, M. (1980a). Relocation of the site of mother-young contact: Maternal transport behavior in Norway rats. *J. Comp. Physiol. Psychol.* **94**, 69–79.

Brewster, J. A., and Leon, M. (1980b). Facilitation of maternal transport by Norway rat pups. *J. Comp. Physiol. Psychol.* **94**, 80–88.

Bridges, R. S. (1975). Long-term effects of pregnancy and parturition upon maternal responsiveness in the rat. *Physiol. Behav.* **14**, 245–250.

Bridges, R. S. (1977). Parturition: Its role in the long term retention of maternal behavior in the rat. *Physiol. Behav.* **18**, 487–490.

Bridges, R. S. (1984). A quantitative analysis of the roles of dosage, sequence and duration of estradiol and progesterone exposure in the regulation of maternal behavior in the rat. *Endocrinology (Baltimore)* **114**, 930–940.

Bridges, R. S., and Dunckel, P. T. (1987). Hormonal regulation of maternal behavior in rats: Stimulation following treatment with ectopic pituitary grafts plus progesterone. *Biol. Reprod.* **37**, 518–526.

Bridges, R. S., and Feder, H. H. (1978). Effects of various progestins and deoxycorticosterone on the hormonal inhibition of maternal behavior in the ovariectomized-hysterectomized primigravid rat. *Horm. Behav.* **10**, 30–39.

Bridges, R. S., and Grimm, C. T. (1982). Reversal of morphine disruption of maternal behavior by concurrent treatment with the opiate antagonist naloxone. *Science* **218**, 166–168.

Bridges, R. S., and Ronsheim, P. M. (1987). Immunoreactive beta-endorphin concentrations in brain and plasma during pregnancy in rats: Possible modulation by progesterone and estradiol. *Neuroendocrinology* **45**, 381–388.

Bridges, R. S., and Russell, D. W. (1981). Steroidal interactions in the regulation of maternal behaviour in virgin female rats: Effects of testosterone, dihydrostestosterone, oestradiol, progesterone and aromatase inhibitor 1,4,-androstatriene-3,17-dione. *J. Endocrinol.* **90**, 31–40.

Bridges, R. S., Zarrow, M. X., Gandelman, R., and Denenberg, V. H. (1972). Differences in maternal responsiveness between lactating and sensitized rats. *Dev. Psychobiol.* **5**, 123–127.

Bridges, R. S., Zarrow, M. S., and Denenberg, V. H. (1973). The role of neonatal androgen in the expression of hormonally induced maternal responsiveness in the adult rat. *Horm. Behav.* **4**, 315–322.

Bridges, R. S., Zarrow, M. X., Goldman, B. D., and Denenberg, V. H. (1974a). A developmental study of maternal responsiveness in the rat. *Physiol. Behav.* **12,** 149–151.

Bridges, R. S., Goldman, B. D., and Bryant, L. P. (1974b). Serum prolactin concentrations and the initiation of maternal behavior in the rat. *Horm. Behav.* **5,** 219–226.

Bridges, R. S., Rosenblatt, J. S., and Feder, H. H. (1978). Stimulation of maternal responsiveness after pregnancy termination in rats: Effects of time of onset of behavioral testing. *Horm. Behav.* **10,** 235–245.

Bridges, R. S., Clifton, D. K., and Sawyer, C. H. (1982). Postpartum luteinizing hormone release and maternal behavior in the rat after late gestational depletion of hypothalamine norepinephrine. *Neuroendocrinology* **34,** 286–291.

Bridges, R. S., DiBase, R., Loundes, D. D., and Doherty, D. C. (1985). Prolactin stimulation of maternal behavior in female rats. *Science* **227,** 782–784.

Bridges, R. S., Dunckel, P. T., and Millard, W. J. (1987). Growth hormone involvement in the stimulation of maternal behavior in rats. *Abstr 726, The Endocrine Society, 69th Annual Meeting,* Indianapolis, IN, June.

Bronstein, P. M., and Hirsch, S. M. (1976). Ontogeny of defensive reactions in Norway rats. *J. Comp. Physiol. Psychol.* **90,** 620–629.

Brooks, P. J., Lund, P. K., Caldwell, J. D., and Pedersen, C. A. (1986). Oxytocinergic cells in the medial preoptic area and perifornical region: An *in situ* hybridization and immunohistochemical study. *Soc. Neurosci. Abstr.* **12,** 1579.

Brown, R. E. (1979). Mammalian social odors: A critical review. *Adv. Study Behav.* **10.**

Brown, R. E. (1986a). Social and hormonal factors influencing infanticide and its suppression in adult male Long-Evans rats (*Rattus norvegicus*). *J. Comp. Psychol.* **100,** 155–161.

Brown, R. E. (1986b). Paternal behavior in the male Long-Evans rat (*Rattus norvegicus*). *J. Comp. Psychol.* **100,** 162–172.

Brown-Grant, K., and Raisman, G. (1972). Reproductive function in the rat following selective destruction of afferent fibres to the hypothalamus from the limbic system. *Brain Res.* **46,** 23–42.

Brunelli, S. A., Shindledecker, R. D., and Hofer, M. A. (1985). Development of maternal behaviors in prepubertal rats at three ages: Age-characteristic patterns of responses. *Dev. Psychobiol.* **18,** 309–326.

Buijs, R. M., de Vries, G. J., and van Leeuwen, F. W. (1985). The distribution and synaptic release of oxytocin in the central nervous system. *In* "Oxytocin: Clinical and Laboratory Studies" (J. A. Amico and A. G. Robinson, eds.), pp. 77–86. Am. Elsevier, New York.

Buntin, J. D., and Tesch, D. (1985). Effects of intracranial prolactin administration on maintenance of incubation readiness, ingestive behavior, and gonadal condition in ring doves. *Horm. Behav.* **19,** 188–203.

Butcher, R. L., Fugo, N. W., and Collins, W. E. (1972). Semicircadian rhythm in plasma levels of prolactin during early gestation in the rat. *Endocrinology (Baltimore)* **90,** 1125–1127.

Caggiula, A. R., Herndon, Jr., J. J., Scanlon, R., Greenstone, D., Bradshaw, W., and Sharp, D. (1979). Dissociation of active from immobility components of sexual behavior in female rats by central 6-hydroxydopamine: Implications for CA involvement in sexual behavior and sensorimotor responsiveness. *Brain Res.* **172,** 505–520.

Cain, D. P. (1974). Olfactory bulbectomy: Neural structures involved in irritability and aggression in the male rat. *J. Comp. Physiol. Psychol.* **86,** 213–220.

Caldwell, J. D., Greer, E. R., Johnson, M. F., Prange, A. J., and Pedersen, C. A. (1987). Oxytocin and vasopressin immunoreactivity in hypothalamic and extrahypothalamic sites in late pregnant and postpartum rats. *Neuroendocrinology* **46,** 39–47.

Calhoun, J. B. (1962). "The Ecology and Sociology of the Norway Rat," Public Health Serv. Publ. No. 1008. U.S. Public Health Serv., Washington, D.C.

Carlson, N. R., and Thomas, G. J. (1968). Maternal behavior of mice with limbic lesions. *J. Comp. Physiol. Psychol.* **66,** 731–737.

Carrer, H. R. (1978). Mesencephalic participation in the control of sexual behavior in the female rat. *J. Comp. Physiol. Psychol.* **92,** 877–887.

Casey, M. L., MacDonald, P. C., and Simpson, E. R. (1985). Endocrinological changes of pregnancy. In "Williams' Textbook of Endocrinology" (J. D. Wilson and D. W. Foster, eds), pp. 422–437. Saunders, Philadelphia, Pennsylvania.

Charton, D., Adrien, J., and Cosnier, J. (1971). Declencheurs chimiques due comportement de lechage des petits par la Ratte parturiente. *Rev. Comp. Anim.* **5,** 89–94.

Christopherson, E. R., and Wagman, W. (1965). Maternal behavior in the albino rat as a function of self-licking deprivation. *J. Comp. Physiol. Psychol.* **60,** 142–144.

Cohen, J., and Bridges, R. S. (1981). Retention of maternal behavior in nulliparous and primiparous rats: Effects of duration of previous maternal experience. *J. Comp. Physiol. Psychol.* **95,** 450–459.

Conrad, L. C. A., and Pfaff, D. W. (1976). Efferents from medial basal forebrain and hypothalamus in the rat. I. An autoradiographic study of the medial preoptic area. *J. Comp. Neural.* **169,** 185–220.

Cosnier, J., and Couturier, C. (1966). Compartement maternal provoqué chez les rattes adultes castrées. *C. R. Seances Soc. Biol. Ses Fil.* **160,** 798–791.

Croskerry, P. G., Smith, G. K., and Leon, M. (1978). Thermoregulation and the maternal behaviour of the rat. *Nature (London)* **273,** 299–300.

Cross, B. A. (1952). Nursing behaviour and the milk ejection reflex in rabbits. *J. Endocrinol.* **8,** xiii–xiv.

Cruz, M. L., and Beyer, C. (1972). Effect of septal lesions on maternal behavior and lactation in the rabbit. *Physiol. Behav.* **9,** 361–365.

Csapo, A. I. (1969). The four direct regulatory factors of myometrial function. *Ciba Found. Study Group* **34**, 13–42.

Dahlström, A., and Fuxe, K. (1964). Evidence for the existence of monoamines in the cell bodies of brain stem neurons. I. Demonstration of monoamines in the cell bodies of brainstem neurons. *Acta Physiol. Scand.* **62**, Suppl. 323, 1–55.

D'Amato, F. R. (1987). Infanticide by genetically deaf mice: Possible evidence for an inhibiting function of pups' ultrasonic calls. *Aggressive Behav.* **13**, 25–28.

Davis, C. D. (1939). The effect of ablations of the neocortex on mating, maternal behavior and the production of pseudopregnancy in the female rat and on copulatory activity in the male. *Am. J. Physiol.* **127**, 374–380.

Davis, D. E., and Hall, O. (1951). The seasonal reproductive condition of female Norway (Brown) rats in Baltimore, Maryland. *Physiol. Zool.* **24**, 9–20.

Day, C. S. D., and Galef, B. G. (1977). Pup cannibalism: One aspect of maternal behavior in golden hamsters. *J. Comp. Physiol. Psychol.* **91**, 1179–1189.

Deis, R. P. (1968). The effect of an exteroceptive stimulus on milk ejection in lactating rats. *J. Physiol. (London)* **197**, 37–46.

Denenberg, V. H., Hudgens, G. A., and Zarrow, M. X. (1966). Mice reared with rats. *Psychol. Rep.* **18**, 455–456.

Doerr, H. K., Siegel, H. I., and Rosenblatt, J. S. (1981). Effects of progesterone withdrawal and estrogen on maternal behavior in nulliparous rats. *Behav. Neural Biol.* **32**, 35–44.

Dogterom, J., Van Wimersma Greidanus, Tj. B., and Swabb, D. F. (1977). Evidence for the release of vasopressin and oxytocin into the cerebrospinal fluid: Measurements in plasma and CSF of intact and hypophysectomized rats. *Neuroendocrinology* **24**, 108–118.

Drago, F., Canonico, P. L., Beitetti, R., and Scapagnini, U. (1980). Systemic and intraventricular prolactin induces excessive grooming. *Eur. J. Pharmacol.* **65**, 457–458.

Drewett, R. F., and Trew, A. M. (1978). The milk ejection of the rat, as a stimulus and a response to the litter. *Anim. Behav.* **26**, 982–987.

Dubois-Dauphin, M., Armstrong, W. E., Tribollet, E., and Dreifuss, J. J. (1985a). Somatosensory systems and the milk-ejection reflex in the rat. I. Lesions of the mesencephalic lateral tegmentum disrupt the reflex and damage mesencephalic somatosensory connections. *Neuroscience* **15**, 1111–1129.

Dubois-Dauphin, M., Armstrong, W. E., Tribollet, E., and Dreifuss, J. J. (1985b). Somatosensory systems and the milk-ejection reflex in the rat. II. The effects of lesions in the ventroposterior thalamic complex, dorsal columns and lateral cervical nucleus-dorsolateral funiculus. *Neuroscience* **15**, 1131–1140.

Dudley, C. A., Jamison, T. S., and Moss, R. L. (1982). Inhibition of lordosis behavior in the female rat by intraventricular infusion of prolactin and by chronic hyperprolactinemia. *Endocrinology (Baltimore)* **110**, 677–679.

Eisenberg, J. F. (1981). "The Mammalian Radiations: An Analysis of Trends in Evolution, Adaptation, and Behavior." Univ. of Chicago Press, Chicago, Illinois.

Elwood, R. W., ed. (1983). "Parental Behaviour of Rodents." Wiley, New York.

Emanuele, N. V., Metcalfe, L., Wallock, L., Tentler, J., Hagen, T. C., Beer, C. T., Martinson, D., Gout, P. W., Kirsteins, L., and Lawrence, A. M. (1986). Hypothalamic prolactin: Characterization by radioimmunoassay and bioassay and response to hypophysectomy and restraint stress. *Neuroendocrinology* **44**, 217–221.

Emanuele, N. V., Metcalfe, L., Wallock, L., Tentler, J., Hagen, T. C., Beer, C. T., Martinson, D., Gout, P. W., Kirsteins, L., and Lawrence, A. M. (1987). Extrahypothalamic brain prolactin: Characterization and evidence for independence from pituitary prolactin. *Brain Res.* **42**, 255–262.

Epstein, A. N. (1971). The lateral hypothalamic syndrome: Its implications for the physiological psychology of hunger and thirst. *Prog. Physiol. Psychol.* **4**, 263–317.

Erskine, M. S., Denenberg, V. H., and Goldman, B. D. (1978a). Aggression in the lactating rat: Effects of intruder age and test arena. *Behav. Biol.* **23**, 52–66.

Erskine, M. S., Barfield, R. J., and Goldman, B. D. (1978b). Intraspecific fighting during late pregnancy and lactation in rats and effects of litter removal. *Behav. Biol.* **23**, 206–218.

Erskine, M. S., Barfield, R. J. and Goldman, B. D. (1980a). Postpartum aggression in rats. I. Effects of hypophysectomy. *J. Comp Physiol. Psychol.* **94**, 484–494.

Erskine, M. S., Barfield, R. J., and Goldman, B. D. (1980b). Postpartum aggression in rats. II. Dependence on maternal sensitivity to young and effects of experience with pregnancy and parturition. *J. Comp. Physiol. Psychol.* **94**, 495–505.

Ewer, R. F. (1968). "Ethology of Mammals," Chapter 10. Elek Science, London.

Fahrbach, S. E., and Pfaff, D. W. (1986). Effects of preoptic area implants of dulute estradiol on the maternal behavior of ovariectomized, nulliparous rats. *Horm. Behav.* **20**, 354–363.

Fahrbach, S. E., Morrell, J. I., and Pfaff, D. W. (1984). Oxytocin induction of short-latency maternal behavior in nulliparous, estrogen-primed female rats. *Horm. Behav.* **18**, 267–286.

Fahrbach, S. E., Morrell, J. I., and Pfaff, D. W. (1985). Possible role for endogenous oxytocin in estrogen-facilitated maternal behavior in rats. *Neuroendocrinology* **40**, 526–532.

Fahrbach, S. E., Morrell, J. I., and Pfaff, D. W. (1986). Effect of varying the duration of pre-test habituation on oxytocin induction of short-latency maternal behavior. *Physiol. Behav.* **37**, 135–139.

Fallon, J. H., and Loughlin, S. E. (1985). Substantia nigra. *In* "The Rat Nervous System" (G. Paxinos, ed.), Vol. 1, pp. 353–374. Academic Press, Orlando, Florida.

Feder, H. H. (1985). Peripheral plasma levels of gonadal steroids in adult male and adult, nonpregnant female mammals. *In* "Handbook of Behavioral Neurobiology: Reproduction" (N. Adler, D. Pfaff, and R. W. Goy, eds.), pp. 299–370. Plenum, New York.

Ferreira, A., and Hansen, S. (1986). Sensory control of maternal aggression in *Rattus norvegicus*. *J. Comp. Psychol.* **100** (2), 173–177.

Ferreira A., Dahlof, L. G., and Hansen, S. (1987). Olfactory mechanisms in the control of maternal aggression, appetite, and fearfulness: Effect of lesions to olfactory receptors, mediodorsal thalamic nucleus, and insular prefrontal cortex. *Behav. Neurosci.* **101**, 709–717.

Festing, M. F. W. (1981). Inbred strains of rats. *Behav. Genet.* **11**, 431–435.

Findlay, A. L. R. (1966). Sensory discharges from lactating mammary glands. *Nature (London)* **211**, 1183–1184.

Findlay, A. L. R. (1969). Nursing behavior and the condition of the mammary gland in the rabbit. *J. Comp. Physiol. Psychol.* **69**, 115–118.

Findlay, A. L. R., and Roth, L. L. (1970). Long-term dissociation of nursing behavior and the condition of the mammary gland in the rabbit. *J. Comp Physiol. Psychol.* **72**, 341–344.

Findlay, A. L. R., and Tallal, P. A. (1971). Effect of reduced suckling stimulation on the duration of nursing in the rabbit. *J. Comp. Physiol. Psychol.* **76**, 236–241.

Fink, J. S., and Smith, G. P. (1980). Mesolimbic cortical dopamine terminal fields are necessary for normal locomotor and investigatory explorations in rats. *Brain Res.* **199**, 359–384.

Fleischer, S., and Slotnick, B. M. (1978). Disruption of maternal behavior in rats with lesions of the septal area. *Physiol. Behav.* **21**, 189–200.

Fleischer, S., Kordower, J. H., Kaplan, B., Dicker, R., Smerling, R., and Ilgner, J. (1981). Olfactory bulbectomy and gender differences in maternal behaviors of rats. *Physiol. Behav.* **26**, 957–959.

Fleming, A. S. (1986). Psychobiology of rat maternal behavior: How and where hormones act to promote maternal behavior at parturition. *Ann. N.Y. Acad. Sci.* **474**, 234–252.

Fleming, A. S., and Luebke, C. (1981). Timidity prevents the virgin female rat from being a good mother: Emotionality difference between nulliparous and parturient females. *Physiol. Behav.* **27**, 863–868.

Fleming, A. S., and Rosenblatt, J. S. (1974a). Olfactory regulation of maternal behavior in rats. I. Effects of olfactory bulb removal in experienced and inexperienced lactating and cycling females. *J. Comp. Physiol. Psychol.* **86**, 221–232.

Fleming, A. S., and Rosenblatt, J. S. (1974b). Olfactory regulation of maternal behavior in rats. II. Effects of peripherally induced anosmia and lesions of the lateral olfactory tract in pup-induced virgins. *J. Comp. Physiol. Psychol.* **86**, 233–246.

Fleming, A. S., and Rosenblatt, J. S. (1974c). Maternal behavior in the virgin and lactating rat. *J. Comp. Physiol. Psychol.* **86**, 617–622.

Fleming, A. S., Vaccarino, F., Tambosso, L., and Chee, P. (1979). Vomeronasal and olfactory system modulation of maternal behavior in the rat. *Science* **203**, 372–374.

Fleming, A. S., Vaccarino, F., and Luebke, C. (1980). Amygdaloid inhibition of maternal behavior in the nulliparous female rat. *Physiol. Behav.* **25**, 731–743.

Franz, J. J., Leo, R. J., Steuer, M. A., and Kristal, M. B. (1986). Effects of hypothalamic knife cuts and experience on maternal behavior in the rat. *Physiol. Behav.* **38,** 629–640.

Friedman, M. I., Bruno, J. P., and Alberts, J. R. (1981). Physiological and behavioral consequences in rats of water recycling during lactation. *J. Comp. Physiol. Psychol.* **95,** 26–35.

Fuchs, A. R. (1985). Oxytocin in animal parturition. *In* "Oxytocin: Clinical and Laboratory Studies" (J. A. Amico and A. G. Robinson, eds.), pp. 207–235. Am. Elsevier, New York.

Fuchs, A. R., Cubile, L., Dawood, M. Y., and Jorgensen, F. S. (1984). Release of oxytocin and prolactin in rabbits throughout lactation. *Endocrinology (Baltimore)* **114,** 462–469.

Fuchs, A. F., Syromlooi, J., and Rasmussen, A. B. (1987). Oxytocin response to conditioned and nonconditioned stimuli in lactating ewes. *Biol. Reprod.* **37,** 301–305.

Gaffori, O., and Le Moal, M. (1979). Disruption of maternal behavior and appearance of cannibalism after ventral mesencephalic tegmentum lesions. *Physiol. Behav.* **23,** 317–323.

Galef, B. G., and Muskus, P. A. (1979). Olfactory mediation of mother-young contact in Long-Evans rats. *J. Comp. Physiol. Psychol.* **93,** 708–716.

Gandelman, R. (1983). Gondal hormones and sensory function. *Neurosci. Biobehav. Rev.* **7,** 1–17.

Gandelman, R., Zarrow, M. X., Denenberg, V. H., and Myers, M. (1971). Olfactory bulb removal eliminates maternal behavior in the mouse. *Science* **171,** 210–211.

Geisler, G. J., Urca, B., Cannon, J. T., and Liebeskind, J. C. (1979). Response properties of neurons of the lateral cervical nucleus in the rat. *J. Comp. Neurol.* **186,** 65–78.

Gilbert, A. N., Burgoon, D. A., Sullivan, K. A., and Adler, N. T. (1983). Mother-weanling interactions in Norway rats in the presence of a successive litter produced by postpartum mating. *Physiol. Behav.* **30,** 267–271.

Gilbert, A. N., Pelchat, R. J., and Adler, N. T. (1984). Sexual and maternal behaviour at the postpartum oestrus: The role of experience in time-sharing. *Anim. Behav.* **32,** 1045–1053.

Giordano, A. L. (1987). Relationship between nuclear estrogen receptor binding in the preoptic area and hypothalamus and the onset of maternal behavior in female rats. Unpublished Doctoral Dissertation, Rutgers University, Newark, New Jersey.

Glickman, S. E., and Schiff, B. (1967). A biological theory of reinforcement. *Psychol. Rev.* **74,** 81–109.

Goldberger, M. E. (1980). Motor recovery after lesions. *Trends Neurosci.* **3,** 288–291.

Goldstein, J. M., and Siegel, J. (1980). Suppression of attack behavior in cats by stimulation of ventral tegmental area and nucleus accumbens. *Brain Res.* **183,** 181–192.

Goldstein, J. M., and Siegel, J. (1981). Stimulation of ventral tegmental area and nucleus accumbens reduce receptive fields for hypothalamic biting reflex in cats. *Exp. Neurol.* **72**, 239–246.

Gorski, R. A. (1971). Gonadal hormones and the perinatal development of neuroendocrine function. In "Frontiers in Neuroendocrinology." (L. Martini and W. F. Ganong, eds.). Oxford Univ. Press, London and New York.

Gottlieb, G. (1971). Ontogenesis of sensory function in birds and mammals. In "Biopsychology of Development" (E. Tobach, L. R. Aronson, and E. Shaw, eds.), pp. 67–128. Academic Press, New York.

Graber, G. C., and Kristal, M. B. (1977). Uterine distention and the onset of maternal behavior in pseudopregnant but not in cycling rats. *Physiol. Behav.* **19**, 133–137.

Gray, P., and Brooks, P. J. (1984). Effect of lesion location within the medial preoptic-anterior hypothalamic continuum on maternal and male sexual behaviors in female rats. *Behav. Neurosci.* **98**, 703–711.

Gray, P., and Chesley, S. (1984). Development of maternal behavior in nulliparous rats (*Rattus norvegicus*): Effects of sex and early maternal experiences. *J. Comp. Psychol.* **98**, 91–99.

Greene, E. C. (1968). "Anatomy of the Rat." Hafner, New York.

Grégoire, S. E., and Smith, D. E. (1975). Mouse-killing in the rat: Effects of sensory deficits on attack behaviour and stereotyped biting. *Anim. Behav.* **23**, 186–191.

Grimm, C. T., and Bridges, R. S. (1983). Opiate regulation of maternal behavior in the rat. *Pharmacol., Biochem. Behav.* **19**, 609–616

Grosvenor, C. E. (1965). Evidence that exteroceptive stimuli can release prolactin from the pituitary gland of the lactating rat. *Endocrinology (Baltimore)* **76**, 340–342.

Grota, L. J., and Ader, R. (1969). Continuous recording of maternal behaviour in *Rattus norvegicus. Anim. Behav.* **17**, 722–729.

Grota, L. J., and Ader, R. (1974). Behavior of lactating rats in a dual-chambered maternity cage. *Horm. Behav.* **5**, 275–282.

Gubernick, D. J., and Alberts, J. R. (1983). Maternal licking of young: Resource exchange and proximate controls. *Physiol. Behav.* **31**, 593–601.

Gubernick, D. J., and Klopfer, P. H. (1981). "Parental Care in Mammals." Plenum, New York.

Haldar, J., and Sawyer, W. H. (1978). Inhibition of oxytocin release by morphine and its analogs. *Proc. Soc. Exp. Biol. Med.* **157**, 476–480.

Hammer, R. P., and Bridges, R. S. (1987). Preoptic opioids and opiate receptors increase during pregnancy and decrease during lactation. *Brain Res.* **420**, 48–56.

Hansen, S., and Ferreira, A. (1986a). Food intake, aggression and fear behavior in the mother rat: Control by neural systems concerned with milk ejection and maternal behavior. *Behav. Neurosci.* **100**, 64–70.

Hansen, S., and Ferreira, A. (1986b). Effects of Bicucullin infusions in the ventromedial hypothalamus and amygdaloid complex on food intake and affective behavior in mother rats. *Behav. Neurosci.* **100**, 410–415.

Hansen, S., and Gummesson, B. M. (1982). Participation of the lateral midbrain tegmentum in the neuroendocrine control of sexual behavior and lactation in the rat. *Brain Res.* **251**, 319–325.

Hansen, S., and Kohler, C. (1984). The importance of the peripeduncular nucleus in the neuroendocrine control of sexual behavior and milk ejection in the rat. *Neuroendocrinology* **39**, 563–572.

Hansen, S., Stanfield, E. J., and Everitt, B. J. (1981). The effects of lesions of lateral tegmental noradrenergic neurons on components of sexual behaviour and pseudopregnancy. *Neuroscience* **6**, 1105–1117.

Hansen, S., Ferreira, A., and Selart, M. E. (1985). Behavioral similarities between mother rats and benzodiazepine-treated non-maternal animals. *Psychopharmacology* **86**, 344–347.

Hard, E., and Hansen, S. (1985). Reduced fearfulness in the lactating rat. *Physiol. Behav.* **35**, 641–643.

Harlan, R. E., Shivers, B. D., and Pfaff, D. W. (1983). Midbrain microinfusions of prolactin increase the estrogen-dependent behavior, lordosis. *Science* **219**, 1451–1453.

Hart, B. L. (1974). The medial preoptic-anterior hypothalamic area and sociosexual behavior of male dogs: A comparative neuropsychological analysis. *J. Comp. Physiol. Psychol.* **86**, 328–349.

Hart, B. L. (1985). Maternal behavior and mother-young interactions. *In* "The Behavior of Domestic Animals," pp. 132–176. Freeman, New York.

Hatton, J. D., and Ellisman, M. H. (1982). A restructuring of hypothalamic synapses is associated with motherhood. *J. Neurosci.* **2**, 704–708.

Heimer, L., and Larsson, K. (1966–1967). Impairment of mating behavior in male rats following lesions in the preoptic-anterior hypothalamic continuum. *Brain Res.* **3**, 248–263.

Heimer, L., and Larsson, K. (1967). Mating behavior of male rats after olfactory bulb lesions. *Physiol. Behav.* **2**, 207–209.

Heimer, L., Alheid, G. F., and Zaborsky, L. (1985). Basal ganglia. *In* "The Rat Nervous System" (G. Paxinos, ed.), Vol. 1, pp. 37–86. Academic Press, Orlando, Florida.

Henkin, R. I. (1974). Sensory changes during the menstrual cycle. *In* "Biorhythms and Human Reproduction" (M. Ferin, F. Halberg, R. M. Richart, and R. L. Vande Wiele, eds.), pp. 277–285. Wiley, New York.

Herrenkohl, L. R., and Rosenberg, P. A. (1972). Exteroceptive stimulation of maternal behavior in the naive rat. *Physiol. Behav.* **8**, 595–598.

Herrenkohl, L. R., and Rosenberg, P. A. (1974). Effects of hypothalamic deafferentation late in gestation on lactation and nursing behavior in the rat. *Horm. Behav.* **5**, 33–41.

Herrenkohl, L. R., and Sachs, B. D. (1972). Sensory regulation of maternal behavior in mammals. *Physiol. Behav.* **9**, 689–692.

Herron, R. Y. (1933). The effect of high and low female sex hormone concentrations on the two-point threshold of pain and touch and upon tactile sensitivity. *J. Exp. Psychol.* **16,** 324–327.

Hinde, R. A. (1970). "Animal Behaviour: A Synthesis of Ethology and Comparative Psychology," 2nd ed. McGraw-Hill, New York.

Hinde, R. A., and Steel, E. A. (1964). Effect of exogenous hormones on the tactile sensitivity of the canary brood patch. *J. Endocrinol.* **30,** 355–359.

Hofer, M. A. (1981). Parental contributions to the development of their offspring. In "Parental Care in Mammals" (D. J. Gubernick and P. H. Klopfer, eds.), pp. 77–116. Plenum, New York.

Hofer, M. A., Fisher, A., and Shair, H. (1981). Effects of infraorbital nerve section on survival, growth, and suckling behaviors of developing rats. *J. Comp. Physiol. Psychol.* **95,** 123–133.

Holloway, W. R., Dollinger, M. J., and Denenberg, V. H. (1980). Parturition in the rat: Description and assessment. In "Maternal Influences and Early Behavior" (R. W. Bell and W. P. Smotherman, eds.), pp. 1–26. SP Med. Sci. Books, New York.

Hudson, R., and Distel, H. (1982). The pattern of behaviour of rabbit pups in the nest. *Behaviour* **79,** 255–271.

Hughes, A. M., Everitt, B. J., Lightman, S. L., and Todd, K. (1987). Oxytocin in the central nervous system and sexual behaviour in male rats. *Brain Res.* **414,** 133–137.

Hunsaker, D., II., and Shupe, D. (1977). Behavior of the New World marsupials. In "The Biology of Marsupials" (D. Hunsaker, II, ed.), pp. 279–347. Academic Press, New York.

Ichikawa, S., and Fujii, Y. (1982). Effect of prenatal androgen treatment on maternal behavior in the female rat. *Horm. Behav.* **16,** 224–233.

Jacobson, C. D., Terkel, J., Gorski, R. A., and Sawyer, C. H. (1980). Effects of small medial preoptic area lesions on maternal behavior: Retrieving and nest building in the rat. *Brain Res.* **194,** 471–478.

Jacquin, M. F. (1983). Gustation and ingestive behavior in the rat. *Behav. Neurosci.* **97,** 98–109.

Jacquin, M. F., and Zeigler, H. P. (1983). Trigeminal orosensation and ingestive behavior in the rat. *Behav. Neurosci.* **97,** 62–97.

Jacquin, M. F., Semba, R. W., Enfiejian, H., and Egger, M. D. (1982). Trigeminal primary afferents project bilaterally to dorsal horn and ipsilaterally to cerebellum, reticular formation and cuneate, solitary, supratrigeminal and vagal nuclei. *Brain Res.* **246,** 285–291.

Jakubowski, M., and Terkel, J. (1982). Infanticide and caretaking in non-lactating *Mus musculus:* Influence of genotype, family group and sex. *Anim. Behav.* **30,** 1029–1035.

Jakubowski, M., and Terkel, J. (1985). Incidence of pup killing and parental behavior in virgin female and male rats (*Rattus norvegicus*): Differences between Wistar and Sprague-Dawley stocks. *J. Comp. Psychol.* **99,** 93–97.

Jakubowski, M., and Terkel, J. (1986). Establishment and maintenance of maternal responsiveness in postpartum Wistar rats. *Anim. Behav.* **34,** 256–262.

Janik, D. S., and Buntin, J. D. (1985). Behavioral and physiological effects of prolactin in incubating ring doves. *J. Endocrinol.* **105,** 201–209.

Jirik-Babb, P., Manaker, S., Tucker, A. M., and Hofer, M. A. (1984). The role of the accessory and main olfactory systems in maternal behavior of the primiparous rat. *Behav. Neural Biol.* **40,** 170–178.

Johnson, S. K., and Stern, J. M. (1988). The nursing posture requires active tactile input from pups. *Abstr. International Society for Developmental Psychobiology,* Toronto, Canada, Nov. 10–13.

Jus, T. S., and Wakerley, J. B. (1981). Mesencephalic areas controlling pulsatile oxytocin release in suckled rats. *J. Endocrinol.* **91,** 233–244.

Kelly, P. A., Shiu, R. P. C., Robertson, M. C., and Freisen, H. G. (1975). Characterization of rat chorionic mammotropin. *Endocrinology (Baltimore)* **96,** 1187–1195.

Kendrick, K. M., Keverne, E. B., Baldwin, B. A., and Sharman, D. F. (1986). Cerebrospinal fluid levels of acetylcholinesterase, monoamines and oxytocin during labour, parturition, vaginocervical stimulation, lamb separation and suckling in sheep. *Neuroendocrinology* **44,** 149–156.

Kendrick, K. M., Keverne, E. B., and Baldwin, B. A. (1987). Intracerebroventricular oxytocin stimulates maternal behaviour in the sheep. *Neuroendocrinology* **46,** 56–61.

Kenyon, P., Cronin, P., and Keeble, S. (1981). Disruption of maternal retrieving by perioral anesthesia. *Physiol. Behav.* **27,** 313–321.

Kenyon, P., Cronin, P., and Keeble, S. (1983). Role of the infraorbital nerve in retrieving behavior in lactating rats. *Behav. Neurosci.* **97,** 255–269.

Keverne, E. B., Levy, F., Poindron, P., and Lindsay, D. R. (1983). Vaginal stimulation: An important determinant of maternal bonding in sheep. *Science* **219,** 81–83.

Kimble, D. P., Rogers, L., and Hendrickson, C. W. (1967). Hippocampal lesions disrupt maternal, not sexual behavior in the albino rat. *J. Comp. Physiol. Psychol.* **63,** 401–407.

King, H. D. (1939). Life processes in gray Norway rats during 14 years in captivity. *Am. Anat. Mem.* **17,** 1–77.

Kinsley, C. H., and Bridges, R. S. (1986). Opiate involvement in postpartum aggression in rats. *Pharmacol., Biochem. Behav.* **25,** 1007–1011.

Kinsley, C. H., and Bridges, R. S. (1988). Prolactin modulation of the maternal-like behavior displayed by juvenile rats. *Horm. Behav.* **22,** 49–65.

Kirby, H. W., and Horvath, T. (1968). Self-licking deprivation and maternal behavior in the primiparous rat. *Can. J. Psychol.* **22,** 369–375.

Kittrell, E. M., and Satinoff, E. (1988). Diurnal rhythms of body temperature, drinking and activity over reproductive cycles. *Physiol. Behav.* **42,** 477–484.

Klopfer, P. H. (1971). Mother love: What turns it on? *Am. Sci.* **59,** 404–407.

Koe, B. K., and Weissman, A. (1968). The pharmocology of parachlorophenylalanine, a selective depletor of serotonin stores. *Adv. Pharmacol.* **68**, 29–47.

Kolunie, J. M., and Stern, J. M. (1987). Perioral anesthetization reduces maternal aggression in postpartum Norway rats. *Abstr.* Conference on Reproductive Behavior, Tlaxcala, Mexico, June 14–18.

Komisaruk, R. B., and Diakow, C. (1973). Lordosis reflex intensity in rats in relation to the estrous cycle, ovariectomy, estrogen administration and mating behavior. *Endocrinology (Baltimore)* **93**, 548–557.

Komisaruk, R. B., Adler, N. T., and Hutchison, J. (1972). Genital sensory field: Enlargement by estrogen treatment in female rats. *Science* **178**, 1295–1298.

Kordon, C., Blake, C. A., Terkel, J., and Sawyer, C. H. (1973–1974). Participation of serotonin-containing neurons in the suckling-induced rise in plasma prolactin levels in lactating rats. *Neuroendocrinology* **13**, 213–223.

Kow, L. M., and Pfaff, D. W. (1973–1974). Effects of estrogen treatment on the size of receptive field and response threshold of pudendal nerve in the female rat. *Neuroendocrinology* **13**, 299–313.

Krehbiel, D. A., and Le Roy, M. L. (1979). The quality of hormonally stimulated maternal behavior in ovariectomized rats. *Horm. Behav.* **12**, 243–252.

Kristal, M. B. (1980). Placentophagia: A biobehavioral enigma. *Neurosci. Biobehav. Rev.* **4**, 141–150.

Kristal, M. B., Whitney, J. F., and Peters, L. C. (1981). Placenta on pups' skin accelerates onset of maternal behaviour in non-pregnant rats. *Anim. Behav.* **29**, 81–85.

Kristal, M. B., Thompson, A. C., Heller, S. B., and Komisaruk, B. R. (1986a). Placenta ingestion enhances opiate analgesia produced by vaginal/cervical stimulation in rats. *Physiol. Behav.* **36**, 1017–1020.

Kristal, M. B., Thompson, A. C., and Abbott, P. (1986b). Ingestion of amniotic fluid enchances opiate analgesia in rats. *Physiol. Behav.* **38**, 809–815.

Larsson, K. (1975). Sexual impairment of inexperienced male rats following pre- and postpubertal olfactory bulbectomy. *Physiol. Behav.* **14**, 195–199.

Lashley, K. S. (1933). Integrative functions of the cerebral cortex. *Physiol. Rev.* **13**, 1–42.

LeDoux, J. E. (1986). Neurobiology of emotion. *In* "Mind and Brain" (J. E. LeDoux and W. Hirst, eds.), pp. 301–354. Cambridge Univ. Press, London and New York.

Lehrman, D. S. (1961). Hormonal regulation of parental behavior in birds and infrahuman mammals. *In* "Sex and Internal Secretions" (W. C. Young, ed.), 3rd ed., pp. 1268–1382. Williams & Wilkins, Baltimore, Maryland.

Lehrman, D. S., and Brody, P. (1964). Effect of prolactin on established incubation behavior in the ring dove. *J. Comp. Physiol. Psychol.* **57**, 161–165.

Leigh, H., and Hofer, M. (1973). Behavioral and physiological effects of littermate removal on the remaining single pup and mother during the pre-weaning period in rats. *Psychosom. Med.* **35**, 497–508.

Leon, M. (1974). Maternal pheromone. *Physiol. Behav.* **13**, 441–453.

Leon, M., Croskerry, P. G., and Smith, G. K. (1978). Thermal control of mother-young contact in rats. *Physiol. Behav.* **21,** 793–811.

Leon, M., Adels, L., and Coopersmith, R. (1985). Thermal limitation of mother-young contact in Norway rats. *Dev. Psychobiol.* **18,** 85–105.

Le Roy, L. M., and Krehbiel, D. A. (1978). Variations in maternal behavior in the rat as a function of sex and gonadal state. *Horm. Behav.* **11,** 232–247.

Levin, R., and Stern, J. M. (1975). Maternal influences on ontogeny of suckling and feeding rhythms in the rat. *J. Comp. Physiol. Psychol.* **89,** 711–721.

Lincoln, D. W. (1974). Suckling: A time constant in the nursing behavior of the rabbit. *Physiol. Behav.* **13,** 711–714.

Lincoln, D. W., and Wakerley, J. B. (1973). The milk-ejection reflex of the rat: An intermittent function not abolished by surgical levels of anaesthesia. *J. Endocrinol.* **57,** 459–476.

Lincoln, D. W., Hentzen, K., Hin, T., van der Schoot, P., Clarke, G., and Summerlee, A. J. (1980). Sleep: A prerequisite for reflex milk ejection in the rat. *Exp. Brain Res.* **38,** 151–162.

Linkie, D. M., and Niswender, G. D. (1973). Characterization of rat placental luteotrophin: Physiological and physiocochemical properties. *Biol. Reprod.* **8,** 48–57.

Lipton, J. M. (1968). Effects of preoptic lesions on heat-escape responding and colonic temperature in the rat. *Physiol. Behav.* **3,** 165–169.

Lisk, R. D. (1971). Oestrogen and progesterone synergism and elicitation of maternal nestbuilding in the mouse (*Mus musculus*). *Anim. Behav.* **19,** 606–610.

Lisk, R. D., and Reuter, L. A. (1977). *In vivo* progesterone treatment enhances ^{3}H-estradiol retention by neural tissue of the female rat. *Endocrinology* (*Baltimore*) **100,** 1652–1658.

Lisk, R. D., Pretlow, R. A., and Friedman, S. A. (1969). Hormonal stimulation necessary for eliciting of maternal nest building in the mouse. *Anim. Behav.* **17,** 730–738.

Lore, R. (1987). Maternal aggression and infanticide in rats. *Aggressive Behav.* **13,** 287 (abstr.).

Lott, D. (1962). The role of progesterone in the maternal behavior of rodents. *J. Comp. Physiol. Psychol.* **55,** 610–613.

Lott, D., and Fuchs, S. (1962). Failure to induce retrieving by sensitization or the injection of prolactin. *J. Comp. Physiol. Psychol.* **55,** 1111–1113.

Loughlin, S. E., and Fallon, J. H. (1985). Locus coeruleus. *In* "The Rat Nervous System" (G. Paxinos, ed.), Vol. 2, pp. 79–94. Academic Press, New York.

Loundes, D. D., and Bridges, R. S. (1986). Length of prolactin priming differentially affects maternal behavior in female rats. *Biol. Reprod.* **34,** 495–501.

Lyons, W. R. (1958). Hormonal synergism in mammary growth. *Proc. R. Soc. London, Ser. B* **149,** 303–325.

MacDonnell, M. F., and Flynn, J. P. (1966). Sensory control of hypothalamic attack. *Anim. Behav.* **14,** 399–405.

MacIntosh, S. R. (1975). Observations on the structure and innervation of the rat snout. *J. Anat.* 119, 537–546.

MacKinnon, D. A., and Stern, J. M. (1977). Pregnancy duration and fetal number: Effects on maternal behavior of rats. *Physiol. Behav.* 18, 793–797.

MacLean, P. D. (1952). Some psychiatric implications of physiological studies on frontotemporal portion of limbic system (visceral brain). *Electroencephalogr. Clin. Neurophysiol.* 4, 407–418.

MacLean, P. D. (1985). Fiber systems of the forebrain. *In* "The Rat Nervous System" (G. Paxinos, ed.), Vol. 1, pp. 417–440. Academic Press, New York.

MacLeod, R. M. (1975). Regulation of prolactin secretion. *In* "Frontiers in Neuroendocrinology" (L. Martini and W. F. Ganong, eds.), Vol. 4, pp. 169–194. Raven Press, New York.

Marques, D. M. (1979). Roles of the main olfactory and vomeronasal systems in the response of the female hamster to young. *Behav. Neural Biol.* 26, 311–329.

Marshall, J. F., Turner, B. H., and Teitelbaum, P. (1971). Sensory neglect produced by lateral hypothalamic damage. *Science* 174, 523–525.

Marshall, J. F., Berrios, N., and Sawyer, S. (1980). Neostriatal dopamine and sensory inattention. *J. Comp. Physiol. Psychol.* 94, 833–846.

Mayer, A. D., and Rosenblatt, J. S. (1975). Olfactory basis for the delayed onset of maternal behavior in virgin female rats: Experiential effects. *J. Comp. Physiol. Psychol.* 89, 701–710.

Mayer, A. D., and Rosenblatt, J. S. (1977). Effects of intranasal zinc sulfate on open field and maternal behavior in female rats. *Physiol. Behav.* 18, 101–109.

Mayer, A. D., and Rosenblatt, J. S. (1979). Hormonal influences during the ontogeny of maternal behavior in female rats. *J. Comp. Physiol. Psychol.* 93, 879–898.

Mayer, A. D., and Rosenblatt, J. S. (1984). Prepartum changes in maternal responsiveness and nest defense in *Rattus norvegicus*. *J. Comp. Psychol.* 98, 177–188.

Mayer, A. D., and Rosenblatt, J. S. (1987). Hormonal factors influence the onset of maternal aggression in laboratory rats. *Horm. Behav.* 21, 253–267.

Mayer, A. D., Faris, P. L., Komisaruk, B. R., and Rosenblatt, J. S. (1985). Opiate antagonism reduces placentophagia and pup cleaning by parturient rats. *Pharmacol., Biochem. Behav.* 22, 1035–1044.

Mayer, A. D., Carter, L., Jorge, W. A., Mota, M. J., Tannu, S. M., and Rosenblatt, J. S. (1987a). Mammary stimulation and maternal aggression in rodents: Thelectomy fails to reduce pre- or postpartum aggression in rats. *Horm. Behav.* 21, 501–510.

Mayer, A. D., Reisbick, S., Siegel, H. I., and Rosenblatt, J. S. (1987b). Maternal aggression in rats: Changes over pregnancy and lactation in a Sprague-Dawley strain. *Aggressive Behav.* 13, 29–43.

McCullough, J., Quadagno, D. M., and Goldman, B. D. (1974). Neonatal gonadal hormones: Effects on maternal and sexual behavior in the male rat. *Physiol. Behav.* 12, 183–188.

McGill, T. E., Dewsbury, D. A., and Sachs, B. D., eds. (1978). "Sex and Behavior: Status and Prospectus." Plenum, New York.

McLean, B. K., and Nikotovich-Winer, M. B. (1973). Corpus luteum function in the rat: A critical period for luteal activation and the control of luteal maintenance. *Endocrinology (Baltimore)* 93, 316–322.

Meites, J. (1966). Control of mammary growth and lactation. In "Neuroendocrinology" (L. Martini and W. F. Ganong, eds.) Vol. 1, pp. 669–707. Academic Press, New York.

Mena, F., and Grosvenor, C. E. (1971). Release of prolactin in rats by exteroceptive stimulation: Sensory stimuli involved. *Horm. Behav.* 2, 107–116.

Merzenich, M. M., Kaas, J. H., Wall, J. T., Nelson, R. J., Sur, M., and Felleman, D. J. (1983). Topographic reorganization of somatosensory cortical areas 3B and 1 in adult monkeys following restricted deafferentation. *Neuroscience* 8, 33–55.

Miaskowski, C., Ong, G. L., and Haldar, J. (1987). Pregnancy and lactation alter spinal cord levels of oxytocin and vasopressin. *Soc. Neuroci. Abstr.* 13, 664.

Miceli, M. O., and Malsbury, C. W. (1982). Sagittal knife cuts in the near and far lateral preoptic area-hypothalamus disrupt maternal behaviour in female hamsters. *Physiol. Behav.* 28, 857–867.

Miceli, M. O., Fleming, A. S., and Malsbury, C. W. (1983). Disruption of maternal behavior in virgin and postparturient rats following sagittal plane knife cuts in the preoptic area-hypothalamus. *Behav. Brain Res.* 9, 337–360.

Modney, B. K., Yang, Q. Z., and Hatton, G. I. (1987). Pup-induced maternal behavior in virgin rats is associated with increased frequency of dye-coupling among supraoptic nucleus neurons. *Soc. Neurosci. Abstr.* 13, 1592.

Mogenson, G. J., and Yim, C. Y. (1981). Electrophysiological and neuropharmacological-behavior studies of the nucleus accumbens: Implications for its role as a limbic-motor interface. In "The Neurobiology of the Nucleus Accumbens" (R. B. Chronister and J. F. De France, eds.), pp. 210–229. Haer Institute, Brunswick, Maine.

Moll, J., and DeWied, D. (1962). Observations on the hypothalamohypophyseal system of the posterior lobectomized rat. *Gen. Comp. Endocrinol.* 2, 215–228.

Moltz, H., and Wiener, E. (1966). Ovariectomy: Effects on the maternal behavior of the primiparous and multiparous rat. *J. Comp. Physiol. Psychol.* 62, 383–387.

Moltz, H., Robbins, D., and Parks, M. (1966). Caesarian delivery and the maternal behavior of primiparous and multiparous rats. *J. Comp. Physiol. Psychol.* 61, 455–460.

Moltz, H., Geller, D., and Levin, R. (1967). Maternal behavior in the totally mammectomized rat. *J. Comp. Physiol. Psychol.* 64, 225–229.

Moltz, H., Levin, R., and Leon, M. (1969a). Differential effects of progesterone on the maternal behavior of primiparous and multiparous rats. *J. Comp. Physiol. Psychol.* 67, 36–50.

Moltz, H., Levin, R., and Leon, M. (1969b). Prolactin in the postpartum rat: Synthesis and release in the absence of suckling stimulation. *Science* 163, 1083–1084.

Moltz, H., Lubin, M., Leon, M., and Numan, M. (1970). Hormonal induction of maternal behavior in the ovariectomized nulliparous rat. *Physiol. Behav.* **5,** 1373–1377.

Monaghan, E. P., and Breedlove, S. M. (1987). Evidence for oxytocin innervation of perineal motor neurons in rats. *Soc. Neurosci. Abstr.* **13,** 55.

Moore, C. L. (1981). An olfactory basis for maternal discrimination of sex of offspring in rats (*Rattus norvegicus*). *Anim. Behav.* **29,** 383–386.

Moore, C. L. (1982). Maternal behavior of rats is affected by hormonal condition of pups. *J. Comp. Physiol. Psychol.* **96,** 123–129.

Moore, C. L. (1984). Maternal contributions to the development of masculine sexual behavior in rats. *Dev. Psychobiol.* **17,** 347–356.

Moore, C. L. (1985). Sex differences in urinary odors produced by young laboratory rats (*Rattus norvegicus*). *J. Comp. Psychol.* **99,** 76–80.

Moore, C. L., and Chadwick-Dias, A. M. (1986). Behavioral responses of infant rats to maternal licking: Variations with age and sex. *Dev. Psychobiol.* **19,** 427–438.

Moore, C. L., and Rogers, S. (1984). Contribution of self-grooming to onset of puberty in male rats. *Dev. Psychobiol.* **17,** 243–253.

Moore, C. L., and Samonte, B. R. (1986). Preputial glands of infant rats (*Rattus norvegicus*) provide chemosignals for maternal discrimination of sex. *J. Comp. Psychol.* **100,** 76–80.

Moore, W. T., and Hampton, J. K. (1974). Effects of parachlorophenylalanine on pregnancy in the rat. *Biol. Reprod.* **11,** 280–287.

Moore, W. W., Woehler, T. R., and Tarry, K. (1965). Effects of lesions of the amygdala on adrenal, thyroid and ovarian function. *Anat. Rec.* **151,** 390.

Moretto, D., Paclik, L., and Fleming, A. (1986). The effects of early rearing environments on maternal behavior in adult female rats. *Dev. Psychobiol.* **19,** 581–592.

Morishige, W. K., Pepe, G. J., and Rothchild, I. (1973). Serum luteinizing hormone (LH), prolactin and progesterone levels during pregnancy in the rat. *Endocrinology (Baltimore)* **92,** 1527–1530.

Murphy, M. R., and Schneider, G. E. (1970). Olfactory bulb removal eliminates mating behavior in the male golden hamster. *Science* **167,** 302–304.

Murphy, M. R., MacLean, P. D., and Hamilton, S. C. (1981). Species-typical behavior of hamsters deprived from birth of the neocortex. *Science* **213,** 459–461.

Nadaud, D., Simon, H., Herman, J. P., and Le Moal, M. (1984). Contributions of the mesencephalic dopaminergic system and the trigeminal sensory pathway to the ventral tegmental aphagia syndrome in rats. *Physiol. Behav.* **33,** 879–887.

Nakamura, K., and Nakamura, K. (1976). Behavioral and neurochemical changes following administration of 6-hydroxydopamine into the ventral tegmental area of the midbrain. *Jpn. J. Pharmacol.* **26,** 269–273.

Nauta, W. J. H. (1958). Hippocampal projections and related neural pathways to the midbrain in the cat. *Brain Res.* **81,** 319–340.

Noonan, M., and Kristal, M. B. (1979). Effects of medial preoptic lesions on placentophagia and on the onset of maternal behavior in the rat. *Physiol. Behav.* **22,** 1197–1202.

Numan, M. (1974). Medial preoptic area and maternal behavior in the female rat. *J. Comp. Physiol. Psychol.* **87,** 746–759.

Numan, M. (1978). Progesterone inhibition of maternal behavior in the rat. *Horm. Behav.* **11,** 209–231.

Numan, M. (1985). Brain mechanisms and parental behavior. *In* "Handbook of Behavioral Neurobiology: Reproduction" (N. Adler, D. Pfaff, and R. W. Goy, eds.), pp. 537–605. Plenum, New York.

Numan, M., and Callahan, E. C. (1980). The connections of the medial preoptic region and maternal behavior in the rat. *Physiol. Behav.* **25,** 653–665.

Numan, M., and Corodimas, K. P. (1985). The effects of paraventricular hypothalamic lesions on maternal behavior in rats. *Physiol. Behav.* **35,** 417–425.

Numan, M., and Nagle, D. S. (1983). Preoptic area and substantia nigra interact in the control of maternal behavior in the rat. *Behav. Neurosci.* **97,** 120–139.

Numan, M., and Smith, H. G. (1984). Maternal behavior in rats: Evidence for the involvement of preoptic projections to the ventral tegmental area. *Behav. Neurosci.* **98,** 712–727.

Numan, M., Rosenblatt, J. S., and Komisaruk, B. R. (1977). Medial preoptic area and onset of maternal behavior in the rat. *J. Comp. Physiol. Psychol.* **91,** 146–164.

Numan, M., Morrell, J. I., and Pfaff, D. W. (1985). Anatomical identification of neurons in selected brain regions associated with maternal behavior deficits induced by knife cuts of the lateral hypothalamus in rats. *J. Comp. Neurol.* **237,** 552–564.

Numan, M., Corodimas, K. P., Numan, M. J., Factor, E. M., and Piers, W. D. (1988). Axon-sparing lesions of the preoptic region and substantia innominata disrupt maternal behavior in rats. *Behav. Neurosci.* **102,** 381–396.

Obias, M. D. (1957). Maternal behavior of hypophysectomized gravid albino rats and development and performance of their progeny. *J. Comp. Physiol. Psychol.* **50,** 120–124.

Okon, E. E. (1972). Factors affecting ultrasound production in infant rodents. *J. Zool.* **168,** 139–148.

Orpen, G. B., Furman, N., Wong, P. Y., and Fleming, A. S. (1987). Hormonal influences on the duration of postpartum maternal responsiveness in the rat. *Physiol. Behav.* **40,** 307–315.

Oswalt, G. L., and Meier, G. W. (1975). Olfactory, thermal and tactual influences on infantile ultrasonic vocalization in rats. *Dev. Psychobiol.* **8,** 129–135.

Paisley, A. C., and Summerlee, J. S. (1983). Suckling and arousal in the rabbit: Activity of neurones in the cerebral cortex. *Physiol. Behav.* **31,** 471–475.

Pan, J. T., and Gala, R. R. (1985). Central nervous system regions involved in the estrogen-induced afternoon prolactin surge. II. Implantation studies. *Endocrinology (Baltimore)* **117,** 388–395.

Panerai, A. E., Casanueva, F., Martini, A., Mantegazza, P., and DiGuilio, A. M. (1981). Opiates act centrally on GH and PRL release. *Endocrinology (Baltimore)* **108,** 2400–2402.

Pedersen, C. A., and Prange, A. J. (1979). Induction of maternal behavior in virgin rats after intracerebroventricular administration of oxytocin. *Proc. Natl. Acad. Sci. U.S.A.* **756,** 6661–6665.

Pedersen, C. A., Ascher, J. A., Monroe, Y. L., and Prange, A. J. (1982). Oxytocin induces maternal behavior in virgin female rats. *Science* **216,** 648–650.

Pedersen, C. A., Caldwell, J. D., Johnson, M. F., and Prange, A. J., Jr. (1985). Oxytocin anti-serum delays onset of ovarian steroid-induced maternal behaviour. *Neuropeptides* **6,** 175–182.

Pedersen, P. E., Williams, C. L., and Blass, E. M. (1982). Activation and odor conditioning of suckling behavior in 3-day-old albino rats. *J. Exp. Psychol.* **8,** 329–341.

Perlow, M. J. (1982). Cerebrospinal fluid prolactin: A daily rhythm and response to acute perturbation. *Brain Res.* **243,** 382–385.

Peters, L. C. (1980). The suppression of filicide in mother rats. Unpublished Doctoral Dissertation, State University of New York at Buffalo.

Peters, L. C., and Kristal, M. B. (1983). Suppression of infanticide in mother rats. *J. Comp. Psychol.* **97,** 167–177.

Pfaff, D. W. (1968). Autoradiographic localization of radioactivity in rat brain after injection of tritiated sex hormones. *Science* **161,** 1355–1356.

Pfaff, D. W. (1973). Luteinizing hormone releasing factor (LRF) potentiates lordosis behavior in hypophysectomized ovariectomized female rats. *Science* **182,** 1148–1149.

Pfaff, D. W. (1976). The neuroanatomy of sex hormone receptors in the vertebrate brain. *In* "Neuroendocrine Regulation of Fertility" (T. C. Anand Kumar, ed.), pp. 30–45. Karger, Basel.

Pfaff, D. W. (1980). "Estrogens and Brain Function." Springer, New York.

Pfaff, D. W., and Keiner, M. (1973). Atlas of estradiol-concentrating cells in the central nervous system of the female rat. *J. Comp. Neurol.* **151,** 121–158.

Pfaff, D. W., and Modianos, D. (1985). Neural mechanisms of female reproductive behavior. *In* "Handbook of Behavioral Neurobiology: Reproduction" (N. Adler, D. Pfaff, and R. Goy, eds.), pp. 423–493. Plenum, New York.

Pfister, J. F., Cramer, C. P., and Blass, E. M. (1986). Weaning in rats: Extended by continuous living with dams and their 16–21-day-old litters. *Anim. Behav.* **34,** 515–520.

Pietras, R. J., and Moulton, D. G. (1974). Hormonal influences on odor detection in rats: Changes associated with the estrous cycle, pseudopregnancy, ovariectomy, and administration of testosterone propionate. *Physiol. Behav.* **12,** 475–491.

Pissonier, D., Thiery, J. C., Fabre-Nys, C., Poindron, P., and Keverne, E. B. (1985). The importance of olfactory bulb noradrenalin for maternal recognition in sheep. *Physiol. Behav.* **35,** 361–363.

Plume, S., Fogarty, C., Grota, L. J., and Ader, R. (1968). Is retrieving a measure of maternal behavior in the rat. *Psychol. Rep.* **23**, 627–630.

Poindron, P., and Le Neindre, P. (1980). Endocrine and sensory regulation of maternal behavior in the ewe. *Adv. Study Behav.* **11**, 75–119.

Pollak, E. I., and Sachs, B. D. (1975). Male copulatory behavior and female maternal behavior in neonatally bulbectomized rats. *Physiol. Behav.* **14**, 337–343.

Poulain, D. A., and Wakerley, J. B. (1986). Afferent projections from the mammary glands to the spinal cord in the lactating rat. II. Electrophysiological responses of spinal neurons during stimulation of the nipples, including suckling. *Neuroscience* **19**, 511–521.

Powers, J. B., and Valenstein, E. S. (1972). Sexual receptivity: Facilitation by medial preoptic lesions in female rats. *Science* **175**, 1003–1005.

Quadagno, D. M., and Rockwell, J. (1972). The effect of gonadal hormones in infancy on maternal behavior in the adult rat. *Horm. Behav.* **3**, 55–62.

Quadagno, D. M., McCullough, J., Ho, G. K., and Spevak, A. M. (1973). Neonatal gonadal hormones: Effect on maternal and sexual behavior in the female rat. *Physiol. Behav.* **11**, 251–254.

Qureshi, G. A., Hansen, S., and Sodersten, P. (1987). Offspring control of cerebrospinal fluid GABA concentrations in lactating rats. *Neurosci. Lett.* **75**, 85–88.

Raisman, G., and Field, P. M. (1971). Anatomical considerations relevant to the interpretation of neuroendocrine experiments. *In* "Frontiers in Neuroendocrinology" (L. Martini and W. F. Ganong, eds.), pp. 3–44. Oxford Univ. Press, London and New York.

Reisbick, S., Rosenblatt, J. S., and Mayer, A. D. (1975). Decline of maternal behavior in the virgin and lactating rat. *J. Comp. Physiol. Psychol.* **89**, 722–732.

Rheingold, H. L., ed. (1963). "Maternal Behavior in Mammals." Wiley, New York.

Rhodes, C. H., Morrell, J. I., and Pfaff, D. W. (1982). Estrogen-concentrating neurophysin-containing hypothalamic magnocellular neurons in the vasopressin-deficient (Brattleboro) rat: A study combining steroid autoradiography and immunocytochemistry. *J. Neurosci.* **2**, 1718–1724.

Richards, M. P. M. (1969). Effects of oestrogen and progesterone on the nest building in the golden hamster. *Anim. Behav.* **17**, 356–361.

Richter, C. P. (1956). Self-regulatory functions during gestation and lactation. *In* "Gestation: Transactions of the Third Conference" (C. A. Villee, ed.), pp. 11–93. Madison Printing, Madison, New Jersey.

Riddle, O., Lahr, E. L., and Bates, R. W. (1942). The role of hormones in the initiation of maternal behavior in rats. *Am. J. Physiol.* **137**, 299–317.

Risser, J. M., and Slotnick, B. M. (1987). Nipple attachment and survival in neonatal olfactory bulbectomized rats. *Physiol. Behav.* **40**, 545–549.

Roberts, W. W. (1980). (^{14}C) Deoxyglucose mapping of first-order projections activated by stimulation of lateral hypothalamic sites eliciting gnawing, eating and drinking in rats. *J. Comp. Neurol.* **194**, 617–638.

Robinson, A. G., Verbalis, J. G., Amico, J. A., and Seif, M. (1981). Recent advances in neurophypophyseal research. *Int. Rev. Physiol.* **24**, 1–40.

Rodriguez-Sierra, J., and Rosenblatt, J. S. (1977). Does prolactin play a role in estrogen-induced maternal behavior in rats: Apomorphine reduction of prolactin release. *Horm. Behav.* **9**, 1–7.

Rose, J. D. (1986). Functional reconfiguration of midbrain neurons by ovarian steroids in behaving hamsters. *Physiol. Behav.* **37**, 633–647.

Rose, J. D., and Bieber, S. L. (1984). Joint and separate effects of estrogen and progesterone on response of midbrain neurons to lordosis-controlling somatic stimuli in the female golden hamster. *J. Neurophysiol.* **51**, 1040–1054.

Rosenberg, K. M. (1974). Effects of pre- and postpubertal castration and testosterone on pup-killing behavior in the male rat. *Physiol. Behav.* **13**, 159–161.

Rosenberg, K. M., and Sherman, G. F. (1975). The role of testosterone in the organization, maintenance and activation of pup-killing behavior in the male rat. *Horm. Behav.* **6**, 173–179.

Rosenberg, K. M., Denenberg, V. H., Zarrow, M. X., and Frank, B. L. (1971). Effects of neonatal castration and testosterone on the rat's pup-killing behavior and activity. *Physiol. Behav.* **7**, 363–368.

Rosenberg, P. A., and Herrenkohl, L. R. (1976). Maternal behavior in male rats: Critical times for the suppressive action of androgens. *Physiol. Behav.* **16**, 293–297.

Rosenberg, P. A., Halaris, A., and Moltz, H. (1977). Effects of central norepinephrine depletion on the initiation and maintenance of maternal behavior in the rat. *Pharmacol., Biochem. Behav.* **6**, 21–24.

Rosenblatt, J. S. (1967). Nonhormonal basis of maternal behavior in the rat. *Science* **156**, 1512–1514.

Rosenblatt, J. S. (1970). Views on the onset and maintenance of maternal behavior in the rat. *In* "Development and Evolution of Behavior" (E. Tobach, D. S. Lehrman, and J. S. Rosenblatt, eds.), pp. 489–515. Freeman, San Francisco, California.

Rosenblatt, J. S. (1975). Selective retrieving by maternal and nonmaternal female rats. *J. Comp. Physiol. Psychol.* **88**, 678–686.

Rosenblatt, J. S., and Lehrman, D. S. (1963). Maternal behavior of the laboratory rat. *In* "Maternal Behavior in Mammals" (H. Rheingold, ed.) pp. 8–57. Wiley, New York.

Rosenblatt, J. S., and Siegel, H. I. (1975). Hysterectomy-induced maternal behavior during pregnancy in the rat. *J. Comp. Physiol. Psychol.* **89**, 685–700.

Rosenblatt, J. S., and Siegel, H. I. (1981). Factors governing the onset and maintenance of maternal behavior among nonprimate mammals: The role of hormonal and nonhormonal factors. *In* "Parental Care in Mammals" (D. J. Gubernick and P. H. Klopfer, eds.), pp. 14–76. Plenum, New York.

Rosenblatt, J. S., Mayer, A. D., and Siegel, H. I. (1985). Maternal behavior among the nonprimate mammals. *In* "Handbook of Behavioral Neurobiology: Reproduction" (N. Adler, D. Pfaff, and R. W. Goy, eds.), pp. 229–298. Plenum Press, New York.

Rosser, A. E., and Keverne, E. B. (1985). The importance of central noradrenergic neurons in the formation of an olfactory memory in the prevention of pregnancy block. *Neuroscience* **15**, 1141–1147.

Rowland, D., Steele, M., and Moltz, H. (1978). Serotonergic mediation of the suckling-induced release of prolactin in the lactating rat. *Neuroendocrinology* **26**, 8–14.

Rubin, B. S., and Bridges, R. S. (1984). Disruption of ongoing maternal responsiveness in rats by central administration of morphine sulfate. *Brain Res.* **307**, 91–97.

Rubin, B. S., Menniti, F. S., and Bridges, R. S. (1983). Intracerebroventricular administration of oxytocin and maternal behavior in rats after prolonged and acute steroid pretreatment. *Horm. Behav.* **17**, 45–53.

Saito, T. R. (1986). Role of the vomeronasal organ in retrieving behavior in lactating rats. *Zool. Sci.* **3**, 919–920.

Sawchenko, P. E., and Swanson, L. W. (1983). The organization and biochemical specificity of afferent projections to the paraventricular and supraoptic nuclei. *Prog. Brain Res.* **60**, 19–29.

Schachter, B. S., Durgerian, S., Harlan, R. E., Pfaff, D. W., and Shivers, B. D. (1984). Prolactin mRNA exists in rat hypothalamus. *Endocrinology (Baltimore)* **114**, 1947–1949.

Schlein, P. A., Zarrow, M. X., Cohen, H. A., Denenberg, V. H., and Johnson, N. P. (1972). The differential effect of anosmia on maternal behaviour in the virgin and primiparous rat. *J. Reprod. Fertil.* **30**, 139–142.

Schwartz, E., and Rowe, F. A. (1976). Olfactory bulbectomy: Influences on maternal behavior in primiparous and multiparous rats. *Physiol. Behav.* **17**, 879–883.

Seegal, R. F., and Denenberg, V. H. (1974). Maternal experience prevents pup-killing in mice induced by peripheral anosmia. *Physiol. Behav.* **13**, 339–341.

Seitz, P. F. D. (1954). The effects of infantile experiences upon adult behavior in animal subjects. I. Effects of litter-size during infancy upon adult behavior in the rat. *Am. J. Psychiatry* **110**, 916–927.

Shepherd, G. M. (1988). "Neurobiology." Oxford Univ. Press, London and New York.

Shivers, B. D., Harlan, R. E., and Pfaff, D. W. (1983). Immunocytochemical mapping of immunoreactive prolactin in female rat brain. *Soc. Neurosci. Abstr.* **13**, 1018.

Siegel, H. I. (1974). Hormonal basis of the onset of maternal behavior in the laboratory rat. Unpublished Doctoral Dissertation, Rutgers University, Newark, New Jersey.

Siegel, H. I., and Rosenblatt, J. S. (1975a). Hormonal basis of hysterectomy-induced maternal being during pregnancy in the rat. *Horm. Behav.* **6**, 211–222.

Siegel, H. I., and Rosenblatt, J. S. (1975b). Estrogen-induced maternal behavior in hysterectomized-ovariectomized virgin rats. *Physiol. Behav.* **14**, 465–471.

Siegel, H. I., and Rosenblatt, J. S. (1975c). Progesterone inhibition of estrogen-induced maternal behavior in hysterectomized-ovariectomized virgin rats. *Horm. Behav.* **6**, 223–230.

Siegel, H. I., Ahdieh, H. B., and Rosenblatt, J. S. (1978). Hysterectomy-induced facilitation of lordosis behavior in the rat. *Horm. Behav.* **11**, 273–278.

Simerly, R. B., and Swanson, L. W. (1986). The organization of neural inputs to the medial preoptic nucleus of the rat. *J. Comp. Neurol.* **246**, 312–342.

Slijper, E. J. (1979). "Whales." Cornell Univ. Press, Ithaca, New York.

Slotnick, B. M. (1967a). Intercorrelations of maternal activities in the rat. *Anim. Behav.* **15**, 267–269.

Slotnick, B. M. (1967b). Disturbances of maternal behavior in the rat following lesions of the cingulate cortex. *Behaviour* **29**, 204–236.

Slotnick, B. M. (1969). Maternal behavior deficits following forebrain lesions in the rat. *Am. Zool.* **9**, 1068 (abstr.).

Slotnick, B. M. (1975). Neural and hormonal basis of maternal behavior in the rat. *In* "Hormonal Correlates of Behavior: An Organismic View" (B. E. Elefthériou and R. L. Sprott, eds.), Vol. 2, pp. 585–656. Plenum, New York.

Slotnick, B. M., and Gutman, L. A. (1977). Evaluation of intranasal zinc sulfate treatment on olfactory discrimination in rats. *J. Comp. Physiol. Psychol.* **91**, 942–950.

Slotnick, B. M., and Nigrosh, B. J. (1975). Maternal behavior of mice with cingulate cortical, amygdala, or septal lesions. *J. Comp. Physiol. Psychol.* **88**, 118–127.

Slotnick, B. M., Carpenter, M. L., and Fusco, R. (1973). Initiation of maternal behavior in pregnant nulliparous rats. *Horm. Behav.* **4**, 1–2, 53–59.

Smith, M. O., and Holland, R. C. (1975). Effects of lesions of the nucleus accumbens on lactation and postpartum behavior. *Physiol. Psychol.* **3**, 331–336.

Smith, M. S. (1978). The relative contribution of suckling and prolactin to the inhibition of gonadotropin secretion during lactation in the rat. *Biol. Reprod.* **19**, 370–374.

Smith, M. S., and Neill, J. D. (1976). Termination at midpregnancy of the two daily surges of plasma prolactin initiated by mating in the rat. *Endocrinology (Baltimore)* **98**, 696–701.

Smith, M. S., Freeman, M. E., and Neill, J. D. (1975). The control of progesterone secretion during the estrous cycle and early pseudopregnancy in the rat: prolactin, gonadotropin and steroid levels associated with rescue of the corpus luteum of pseudopregnancy. *Endocrinology (Baltimore)* **96**, 219–226.

Smith, R. L. (1973). The ascending fibers from the principal trigeminal sensory nucleus in the rat. *J. Comp. Neurol.* **148**, 423–446.

Smotherman, W. P., Bell, R. W., Starzec, J., Elias, J., and Zachman, T. A. (1974). Maternal responses to infant vocalizations and olfactory cues in rats and mice. *Behav. Biol.* **12**, 55–66.

Soloff, M. S., Alexandrova, M., and Fernstrom, M. J. (1979). Oxytocin receptors: triggers for parturition and lactation? *Science* **204**, 1313–1314.

Spencer-Booth, Y. (1970). The relationship between mammalian young and con-specifics other than mothers and peers: A review. *Adv. Study Behav.* **3**, 120–194.

Stamm, J. S. (1955). The function of the median cerebral cortex in maternal behavior in rats. *J. Comp. Physiol. Psychol.* **48**, 347–356.

Steele, M., Rowland, D., and Moltz, H. (1979). Initiation of maternal behavior in the rat: Possible involvement of limbic norepinephrine. *Pharmacol., Biochem. Behav.* **11**, 123–130.

Stern, J. M. (1977). Effects of ergocryptine on postpartum maternal behavior, ovarian cyclicity and food intake in rats. *Behav. Biol.* **21**, 134–140.

Stern, J. M. (1983). Maternal behavior priming in virgin and Caesarean-delivered Long-Evans rats: Effects of brief contact or continuous exteroceptive pup stimulation. *Physiol. Behav.* **31**, 757–763.

Stern, J. M. (1985). Parturition experience influences initial pup preferences at later onset of maternal behavior in primiparous rats. *Physiol. Behav.* **35**, 25–31.

Stern, J. M. (1986). Licking, touching, and suckling: Contact stimulation and maternal psychobiology in rats and women. *Ann. N.Y. Acad. Sci.* **474**, 95–107.

Stern, J. M. (1987a). Peripartum onset of maternal behavior in Long-Evans rats in response to varying pup stimuli: Role of parturition experience. *Abst., Conf. Reprod. Behav., 1987.* Tlaxcala, Mexico, June 14–18.

Stern, J. M. (1987b). Pubertal decline in maternal responsiveness in Long-Evans rats: Maturational influences. *Physiol. Behav.* **41**, 93–99.

Stern, J. M. (1989). A revised view of the multisensory control of maternal behaviour in rats: critical role of tactile inputs. *In* "Ethoexperimental Analysis of Behaviour" (R. J. Blanchard, D. C. Blanchard, S. Parmigiani, and P. F. Brain, eds.). Nijhoff, The Hague.

Stern, J. M., and Johnson, S. K. (1989). Perioral somatosensory determinants of nursing behavior in Norway rats. *J. Comp. Psychol.* (in press).

Stern, J. M., and Kolunie, J. M. (1987). Perioral stimulation from pups: Role in Norway rat maternal retrieval, licking, crouching and aggression. *Soc. Neurosci. Abstr.* **13**, 400.

Stern, J. M., and Kolunie, J. M. (1989). Perioral anesthesia disrupts maternal behavior during early lactation in Long-Evans rats. *Behav. Neural Biol.* (in press).

Stern, J. M., and Levin, R. (1976). Food availability as a determinant of the rats' circadian rhythm in maternal behavior. *Dev. Psychobiol.* **9**, 137–148.

Stern, J. M., and Levine, S. (1972). Pituitary-adrenal activity in the post-partum rat in the absence of suckling stimulation. *Horm. Behav.* **3**, 237–246.

Stern, J. M., and MacKinnon, D. A. (1976). Postpartum, hormonal, and nonhormonal induction of maternal behavior in rats: Effects on T-maze retrieval of pups. *Horm. Behav.* **7**, 305–316.

Stern, J. M., and MacKinnon, D. A. (1978). Sensory regulation of maternal behavior in rats: Effects of pup age. *Dev. Psychobiol.* **11**, 579–586.

Stern, J. M., and McDonald, C. (1989). Ovarian hormone-induced short-latency maternal behavior in ovariectomized virgin Long-Evans rats. *Horm. Behav.* (in press).

Stern, J. M., and Rogers, L. (1988). Experience with younger siblings facilitates maternal responsiveness in pubertal Norway rats. *Dev. Psychobiol.* **21,** 575–590.

Stern, J. M., and Siegel, H. I. (1978). Prolactin release in lactating, primiparous and multiparous thelectomized, and maternal virgin rats exposed to pup stimuli. *Biol. Reprod.* **19,** 177–182.

Stern, J. M., and Strait, T. (1983). Reproductive success, postpartum maternal behavior, and masculine sexual behavior of neonatally androgenized female hamsters. *Horm. Behav.* **17,** 208–224.

Stern, J. M., Erskine, M., and Levine, S. (1973). Dissociation of open-field behavior and pituitary-adrenal function. *Horm. Behav.* **4,** 149–162.

Stern, J. M., Thomas, D. A., Rabii, J., and Barfield, R. J. (1984). Do pup ultrasonic cries provoke prolactin secretion in lactating rats? *Horm. Behav.* **18,** 86–94.

Stern, J. M., Konner, M., Herman, T., and Reichlin, S. (1986). Nursing behaviour, prolactin and postpartum amenorrhoea during prolonged lactation in American and !Kung mothers. *Clin. Endocr.* **25,** 247–258.

Stern, J. M., Johnson, S., and McDonald, C. (1987). Prepartum infraorbital denervation severely disrupts parturition behavior in rats. *Int. Soc. Dev. Psychobiol. Abstr.,* New Orleans, LA, Nov. 12–15.

Steuer, M. A., Thompson, C., Doerr, J. S., Youakim, M., and Kristal, M. B. (1987). Induction of maternal behavior in rats: Effects of pseudopregnancy termination and placenta-smeared pups. *Behav. Neurosci.* **101,** 219–227.

Stinus, L., Gaffori, O., Simon, H., and Le Moal, M. (1978). Disappearance of hoarding and disorganization of eating behavior after ventral mesencephalic tegmentum lesions in rats. *J. Comp. Physiol. Psychol.* **92,** 289–296.

Stone, C. P. (1925). Preliminary note on maternal behavior of rats living in parabiosis. *Endocrinology (Baltimore)* **9,** 505–512.

Stone, C. P. (1938). Effects of cortical destruction on reproductive behavior and maze learning in albino rats. *J. Comp. Psychol.* **26,** 217–236.

Stoneham, M. D., Everitt, B. J., Hansen, S., Lightman, L. L., and Todd, K. (1985). Oxytocin and sexual behaviour in the male rat and rabbit. *J. Endocrinol.* **107,** 97–106.

Sturman-Hulbe, M., and Stone, C. P. (1929). Maternal behavior in the albino rat. *J. Comp. Psychol.* **9,** 203–237.

Svare, B. B. (1977). Maternal aggression in mice: Influence of the young. *Biobehav. Rev.* **1,** 151–167.

Svare, B. R. (1981). Maternal aggression in mammals. *In* "Parental Care in Mammals" (D. J. Gubernick and P. H. Klopfer, eds.), pp. 179–210. Plenum, New York.

Swanson, L. W., and Sawchenko, P. E. (1980). Paraventricular nucleus: A site for the integration of neuroendocrine and autonomic mechanisms. *Neuroendocrinology* **31,** 410–417.

Swanson, L. W., Kucharczyk, J., and Mogenson, G. J. (1978). Autoradiographic evidence for pathways from the medial preoptic area to the midbrain involved in the drinking response to angiotensin II. *J. Comp. Neurol.* **178,** 645–660.

Swanson, L. W., Mogenson, G. J., Gerfen, C. R., and Robinson, P. (1984). Evidence for a projection from the lateral preoptic area and substantia innominata to the 'mesencephalic locomotor region' in the rat. *Brain Res.* 295, 161–178.

Swanson, L. W., Mogenson, G. J., Simerly, R. B., and Wu, M. (1987). Anatomical and electrophysiological evidence for a projection from the medial preoptic area to the 'mesencephalic and subthalamic locomotor regions' in the rat. *Brain Res.* 405, 108–122.

Szechtman, H., Siegel, H. I., Rosenblatt, J. S., and Komisaruk, B. R. (1977). Tail-pinch facilitates onset of maternal behavior. *Physiol. Behav.* 19, 807–809.

Taleisnik, S., and Deis, R. P. (1964). Influence of cerebral cortex in inhibition of oxytocin release induced by stressful stimuli. *Am. J. Physiol.* 207, 1394–1398.

Tasker, J. G., Theodosis, D. T., and Poulain, D. A. (1986). Afferent projections from the mammary glands to the spinal cord in the lactating rat. I. A neuroanatomical study using the transganglionic transport of horseradish peroxidase-wheatgerm agglutinin. *Neuroscience* 19, 495–509.

Teicher, M. H., and Blass, E. M. (1976). Suckling in newborn rats: Eliminated by nipple lavage, reinstated by pup saliva. *Science* 193, 422–425.

Teitelbaum, P., Schallert, T., and Whishaw, I. Q. (1983). Sources of spontaneity in motivated behavior. *In* "Handbook of Behavioral Neurobiology: Motivation" (E. Satinoff and P. Teitelbaum, eds.), pp. 23–65. Plenum, New York.

Terkel, J. (1972). A chronic cross-transfusion technique in freely behaving rats by use of a single heart catheter. *J. Appl. Physiol.* 33, 519–522.

Terkel, J., and Rosenblatt, J. S. (1968). Maternal behavior induced by maternal blood plasma injected into virgin rats. *J. Comp. Physiol. Psychol.* 65, 479–482.

Terkel, J., and Rosenblatt, J. S. (1971). Aspects of nonhormonal maternal behavior in the rat. *Horm. Behav.* 2, 161–171.

Terkel, J., and Rosenblatt, J. S. (1972). Humoral factors underlying maternal behavior at parturition: Cross transfusion between freely moving rats. *J. Comp. Physiol. Psychol.* 80, 365–371.

Terkel, J., Damassa D. A., and Sawyer, C. H. (1979a). Ultrasonic cries from infant rats stimulate prolactin release in lactating mothers. *Horm. Behav.* 12, 95–102.

Terkel, J., Bridges, R. S., and Sawyer, C. H. (1979b). Effects of transecting lateral neural connections of the medial preoptic area on maternal behavior in the rat: Nest building, pup retrieval and prolactin secretion. *Brain Res.* 169, 369–380.

Terlecki, L. J., and Sainsbury, R. S. (1978). Effects of fimbria lesions on maternal behavior in the rat. *Physiol. Behav.* 21, 89–97.

Tindal, J. S., and Blake, L. a. (1984). Central inhibition of milk ejection in the rabbit: Involvement of hippocampus and subiculum. *J. Endocrinol.* 100, 125–129.

Tindal, J. S., and Knaggs, G. S. (1975). Further studies on the afferent path of the milk-ejection reflex in the brain stem of the rabbit. *J. Endocrinol.* 66, 107–113.

Tindal, J. S., and Knaggs, G. S. (1977). Pathways in the forebrain of the rat concerned with the release of prolactin. *Brain Res.* 119, 211–221.

Tindal, J. S., Knaggs, G. S., and Turvey, A. (1967a). Central nervous control of prolactin secretion in the rabbit: Effect of local oestrogen implants in the amygdaloid complex. *J. Endocrinol.* **37,** 279–287.

Tindal, J. S., Knaggs, G. S., and Turvey, A. (1967b). The afferent path of the milk-ejection reflex in the brain of the guinea pig. *J. Endocrinol.* **38,** 337–349.

Tindal, J. S., Knaggs, G. S., and Turvey, A. (1969). The afferent path of the milk-ejection reflex in the brain of the rabbit. *J. Endocrinol.* **43,** 663–671.

Tork, I. (1985). Raphe nuclei and serotonin containing systems. In "The Rat Nervous System" (G. Paxinos, ed.), Vol. 2, pp. 43–78. Academic Press, New York.

Vanderwolf, C. H. (1985). The role of the cerebral cortex and ascending activating systems in the control of behavior. In "Handbook of Behavioral Neurobiology: Motivation" (E. Satinoff and P. Teitelbaum, eds.), pp. 67–104. Plenum, New York.

Vanderwolf, C. H., Kolb, B., and Cooley, R. K. (1978). The behavior of the rat after removal of the neocortex and hippocampal formation. *J. Comp. Physiol. Psychol.* **92,** 156–175.

van Leengoed, E., Kerker, E., and Swanson, H. H. (1987). Inhibition of post-partum maternal behavior in the rat by injecting an oxytocin antagonist into the cerebral ventricles. *J. Endocrinol.* **112,** 275–282.

Vince, M. A. (1987). Tactile communication between ewe and lamb on the onset of suckling. *Behaviour* **101,** 156–176.

Voci, V. E., and Carlson, N. R. (1973). Enhancement of maternal behavior and nest building following systemic and diencephalic administration of prolactin and progesterone in the mouse. *J. Comp. Physiol. Psychol.* **83,** 388–393.

Voith, V. L. (1982). The role of the medial preoptic-anterior hypothalamic continuum in maternal and sexual behavior of the female cat. Unpublished Doctoral Dissertation, University of California, Davis.

Voloschin, L. M., and Dottaviano, E. J. (1976). The channeling of natural stimuli that evoke the ejection of milk in the rat. Effect of transections in the midbrain and hypothalamus. *Endocrinology (Baltimore)* **99,** 49–58.

Voloschin, L. M., and Tramezzani, J. H. (1979). Milk ejection reflex linked to slow wave sleep in nursing rats. *Endocrinology (Baltimore)* **105,** 1201–1207.

Voloschin, L. M., and Tramezzani, J. H. (1984). Relationship of prolactin release in lactating rats to milk ejection, sleep state, and ultrasonic vocalization by the pups. *Endocrinology (Baltimore)* **114,** 618–623.

Voogt, J. L., Sar, M., and Meites, J. (1969). Influence of cycling, pregnancy, labor and suckling on corticosterone-ACTH levels. *Am. J. Physiol.* **216,** 655–658.

Wall, P. D., and Egger, M. D. (1971). Formation of new connections in adult rat brains after partial deafferentation. *Nature (London)* **232,** 542–544.

Walsh, R. J., Posner, B. I., Brawer, J. R., and Kopriwa, B. M. (1978). Prolactin binding sites in the rat brain. *Science* **201,** 1041–1043.

Walsh, R. J., Posner, B. I., and Patel, B. (1984). Binding and uptake of

[I^{125}]Iodoprolactin by epithelial cells of the rat choroid plexus: An *in vivo* autoradiographic analysis. *Endocrinology (Baltimore)* 114, 1496–1505.

Walsh, R. J., Slaby, F. J., and Posner, B. (1987). A receptor-mediated mechanism for the transport of prolactin from blood to cerebrospinal fluid. *Endocrinology (Baltimore)* 120, 1846–1950.

Wamboldt, M. Z., and Insel, T. r. (1987). The ability of oxytocin to induce short latency maternal behavior is dependent on peripheral anosmia. *Behav. Neurosci.* 101, 439–441.

Warembourg, M. (1980). Uptake of ^{3}H-labeled synthetic progestin by rat brain and pituitary: A radioautography study. *Neurosci. Lett.* 9, 329–332.

Welker, W. I. (1964). Analysis of sniffing of the albino rat. *Behaviour* 22, 223–244.

Welle, S. L., and Coover, G. D. (1978). Deficits in mouse-killing following trigeminal lesions in the rat. *Physiol. Psychol.* 6, 332–339.

Wiesner, B. P., and Sheard, N. M. (1933). "Maternal Behavior in the Rat." Oliver & Boyd, London.

Wise, D. A., and Pryor, T. L. (1977). Effects of ergocornine and prolactin on aggression in the postpartum golden hamster. *Horm. Behav.* 8,30–39.

Woodside, B., and Leon, M. (1980). Thermoendocrine influences on maternal nesting behavior in rats. *J. Comp. Physiol. Psychol.* 94, 41–60.

Wysocki, C. J., Bean, N. J., Beauchamp, G. K., Labov, J. B., Lepri, J. J., and Wysocki, L. M. (1986). Role of the vomeronasal organ in social and reproductive behaviors. *Abstr, Conf. Reprod. Behav., 1986*, Montreal, Canada, June.

Yamada, Y. (1975). Effects of iontophoretically applied prolactin on unit activity of the rat brain. *Neuroendocrinology* 18, 263–271.

Yamaguchi, K., Akaishi, T., and Negoro, H. (1979). Effect of estrogen treatment on plasma oxytocin and vasopressin in ovariectomized rats. *Endocrinol. Jpn.* 26, 197–205.

Yang, Q. Z., and Hatton, G. I. (1987). Dye coupling incidence among supraoptic nucleus neurons is increased by lateral olfactory tract stimulation in slices from lactating, but not virgin or male rats. *Soc. Neurosci. Abstr.* 13, 1953.

Yeo, J. A. G., and Keverne, E. B. (1986). The importance of vaginal-cervical stimulation for maternal behavior in the rat. *Physiol. Behav.* 37, 23–26.

Zarrow, M. X., Sawin, P. B., Ross, S., and Denenberg, V. H. (1962). Maternal behavior and its endocrine basis in the rabbit. *In* "Roots of Behavior" (E. L. Bliss, ed.), pp. 187–197. Harper & Row, New York.

Zarrow, M. X., Gandelman, R., and Denenbert, V. H. (1971). Prolactin: Is it an essential hormone for maternal behavior in mammals? *Horm. Behav.* 2, 343–354.

Zarrow, M. X., Schlein, P. A., Denenberg, V. H., and Cohen, H. A. (1972). Sustained corticosterone release in lactating rats following olfactory stimulation from the pups. *Endocrinology (Baltimore)* 91, 191–196.

Zarrow, M. X., Johnson, N. P., Denenberg, V. H., and Bryant, L. P. (1973). Maintenance of lactational diestrum in the postpartum rat through tactile stimulation in the absence of suckling. *Neuroendocrinology* **11,** 150–155.

Zeigler, H. P., and Karten, H. J. (1974). Central trigeminal structures and the lateral hypothalamic syndrome. *Science* **186,** 636–638.

Zeigler, H. P., Semba, K., and Jacquin, M. F. (1984). Trigeminal reflexes and ingestive behavior in the rat. *Behav. Neurosci.* **98,** 1023–1038.

Zeigler, H. P., Jacquin, M. F., and Miller, M. G. (1985). Trigeminal orosensation and ingestive behavior in the rat. *Prog. Psychobiol. Physiol. Psychol.* **11,** 63–196.

Zucker, I. (1969). Hormonal determinants of sex differences in saccharine preference, food intake, and body weight. *Physiol. Behav.* **4,** 595–602.

CHAPTER FOUR

Effects of Sexual Behavior on Gonadal Function in Rodents

T. O. Allen and N. T. Adler

I. Interactions between Behavior and Hormones

Successful reproduction depends upon the integration of behavioral and physiological events. A central concern of reproductive physiology has been the nature and timing of endocrine secretions, and it is well recognized that the expression of copulatory behavior depends on the hormonal condition of the organism. For example, from the classical work of Young (1961) and of Beach (McGill et al., 1978), as well as of their co-workers, it is known that the ovarian hormones estrogen and progesterone stimulate female mating behavior in many mammals and that, analogously, androgen stimulates copulatory behavior in males.

To present a full psychobiological picture of hormone–behavior interaction, however, requires more than the presentation of a catalog of endocrine-behavior contingencies. Hormones do not directly release reproductive behavior patterns: gonadal hormones, at least in the adult, act in a "permissive" way (Ingle, 1954) to allow full expression of copulatory events when the external physical and social environments are sufficient (Komisaruk and Diakow, 1973). During copulation in rats, for example, the hormonally primed female solicits the male; he responds by mounting the female (McClintock and Adler, 1978). The mounting in turn elicits the lordosis reflex from the female, which allows the male to insert his penis into the vaginal orifice, and thus permits the progression of mating (Pfaff *et al.*, 1973). These few examples, and the very large literature from which they were drawn, illustrate the general phenomenon that *behavior* is the joint product of endogenous physiological mechanisms and exogenous (behavioral) stimulation.

Reproductive physiology is also jointly controlled by endogenous and exogenous (behavioral) events. That is, just as behavior is controlled by physiological and environmental (e.g., behavioral) stimulation, so too endocrine secretions are regulated by the interaction of endogenous and exogenous influences. In this chapter, we will discuss the consequences of sexual behavior patterns on gonadal hormones. Although we will be concerned with this general paradigm, we will concentrate our specific examples on behavior–hormone actions in rodents.

The study of behavioral influences on physiology impels us to consider both endocrine and neural mediators (Adler, 1981). The "final common path" in this sequence is the gonad. The gonad and its secretions influence (and are influenced by) the pituitary, which is in turn controlled by the central nervous system (CNS). Exogenous influences—as from the light cycle or behaviorally derived stimulation—are relayed to the CNS (and thus to the gonad). As will be discussed in greater detail later, pregnancy in the female rat depends on the behavioral induction of progesterone secretion. This stimulation is relayed by the pelvic nerves which innervate the cervix (Reiner *et al.*, 1981). Activity in this peripheral nervous pathway mediates the copulation-induced surges of prolactin (Smith *et al.*, 1976).

II. Effects of Behavior on the Rate of Sexual Development

In the remainder of the chapter, we will describe the influence of sexual behavior on the secretion of hormones in male and female mammals. In the three sections that follow, we examine a number of reproductive events in the female that are influenced by behavior. Our endpoint will be the control of ovarian steroid hormones in three general reproductive epochs: (1) the induction of puberty, (2) the induction of estrus in adult females, and (3) the induction of progestational hormone secretion. Another important reproductive event, ovulation and its behavioral control, has been reviewed elsewhere (Allen and Adler, 1984).

A. Acceleration of the Onset of Puberty

Puberty in female rodents (measured by a number of different physiological and morphological parameters including uterine weight, day of vaginal opening, and day of first vaginal estrus) reflects the operation of a functional hypothalamo–pituitary–ovarian axis. This maturation involves a changing relationship between negative and positive feedback systems, resulting in the elicitation of a surge of pituitary luteinizing hormone (LH) in response to ovarian steroid feedback (see Ramaley, 1979, for review).

Although puberty (and its underlying neuroendocrine mechanism) in female mammals can occur by virtue of an endogenously organized maturation of their neuroendocrine systems, the *rate* of maturation can be accelerated by socially derived stimulation. If, for example, an immature female mouse is caged with an adult male, puberty can be advanced by 20 days (Vandenbergh, 1967). The stimulus for this acceleration is quite potent, since only 3 days of exposure to males is sufficient to induce a first estrus in young females (Bronson and Stetson, 1973).

This effect of the male, however, depends on changes within the central neuroendocrine apparatus of the female, not just the release of ovarian hormones. Thus, "immature" female mice can respond to the presence of mature males with an enhanced secretion of estrogen (as measured by heightened uterine growth); however, this enhanced estrogen is not effective in eliciting ovulation until a later developmental stage, when the *ovulatory responsiveness* of the neuroendocrine system to ovarian steroids has matured (Bronson, 1975, 1981).

I. AFFERENT SIGNALS

Puberty can be accelerated by the introduction of mature male conspecifics. A female whose puberty has been accelerated in this way is, in fact, fully mature reproductively: she will mate and produce young (Bronson and Stetson, 1973; Vandenbergh et al., 1972).

There are at least two factors provided by the male that are relevant for the induction of puberty; one is the olfactory cues he provides, and the other is the tactile contact between the partners. When both of these cues are present simultaneously, a synergistic acceleration of puberty in the recipient female ensues (Bronson and Maruniak, 1975; Drickamer, 1974a, 1975). As we shall see in the following paragraphs though, it seems that the olfactory channel is the more critical.

a. Male Odor The influence of male odors on the onset of female ovarian competence has been well documented in mice (Bronson and Desjardin, 1974; Bronson and Stetson, 1973; Colby and Vandenbergh, 1974; Cowley and Wise, 1972; Drickamer and Murphy, 1978; Fullerton and Cowley, 1971; Kennedy and Brown, 1970; Lombardi and Vandenbergh, 1977; Lombardi *et al.*, 1976; Vandenbergh, 1967, 1979; Vandenbergh *et al.*, 1972). Female *Microtus* housed with males also display accelerated onset of puberty (Hasler and Nalbandov, 1974). It seems that one important source of this olfactory cue is urine and that it is androgen dependent since gonadectomy of the male mouse eliminates the urine-based accelerating signal (Colby and Vandenbergh, 1974; Drickamer and Murphy, 1978; Vandenbergh, 1979).

Receptor mediation. Tactile cues are not necessary since a male caged behind wire mesh was effective in accelerating first estrus (Fullerton and Cowley, 1971; Kennedy and Brown, 1970; Vandenbergh, 1969). Visual and auditory cues were also expendable: either male urine or bedding from the male's cage was effective (Colby and Vandenbergh, 1974; Cowley and Wise, 1972; Vandenbergh, 1969; Vandenbergh *et al.*, 1972). Drickamer and Assman (1981) did, however, find that *direct contact* with the male urine was necessary to induce early puberty.

Thus the olfactory system seems to be the primary receptive channel: olfactory bulbectomy or peripheral anosmia retards sexual maturation in female mice and rats (Kling, 1964; Lamond, 1958; McClella and Cowley, 1972; Orhbach and Kling, 1966; Vandenbergh, 1973). It also seems that the *accessory olfactory system* is the critical sensory pathway, rather than the main olfactory system (Kameko *et al.*, 1980).

Pituitary and CNS mediation. The accessory olfactory system (i.e., the vomeronasal system) and the main olfactory system have quite distinct and separate central targets (Raisman, 1972; Scalia and Winans, 1975; Winans and Scalia, 1970). The accessory olfactory system (the afferent channel for the acceleration of puberty) projects centrally to the amygdala (medial and posteriomedialcortical nuclei) and the bed nucleus of the stria terminalis (Raisman, 1972; Scalia and Winans, 1975). These areas in turn connect to hypothalamic areas (medial preoptic area, ventromedial nucleus, premammillary nucleus) related to the control of reproductive behavior and of anterior pituitary hormone secretion (de Olmos, 1972; Kevetter and Winans, 1981; Krettek and Price, 1978; Leonard and Scott, 1971). Precocious puberty can be induced in female rodents by irritative foci of either the corticomedial amygdala or the stria terminalis (Elwers and Critchlow, 1960, 1961; Velasco, 1972). Lesions of the medial amygdala with platinum electrodes delay puberty in female rats (Velasco, 1972).

Upon presentation of a male's odor, there is an early (within 1 to 3 hours) elevation of LH. Two peaks of estradiol follow: one after 12 hours and the other after 36 hours of exposure to the male. The initial surge of LH and the early rise in estradiol are not alone sufficient to induce puberty. To induce proestruslike changes in the uteri of the females, 36 to 48 hours of exposure to the male is needed. During this later period, progesterone, LH, and follicle-stimulating hormone (FSH) show peaks at 56, 60, and 72 hours, respectively, after exposure to the male (Bronson and Desjardin, 1974; Bronson and Maruniak, 1976; Bronson and Stetson, 1973).

b. Tactile Stimulation Although tactile cues may not be sufficient to induce the early onset of puberty in the absence of olfactory input, they *can* augment the olfactory effect. While this primary olfactory stimulus requires a gonadally intact male for its source, a castrated male *is* capable of providing the necessary tactile cues (Bronson and Maruniak, 1975). The exact tactile signal is unknown; however, it may involve somatosensory cues delivered during chasing and/or mounting of the female (Drickamer, 1974a, 1975).

Receptor mediation. The somatosensory signals involved in the release of LH have been more fully investigated in the rat. Mounts (without intromission) stimulate the female rat's perineum, flanks, the rump (Adler et al., 1977; Pfaff et al., 1977). In addition, the male's penis is thrust several times around the vaginal orifice before finally achieving insertion (Adler and Bermant, 1966; Bermant, 1965). The somatosensory information derived from these mounts-without-intromission is relayed to the spinal cord via peripheral nerves, including the pudendal, genitofemoral, and posterior cutaneous (of the thigh) nerves (Komisaruk et al., 1972; Kow and Pfaff, 1976; Reiner et al., 1981).

The pelvic nerve which innervates the vaginocervical area—but not the perineum (Komisaruk et al., 1972)—can be dispensed with: females whose pelvic nerves have been cut continue to show a significant elevation in LH release following coitus (Spies and Niswender, 1971).

Pituitary and CNS mediation. Once the peripheral nervous system receives the stimuli of mounting, the CNS relays this input and translates it into an endocrine signal. Copulatory stimulation from the mounting male enters the spinal cord and probably travels in the anterolateral columns: transection of these columns severely reduces the lordosis response to mounting (Kow et al., 1977). Midbrain areas are involved as relay stations. Unit activity of cells in the midbrain reticular formation and in the nucleus reticularis gigantocellularis changes in response to genital stimulation (Hornby and Rose, 1976; Kawakami and Kubo, 1971).

B. Delay in the Onset of Puberty

While the stimulation derived from the presence of adult males can *accelerate* the onset of puberty, the stimulation provided by the presence of females can *delay* the onset of puberty (Bujalska, 1973; Christian *et al.*, 1965; Cowley and Wise, 1972; Drickamer, 1974b, 1977; Drickamer *et al.*, 1978; Fullerton and Cowley, 1971; McIntosh and Drickamer, 1977; Stiff *et al.*, 1974; Vandenbergh *et al.*, 1972).

1. AFFERENT SIGNAL

The emission of the puberty-delaying cue is hormone dependent, but the hormones are not ovarian. Ovariectomy of adult females does not eliminate the puberty delaying signal from their urine (Drickamer *et al.*, 1978), but adrenalectomy does (Drickamer and McIntosh, 1980). Delay of puberty can occur in a female which is singly caged *if* she is exposed to the urine of other group-housed females. Hence direct contact between females is not necessary to induce production of the delaying chemosignal.

2. RECEPTOR MEDIATION

In the young recipient females, neither auditory nor visual contact with others is necessary: either urine (Cowley and Wise, 1972) or bedding (Drickamer, 1974b) from grouped females is sufficient to delay puberty. In fact, direct contact with urine is not necessary to delay puberty (Drickamer and Assman, 1981). Therefore, it is not mediated through the vomeronasal system, but more probably through the main olfactory system. Hence, the two chemosignals affecting puberty in young female mice operate via different sensory channels in the recipient females.

3. PITUITARY AND CNS MEDIATION

The main olfactory system affects pituitary secretion most probably via its projection to the anteriocortical and posterolateralcortical nuclei of the amygdala (Scalia and Winans, 1975). The anterior cortical amygdala in turn projects to the bed nucleus of the stria terminalis, to the ventromedial nucleus, and to other hypothalamic sites (de Olmos, 1972; Kevetter and Winans, 1981). Hence, the distinct neuroanatomical projections of the accessory and main olfactory systems converge in the hypothalamus and affect pituitary secretion of LH in opposite ways.

C. Functional (and Dysfunctional) Integration

In several species of rodents, the phenomenon of socially induced changes in the rate of reproductive maturation is, perhaps, part of a regulatory mechanism that operates on the level of the population.

Because of all of the inputs, this regulation can be quite complex. There is, for example, the question of the *balance* between accelerating and inhibiting factors. When both the stimuli that delay and the stimuli that accelerate puberty are simultaneously present, as they would be in a natural population, the delaying effect dominates (Drickamer, 1982). In this way, signals that depend on sex-specific social densities may affect the growth rate of the population—for example, by delaying the sexual maturation of the next generation.

There is a second level of complexity in the puberty-accelerating effect because the *production of the male's accelerating odor* is itself under the control of social stimuli: social subordination, known to suppress gonadal function in male mice, was found to reduce the activity of the androgen-dependent odor (Lombardi and Vandenbergh, 1977).

The effects of social stimulation on the advancement or delay of puberty in female rodents represents a natural biological event in the natural history of those species. There may also be effects of behavior and lifestyle on the timing of puberty in humans, even though these effects may not be biologically adaptive. In women, the attainment of puberty (and the subsequent continuation of menstrual cyclicity) requires a minimal weight for height (Frisch and McArthur, 1974). Are there implications of this nutritional requirement for women who engage in activities that result in lower weight? The answer seems to be yes. Young female ballet dancers who, for example, attend professional schools of dancing or who dance in companies in which thinness is much admired, restrict their food intake and are highly active. This lifestyle affects their reproductive potential; in a population of young ballet dancers, there was a relatively late mean age of menarche (Frisch *et al.*, 1980).

These data are, of course, correlational; and while it is not logically possible to conclude from these data whether it is professional activity or the rate of maturation that is the independent variable, several factors lead to the conclusion that the choice of lifestyle affects reproductive development. Most of the dancers, for example, began their training at young ages; and the occurrence of irregular cycles after puberty suggests that continued, intense physical training coupled with low food intake contribute to an ongoing effect on reproduction (Frisch *et al.*, 1980). It is clearly not likely that dancing per se had any evolutionary influence on our species. It is more likely that the onset of puberty is a basic biological phenomenon that can be modulated by social factors and type of lifestyle. Through our recent evolution, this biological potential remained and is expressed as a function of particular cultural and social variables, whether or not they are reproductively functional in the biological sense.

III. Induction of Estrus in Adult Females

In the previous section, we discussed the way in which behaviorally derived stimulation can modify the rate of sexual development. For female mammals, the endpoint of this development is the ability to sustain ovarian (estrous) cyclicity. Once maturity has been reached, bouts of ovarian cyclicity are broken by the interposition of other reproductive stages: pregnancy, maternity, and anestrus (as in the dormant periods between breeding seasons). The length and frequency of estrous cycles in the adult female can themselves be influenced by behavioral stimulation. In this section, we discuss two social effects on estrous cyclicity of adult mice: the Bruce effect and the Whitten effect.

A. Afferent Signal

I. WHITTEN EFFECT

Adult female mice respond to the presence of a male with an acceleration of their estrous cycles. That is, single females show shorter estrous cycles when housed in the presence of a male (Whitten, 1957, 1958, 1959). [The stimulus from the male is most effective during the early part of the cycle, the stage of follicular growth (Whitten, 1958).] This accelerating effect of a male is highlighted against the inhibitory background provided by housing females together. That is, when female mice live in the same cage, there is a basic inhibitory influence operating on their estrous cycles, an effect which is removed when the individual female is isolated from the isosexual group (Marsden and Bronson, 1965).

If all females are in the stage of anestrus induced by female grouping, stimulation by the male will induce estrus concurrently in the females, resulting in an apparent "estrus synchrony" (Whitten, 1958). A peak number of copulatory plugs is found 3 days after pairing with a male (Chipman and Fox, 1966a; Lamond, 1959; Whitten, 1956a, 1959). The use of the term "synchrony" for this effect can be somewhat misleading; what seems to be happening is a simultaneous disinhibition in the females exposed to the male (Whitten, 1958).

One source of the estrus-inducing factor is male urine. Exposure of female mice to bedding soiled by a male or to as little as .01 ml of male urine induced estrus in females earlier than expected (Bronson, 1971; Bronson and Whitten, 1963; Chipman and Albrecht, 1974; Whitten, 1956b).

There have been some studies attempting to isolate the components in the urine which produce the biological effect. Male urine taken directly from the bladder, i.e., containing no glandular secretions, is a sufficient stimulus for inducing estrus (Bronson and Whitten, 1963).

A second source of the estrus-accelerating factor is the preputial gland secretion. That is, even though bladder urine is sufficient, preputial secretions alone are also sufficient to accelerate estrus, and urine from preputialec-tomized males is less effective than that from intact males (Chipman and Albrecht, 1974; Gaunt, 1967).

The estrus-accelerating components of male urine are androgen dependent. Estrus-accelerating properties are absent from the urine of castrates and present in the urine of intact males and androgen-treated castrates of either sex (Bronson, 1971; Bronson and Whitten, 1963; Bruce, 1965; Chipman and Albrecht, 1974; Marsden and Bronson, 1964, 1965; Scott and Pfafff, 1970).

When housed in isolation or in all-female groups, the females in some species of *Microtus* do not spontaneously display either behavioral or vaginal estrus (Cluclow and Mallory, 1970). The presence of an intact male, even when no physical contact is allowed, can induce estrus in these females (Gray *et al.*, 1974; Hasler and Conaway, 1973; Richmond and Conaway, 1969). This effect may be similar to the well known Whitten effect displayed by mice (Whitten, 1956a, 1957, 1958, 1959).

2. BRUCE EFFECT

The Bruce effect, first described in mice, is a blockade of pregnancy upon exposure to the odors of a strange male; the female returns to estrus and will mate if contact with the male is allowed (Bruce, 1959, 1960, 1961, 1963a,b, 1966; Bruce and Parrott, 1960; Chipman and Fox, 1966b; Godowicz, 1970). Although the more fundamental mechanism may be the induction of estrus in a female that would not normally ovulate (e.g., one that is pregnant), investigators have usually focused on the progestational consequences of the male's odor rather than on the correlated induction of a new estrus.

The male urinary chemosignals mediating the Bruce effect are androgen dependent. Exposure of recently inseminated females to other females, castrated males, or ovariectomized females did not produce a significant elevation of pregnancy termination when compared to control levels. On the other hand, androgen administration to castrated males (or to females) rendered the urine of these hormonally treated animals capable of blocking pregnancy (Bronson and Whitten, 1963; Bruce, 1959, 1960, 1965; Dominic, 1965; Godowicz, 1970). Conversely, prior treatment of an intact male with the antiandrogen cyproterone acetate abolished the capacity of his urine to block pregnancy (Bloch and Wyss, 1973). The pregnancy-blocking chemosignal is not contained in the preputial gland, for removal of this gland did not impair the capacity of male urine to block pregnancy (Parkes and Bruce, 1961).

Microtus females are susceptible to pregnancy blockade when exposed to strange males (*M. agrestis*, Cluclow and Clarke, 1968; *M. pennsylvanicus*, Cluclow and Langford, 1971). The Bruce effect in voles differs somewhat from that in mice, however; *Microtus* females respond to the presence of a strange male even in the later stages of pregnancy (Kenney *et al.*, 1977; Stehn and Richmond, 1975) whereas female mice are only susceptible in the early stages.

The reproductive advantage of the Bruce effect is obvious for the male, but less so for the female. By returning the female to estrus when he arrives in the vicinity, the strange male increases his chances of siring a litter. For the female, the advantage may lie in the greater ability of the dominant male to induce pregnancy blockade (Huck, 1982). Pregnancy block may have evolved as a female strategy to increase the probability of mating with dominant males and optimization of paternal investment (Schwagmeyer, 1979).

B. *Receptor Mediation*

Tactile contact with the male is not necessary to produce the Whitten effect. The majority of females will attain estrus 3 days after exposure to a caged male (Whitten, 1956b, 1958). In mice, the effect is mediated through the main olfactory system (Bruce and Parrott, 1960; Lamond, 1959; Mody, 1963; Vandenbergh, 1973; Whitten, 1956a). Females placed downwind of males show an increase in the incidence of estrus comparable to that of females allowed access to the male's urine (Whitten *et al.*, 1968).

In a different species (*M. ochrogaster*), male-induced estrus is probably mediated by the olfactory system, since females whose main and accessory olfactory bulbs had been completely removed failed to display estrus after exposure to males (Horton and Shepherd, 1979; Richmond and Stehn, 1976). However, airborne olfactory cues from males were not sufficient to activate female reproduction in *M. ochrogaster*. At least in this species, the accessory olfactory system rather than the main olfactory system may be the critical component (Carter *et al.*, 1980).

The Bruce effect is mediated through the olfactory system; the response is abolished by bulbectomy of the female (Bruce and Parrott, 1960). Exposure to the soiled cages or urine of strange males is effective (Bronson and Eleftheriou, 1963; Bruce, 1959, 1960, 1961; Bruce and Parkes, 1960; Chipman and Fox, 1966b; Chipman *et al.*, 1966; Dominic, 1964, 1965, 1966a,b;

Parkes and Bruce, 1962). In mice, in contrast to the Whitten effect, the Bruce effect is dependent on an intact accessory olfactory system (Bellringer et al., 1980).

C. Pituitary and CNS Mediation

Both the Whitten and Bruce effects are characterized by increased gonadotropin secretion. When testing for the Whitten effect, increases in both FSH and LH occur upon exposure to a male (Bronson and Desjardin, 1969; Ryan and Schwartz, 1980).

When testing for the Bruce effect, investigators note the failure of blastocyst implantation with a return to estrus (Bruce, 1960; Parkes and Bruce, 1961). The absence of ovum implantation is due to lack of progestational changes in the uterus, caused by a failure of corpus luteum function (Bruce, 1960; Dominic, 1966c). This ovarian deficit is caused by a reduction in the release of pituitary prolactin (PRL) (Dominic, 1966c), correlated with an increased release of gonadotropin in the female (Bronson, 1971; Chapman et al., 1970; Hoppe and Whitten, 1972). Treatment with exogenous PRL or elevations in endogenous PRL protects females against pregnancy block (Bronson et al., 1969; Bruce and Parkes, 1960, 1961; Dominic, 1966c).

IV. Induction of Progestational Hormone Secretion

Once a female mammal has successfully ovulated, one of two neuroendocrine sequences can follow—she can proceed through the luteal phase of one estrous cycle and return to the follicular phase of another; or her ova are fertilized and she enters the prolonged luteal phase characteristic of pregnancy. In the females of some species (e.g., in some primates), the spontaneous luteal phase of each ovarian cycle is long and there is enough progesterone to permit uterine implantation of a fertilized ovum. If no ovum implants, the uterus, which has been developing under the influence of progesterone, sheds its inner lining (the endometrium), and another ovarian cycle begins.

Unlike the ovarian cycle of most primates, however, the estrous cycle of some species (the rat, the hamster, and several species of mice) does not spontaneously include a luteal phase of any functional consequence; in these species, if the female is to become pregnant some event must trigger the progestational state. Often it is some aspect of the male's copulation that triggers the progestational state underlying pregnancy (Adler, 1978; Dewsbury, 1972, 1975; Diakow, 1975; Diamond, 1970; Diamond and Yanagimachi, 1968; McGill, 1970). The progestational state is a physiological response that can either occur spontaneously or be induced by mating, depending on the natural history of the species and/or experimental manipulation.

In this section, we will focus on how the male rat's behavior facilitates the induction of pregnancy in the female and we will describe some of the mechanisms that produce these adaptive behavioral patterns. For over half a century it has been known that mating induced the progestational state underlying pregnancy in the female rat (Long and Evans, 1922). The total pattern of stimulation received by the female during coitus, however, is quite complex; it consists of tactile, visual, olfactory, and auditory cues, and consequently, the behavioral induction of progesterone secretion could depend on one or on many of these stimuli (Diakow, 1974).

One of the purposes of this section is to examine the specific characteristics of the mating situation which are necessary and sufficient for the induction of the progestational state. We will also discuss the behavioral adaptations which facilitate the induction of the progestational state as well as the CNS mechanisms which generate this state.

A. Afferent Signal

Since one of the most consistent features of the rat's copulatory pattern is the series of multiple intromissions preceding each ejaculation (Adler, 1978; Dewsbury, 1975), the question arises: what function do these multiple intromissions possess for successful reproduction?

In several experiments, the incidence of pregnancy was determined for females that received a normal complement of intromissions preceding the male's ejaculation (high-intromission group); these data were compared to the incidence of inducing pregnancy in a group of females that were permitted only a reduced number of preejaculatory intromissions (low-intromission group) (Adler, 1969; Wilson et al., 1965). About 20 days later, the females in both groups were sacrificed and their uteri examined for the presence of viable fetuses. In one study (Wilson et al., 1965), approximately 90% of the females in the high-intromission group were pregnant while only 20% of the females in the low-intromission group were pregnant. Thus, multiple intromissions appear to be necessary for the induction of pregnancy in rats.

Since the female rat does not spontaneously produce a progestational state as part of her estrous cycle, while she does ovulate spontaneously, it seemed more likely that the multiple copulatory intromissions facilitated pregnancy by invoking progesterone secretion rather than by affecting ovulation. We hypothesized that the stimulation derived from multiple intromissions triggers a neuroendocrine reflex which results in the secretion of progesterone. This interpretation was supported by progesterone assays (Adler *et al.*, 1970): within 24 hours after mating, females in the high-intromission group had higher, sustained elevations of progesterone in their peripheral blood than did females in the low-intromission group. The majority of the females in the high-intromission group also displayed the nocturnal surges of prolactin characteristic of pregnancy (Butcher *et al.*, 1972; Freeman and Neill, 1972; Freeman *et al.*, 1974), while those in the low-intromission group did not (Terkel and Sawyer, 1978).

Later in this section, we will examine the neural and hormonal links in the neuroendocrine reflex leading to pregnancy induction and discuss how the copulatory stimulation is transduced through these various stages. The major point for now is that stimulation from the male rat's copulatory intromissions initiates a neuroendocrine reflex in the female rat; the result of this reflex is repeated surges of PRL which in turn induce an elevated concentration of progesterone in the female.

Although there was a difference in the number of intromissions received by females in the high- and low-intromission groups, all of them received an ejaculation. Thus, although multiple intromissions might have been *necessary* for the induction of the progestational state, they may not have been *sufficient*. Therefore, we performed a series of experiments to evaluate the role of the ejaculation, the vaginal plug, and the mounts. When the vaginal orifice was sutured closed, females receiving mounts alone (over 40) did not become progestational (Adler, 1969). Therefore, vaginocervical stimulation was necessary. In another study (Adler, 1974), males were treated with isemelin (guanethidine sulfate). This permitted them to copulate but prevented the formation of a vaginal plug, the enzymatic coagulate of seminal-vesicle and prostate fluid in the semen. Females that copulated with treated males received multiple copulatory intromissions prior to ejaculation but did not receive the vaginal plug. Despite the absence of the vaginal plug, these females entered the progestational state.

In another study, we found that with increasing numbers of intromissions (without ejaculation), the proportion of females becoming progestational increased (Adler, 1969) (Fig. 1). With less than 4 intromissions, fewer than 10% of the females were progestational; with 13–16 intromissions, approximately 85% of the females became progestational. The occurrence of multiple copulatory intromissions is necessary and sufficient to trigger the hormonal state of pregnancy, even though the deposition of a vaginal plug adds an increment to the stimulation (Brown-Grant, 1977; Chester and Zucker, 1970).

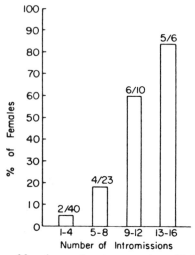

FIGURE 1.   Number of female rats showing cessation of behavioral receptivity and therefore, by inference, secretion of progesterone after different numbers of intromissions. From Adler, 1969. Copyright 1969 by the American Psychological Association. Reprinted by permission of the publisher and author.

Although there are many different species that have induced progestational changes, the necessary and sufficient behavioral events that trigger these physiological events vary. In some, a single ejaculation (with its antecedent mounts and/or intromissions) is sufficient (e.g., rats, Adler, 1969; mice, Land and McGill, 1967; McGill and Coughlin, 1970; cactus mice, Dewsbury and Estep, 1975). In others, multiple ejaculatory series are required (peromyscus, Dewsbury, 1979; Dewsbury and Lanier, 1976). The ecological causes for each pattern (and the specific evolutionary path) remain to be worked out (Dewsbury, 1975).

Along with comparative studies and work on the relationship between behavior and pregnancy induction, there has been an increasing amount of attention devoted to tracing the effects of the behavioral stimulus through successive levels of the female's neuroendocrine system to determine the mechanism by which copulatory intromissions trigger the secretion of the gestational hormones.

Mechanical stimulation provided by the experimenter may fail to provide an essential feature of the male's natural copulatory stimulation. That is, the experimental stimulation does not mimic what Diamond (1970) has described as the "species-specific vaginal code." This term highlights a major biological principle: what may seem at first sight to be minor variations in the physical nature of the stimulus are in fact important biological parameters leading to profound effects on the organism. For example, penile morphology also plays a role in induction of progestional hormone secretion. Males with smaller penes and a decreased number of penile spines are less capable of inducing pregnancy (Baumgardner and Dewsbury, 1980; Sachs, 1982).

B. Receptor Mediation

From the studies that we discussed in the preceding paragraphs, it seems that stimulation of the vaginocervical area is the definitive component for the induction of pregnancy in the female rat.

Manual tapping or electrical stimulation of the cervix provide adequate stimuli for inducing the progestational state (Carlson and DeFeo, 1965; Friedgood and Bevin, 1938; Greep and Hisaw, 1938; Haterius, 1933; Kollar, 1953; Long and Evans, 1922; Meyer et al., 1929; Shelesnyak, 1931; Slonaker, 1929). Desensitization of the genital area with the local anesthetic lidocaine just prior to copulation prevented the occurrence of pregnancy (Adler, 1974; Long and Evans, 1922). Generally, the greater the strength of the stimulus, the greater the probability of inducing pseudopregnancy (Castro-Vazquez and Carreno, 1981).

From the anatomical and electrophysiological study of the peripheral genital nervous system, we confirmed earlier findings that it was the pelvic nerve that innervated the vaginocervical area (Komisaruk et al., 1972; Reiner et al., 1981). When the pelvic nerves had been cut, female rats did not become progestational following mechanical stimulation of the cervix (Carlson and DeFeo, 1965; Kollar, 1953; Spies and Niswender, 1971; Spies et al., 1971). The data reviewed in this section support the notion that the pelvic nerve is the afferent channel for the induction of progesterone secretion.

An alternative channel may involve the vagus nerve; abdominal vagotomy blocks the induction of pseudopregnancy and its concomitant surges of prolactin in response to vaginocervical stimulation (Burden et al., 1981).

C. Pituitary Mediation

A variety of studies have documented the nature of the next stage in the neuroendocrine reflex—pituitary secretion. Radioimmunoassays reveal changes in FSH, LH, PRL, and melanocyte-stimulating hormone (MSH) within the first few minutes after stimulation of the vaginocervical area (Alonso and Deis, 1973–1974; Amenomori et al., 1970; Blake and Sawyer, 1972;

Davidson *et al.*, 1973; Linkie and Niswender, 1972; Moss, 1974; Moss and Cooper, 1973; Moss *et al.*, 1973, 1977; Rodgers, 1971; Spies and Niswender, 1971; Taleisnik and Tomatis, 1970; Taleisnik *et al.*, 1966).

Precisely which of these many pituitary hormonal responses mediate(s) the male's copulatory stimulation of progesterone secretion? Both PRL and LH plays a role in regulating progesterone secretion. Although LH is important for the maintenance of progesterone secretion in the later stages of pregnancy (Madhwah Raj and Moudgal, 1970; Morishige and Rothchild, 1974), the initial luteotropic stimulus seems to be PRL (McLean and Nikitovitch-Winer, 1973; Smith *et al.*, 1975).

This increase is, however, not a tonic elevation but rather takes the form of two daily surges (Butcher *et al.*, 1972; Freeman and Neill, 1972; Freeman *et al.*, 1974) (Fig. 2). The so-called diurnal surge begins while the lights are on, reaches maximal levels as the lights go off, and returns to basal levels by midnight. The other daily PRL peak, the nocturnal surge, begins while the lights are off, reaches its maximum as the lights come on, and decreases to baseline by the middle of the day. These surges are labeled diurnal and nocturnal on the basis of when they begin rather than when they peak.

A rhythmic, bicircadian surge of PRL suggests that a biological oscillator or clock is at work. With such clocks, rhythmicity per se is not actively triggered by environmental cues, but is generated by a "self-sustaining biological oscillator" (Aschoff, 1960).

Further evidence for the copulatory triggering of a "prolactin clock" was provided by Smith and Neill (1976a) when they showed that, while the time of mating could be experimentally varied, the surges of PRL occurred at the same time of day, regardless of when the stimulation occurred. In one experiment, even though cervical stimulation was delivered at different times of day (1900, 2400, or 0400 hours), the PRL surge always occurred between 0300 and 0700 hours. For the three times of stimulation in this study, the latencies from cervical stimulation to the respective PRL surge was 8, 5, and 3 hours, respectively.

The environmental rhythm entrains the biological rhythm by continually modifying the phase of the biological oscillator. In the absence of an external cycle (generally light), the biological rhythm "free runs" and will gradually go out of phase with the solar day. True circadian rhythms persist in constant photic conditions [either constant light (LL) and/or constant dark (DD)].

Three studies examined the pattern of plasma PRL secretion under

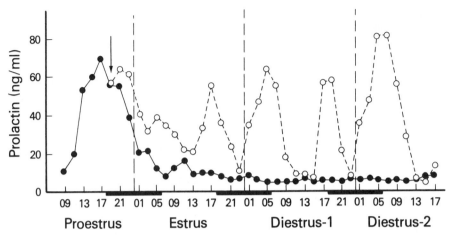

F I G U R E 2. Serum prolactin levels during the estrous cycle (●) and early pseudopregnancy (○). The numbers along the bottom abcissa represent the time of day in terms of a 24-hour clock. The dark bars represent the dark cycle (1800–0600 hours) and the dashed line bisecting each bar represents midnight. Cervical stimulation was performed at 1900 hours on proestrous and each point represents the mean of 5–6 decapitated animals. From Smith et al., 1975. Copyright 1975 by The Endocrine Society. Reprinted by permission of the publisher and authors.

constant photic conditions (Bethea and Neill, 1979; Pieper and Gala, 1979; Yogev and Terkel, 1980). These investigators compared the PRL surges in female rats housed under light–dark cycles with those in females housed under constant conditions of light or dark. They found that animals housed in DD or blinded animals continued to show PRL surges following cervical stimulation. These surges were not diminished in size and were still circadian in timing. In two of these studies, the PRL surges were not synchronized between animals, i.e., the rhythm free runs (Bethea and Neill, 1979; Pieper and Gala, 1979). In the third study, the surges were synchronized; however, nonphotic timing cues may have entrained PRL secretion (Yogev and Terkel, 1980). Constant light, LL, reduced the magnitude of the PRL surges and altered their periodicity (Bethea and Neill, 1979; Pieper and Gala, 1979; Yogev and Terkel, 1980). However, constant light not only removes periodic photic cues, but may additionally suppress the expression of periodic hormonal events, e.g., ovulation in rats (Dempsey and Searles, 1943).

The oscillatory basis of the PRL surges is also demonstrated by the fact that the surges can be advanced by 6 hours when the light cycle is phase shifted by this amount (Bethea and Neill, 1979; Freeman et al., 1974; Pieper and Gala, 1979; Yogev and Terkel, 1980). The results of all of these experiments imply that cervical stimulation activates an endogenous oscillator (perhaps in the CNS). At least initially, this oscillator requires neither the support of ovarian steroids nor the temporal information from environmental light cycles. Of course, like other biological rhythms, the environmental LD cycle does have an influence on the endogenous oscillator. The light cycle entrains the approximately 24-hour endogenous oscillator so that PRL secretions occur at a precise 24-hour periodicity.

In the previous paragraphs, we have seen that vaginocervical stimulation is transduced by the nervous system into a pituitary response (PRL surges) which in turn stimulates the ovary. Much of the physiological analysis that follows will be devoted to unraveling the way in which the graded stimulation provided by a series of copulatory intromissions is converted into the state of pregnancy. At what stage of the neuroendocrine axis—CNS, pituitary, ovary—is the graded stimulus of mating converted to the digital state of pregnancy? In more colloquial terms, at what stage of the neuroendocrine reflex is it possible for a female rat to be a little bit pregnant?

As we mentioned in the previous section, Terkel and Sawyer (1978) found that there is a relationship between the number of intromissions a female rat receives and the *probability* that she will display the PRL surges characteristic of pregnancy. That is, the majority of females in the high-intromission group displayed the surges, while only 6% of females in the low-intromission group did so. There was, however, no correlation between the magnitude of the surges and the number of intromissions. With reference to the pituitary, at least, the female rat cannot be a little bit pregnant—the surges are all or none.

D. CNS Mediation

I. STORAGE OF THE STIMULATION

One of the major functions of the CNS in the pregnancy system is storage of the stimulation received during mating. Since a number of discrete intromissions, spaced over several minutes, are required to trigger the progestational state, there must be a storage mechanism which integrates the stimulation from each brief (250-ms) penile intromission. To determine the limits of this storage, we performed an experiment in which the rate of stimulation was manipulated (Edmonds *et al.*, 1972). For different groups of female rats, the number of intromissions was varied (2, 5, or 10). Within each of these groups, the females were placed into subgroups which were defined

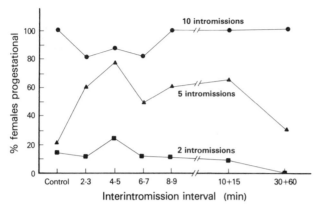

FIGURE 3. Effects of different interintromission intervals on the induction of progestational hormone secretion in female rats. From Edmonds *et al.*, 1972, in Adler, 1974. Copyright 1974 by Plenum Press. Reprinted by permission of the publisher and author.

by the intervals between successive intromissions. The interintromission-interval values ranged from the control rate of *ad libitum* copulation (approximately 40 sec) up to 1 hour. The results are summarized in Fig. 3.

With 10 intromissions, 100% of the females became progestational, even if the intromissions were distributed at the rate of one every half hour. Even when females received fewer than 10 intromissions, one could impose interintromission intervals that were longer than normal without reducing the effectiveness of the copulatory stimulation. When female rats received a total of 5 intromissions at the rate of one every 4 or 5 minutes, the stimulation was more effective in inducing the progestational state than when females received copulatory stimulation at the "control" rate.

In fact, the true "ad libitum" rate of copulation is longer than one intromission every 40 seconds, for when the female is allowed to control the pacing of the copulation, the interval between intromissions is longer (Bermant, 1961; Gilman *et al.*, 1979; McClintock and Adler, 1978; Pierce and Nuttal, 1961). With a limited number of intromissions (5), more females became progestational when they were allowed to pace the copulation (Gilman *et al.*, 1979).

The ability of copulatory intromissions to stimulate the progestational state—even when these intromissions are widely spaced—points to an exquisitely adapted form of neuroendocrine integration by which species-specific stimuli (the multiple intromissions) are stored by an adaptively specialized neural mechanism. For example, one intromission delivered every half hour provided a density of stimulation of only one part per 7200 (that is one 250-ms intromission per half hour).

In another variant of this storage experiment (Edmonds *et al.*, 1972), we compared the rate of inducing progestational responses in female rats that received eight consecutive intromissions (with about 1 minute between successive intromissions) with females that received two batches of four intromissions each, the two bursts of intromissions being separated by intervals that could be as long as 6 hours. With even 4 hours between the two blocks of four intromissions, female rats showed no diminution of the progestational response.

The phenomenon of delayed pseudopregnancy (Greep and Hisaw, 1938) is another example of storage of copulatory stimulation (Zeilmaker, 1965). In this paradigm, a female rat receives vaginocervical stimulation hours or days before ovulation is to occur. Under these conditions, the progestational state occurs but is delayed until after the succeeding estrus. Following Everett's (1967) postulation that the central neuroendocrine system retains a trace of the stimulus until the ovary contains corpora lutea that are competent to respond to the stimulation, Neill and his co-workers proposed that the PRL surges could bridge the gap between cervical stimulation and the delayed ovarian response (Freeman *et al.*, 1974). However, Beach *et al.* (1978) found that PRL levels surged during the delay interval only when the pseudopregnancy was initiated by direct electrical stimulation of the dorsomedial–ventromedial hypothalamus. It is not *necessary* for PRL surges to occur during the delay interval since, under normal conditions, even the surges of PRL are postponed until after the next estrus (Beach *et al.*, 1975; de Greef and Zeilmaker, 1976). The PRL surges therefore represent the "hands of the clock" rather than the clock itself.

These data are consistent with a dual mechanism for processing cervical stimulation. First, the stimulation is received and stored in the CNS, then it activates a clock which generates the twice-daily PRL surges characteristic of pregnancy. Developmentally, these two aspects of the neural processing are separable. The ability to retain the information from cervical stimulation develops by day 23 in the rat, but the ability to express the information as PRL surges does not develop until day 25 (Smith and Ramaley, 1978).

2. METHODOLOGICAL APPROACHES

Investigators have used three basic approaches to determine what parts of the CNS are critically involved in translating the neural signals from vaginocervical stimulation into the endocrine signals of twice-daily PRL surges. Each of these approaches, stimulation, lesion–disconnection, and recording, has its advantages and disadvantages. When the results of one are assessed with regard to the results derived from the others, a coherent picture of the role of the CNS in processing vaginocervical stimulation emerges.

a. Stimulation Studies Although the progestational state is normally induced following ovulation, these two neuroendocrine events have at least partially separable neural mechanisms. The progestational state (and/or PRL surges) can be elicited without a concomitant ovulation if one stimulates portions of the medial hypothalamus, including the anterior hypothalamus–paraventricular nuclei, the dorsomedial-ventromedial nuclei, and the pre-mamillary complex (Beach *et al.*, 1978; Everett, 1968; Everett and Quinn, 1966; Freeman and Banks, 1980; Gunnet *et al.*, 1981; Quinn and Everett, 1967) (Fig. 4). On the other hand, neither the progestational state nor increased PRL release followed stimulation of the preoptic area (Clemens *et al.*, 1971; Everett, 1962; Everett and Quinn, 1966; Kalra *et al.*, 1971)—despite the fact that ovulation does follow this kind of stimulation.

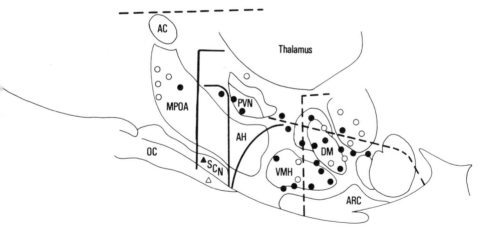

F I G U R E 4. Sagittal brain section depicting areas in the hypothalamus involved with induction of the progestational state. _____, knife cuts which block PSP or PRL surges in response to vaginocervical stimulation; ---, knife cuts which failed to block PSP or PRL surges in response to vaginocervical stimulation; ▲, lesions blocking PRL surges in reaction to vaginocervical stimulation; △, lesions which failed to block PRL surges in reaction to vaginocervical stimulation; •, sites in which electrical (or electrochemical) stimulation invoked PSP or PRL surges; ○, sites in which stimulation failed to invoke PSP or PRL surges. From Allen and Adler, 1984. Copyright 1984 by Plenum Press. Reprinted by permission of the publisher and authors.

b. Lesion and Disconnection Studies   Bilateral electrolytic lesions of the medial preoptic area (MPOA) induce repeated bouts of pseudopregnancy (Brown-Grant *et al.*, 1977; Clemens and Bennett, 1977; Clemens *et al.*, 1976). In contrast to the experiments in which MPOA stimulation was ineffective, the lesions of this area induced the progestational state, characterized by daily nocturnal surges of PRL (Freeman and Banks, 1980).

Midbrain areas act as relay stations: lesions of ventromedial and lateral areas of midbrain (Kawakami and Arita, 1982) or of the ventral noradrenergic bundle (Hansen *et al.*, 1980) prevent pseudopregnancy in response to vaginocervical stimulation. These lesion studies imply that the MPOA is an inhibitory area, whereas the ventral noradrenergic bundle is a facilatatory structure.

Although lesioning a CNS structure can show that it is involved (as a facilitator or inhibitor) in a specific neuroendocrine function, the relationship between the lesioned structure and other CNS sites is determined by other techniques, e.g., surgically disconnecting areas of the brain that are normally connected. Severing the posterior and lateral connections to the mediobasal hypothalamus (MBH) did not interfere with the induction of pseudopregnancy (Arai, 1969). However, when the anterior connections to the hypothalamus had been severed in animals that were primed with LH, the progestational response could not be elicited by cervical stimulation (Carrer and Taleisnik, 1970). Retrochiasmatic cuts separating the anterior hypothalamus (AH) and the preoptic area (POA) prevented both the nocturnal and the diurnal surge of PRL that normally follow cervical stimulation (Freeman *et al.*, 1974).

This anterior deafferentation also disrupted the ongoing progestational state that had been initiated prior to making the lesions (Carrer and Taleisnik, 1970; Velasco *et al.*, 1974), with both surges of PRL being abolished (Kato *et al.*, 1978).

More medial deafferentation, separating the AH from the median emminence (ME), did not disrupt a previously initiated pregnancy (Velasco *et al.*, 1974); in this case, the lesion abolished the diurnal PRL surge while only reducing the magnitude of the nocturnal surge (Kato *et al.*, 1978), giving further evidence that the nocturnal surge is the one necessary for pregnancy. These data also demonstrate that the diurnal and nocturnal peaks of PRL are controlled from different areas of the hypothalamus.

At this point, it may be possible to put these various pieces of anatomical information together into a working hypothesis describing the way in which the CNS triggers the progestational state. The suprachiasmatic nucleus

(SCN) is a major circadian pacemaker in the rat. Although cervical stimulation normally induces twice-daily surges of PRL, lesion of the SCN blocked the response (Bethea and Neill, 1980; Yogev and Terkel, 1980). Bethea and Neill postulated that (1) the SCN is a circadian pacemaker for PRL surges in the female rat; (2) the MPOA acts as a tonic inhibitory area which blocks the SCN's oscillatory input on PRL secretion; and (3) the MPOA's inhibition is turned off when the female rat receives vaginocervical stimulation, thereby allowing the SCN to drive the PRL surges.

When it was first discovered that retrochiasmatic cuts abolished the PRL surges, it was thought that the MPOA itself generated the PRL secretory rhythm. It seems more likely now that the retrochiasmatic cuts removed the influence of *both* the SCN (stimulatory) and the MPOA (inhibitory) on the PRL-regulating neural elements. Other data also fit with this model. Lesions of the MPOA would produce the progestational state because of the SCN's facilitatory output is no longer opposed. Stimulation of the MPOA reduces PRL output because it increases the inhibition by the MPOA.

c. Recording Studies Although lesion and disconnection studies can demonstrate what the system is capable of doing, to understand what normally occurs when a female goes into the progestational state it is also necessary to record from (assay) the CNS.

Electrophysiological recording has identified brain areas responsive to cervical stimulation. Areas that receive copulatory information in female rats include the anterior, lateral, ventromedial, and preoptic areas of the hypothalamus, the medial and basolateral complexes of the amygdala, the nucleus reticularis gigantocellularis of the medulla, the dorsal midbrain, the dorsal hippocampus, the lateral septum, and the reticular formation (Barraclough, 1960; Barraclough and Cross, 1963; Blake and Sawyer, 1972; Cross and Silver, 1965; Haller and Barraclough, 1970; Hornby and Rose, 1976; Kawakami and Kubo, 1971; Lincoln, 1969; Malsbury *et al.*, 1973; Margharita *et al.*, 1965; Ramirez *et al.*, 1967; Terasawa and Sawyer, 1969; Wüttke, 1974). These areas responsive to the vaginocervical stimulation may be involved with the storage of the stimulation.

We are using the [^{14}C]deoxyglucose (2DG) method of neural mapping to investigate brain areas responsive to vaginocervical stimulation (Allen *et al.*, 1981). A major advantage of the 2DG method over other techniques is the ability to assay concurrently a virtually unlimited number of brain areas in an individual animal, without invading or destroying the integrity of the brain. This method is based on the fact that nerve cells preferentially use glucose for energy, and on the assumption that the functional activity and energy requirements of neural tissue are coupled. [^{14}C]deoxyglucose is taken up by active cells and metabolized to deoxyglucose-6-phosphate. It is not metabolized further and is essentially trapped in the cell. The rate of glucose utilization, and hence the activity of that structure, can be estimated by autoradiographic measurement of the [^{14}C]deoxyglucose-6-phosphate concentration (for a description of the procedure, see Sokoloff et al., 1977). Areas of high metabolic activity are directly visualized as areas of high optical density. Histological staining of the tissue after it has been autoradiographed permits precise identification of anatomical structures.

In order to compare the ^{14}C concentration of a gray-matter structure between animals in these groups, we needed to control for individual variation in the brain's metabolism of glucose. Accordingly, the average concentration of ^{14}C in fiber tracts of the brain was used as an internal baseline. In a direct comparison, Sharp et al. (1983) have determined that the use of optical density ratios was equivalent to the determination of local cerebral glucose utilization values when animals are in similar physiological states.

Eight of the forty-four CNS structures increased their metabolic activity by 20% or more in response to vaginocervical stimulation. The MPOA showed the greatest increase (37% over control values). This increase was statistically significant ($p = .014$). On the autoradiograms, this area is visible as a bilateral wedge of darkening, lateral to the third ventricle and extending from the anterior commissure to the optic chiasm (Fig. 5).

In addition to heightened activity in the MPOA, the stimulated females also showed statistically significant increases in activity in the mesencephalic reticular formation, nucleus of the stria terminalis, and dorsal raphe. Although the locus coeruleus, lateral preoptic area, anterior hypothalamus, and habenula showed elevated uptake of 2DG, these increases were not statistically significant.

The other 36 nuclei exhibited less than 18% difference in relative ^{14}C concentration between stimulated and unstimulated animals. None of these comparisons was statistically significant, except for the 13% increase over control values in the globus pallidus ($p = .014$).

FIGURE 5. (A) [^{14}C]2-deoxyglucose autoradiograph from a female rat which received vaginocervical stimulation. The preoptic area is visible as the bilateral wedge of darkening extending from the anterior commissure (AC) to the optic chiasm (OC). Average relative ^{14}C concentration in medial preoptic area in this animal was 2.15. (B) Autoradiograph from female which did not receive the stimulation. The medial preoptic area is pale. Average relative ^{14}C concentration in this animal was 1.70. (C) Cresyl violet-stained section. This is the same section from which the autoradiograph

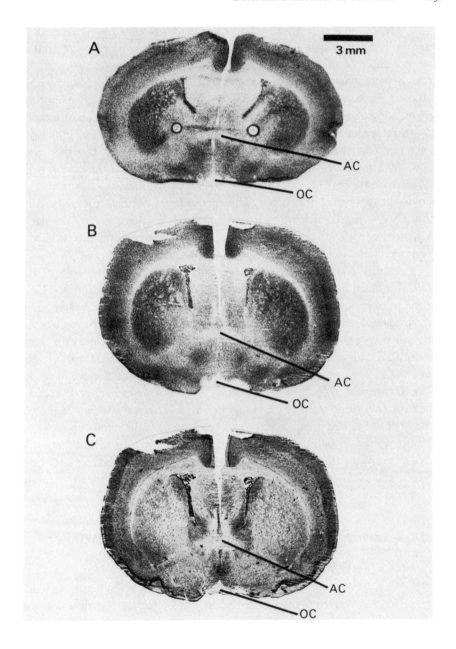

in (B) was taken. The stained sections were used to confirm the anatomical identification in the autoradiographs; note the medial preoptic area here. From Allen *et al.*, 1981. Copyright 1981 by the American Association for the Advancement of Science. Reprinted by permission of the publisher and authors.

These findings provide a map of brain areas which are activated by vaginocervical stimulation. At least some of these areas may be involved in the induction of the progestational state. The CNS areas that showed increased metabolism following vaginocervical stimulation have been implicated in the control of reproduction by experiments that used recording, lesion, and/or stimulation techniques. The 2DG method of autoradiography can therefore validate the results of other experimental methods. However, several brain structures implicated in reproductive physiology by other techniques did not increase their metabolic activity in this study. At least in part, the discrepancy may have been due to 2DG autoradiography's assessment of activity in aggregates of neurons. This method may not detect increases in a diffusely organized nucleus, such as the nucleus reticularis gigantocellularis, in which only 4% of the genitally sensitive neurons respond specifically to vaginocervical stimulation (Hornby and Rose, 1976).

At this point we can only speculate about the specific role of the MPOA in the induction of the progestational state. One function of the CNS in the induction of the progestational state is that of a storage system. This system integrates the brief stimulation from each intromission, even when the intromissions are widely spaced (Edmonds et al., 1972), and delays expression of the PRL surges until the appropriate time (Smith and Neill, 1976a). The MPOA fulfills a similar timing function in another reproductive event, ovulation. The neural trigger for spontaneous ovulation arises here and precedes the ovulatory surge of LH by a number of hours (Everett and Sawyer, 1950). The MPOA has been characterized as a receiving area for exteroceptive stimulation related to ovulation (Blake and Sawyer, 1972). In this schema, the MPOA evaluates the amount of copulatory stimulation, then at a later time (Smith and Neill, 1976a) is inhibited to allow the progestationally relevant nocturnal surge of PRL (Bethea and Neill, 1980; Freeman and Banks, 1980).

V. Behavioral Control of Testicular Hormone Secretion

A. Induction of Testosterone Release

We have devoted the bulk of this chapter to the ways in which behavior can affect the secretion of ovarian hormones in female rodents. We will conclude this discussion with the homologous phenomenon: behavioral control of testosterone secretion in the male.

As in female rodents, copulation affects endocrine events in males. After initiation of sexual contact, androgen levels are elevated in rats (Kamel and Frankel, 1978a,b; Kamel et al., 1975, 1977; Purvis and Haynes, 1974; Taylor *et al.*, 1983), in voles (Charlton *et al.*, 1975), and in mice (Batty, 1978; Macrides *et al.*, 1975; Quadagno *et al.*, 1976).

The gonadal response in the male can occur very quickly after the female is introduced (before intromission and ejaculation). This endocrine effect may thus depend upon the *presence* of the female (and a neuroendocrine representation of a heightened motivational state) without requiring copulation (Batty, 1978; Graham and Desjardin, 1980; Kamel and Frankel, 1978b; Kamel *et al.*, 1975; Macrides *et al.*, 1975; Taylor *et al.*, 1983).

B. Afferent Signal

Maruniak and Bronson (1976) identified a signal in the female's urine which is responsible for LH release in the male (Kamel *et al.*, 1977). This substance is not dependent on the ovary, but rather is from the pituitary (Johnston and Bronson, 1982).

C. Receptor Mediation

When no tactile contact could occur, males caged in proximity to estrous females still showed elevated levels of circulating testosterone (Batty, 1978; Purvis and Haynes, 1974; Taylor *et al.*, 1983). Wysocki *et al.* (1983) attribute the female-induced rise in androgen to the vomeronasal system: in the absence of a vomeronasal organ, testosterone and dihydrotestosterone remained at basal levels following exposure to a nonreactive female.

D. Pituitary and CNS Mediation

The behavioral induction of androgen secretion seems to be mediated by surges in LH secretion (Kamel *et al.*, 1975; Taleisnik *et al.*, 1966). The LH surge precedes the testosterone rise (Coquelin and Bronson, 1980; Coquelin and Desjardin, 1982; Kamel and Frankel, 1978b; Kamel *et al.*, 1977). The LH response is seen in mating males prior to intromissions and ejaculation (Kamel and Frankel, 1978b), and even in sexually sated males (Bronson and Desjardin, 1982). Hence, the release of this pituitary hormone seems controlled by a neuroendocrine reflex, triggered by the odor of a female.

Despite the positive evidence for concluding that LH mediates the behavioral initiation of androgen secretion, a number of studies have failed to find postcoital rises in LH in male rodents (Balin and Schwartz, 1976; Quadagno *et al.*, 1979; Spies and Niswender, 1971). It is, however, possible that in these studies samples were taken too late (i.e., postejaculation) to see a mating-stimulated rise in LH, and/or these samples may have been taken during the period when the male is refractory to the female (Coquelin and Bronson, 1981).

Investigators have begun to examine the CNS components responsible for androgen secretion in response to stimulation from a female. Although the MPOA is critical for his sexual behavior, the female-elicited alterations of hormones in the male may be spared following lesions to this structure. It remains to be determined how much of the neuroendocrine response depends on the MPOA, since lesions can produce somewhat different results even when the experiments are performed in the same laboratory. In one study, for example, lesions of the MPOA abolished sexual behavior and the female-induced elevation in testosterone without affecting the female-elicited rise in LH (Kamel and Frankel, 1978a). In another study, both the LH *and* the testosterone responses were retained in lesioned males (Ryan and Frankel, 1978).

Although there has been a renewal of interest in the effects of behavior on reproductive physiology in male mammals, there is still a need to relate these effects to more functional considerations. Just how important are these increases in androgen levels? That is, do they affect social organization, dominance, or other population parameters? A kind of physiological behavioral ecology seems needed.

VI. Summary

In this chapter, we have examined some of the machinery of the mammalian body, reviewing studies on the neuroendocrine organization of reproduction in that class.

While endogenous feedback among brain, pituitary, and gonad provides the basic framework for reproduction, we have concentrated on the ways in which behavior, and the stimulation derived from behavior, coordinate the physiological events. Our thesis is that such behavioral influences are not merely added on to an autonomous neuroendocrine apparatus but rather provide the fine-tuning by which members of a particular species are adapted to their local environments. The physiological events must be coordinated with reproductive behavior patterns for the sexually reproducing species to continue.

References

Adler, N. T. (1969). Effects of the male's copulatory behavior in successful pregnancy of the female rat. *J. Comp. Physiol. Psychol.* **69**, 613–622.

Adler, N. T. (1974). The behavioral control of reproductive physiology. In "Repro-ductive Behavior" (W. Montagna and W. A. Sadler, eds.). Plenum, New York.

Adler, N. T. (1978). Social and environmental control of reproductive processes in animals. In "Sex and Behavior: Status and Prospectus" (T. McGill, D. Dews-bury, and B. Sachs, eds.). Plenum, New York.

Adler, N. T., ed. (1981). "Neuroendocrinology of Reproduction: Physiology and Behavior." Plenum, New York.

Adler, N. T., and Bermant, G. (1966). Sexual behavior of male rats: Effects of reduced sensory feedback. *J. Comp. Physiol. Psychol.* **61**, 240–243.

Adler, N. T., Resko, J. A., and Goy, R. W. (1970). The effect of copulatory behavior on hormonal change in the female rat prior to implantation. *Physiol. Behav.* **5**, 1003–1007.

Adler, N. T., Davis, P. G., and Komisaruk, B. K. (1977). Variation in the size and sensitivity of a genital sensory field in relation to the estrous cycle in rats. *Horm. Behav.* **9**, 334–344.

Allen, T. O., and Adler, N. T. (1984). Neuroendocrine consequences of sexual behav-ior. In "Handbook of Behavioral Neurobiology" (N. Adler, D. Pfaff, and R. Goy, eds.), Vol. 8. Plenum, New York.

Allen, T. O., Adler, N. T., Greenberg, J. H., and Reivich, M. (1981). Vaginocervical stimulation selectively increases metabolic activity in the rat brain. *Science* **211**, 1070–1072.

Alonso, N., and Deis, R. P. (1973–1974). Hypophysial and serum concentration of prolactin in the rat after vaginal stimulation. *Neuroendocrinology* **13**, 63–68.

Amenomori, Y., Chen, C. L., and Meites, J. (1970). Serum prolactin levels in rats during different reproductive state. *Endocrinology (Baltimore)* **86**, 506–510.

Arai, Y. (1969). Effect of hypothalamic deafferentation on induction of pseudopreg-nancy by vaginal-cervical stimulation in the rat. *J. Reprod. Fertil.* **19**, 573–575.

Aschoff, J. (1960). Exogenous and endogenous components in circadian rhythms. *Cold Spring Harbor Symp. Quant. Biol.* **25**, 11–28.

Balin, M., and Schwartz, N. (1976). Effects of mating on serum LH, FSH, and prolactin, and accessory tissue weight in male rats. *Endocrinology (Baltimore)* **98**, 522–526.

Barraclough, C. A. (1960). Hypothalamic activity associated with stimulation of vagi-nal cervix in rats. *Anat. Rec.* **136**, 159.

Barraclough, C. A., and Cross, B. A. (1963). Unit activity in the hypothalamus of the cyclic female rat: Effect of genital stimuli and progesterone. *Endocrinology (Bal-timore)* **26**, 339–359.

Batty, J. (1978). Acute changes in plasma testosterone levels and their relation to measures of sexual behavior in the male house mouse (*Mus musculus*). *Anim. Behav.* **26**, 349–357.

Baumgardner, D. J., and Dewsbury, D. A. (1980). Pseudopregnancy in female rats: Effects of hormonal manipulations of the male. *Physiol. Behav.* **24**, 693–697.

Beach, J. E., Tyrey, L., and Everett, J. W. (1975). Serum prolactin and LH in early phases of delayed vs direct pseudopregnancy in the rat. *Endocrinology (Baltimore)* **96,** 1241–1246.

Beach, J. E., Tyrey, L., and Everett, J. W. (1978). Prolactin secretion preceding delayed pseudopregnancy in rats after electrical stimulation of the hypothalamus. *Endocrinology (Baltimore)* **103,** 2247–2251.

Bellringer, J., Pratt, H., and Keverne, E. (1980). Involvement of the vomeronasal organ and prolactin in pheromonal induction of delayed implantation in mice. *J. Reprod. Fertil.* **59,** 223–228.

Bermant, G. (1961). Response latencies of female rats during sexual intercourse. *Science* **133,** 1771.

Bermant, G. (1965). Sexual behavior of male rats: Photographic analysis of the intromission response. *Psychon. Sci.* **2,** 65–66.

Bethea, C. L., and Neill, J. D. (1979). Prolactin secretion after cervical stimulation of rats maintained in constant dark or constant light. *Endocrinology (Baltimore)* **104,** 870–876.

Bethea, C. L., and Neill, J. D. (1980). Lesions of the suprachiasmatic nuclei abolish the cervically stimulated prolactin surges in the rat. *Endocrinology (Baltimore)* **107,** 1–5.

Blake, C. A., and Sawyer, C. H. (1972). Effects of vaginal stimulation on hypothalamic multiple-unit activity and pituitary LH release in the rat. *Neuroendocrinology* **10,** 358–370.

Bloch, S., and Wyss, H. (1973). Antiandrogen (cyproterone acetate) inhibits pregnancy block in mice caused by the presence of strange males (Bruce effect). *J. Endocrinol.* **59,** 365–366.

Bronson, F. H. (1971). Rodent pheromones. *Biol. Reprod.* **4,** 344–357.

Bronson, F. H. (1975). A developmental comparison of steroid-induced and male-induced ovulation in young mice. *Biol. Reprod.* **12,** 431–437.

Bronson, F. H. (1981). The regulation of luteinizing hormone secretion by estrogen: Relationships among negative feedback, surge potential, and male stimulation in juvenile, peripubertal and adult female mice. *Endocrinology (Baltimore)* **108,** 506–516.

Bronson, F. H., and Desjardin, C. (1969). Release of gonadotrophin in ovariectomized mice after exposure to males. *J. Endocrinol.* **44,** 293–297.

Bronson, F. H., and Desjardin, C. (1974). Circulating concentrations of FSH, LH, estradiol and progesterone associated with acute, male-induced puberty in female mice. *Endocrinology (Baltimore)* **94,** 1658–1668.

Bronson, F. H., and Desjardin, C. (1982). Endocrine responses to sexual arousal in male mice. *Endocrinology (Baltimore)* **111,** 1286–1291.

Bronson, F. H., and Elefthériou, B. (1963). The influence of strange males on implantation in deer mice. *Gen. Comp. Endocrinol.* **3,** 515–519.

Bronson, F. H., and Maruniak, J. A. (1975). Male-induced puberty in female mice: Evidence for a synergistic action of social cues. *Biol. Reprod.* **13,** 94–98.

Bronson, F. H., and Maruniak, J. A. (1976). Differential effects of male stimuli on follicle-stimulating hormone, luteinizing hormone, and prolactin secretion in prepubertal female mice. *Endocrinology (Baltimore)* **98**, 1101–1108.

Bronson, F. H., and Stetson, M. (1973). Gonadotrophin release in prepubertal female mice following male exposure: A comparison with the adult cycle. *Biol. Reprod.* **9**, 449–459.

Bronson, F. H., and Whitten, W. K. (1963). Oestrus-accelerating pheromone of mice: Assay, androgen dependency, and presence in bladder urine. *J. Reprod. Fertil.* **15**, 131–134.

Bronson, F. H., Elefthériou, B., and Dezell, H. E. (1969). Strange male pregnancy block in deermice: Prolactin and adrenocortical hormones. *Biol. Reprod.* **1**, 302–306.

Brown-Grant, K. (1977). The induction of pseudopregnancy and pregnancy by mating in albino rats exposed to constant light. *Horm. Behav.* **8**, 62–76.

Brown-Grant, K., Murray, M. A., Raisman, G., and Sood, M. C. (1977). Reproductive function in male and female rats following extra- and intra-hypothalamic lesions. *Proc. R. Soc. London* **198**, 267–278.

Bruce, H. (1959). An exteroceptive block to pregnancy in the mouse. *Nature (London)* **184**, 105.

Bruce, H. (1960). A block to pregnancy in the mouse caused by proximity to strange males. *J. Reprod. Fertil.* **1**, 96–103.

Bruce, H. (1961). Time relations in the pregnancy-block induced in mice by strange males. *J. Reprod. Fertil.* **2**, 138–142.

Bruce, H. (1963a). A comparison of olfactory stimulation and nutritional stress as pregnancy-blocking agents in mice. *J. Reprod. Fertil.* **6**, 221–227.

Bruce, H. (1963b). Olfactory block to pregnancy among grouped mice. *J. Reprod. Fertil.* **6**, 451–460.

Bruce, H. (1965). The effect of castration on the reproductive pheromones of male mice. *J. Reprod. Fertil.* **2**, 138–142.

Bruce, H. (1966). Smell as an exteroceptive factor. *J. Anim. Sci.* **25**, 83–89.

Bruce, H., and Parkes, A. (1960). Hormonal factors in exteroceptive block to pregnancy in mice. *J. Endocrinol.* **20**, 29–30.

Bruce, H., and Parkes, A. (1961). An olfactory block to implantation in mice. *J. Reprod. Fertil.* **2**, 195–196.

Bruce, H., and Parrott, D. (1960). The role of olfactory sense in pregnancy-block by strange males. *Science* **131**, 1526.

Bujalska, G. (1973). The role of spacing behavior among females in the regulation of reproduction in the bank vole. *J. Reprod. Fertil., Suppl.* **19**, 465–474.

Burden, H. W., Lawrence, I. E., Louis, T. M., and Hodson, C. A. (1981). Effects of abdominal vagotomy on the estrous cycle of the rat and the induction of pseudopregnancy. *Neuroendocrinology* **33**, 218–222.

Butcher, R. L., Fugo, H. W., and Collins, W. E. (1972). Semicircadian rhythm in plasma levels of prolactin during early gestation in the rat. *Endocrinology (Baltimore)* **90**, 1125–1127.

Carlson, R. R., and DeFeo, V. J. (1965). Role of the pelvic nerve vs. the abdominal sympathetic nerves in the reproductive function of the female rat. *Endocrinology (Baltimore)* **77**, 1014–1022.

Carrer, H. F., and Taleisnik, S. (1970). Induction and maintenance of pseudopregnancy after interruption of preoptic hypothalamic connections. *Endocrinology (Baltimore)* **86**, 231–236.

Carter, C. S., Getz, L., Gavish, L., McDermott, J., and Arnold, P. (1980). Male-related phermones and the activation of female reproduction in the prairie vole *(M. ochrogaster)*. *Biol. Reprod.* **23**, 1038–1045.

Castro-Vazquez, A., and Carreno, N. B. (1981). A study of the sensory requirement for eliciting the pseudopregnancy response. *Brain Res.* **230**, 205–220.

Chapman, V. M., Desjardin, C., and Whitten, W. K. (1970). Pregnancy block in mice: Changes in pituitary LH and LTH and plasma progestin levels. *J. Reprod. Fertil.* **21**, 333–337.

Charlton, H. M., Naftolin, F., Sood, M. C., and Worth, R. W. (1975). The effect of mating upon LH release in male and female voles of the species, *Microtus agretis*. *J. Reprod. Fertil.* **42**, 171–174.

Chester, R. V., and Zucker, I. (1970). Influence of male copulatory behavior on sperm transport, pregnancy, and pseudopregnancy in female rats. *Physiol. Behav.* **5**, 35–43.

Chipman, R., and Albrecht, E. (1974). The relationship of the male preputial gland to the acceleration of estrus in the latoratory mouse. *J. Reprod. Fertil.* **38**, 91–96.

Chipman, R., and Fox, K. (1966a). Oestrus synchronization and pregnancy blocking in wild house mice. *J. Reprod. Fertil.* **12**, 233–236.

Chipman, R., and Fox, K. (1966b). Factors in pregnancy block: Age and reproductive background of females and number of strange males. *J. Reprod. Fertil.* **12**, 399–403.

Chipman, R., Holt, J., and Fox, K. (1966). Pregnancy failure in lab mice after multiple short-term exposure to strange males. *Nature (London)* **210**, 653.

Christian, J., Lloyd, J., and Davis, D. (1965). The role of endocrines in self regulation of mammalian populations. *Recent Prog. Horm. Res.* **21**, 501–578.

Clemens, J. A., and Bennett, D. R. (1977). Do aging changes in the POA contribute to the loss of cyclic endocrine functions? *J. Gerontol.* **32**, 19–24.

Clemens, J. A., Shaar, C. J., Kleber, J. W., and Tandy, W. A. (1971). Areas of the brain stimulatory to LH and FSH secretion. *Endocrinology (Baltimore)* **88**, 180–184.

Clemens, J. A., Smalstig, E. B., and Sawyer, B. D. (1976). Studies on the role of the preoptic area in the control of reproductive function in the rat. *Endocrinology (Baltimore)* **99**, 728–735.

Cluclow, F., and Clarke, J. (1968). Pregnancy-block in *Microtus agrestis*, an induced ovulator. *Nature (London)* **219**, 511.

Cluclow, F., and Langford, P. (1971). Pregnancy-block in the meadow vole, *Microtus pennsylvanicus*. *J. Reprod. Fertil.* **24**, 275–277.

Clulow, F. V., and Mallory, F. (1970). Oestrus and induced ovulation in the meadow vole, *Microtus pennsylvanicus*. *J. Reprod. Fertil.* **23**, 341–343.

Colby, D. R., and Vandenbergh, J. G. (1974). Regulatory effects of urinary pheromones on puberty in the mouse. *Biol. Reprod.* **11**, 268–279.

Coquelin, A., and Bronson, F. H. (1980). Secretion of luteinizing hormone in male mice: Factors that influence release during sexual encounters. *Endocrinology (Baltimore)* **106**, 1224–1229.

Coquelin, A., and Bronson, F. H. (1981). Episodic release of luteinizing hormone in male mice: Antagonism by a neural refractory period. *Endocrinology (Baltimore)* **109**, 1605–1610.

Coquelin, A., and Desjardin, C. (1982). Luteinizing hormone and testosterone secretion in young and old male mice. *Am. J. Physiol.* **243**, 257E–263.

Cowley, J. J., and Wise, D. R. (1972). Some effects of mouse urine on neonatal growth and reproduction. *Anim. Behav.* **20**, 499–506.

Cross, B. A., and Silver, I. A. (1965). Effect of luteal hormone on the behavior of hypothalamic neurons in pseudopregnant rats. *J. Endocrinol.* **31**, 251–263.

Davidson, J., Smith, E., and Bowers, C. (1973). Effects of mating on gonadotrophin release in female rat. *Endocrinology (Baltimore)* **93**, 1185–1192.

de Greef, W. J., and Zeilmaker, G. H. (1976). Prolactin and delayed pseudopregnancy in the rat. *Endocrinology (Baltimore)* **98**, 305–310.

Dempsey, E., and Searles, H. (1943). Environmental modification of certain endocrine phenomena. *Endocrinology (Baltimore)* **32**, 119–128.

de Olmos, J. S. (1972). The amygdaloid projection field in the rat as studied with the cupric-sulver method. *In* "The Neurobiology of the Amygdala" (B. Eleftheriou, ed.), pp. 145–204. Plenum, New York.

Dewsbury, D. A. (1972). Patterns of copulatory behavior in male mammals. *Q. Rev. Biol.* **47**, 1–33.

Dewsbury, D. A. (1975). Diversity and adaptation in rodent copulatory behavior. *Science* **190**, 949–954.

Dewsbury, D. A. (1979). Copulatory behavior of deer mice *(Peromyscus maniculatus)*. III. Effects of pregnancy initiation. *J. Comp. Physiol. Psychol.* **93**, 178–188.

Dewsbury, D. A., and Estep, D. Q. (1975). Pregnancy in cactus mice: effects of prolonged copulation. *Science* **187**, 552–553.

Dewsbury, D. A., and Lanier, D. L. (1976). Effects of variations in copulatory behavior on pregnancy in two species of *Peromyscus*. *Physiol. Behav.* **17**, 921–924.

Diakow, C. (1974). Male-female interactions and the organization of mammalian mating patterns. *Adv. Study Behav.* **5**, 227–268.

Diakow, C. (1975). Motion picture analysis of rat mating behavior. *J. Comp. Physiol. Psychol.* **88**, 704–712.

Diamond, M. (1970). Intromission pattern and species vaginal code in relation to induction of pseudopregnancy. *Science* **169**, 995–997.

Diamond, M., and Yanagimachi, R. (1968). Induction of pseudopregnancy in the golden hamster. *J. Reprod. Fertil.* **17**, 165–168.

Dominic, C. (1964). The source of the male odor causing pregnancy block in mice. *J. Reprod. Fertil.* **8**, 266–267.

Dominic, C. (1965). The origin of the pheromones causing pregnancy block in mice. *J. Reprod. Fertil.* **10**, 469–472.

Dominic, C. (1966a). Block to pseudopregnancy in mice caused by exposure to male urine. *Experientia* **22**, 534–535.

Dominic, C. (1966b). Observation of the reproductive pheromones of mice. I. Source. *J. Reprod. Fertil.* **11**, 407–414.

Dominic, C. (1966c). Observations of the reproductive pheromones of mice. II. Neuroendocrine mechanisms involved in the olfactory block to pregnancy. *J. Reprod. Fertil.* **11**, 415–421.

Drickamer, L. C. (1974a). Contact stimulation, androgenized females, and accelerated sexual maturation in female mice. *Behav. Biol.* **12**, 101–110.

Drickamer, L. C. (1974b). Sexual motivation of female house mice: social inhibition. *Dev. Psychobiol.* **7**, 257–265.

Drickamer, L. C. (1975). Contact stimulation and accelerated sexual maturation of female mice. *Behav. Biol.* **15**, 113–115.

Drickamer, L. C. (1977). Delay of sexual maturation in female house mice by exposure to grouped females or urine from grouped females. *J. Reprod. Fertil.* **51**, 77–81.

Drickamer, L. C. (1982). Delay and acceleration of puberty in female mice by urinary chemosignals from other females. *Dev. Psychobiol.* **15**, 433–445.

Drickamer, L. C., and Assman, S. M. (1981). Acceleration and delay of puberty in female housemice: Methods of delivery of the urinary stimulus. *Dev. Psychobiol.* **14**, 487–497.

Drickamer, L. C., and McIntosh, T. K. (1980). Effects of adrenalectomy on the presence of a maturation-delaying pheromone in the urine of female mice. *Horm. Behav.* **14**, 146–152.

Drickamer, L. C., and Murphy, R. (1978). Female mouse maturation: effects of excreted and bladder urine from juvenile and adult males. *Dev. Psychobiol.* **11**, 63–72.

Drickamer, L. C., McIntosh, T. K., and Rose, E. (1978). Effects of ovariectomy on the presence of a maturation delaying phermone in the urine of female mice. *Horm. Behav.* **11**, 135–138.

Edmonds, S., Zoloth, S. R., and Adler, N. T. (1972). Storage of copulatory stimulation in the female rat. *Physiol. Behav.* **8**, 161–164.

Elwers, M., and Critchlow, V. (1960). Precocious ovarian stimulation following hypothalamic and amygdaloid lesions in rats. *Am. J. Physiol.* **198**, 381–385.

Elwers, M., and Critchlow, V. (1961). Precocious ovarian stimulation following interruption of stria terminalis. *Am. J. Physiol.* **201**, 281–284.

Everett, J. W. (1962). Absence of pseudopregnancy in rats after ovulation-inducing stimulation of the preoptic brain. *Physiologist* **5**, 137.

Everett, J. W. (1967). Provoked ovulation or long-delayed pseudopregnancy from coital stimulation in barbiturate blocked rats. *Endocrinology (Baltimore)* **80**, 145–154.

Everett, J. W. (1968). Delayed pseudopregnancy in the rat, a tool for the study of central neural mechanisms in reproduction. *In* "Perspectives in Reproduction and Sexual Behavior" (M. Diamond ed.). Indiana Univ. Press, Bloomington.

Everett, J. W., and Quinn, D. L. (1966). Differential hypothalamic mechanisms inciting ovulation and pseudopregnancy in the rat. *Endocrinology (Baltimore)* **78**, 141–150.

Everett, J. W., and Sawyer, C. (1950). A 24-hour periodicity in the LH-release apparatus of female rats, disclosed by barbiturate sedation. *Endocrinology (Baltimore)* **47**, 198–218.

Freeman, M. E., and Banks, J. A. (1980). Hypothalamic sites which control the surges of prolactin secretion induced by cervical stimulation. *Endocrinology (Baltimore)* **106**, 668–673.

Freeman, M. E., and Neill, J. D. (1972). The pattern of prolactin secretiopn during pseudopregnancy in the rat: A daily nocturnal surge. *Endocrinology (Baltimore)* **90**, 1292–1294.

Freeman, M. E., Smith, M. S., Nazian, S. J., and Neill, J. D. (1974). Ovarian and hypothalamic control of the daily surges of prolactin secretion during pseudopregnancy in the rat. *Endocrinology (Baltimore)* **94**, 875–882.

Friedgood, H. B., and Bevin, S. (1938). Relationship of cervical sympathetics to pseudopregnancy in the rat. *Am. J. Physiol.* **123**, 71.

Frisch, R., and McArthur, J. (1974). Menstual cycle: Fatness as a determinant of minimum weight for height necessary for their maintenance of onset. *Science* **185**, 949–951.

Frisch, R., Wyshak, G., and Vincent, L. (1980). Delayed menarche and amenorrhea in ballet dancers. *N. Engl. J. Med.* **303**, 17–19.

Fullerton, C., and Cowley, J. (1971). The differential effect of the presence of male and female mice on the growth and development of young. *J. Genet. Psychol.* **119**, 89–97.

Gaunt, S. (1967). Classification and effect of the preputial pheromone in the mouse. *Am. Zool.* **7**, 713.

Gilman, D. P., Mercer, L. F., and Hitt, J. C. (1979). Influence of female copulatory behavior on the induction of pseudopregnancy in the female rat. *Physiol. Behav.* **22**, 675–678.

Godowicz, B. (1970). The influence of genotype of the male on pregnancy block in inbred mice. *J. Reprod. Fertil.* **23**, 237–241.

Graham, M. J., and Desjardin, C. (1980). Classical conditioning: induction of luteinizing hormone and testosterone secretion in anticipation of sexual activity. *Science* **210**, 1039–1041.

Gray, G. D., Davis, H. N., Zerylnick, M., and Dewsbury, D. A. (1974). Oestrus and induced ovulation in montane voles. *J. Reprod. Fertil.* **38**, 193–196.

Greep, R. O., and Hisaw, R. L. (1938). Pseudopregnancies from electrical stimulation of the cervix in the diestrum. *Proc. Soc. Exp. Biol. Med.* **39**, 359–360.

Gunnet, J. W., Mick, C., and Freeman, M. E. (1981). The role of the dorsomedial-ventromedial area of the hypothalamus in the control of prolactin secretion induced by cervical stimulation. *Endocrinology (Baltimore)* **109**, 1846–1850.

Haller, E., and Barraclough, C. (1970). Alterations in unit activity of hypothalamic ventromedial nuclei by stimuli which affect gonadotropic hormone secretion. *Exp. Neurol.* **29**, 111–120.

Hansen, S., Stanfield, E. J., and Everitt, B. J. (1980). The role of the ventral bundle noradrenergic neurons in sensory components of sexual behavior and coitus-induced pseudopregnancy. *Nature (London)* **286**, 151–154.

Hasler, M. J., and Conaway, C. H. (1973). The effects of males on the reproductive state of female *Microtus ochrogaster*. *Biol. Reprod.* **9**, 426–436.

Hasler, M. J., and Nalbandov, A. (1974). The effect of weaning and adult males on sexual maturation in female voles (*Microtus ochrogaster*). *Gen. Comp. Endocrinol.* **23**, 237–238.

Haterius, H. O. (1933). Partial sympathectomy and induction of pseudopregnancy. *Am. J. Physiol.* **103**, 97–103.

Hoppe, P., and Whitten, W. (1972). Pregnancy block: Imitation by administered gonadotropin. *Biol. Reprod.* **7**, 254–259.

Hornby, J. B., and Rose, J. D. (1976). Responses of caudal brain stem neurons to vaginal and somatosensory stimulation in the rat and evidence of genital-nociceptive interactions. *Exp. Neurol.* **51**, 363–376.

Horton, L. W., and Shepherd, B. A. (1979). Effects of olfactory bulb ablation on estrus-induction and frequency of pregnancy. *Physiol. Behav.* **22**, 847–850.

Huck, U. W. (1982). Pregnancy block in laboratory mice as a function of male social status. *J. Reprod. Fertil.* **66**, 181–184.

Ingle, D. J. (1954). Permissibility of hormone action: A review. *Acta Endocrinol. (Copenhagen)* **17**, 172–186.

Johnston, R. E., and Bronson, F. H. (1982). Endocrine control of female mouse odors that elicit luteinizing hormone surges and attraction in males. *Biol. Reprod.* **27**, 1174–1180.

Kalra, S., Ajika, K., Krulich, L., Fawcett, C., Quijada, M., and McCann, S. (1971). Effects of hypothalamic and preoptic electrochemical stimulation on gonado-tropin and prolactin release in proestrous rats. *Endocrinology (Baltimore)* **88**, 1150–1158.

Kameko, W., Debski, E., Wilson, M., and Whitten, W. (1980). Puberty acceleration in mice. 2. Evidence that the vomeronasal organ is a receptor for the primer pheromone in male mouse urine. *Biol. Reprod.* **22**, 873–878.

Kamel, F., and Frankel, A. I. (1978a). The effect of medial preoptic area lesions on sexually stimulated hormone release in the male rat. *Horm. Behav.* **10**, 10–21.

Kamel, F., and Frankel, A. I. (1978b). Hormone release during mating in the male rat: Time course, relation to sexual behavior, and interaction with handling procedures. *Endocrinology (Baltimore)* 103, 2172–2179.

Kamel, F., Mock, E., Wright, W., and Frankel, A. (1975). Alterations in plasma concentrations of testosterone, LH and prolactin associated with mating in the male rat. *Horm. Behav.* 6, 277–288.

Kamel, F., Wright, W., Mock, E., and Frankel, A. (1977). The influence of mating and related stimuli on plasma levels of leuteinizing hormone, follicle stimulating hormone, prolactin, and testosterone in the male rat. *Endocrinology (Baltimore)* 101, 421–429.

Kato, H., Velasco, M. E., and Rothchild, I. (1978). Effects of medial hypothalamic deafferentation on prolactin secretion in pseudopregnant rats. *Acta Endocrinol. (Copenhagen)* 89, 425–431.

Kawakami, M., and Arita, J. (1982). Midbrain pathways for the initiation and maintenance of the nocturnal prolactin surge in pseudopregnant rats. *Endocrinology (Baltimore)* 110, 1977–1982.

Kawakami, M., and Kubo, K. (1971). Neuro-correlate of limbic-hypothalmo-pituitary-gonadal axis in the rat: Change in limbic-hypothalmic unit activity induced by vaginal and electrical stimulation. *Neuroendocrinology* 7, 65–89.

Kennedy, J., and Brown, K. (1970). The effects of male odor in infancy on the maternal behavior and reproduction of female mice. *Dev. Psychobiol.* 3, 179–189.

Kennedy, A., Evans, R., and Dewsbury, D. (1977). Postimplantation pregnancy disruption in *Microtus ochrogaster, M. pennsylvanicus* and *Peromyscus maniculatus. J. Reprod. Fertil.* 49, 365–367.

Kevetter, G. A., and Winans, S. S. (1981). Connections of the corticomedial amygdala in the golden hamster. I. Efferents of the "vomeronasal amygdala." *J. Comp. Neurol.* 197, 81–98.

Kling, A. (1964). The effects of rhinencephalic lesions on endocrine and somatic developmant in the rat. *Am. J. Physiol.* 206, 1395–1400.

Kollar, E. J. (1953). Reproduction in the female rat after pelvic nerve neurectomy. *Anat. Rec.* 115, 641–658.

Komisaruk, B. R., and Diakow, C. (1973). Lordosis reflex intensity in rats in relation to the estrous cycle, ovariectomy, estrogen administration and mating behavior. *Endocrinology (Baltimore)* 93, 548–557.

Komisaruk, B. R., Adler, N. T., and Hutchison, J. B. (1972). Genital sensory field: Enlargement by estrogen treatment in female rats. *Science* 178, 1295–1298.

Kow, L. M., and Pfaff, D. W. (1976). Sensory requirements for the lordosis reflex in female rats. *Brain Res.* 101, 47–66.

Kow, L. M., Montgomery, M. O., and Pfaff, D. W. (1977). Effects of spinal cord transections on lordosis reflex in female rats. *Brain Res.* 123, 75–88.

Krettek, J., and Price, J. (1978). Amygdaloid projections to subcortical structures within the basal forebrain and brainstem in the rat and cat. *J. Comp. Neurol.* 178, 225–253.

Lamond, D. (1958). Infertility associated with the extirpation of the olfactory bulbs in female albino mice. *Aust. J. Exp. Biol. Med. Sci.* **36**, 103–108.

Lamond, D. (1959). The effect of stimulation derived from other animals of the same species on estrus cycles in mice. *J. Endocrinol.* **18**, 343–349.

Land, R. B., and McGill, T. E. (1967). The effect of the mating pattern of the mouse on the formation of corpora lutea. *J. Reprod. Fertil.* **13**, 121–125.

Leonard, C. M., and scott, J. W. (1971). Origin and distribution of the amygdalofugal pathways in the rat: An experimental neuroanatomical study. *J. Comp. Neurol.* **141**, 313–330.

Lincoln, D. W. (1969). Response of hypothalamic units to stimulation of the vaginal cervix: Specific vs. non-specific effects. *J. Endocrinol.* **43**, 683–684.

Linkie, D. M., and Niswender, G. D. (1972). Serum levels of prolactin, luteinizing hormone and follicle stimulating hormone during pregnancy in the rat. *Endocrinology (Baltimore)* **90**, 632–637.

Lombardi, J., and Vandenbergh, J. (1977). Pheromonally-induced sexual maturation in females. Regulation by the social environment of the male. *Science* **196**, 545–546.

Lombardi, J., Vandenbergh, J., and Whitsett, M. (1976). Androgen control of the sexual maturation and pheromone in house mouse urine. *Biol. Reprod.* **15**, 179–186.

Long, J. A., and Evans, H. Mc. (1922). The oestrous cycle in the rat and its associated phenomena. *Mem. Univ. Calif.* **6**, 1–148.

Macrides, F., Bartke, A., and Dalterio, S. (1972). Strange females increase plasma testosterone levels in male mice. *Science* **189**, 1104–1106.

Madhwa Raj, H. G., and Moudgal, N. R. (1970). Hormonal control of gestation in the intact rat. *Endocrinology (Baltimore)* **86**, 874–889.

Malsbury, C. W., Kelley, D. B., and Pfaff, D. W. (1973). Responses of single units in the dorsal midbrain to somatosensory stimulation in female rats. *Int. Congr. Ser.—Excerpta Med.* **273**, 205–209.

Margharita, G., Albritton, D., MacInnes, R., Hayward, J., and Gorski, R. A. (1965). Electroencephalographic changes in ventromedial hypothalmus and amygdala induced by vaginal and other peripheral stimuli. *Exp. Neurol.* **13**, 96–108.

Marsden, H., and Bronson, F. (1964). Estrous synchrony in mice: Alteration by exposure to male urine. *Science* **144**, 3625.

Marsden, H., and Bronson, F. (1965). Strange male block to pregnancy: Its absence in inbred mouse strains. *Nature (London)* **207**, 878.

Maruniak, J., and Bronson, F. (1976). Gonadotropic responses of male mice to female urine. *Endocrinology (Baltimore)* **99**, 963–969.

McClella, R., and Cowley, J. (1972). The effects of lesions of olfactory bulbs on the growth and behavior of mice. *Physiol. Behav.* **9**, 319–324.

McClintock, M. K. (1978). Estrous synchrony and its mediation by airborne chemical communication. *Horm. Behav.* **10**, 264–276.

McClintock, M. K., and Adler, N. (1978). The role of the female during copulation in the wild and domestic Norway rat. *Behaviour* **67,** 67–96.

McGill, T. E. (1970). Induction of luteal activity in female house mice. *Horm. Behav.* **11,** 211–222.

McGill, T. E., and Coughlin, R. C. (1970). Ejaculatory reflex and luteal activity induction in mus musculus. *J. Reprod. Fertil.* **21,** 215–220.

McGill, T. E., Dewsbury, D., and Sachs, B., eds. (1978). "Sex and Behavior: Status and Prospectus." Plenum, New York.

McIntosh, T., and Drickamer, L. (1977). Excreted urine, bladder urine and the delay of sexual maturation in female house mice. *Anim. Behav.* **25,** 999–1004.

McLean, B. K., and Nikitovitch-Winer, M. B. (1973). Corpus luteum function in the rat: A critical period for luteal activation and the control of luteal maintenance. *Endocrinology (Baltimore)* **93,** 316–322.

Meyer, R. K., Leonard, S. L., and Hisaw, F. L. (1929). Effect of anesthesia on artificial production of pseudopregnancy in the rat. *Proc. Soc. Exp. Biol. Med.* **27,** 340–342.

Mody, J. (1963). Structural changes in the ovaries of IF mice due to age and various other states: Demonstrations of spontaneous peudopregnancy in grouped virgins. *Anat. Rec.* **145,** 439–447.

Morishige, W. K., and Rothchild, I. (1974). A paradoxical inhibiting effect of ether on prolactin release in the rat: Comparison with effect of ether on LH and FSH. *Neuroendocrinology* **16,** 93–107.

Moss, R., and Cooper, K. (1973). Temporal relationship of spontaneous and coitus-induced release of luteinizing hormone in the normal cyclic rat. *Endocrinology (Baltimore)* **92,** 1748–1753.

Moss, R., Cooper, K., and Danhof, I. (1973). Coitus-induced release of LH in normal cyclic female rats. *Fed. Proc., Fed. Am. Soc. Exp. Biol.* **32,** 239.

Moss, R., Dudley, C., and Schwartz, N. (1977). Coitus-induced release of luteinizing hormone in the proestrous rat: Fantasy or fact? *Endocrinology (Baltimore)* **100,** 394–397.

Ohrbach, J., and Kling, A. (1966). The effects of sensory deprivation on the onset on puberty, mating, fertility and gonadal weight in rats. *Brain Res.* **3,** 141–149.

Parkes, A., and Bruce, H. (1961). Olfactory stimuli in mammalian reproduction. *Science* **134,** 1049–1054.

Parkes, A., and Bruce, H. (1962). Pregnancy block in female mice placed in boxes soiled by males. *J. Reprod. Fertil.* **4,** 303–308.

Pfaff, D., Lewis, C., Diakow, C., and Keiner, M. (1973). Neurophysiological analyses of mating behavior responses as hormone sensitive reflexes. *Prog. Physiol. Psychol.* **5.**

Pfaff, D., Montgomery, M., and Lewis, C. (1977). Somatosensory determinants of lordosis in female rats: Behavioral definition of the estrogen effect. *J. Comp. Physiol. Psychol.* **91,** 134–145.

Pieper, D. R., and Gala, R. R. (1979). The effect of light on the prolactin surges of pseudopregnant and ovariectomized estrogenized rats. *Biol. Reprod.* 20, 727–732.

Pierce, J. T., and Nuttal, R. L. (1961). Self-paced sexual behavior in the female rat. *J. Comp. Physiol. Psychol.* 54, 310–313.

Purvis, K., and Haynes, N. (1974). Short term effects of copulation, human chorionic gonadotropin injection and nontactile association with a female on testosterone levels in the male rat. *J. Endocrinol.* 60, 429–439.

Quadagno, D., Albelda, S., McGill, T., and Kaplan, L. (1976). Intracranial cycloheximide: Effect on male mouse sexual behavior and plasma testosterone. *Pharmacol., Biochem. Behav.* 4, 185–189.

Quadagno, D., McGill, T., Yellon, S., and Goldman, B. (1979). Neither non-contact exposure nor mating affect serum LH and FSH in male B6D2F1 house mice. *Physiol. Behav.* 22, 191–192.

Quinn, D. L., and Everett, J. W. (1967). Delayed pseudopregnancy induced by selective hypothalamic stimulation. *Endocrinology (Baltimore)* 80, 155–162.

Raisman, G. (1972). An experimental study of the projection of the amygdala to the accessory olfactory bulb and its relationship to the concept of a dual olfactory system. *Exp. Brain Res.* 14, 395–408.

Ramaley, J. (1979). Development of gonadotropin regulation in the prepubertal mammal. *Biol. Reprod.* 20, 1–34.

Ramirez, V. D., Komisaruk, B. R., Whitmoyer, D. I., and Sawyer, C. H. (1967). Effects of hormones and vaginal stimulation on the EEG and hypothalamic units in rats. *Am. J. Physiol.* 212, 1376–1384.

Reiner, P., Woolsey, J., Adler, N., and Morrison, A. (1981). Appendix: A gross anatomical study of the peripheral nerves associated with reproductive function in the female albino rat. *In* "Neuroendocrinology of Reproduction" (N. T. Adler, ed.). Plenum, New York.

Richmond, M., and Conaway, C. H. (1969). Induced ovulation and oestrus in *Microtus ochrogaster. J. Reprod. Fertil.* 6, 357–376.

Richmond, M., and Stehn, R. (1976). Olfaction and reproductive behavior in microtine rodents. *In* "Mammalian Olfaction, Reproductive Processes, and Behavior" (R. L. Doty, ed.). Academic Press, New York.

Rodgers, C. (1971). Influence of copulation on ovulation in the cycling rat. *Endocrinology (Baltimore)* 88, 433–436.

Ryan, E., and Frankel, A. (1978). Studies on the role of medial preoptic area in sexual behavior and hormonal response to sexual behavior in the mature male laboratory rat. *Biol. Reprod.* 19, 971–983.

Ryan, K., and Schwartz, N. (1980). Changes in serum hormone levels associated with male-induced ovulation in group-housed adult female mice. *Endocrinology (Baltimore)* 106, 959–966.

Sachs, B. D. (1982). Role of striated penile muscles in penile reflexes, copulation and induction of pregnancy in the rat. *J. Reprod. Fertil.* **66**, 433–443.

Scalia, F., and Winans, S. (1975). The differential projections of the olfactory bulb and accessory olfactory bulb in mammals. *J. Comp. Neurol.* **161**, 31–56.

Schwagmeyer, P. (1979). The Bruce effect: An evaluation of male/female advantages. *Am. Nat.* **114**, 932–938.

Scott, J., and Pfaff, D. (1970). Behavioral and electrophysiological responses of female mice to male urine odors. *Physiol. Behav.* **5**, 407–411.

Sharp., F., Kilduff, T., Bzorgchami, S., Heller, H., and Ryan, A. (1983). The relationship of local cerebral glucose utilization to optical density ratios. *Brain. Res.* **263**, 97–103.

Shelesnyak, M. C. (1931). The induction of pseudopregnancy in the rat by means of electrical stimulation. *Anat. Res.* **49**, 179–183.

Slonaker, J. R. (1929). Pseudopregnancy in the albino rat. *Am. J. Physiol.* **89**, 406–416.

Smith, M. S., and Neill, J. D. (1976a). A critical period for cervically-stimulated prolactin release. *Endocrinology (Baltimore)* **98**, 324–328.

Smith, M. S., and Neill, J. D. (1976b). Termination at midpregnancy of the two daily surges of plasma prolactin initiated by mating in the rat. *Endocrinology (Baltimore)* **98**, 696–701.

Smith, M. S., and Ramaley, J. A. (1978). Development of ability to initiate and maintain prolactin surges induced by uterine cervical stimulation in immature rats. *Endocrinology (Baltimore)* **102**, 351–357.

Smith, M. S., Freeman, M. E., and Neill, J. D. (1975). The control of progesterone secretion during the estrous cycle and early pseudopregnancy in the rat: Prolactin, gonadotropin, and steroid levels associated with rescue of the corpus luteum of pseudopregnancy. *Endocrinology (Baltimore)* **96**, 219–226.

Smith, M. S., McLean, B. K., and Neill, J. D. (1976). Prolactin: The initial luteotrophic stimulus of pseudopregnancy in the rat. *Endocrinology (Baltimore)* **98**, 1370–1377.

Sokoloff, L., Reivich, M., Kennedy, C., Des Rosiers, M. H., Patlak, C. S., Pettigrew, K. D., Sakurada, O., and Shinohara, M. (1977). The [14C] deoxyglucose method for the measurement of local cerebral glucose utilization: Theory, procedure, and normal values in the conscious and anesthetized albino rat. *J. Neurochem.* **28**, 897–916.

Spies, H. G., and Niswender, G. (1971). Levels of prolactin, LH and FSH in the serum of intact and pelvic-neurectomized rats. *Endocrinology (Baltimore)* **88**, 937–943.

Spies, H. G., Forbes, Y. M., and Clegg, M. T. (1971). The influence of coitus, suckling, and prolactin injections on pregnancy in pelvic neurectomized rats. *Proc. Soc. Exp. Biol. Med.* **138**, 470–474.

Stehn, R., and Richmond, M. (1975). Male-induced pregnancy termination in the prairie vole, *Microtus ochrogaster. Science* **187**, 1211–1213.

Stiff, M., Bronson, F., and Stetson, M. (1974). Plasma gonadotropins in prenatal and prepubertal female mice: Disorganization of pubertal cycles in absence of a male. *Endocrinology (Baltimore)* **94**, 492–496.

Taleisnik, S., and Tomatis, M. (1970). Mechanisms that determine the changes in pituitary MSH activity during pseudopregnancy induced by vaginal stimulation in the rat. *Neuroendocrinology* **6**, 368–377.

Taleisnik, S., Caligaris, L., and Astrada, J. (1966). Effect of copulation on the release of pituitary gonadotropins in male and female rat. *Neuroendocrinology* **79**, 49–54.

Taylor, G., Regan, D., and Haller, J. (1983). Sexual experience, androgens and female choice of a mate in laboratory rats. *J. Endocrinol.* **96**, 43–52.

Terasawa, E., and Sawyer, C. (1969). Changes in electrical activity in the rat hypothalamus related to electrochemical stimulation of adenohypophyseal function. *Endocrinology (Baltimore)* **85**, 143–149.

Terkel, J., and Sawyer, C. H. (1978). Male copulatory behavior triggers nightly prolactin surges resulting in successful pregnancy in rats. *Horm. Behav.* **11**, 304–309.

Vandenbergh, J. (1967). The effect of the presence of a male on the sexual maturation of female male. *Endocrinology (Baltimore)* **81**, 345–349.

Vandenbergh, J. (1969). Male odor accelerates female sexual maturation in mice. *Endocrinology (Baltimore)* **84**, 658–660.

Vandenbergh, J. (1973). The effects of central and peripheral anosmia on reproduction in female mice. *Physiol. Behav.* **10**, 257–261.

Vandenbergh, J., Drickamer, L., and Colby, D. (1972). Social and dietary factors in the sexual maturation of female mice. *J. Reprod. Fertil.* **28**, 397–405.

Velasco, M. E. (1972). Opposite effects of platinum and stainless steel lesions of the amygdala on gonadotropin secretion. *Neuroendocrinology* **10**, 301–308.

Velasco, M. E., Castro-Vazquez, A., and Rothchild, I. (1974). Effects of hypothalamic deafferentation on criteria of prolactin secretion during pregnancy and lactation in the rat. *J. Reprod. Fertil.* **41**, 385–395.

Whitten, W. (1956a). Modification of the estrus cycle of the mouse by external stimuli associated with the male. *J. Endocrinol.* **13**, 399–404.

Whitten, W. (1956b). The effect of removal of olfactory bulbs on the gonads of mice. *J. Endocrinol.* **14**, 160–163.

Whitten, W. (1957). The effect of exteroceptive factors on estrus cycle of mice. *Nature (London)* **180**, 1436.

Whitten, W. (1958). Modification of the estrus cycle of the mouse by external stimuli associated with the male. Changes in estrus cycle determined by vaginal smears. *J. Endocrinol.* **17**, 307–313.

Whitten, W. (1959). Occurrence of anestrus in mice caged in groups. *J. Endocrinol.* **18**, 102–107.

Whitten, W., Bronson, F., and Greenstein, J. (1968). Estrus-inducing pheromone of male mice: Transport by movement of air. *Science* **161**, 584–585.

Wilson, J. R., Adler, N., and Leboeuf, B. (1965). The effects of intromission frequency on successful pregnancy in the female rat. *Proc. Natl. Acad. Sci. U.S.A.* **53,** 1392–1395.

Winans, S., and Scalia, F. (1970). Amygdaloid nucleus: New afferent input from the vomeronasal organ. *Science* **170,** 330–332.

Wüttke, W. (1974). Preoptic unit activity and gonadotropin release. *Exp. Brain Res.* **19,** 205–216.

Wysocki, C., Katz, Y., and Bernhard, R. (1983). Male vomeronasal organ mediates female-induced testosterone surges in mice. *Biol. Reprod.* **28,** 917–922.

Yogev, L., and Terkel, J. (1980). Effects of photoperiod, absence of photicues, and suprachiasmatic nucleus lesions on nocturnal prolactin surges in pregnant and pseudopregnant rats. *Neuroendocrinology* **31,** 26–33.

Young, W. C., ed. (1961). "Sex and Internal Secretions," 3rd ed., Vol. 2. Williams & Wilkins, Baltimore, Maryland.

Zeilmaker, G. H. (1965). Normal and delayed pseudopregnancy in the rat. *Acta Endocrinol. (Copenhagen)* **49,** 558–566.

CHAPTER FIVE

Hormonal Regulation of Learning Performance

Tj. B. van Wimersma Greidanus and H. Rigter

I. Introduction

A. Defining Learning and Memory

Behavior of an organism is regulated by complex influences exerted by earlier experiences on present response tendencies, i.e., by learning. Unfortunately, the concept of learning is ill defined. Usually memory is considered to be the end product of learning. However, the concepts of learning and memory are not strictly separable. Learning is synonymous, in our view, with the formation of memory. Learning requires the operation of supportive functions such as attention and motivation. On the other hand the concept of "memory" is not limited to formation of memory but includes other aspects too (e.g., retrieval and forgetting). In view of this overlap in meaning between the concepts of learning and memory, it is clear that in studies of the hormonal regulation of learning, memory cannot be disregarded and vice versa.

Neuroendocrine studies have shown that hormones affect brain processes mediating behavior, including learning and memory. These findings have helped to establish the discipline of behavioral endocrinology. Extensive evidence that learning and memory are influenced by hormones has been provided by research of the last 25 years and this research is still expanding.

TABLE I Hormonal Effects on Learning Performance[a]

Peptide	Behavioral effect
ACTH, ACTH (1-10), ACTH (4-10), α-MSH	Normalization of deficient acquisition of an active avoidance response (hypox rats)
	Normalization of impaired extinction of an active avoidance response (hypox rats)
	Enhancement of acquisition of active avoidance behavior
	Delay of extinction of active avoidance behavior
	Facilitation of retrieval of inhibitory avoidance responses
	Delay of extinction of conditioned taste aversion
	Delay of extinction of positively rewarded runway responses
[D-Phe⁷]-ACTH (1-10), [D-Phe⁷]-ACTH (4-10)	Impairment of acquisition of active avoidance behavior
	Facilitation of extinction of active avoidance behavior
α-Endorphin	Delay of extinction of active avoidance behavior
	Increased inhibitory avoidance behavior
γ-Endorphin	Facilitation of extinction of active avoidance behavior
	Impairment of inhibitory avoidance behavior
Met-enkephalin, Leu-enkephalin	Impairment of acquisition of active avoidance behavior
	Reduction of experimentally induced amnesia
Prolactin (hyperprolactinemia)	Improvement of acquisition of active avoidance behavior
Vasopressin	Normalization of deficient acquisition of active avoidance behavior; long-term effect (hypox rats)
	Normalization of impaired extinction of active avoidance behavior; long-term effect (hypox rats)
	Long-term delay of extinction of active avoidance behavior
	Long-term enhancement of inhibitory avoidance behavior
	Delay of extinction of an appetitive discrimination task
	Facilitation of retrieval of an inhibitory avoidance response
	Facilitation of retrieval of positively reinforced maze responses
	Prevention and reversal of experimentally induced amnesia
	Delay of postcastration decline in copulatory behavior
	Enhancement of development of tolerance to the analgesic effect of morphine
Oxytocin	Facilitation of extinction of active avoidance behavior
	Attenuation of inhibitory avoidance behavior

TABLE I *(Continued)*

Peptide	Behavioral effect
CCK	Impairment of acquisition of active avoidance behavior
	Faciliation of extinction of active avoidance behavior
	Impairment of acquisition of food-motivated conditioned behavior
	Facilitation of extinction of food-motivated conditioned behavior
	Facilitation of inhibitory avoidance behavior (s.c., or i.c.v. administration)
	Attentuation of experimentally induced retrograde amnesia
Neurotensin	Delay of extinction of an active avoidance response
	Enhancement of inhibitory avoidance behavior
Substance P	Facilitation or attenuation of inhibitory avoidance behavior (depending on brain site of injection)
VIP	Facilitation of extinction of an active avoidance response
	Attentuation of retrieval of an inhibitory avoidance response
Glucocorticosteroids	Facilitation of extinction of active avoidance behavior (intact and hypox rats)
Progesterone	Facilitation of extinction of active avoidance behavior

[a]Only the established effects as discussed and/or mentioned in this chapter on intact rats and/or mice are included, unless otherwise indicated.

Although details concerning the sources and modes of actions of the hormones are not completely understood, the evidence that the brain may serve as target organ for hormones has been widely accepted. Given the tendency to christen hormones according to their known effects it seems appropriate to call some hormones "learning modulatory hormones" or LMHs (McGaugh and Martinez, 1981).

It is likely that learning involves a complex interaction between different neuronal systems and processes. Thus, modulation of any of these systems or processes may finally result in an effect on learning. Hormones may influence learning through one or more effects on arousal, vigilance, attention, motivation, and consolidation (memory formation). Thus, many different actions of each of a variety of hormones may affect learning in some way or another (Table I).

In the following section some models to assess hormonal effects on learning and memory in animals are discussed.

B. Animal Models of Learning and Memory

When exploring pharmacological, including hormonal, influences on learning and memory, the time of administration of the treatment is crucial for the interpretation of the results. In older studies the treatment usually was given before training the animal on some learning task. However, as McGaugh (1961) pointed out, such a schedule of treatment cannot distinguish between effects on memory formation or storage, on the one hand, and on supportive functions which are essential for but not identical with learning (e.g., perception, attention, motivation), on the other. McGaugh introduced a methodological improvement by administering treatments after training but still long before testing for remembering (retrieval) of the learned response. A pharmacological treatment given after training cannot logically affect such processes as attention, arousal, and motivation, and must be assumed to act on some aspects of memory formation.

A key assumption in most theories of learning and memory is that a burgeoning memory "trace" is labile. The lability of the trace is thought to decline with time after the learning experience. This phenomenon is used to dissociate effects on memory formation from effects on other aspects of behavior. Duncan (1949) did the first relevant study. He placed rats in a box and trained them to run to an adjoining chamber to avoid electric shock to the paws. In such a one-way avoidance task a rat's memory of the shock experience is measured by the speed (latency) of its run into the safe chamber on later trials. Duncan administered electroconvulsive shock, an agent to induce memory loss (amnesia), at different time intervals after the daily training trial. If the electroconvulsive shock followed training within 1 hour, a graded loss of the avoidance response was seen, which was presumed to represent amnesia for the response. When the treatment was administered more than 1 hour after training, it produced no amnesia and, therefore, probably no longer interfered with memory formation. The underlying assumption is that the memory trace gradually loses its susceptibility to interference and becomes "fixed." Parallel changes in the posttraining effectiveness of treatments, such as electroconvulsive shock or hormones, are believed to reflect direct effects on memory formation.

Effects of hormones on retrieval of a learned response can be measured by administering the agent shortly before the test trial. An effect on memory retrieval does not exclude an effect on learning. It may be that hormones improve retrieval of a learned response by increasing responsiveness to relevant stimuli. Such an effect would also be helpful at the time of learning.

Thus, not surprisingly some hormones are similarly effective in influencing learning and memory when administered before learning (acquisition), immediately after learning, or shortly before the memory test (Rigter *et al.*, 1974; Rigter, 1982). Although it may be that such hormones have multiple behavioral effects, it is more parsimonious to presume that a single behavioral state is involved in all of the above-mentioned phases.

We would like to consider at this point a few paradigms of learning and memory in animals. In rodents active avoidance tasks are frequently used. One such task is the pole-jumping test. A rat is placed in a box on a grid floor, in the center of which a pole is situated. When a light is turned on, the animal has to jump onto the pole in order to avoid electric shock. Accomplishment of the task is reflected in the ability of the animal to avoid the shock completely by jumping onto the pole during the 5- or 10-second interval between light and the onset of shock. Clear evidence of learning is seen when the rat reacts as soon as the light appears, but refrains from responding at all other times. When the rat has met a certain learning (acquisition) criterion, extinction of the response may be studied. In this phase of the experiment the light is no longer followed by the shock. Nevertheless, the animal continues to jump onto the pole when the light is turned on. Over several trials the number of jumps will gradually decrease until extinction ("unlearning") is complete.

A second task which is often used is the shuttle box. The animal is placed in one compartment of a box with a grid floor, separated from another identical compartment by a small hurdle. Once again a light or other stimulus is used as a warning signal. When the signal is presented the animal must shuttle to the other compartment in order to avoid shock.

Another task, passive or inhibitory avoidance, emphasizes withholding a response rather than performing a response. One example is the step-through test. A rat placed beneath a light on a small ledge hanging over empty space finds the situation aversive. At the end of the platform is a dark box with a hole. Within a few seconds a rat, innately preferring the dark, will run into the hole. The rat is, however, given a shock upon entering the box. After this training experience (usually after 24 hours) the animal's retrieval of the experience may be tested by placing it again on the ledge and measuring the time it takes to step through the hole. Memory is reflected in an increased latency to enter the box.

The tasks discussed so far involve avoidance, or aversively motivated, learning. To assess the generality of hormonal effects on learning, one must also consider so-called appetitively motivated responses. Examples are hungry animals learning to find a route to a source of food or sexually deprived animals learning to find access to a partner.

II. Hormones of Pituitary Origin

A. Proopiomelanocortin Fragments and Enkephalins

1. INTRODUCTION

Early this century a pituitary hormone was discovered which induced darkening of frog skin and changes in skin color in several fish species. This hormone, originally called "intermedin" due to its presence in the intermediate lobe of the pituitary, is presently named melancyte-stimulating hormone, MSH. Three types of MSH have been identified, i.e., α-MSH, β-MSH, and γ-MSH. α- and β-MSH structurally resemble corticotropin (ACTH). The three hormones have a common sequence of seven amino acids. γ-MSH appears to have a slightly different amino-acid composition. Although the presence of MSH in the brain had already been established in the 1930s, little attention was paid to its possible central effects until the mid-1950s, when a number of central actions of MSH, ACTH, and their fragments were found. We now know that a family of peptides, including MSH and ACTH, is produced not only by the pituitary but also by the nervous system. It has become apparent that MSH and ACTH and related peptides are derived from a large precursor molecule, generally designated as proopiomelanocortin. Successive cleavage of proopiomelanocortin results in the production of α-, β-, and γ-MSH; α-, β-, and γ-endorphin; ACTH; and corticotropinlike intermediate lobe peptide (CLIP) (see Chapter 8, this volume). All these peptides possess behavioral effects, and in the following paragraphs their influences on learning will be discussed (for review, see de Wied and Jolles, 1982).

2. ACTH, MSH, AND RELATED PEPTIDES

a. Active Avoidance   Hypophysectomy (removal of the pituitary gland) has been found to impair the learning performance of rats in the shuttle-box test (Applezweig and Baudry, 1955; de Wied, 1969). This impairment may be ascribed to one or more hormonal deficits. Treatment with adrenal-maintenance doses of the pituitary hormone ACTH restored acquisition to a considerable extent. α- and β-MSH were also effective (de Wied, 1969). The effect of ACTH was independent of its action on the adrenal glands, because administration of the synthetic glucocorticosteroid dexamethasone failed to reverse the acquisition deficit in hypophysectomized rats (de Wied, 1976). Also, the fragments ACTH (4-10) and ACTH (1-10), neither of which exerts significant adrenocorticotrophic activity, both normalized acquisition of the avoidance response (de Wied, 1969).

Hypophysectomy is known to enhance responsiveness of rats to electric shock. However, it is unlikely that this effect of removal of the pituitary gland accounts for the impairment in acquisition of the shuttle-box response. Importantly, peptides such as ACTH (1-10) do not return pain responsiveness to normal (Gispen et al., 1970, 1973). Therefore, the efficacy of these peptides in improving deficient acquisition of the shuttle-box response in hypophysectomized rats does not appear to be due to an effect on pain responsiveness.

Alternatively, ACTH-related peptides may enhance the motility of hypophysectomized rats, thus increasing the chance that the animal may "shuttle" over the hurdle of the avoidance box. However, measures of ambulation have generally failed to detect significant effects of ACTH-like peptides in hypophysectomized (de Wied, 1968) or normal rats (Weijnen and Slangen, 1970). Therefore, motility cannot account for the efficacy of these peptides in normalizing deficient acquisition of shuttle-box avoidance responding.

We will now address the question of whether ACTH-like peptides facilitate learning. At the outset, it should be made clear that pituitary ACTH has no indispensable role in learning. It has been found for instance that a hormone-replacement therapy consisting of cortisone, testosterone, and thyroxine, but not ACTH, is also capable of restoring deficient acquisition of the shuttle-box response in hypophysectomized rats (de Wied, 1969). The possibility remains, however, that ACTH-like substances *modulate* learning performance. A study by Bohus et al. (1973) renders it likely that these peptides do not affect learning *per se* but rather some supportive function. Administration of ACTH (4-10) to hypophysectomized rats during the first week of training on the shuttle-box task normalized acquisition. At the end of this period of time these rats had learned the response well enough to be able to avoid the shock 8 out of 10 times. In the second week of the study peptide treatment was stopped and performance declined. If the sole effect of the peptide had been an improvement of learning, it is difficult to see why cessation of treatment after learning had occurred would result in a decline in performance. One possible explanation for the effect of ACTH (4-10) is that it transiently improved attention or motivation thus facilitating the expression of the learned response. In support of this view is the finding that hypophysectomized rats responded more slowly to the warning signal than intact rats; using a long signal–shock interval, Bélanger (1958) failed to find impaired learning performance.

As mentioned above, under selected experimental conditions hypophysectomized rats may exhibit normal acquisition of an active avoidance response. Nevertheless, extinction of the response may still be impaired, i.e., proceed at a more rapid rate than in intact animals. Administration of ACTH normalized extinction (de Wied, 1969). This effect may also be interpreted as reflecting a normalization of deficient attention or motivation. However, it should be realized that extinction is a complex behavioral phenomenon. Early on during testing for extinction, continued responding undoubtedly is a measure of the strength of retrieval of the originally learned avoidance reaction. Until the animal fails to respond for the first time and notes that such failure is not punished, its situation is the same as during the later stages of acquisition, when it rarely received shock. However, as responding begins to extinguish and the rat begins to discover that the warning signal is no longer followed by shock, the behavioral measure reflects at least two processes, i.e., retrieval of the original response and acquisition of a new one (ignoring the signal) (Rigter and Crabbe, 1979).

ACTH may also enhance acquisition of an active avoidance response in intact rats, although not surprisingly such an effect is more difficult to establish than after removal of the pituitary gland. Effects have been seen when acquisition was suboptimal, e.g., when low shock levels were used (Mirsky *et al.*, 1953; Bohus and Endröczi, 1965; Ley and Corson, 1971). α-MSH (Stratton and Kastin, 1974) and ACTH (4-10) (Bohus and de Wied, 1981) were similarly effective in facilitating acquisition. However, structurally related peptides have been found to impair acquisition of an active avoidance response. Two examples are [D-Phe^7]ACTH (4-10) and γ_2-MSH (Bohus and de Wied, 1981; Bohus *et al.*, 1983).

In intact rats extinction of an active avoidance response is more sensitive to treatment with peptides than acquisition. ACTH-like peptides have been found to delay extinction (de Wied, 1969; Bohus *et al.*, 1968). [D-Phe^7]ACTH (1-10) facilitated extinction (Bohus and de Wied, 1966). When treatment with the D-peptide was stopped, extinction performance returned to control levels (de Wied, 1969). This latter finding suggests that the peptide did not facilitate learning to ignore the warning signal but rather weakened the attention or motivation needed for continued responding to the signal.

b. Inhibitory Avoidance Hypophysectomy not only impairs acquisition of active avoidance but also of inhibitory avoidance responses (Anderson *et al.*, 1968; Lissák and Bohus, 1972). Thus, the nature of the response seems to be irrelevant. In one study administration of ACTH *after* completion of the learning trial restored later retention of the inhibitory avoidance response (Gold and McGaugh, 1977). Such posttrial effects generally are taken to reflect a direct action on memory formation. The generality of this finding will be discussed below.

In studies with intact rats peptides generally have been administered shortly before the test, 24 hours after the learning experience. In experiments designed to assess a facilitatory effect of peptides on retrieval of the inhibitory avoidance response, a relatively mild shock of short duration has been used. Control animals hardly avoided this shock at all. Injection of peptides such as ACTH (1-10) and (4-10) raised entrance latencies (Greven and de Wied, 1973). Possibly, the peptide-treated rats were more attentive to some remembered feature of the task or were quicker to recognize the motivational value of these features. The possibility that these peptide-treated rats were less active is not likely since ACTH (4-10) did not affect entrance latencies in non-shocked animals (Rigter *et al.*, 1974).

A physiological role for cerebral ACTH-like peptides in mediating adaptive behavior is suggested by studies using antisera (van Wimersma Greidanus *et al.*, 1978). Intracerebroventricular administration of antiserum against ACTH (1-24) interfered with retrieval of an inhibitory avoidance response. A similar effect was observed for antiserum against α-MSH.

Effects on memory formation may be examined by administering a hormone posttrial, i.e., after completion of the single learning trial in inhibitory avoidance experiments. Gold and Van Buskirk (1976) injected rats with 0.03, 0.3, or 3.0 IU of ACTH immediately after the learning trial. The lower doses enhanced and the higher dose impaired performance at the later retrieval test. Furthermore, when administration of the hormone was delayed for 2 hours after the learning trial, its effectiveness in modulating memory formation was lost. Finally, the nature of the effect of ACTH depended on the intensity of the shock delivered during training. The dose of 3.0 IU, which was disruptive in the first study, both enhanced and impaired performance in a second experiment depending on the level of shock. These data have important implications. First, the findings suggest that the effect of a particular dose of the hormone is related to the intensity of the learning experience. Possibly, the nature of the effect of treatment with ACTH depends on the amount of endogenous hormone mobilized at the time of the learning trial. Second, there is an inverted U-shaped dose–response curve relating the dose of ACTH to test performance. Conceivably, the memory-related process upon which the hormone may act has an optimal level of functioning; too much ACTH may interfere with rather than facilitate the hypothetical process. Third, the effectiveness of ACTH decreased as the time between learning and treatment was increased. As discussed in Section I,B such time dependency is strong evidence for an effect on memory formation rather than on some supportive function.

Gold and McGaugh (1977) reported that posttrial injection of corticosterone was ineffective in their inhibitory avoidance task. This finding indicates that ACTH did not exert its effect through its corticotropic properties. This is also suggested by data showing that posttrial injection with the fragment ACTH (4-10), which has virtually no effect on the adrenal glands, similarly improved performance of mice at the retrieval test of an inhibitory avoidance task (Flood *et al.*, 1976). In contrast to the experiments discussed so far, other studies have generally failed to find a posttrial effect of injection with ACTH-like peptides. For instance, van Wimersma Greidanus (1977) reported that administration of ACTH (4-10) immediately after training did not influence performance at the retrieval test.

Also Martinez *et al.* (1979) did not observe a posttrial effect of the ACTH (4-9) analog, Org 2766. Rather, these investigators found this peptide to be effective only when injected in a high dose 1 hour before training. They suggested that Org 2766 acts through sensory, motivational, or attentional variables rather than directly by affecting memory formation.

Data described by Fekete and de Wied (1982), although of a different nature, point in the same direction. Administration of Org 2766 prior to the retrieval test improved performance in much lower doses than for any other ACTH-like peptide mentioned in this review. The time span of effectiveness was also much longer than usual. However, by using long time intervals between training and testing the authors showed that the peptide did not affect memory formation but rather proactively influenced test performance. Finally, it should be noted that, according to most studies, ACTH-like peptides are not able to prevent the experimental induction of amnesia in animals (Rigter *et al.*, 1974; van Wimersma Greidanus, 1982a), as would be expected for treatments which boost memory formation.

In conclusion, although many effects of ACTH-like peptides can be explained by a modulation of some supportive function associated with learning and memory retrieval, it may be necessary to postulate that some of these substance occasionally exert an additional effect on memory formation.

c. Nonshock Tasks   A distinction should be made between appetitively motivated responses and tasks in which aversive stimuli other than shock are used. We will discuss the two sorts of tasks, beginning with those involving aversive stimuli. When drinking of a novel palatable solution, such as sugar water, is followed by an aversive treatment, such as injection with lithium chloride, animals will avoid drinking the solution on subsequent occasions. This phenomenon is called conditioned taste aversion. The avoidance response is subject to extinction when the animal is repeatedly tested under nonpunished conditions. Several studies have established that ACTH-like peptides retard extinction of the conditioned taste aversion (Rigter and Popping, 1976; Hennessy *et al.*, 1980; Kendler *et al.*, 1976).

Leshner and Roche (1977) used another variation of the inhibitory avoidance task. They trained mice on a task in which entry into a compartment was punished by attack of an aggressive mouse of the same strain. ACTH injected immediately after the learning criterion was achieved enhanced performance on the first two retrieval tests. An effect of the hormone was no longer seen at the third test, which was conducted 240 hours after training. It was concluded that ACTH has a short-lived effect on retention of the avoidance response.

Garrud *et al.* (1974) examined the action of ACTH on extinction of a runway response when food was used as the reward. The hormone delayed extinction. This finding has been confirmed by Kastin *et al.* (1974). Bohus *et al.* (1973) trained male rats to run to a sexually receptive female. One group of rats was allowed to copulate; another was prevented from entering the compartment containing the female rat. During extinction the female rat was removed from the compartment. Treatment with ACTH (4-10) delayed extinction of the runway response, but only in those rats allowed to copulate during training. Thus, the effects of ACTH-like peptides in nonshock tasks are quite consistent with the effects seen in shock-motivated tasks.

d. Effects in Man The results of studies in man will be reviewed briefly here (for a more detailed account, see Pigache and Rigter, 1981). Most studies have used the peptides Org 2766 and ACTH (4-10). These substances have been examined in connection with a wide range of "cognitive performance" tasks and in terms of psychophysiology, EEG, and rating scales. Studies using acute subcutaneous or oral administration of Org 2766 or of ACTH (4-10) have failed to detect consistent effects of these peptides on performance of subjects on tests of learning and memory and other tests of cognitive performance, with the exception of tests requiring sustained vigilance. Both peptides facilitated performance under conditions where people start to make errors due to a breakdown of vigilance. A similar story holds for most of the physiological indices. *Subchronic* treatment with Org 2766 may have a positive effect on mood in man, as evidenced by a reduction in self-rated anxiety and depression, an increase in self-rated competence and observer-rated sociability, and an improvement in ward behavior.

3. ENDORPHINS

Data obtained by Kovács and de Wied (1981) indicate that β-endorphin exerts a dose-dependent influence on retrieval of an inhibitory avoidance response. The neuropeptide improved retrieval of the avoidance response when administered either immediately after the learning trial or shortly before the test trial. However, different findings have been reported by others (Martinez and Rigter, 1980; Gorelick et al., 1978). β-Endorphin injected immediately after training in an active or inhibitory avoidance task produced a performance deficit at later tests. Possibly, the direction and strength of endorphin effects on behavior depend on the experimental parameters and doses of the peptide used. It should be noted in this respect that α-endorphin and γ-endorphin, endorphins which may be generated by biotransformation of β-endorphin in the brain, have been reported to affect inhibitory avoidance behavior in an opposite direction (Kovács and de Wied, 1981). α-Endorphin increased inhibitory avoidance behavior, both when injected postlearning or prior to the test. On the other hand, γ-endorphin reduced inhibitory avoidance behavior. However, these effects on inhibitory avoidance behavior have not been confirmed by others (e.g., Martinez and Rigter, 1980), and further experiments are needed to establish whether or not endorphins affect learning. One hypothesis to pursue is that administration of β-endorphin may produce a behavioral change which is a result of the influence of β-endorphin itself and also of α- and γ-type endorphins, depending on the dose and route of administration of β-endorphin. Opposite effects of α- and γ-endorphin have also been observed in tests of extinction of active avoidance behavior (de Wied et al., 1978; Le Moal et al., 1979; also see de Wied and Jolles, 1982).

4. ENKEPHALINS

The two main forms of enkephalin are Met-enkephalin and Leu-enkephalin. There are indications that the two peptides may have different behavioral functions. Intracerebroventricular injection of Met-enkephalin lowered, but Leu-enkephalin raised, pain threshold in a test for low-intensity pain (Leybin et al., 1976). Furthermore, intracerebroventricular administration of a high dose of Met-enkephalin produced in rats a stuporous immobility that could be prevented by the opiate antagonist naloxone. The same dose of Leu-enkephalin, on the other hand, elicited naloxone-insensitive rotational behavior (Chang et al., 1976). Both pentapeptides were able to reduce experimentally induced amnesia in rats in doses which were much lower than those needed to influence pain perception. Met-enkephalin attenuated amnesia both when given before training and before the retrieval test, whereas Leu-enkephalin was only effective when injected before the retrieval test (Rigter, 1978).

The enkephalin effects on amnesia could not be blocked by naloxone, suggesting that opiate receptors were not involved in the mediation of these effects (Rigter et al., 1977). However, in other behavioral studies an interaction between enkephalin and high doses of naloxone has been found. Injection of the enkephalins immediately before training rats on a one-way active avoidance task impaired acquisition. Naloxone blocked the effect of Leu-enkephalin. The effective dose of the antagonist was so high that it is unlikely that the peptide effect was mediated through the classical μ opiate receptor system (Rigter et al., 1980a). In subsequent experiments (Rigter et al., 1980b) the peptide effect was replicated. Additional studies ruled out that Leu-enkephalin interfered with motility or pain perception or reactivity. Rather, it was suggested that the peptide strengthens the tendency of rats to suppress behavior in the presence of cues previously associated with an aversive experience. This effect may be due to an enkephalin-induced increase in fear or arousal. Such an effect may sometimes interfere with and sometimes facilitate learning performance, depending on the characteristics of the task. Thus, the behavioral suppression caused by a Leu-enkephalin analog disrupted acquisition of an active avoidance response but speeded acquisition of an inhibitory avoidance response.

The enkephalin effect on behavior was obtained at low doses and was observed within a few minutes after administration. These findings suggest that this particular enkephalin effect may be initiated at a peripheral site. This hypothesis is supported by other data. The action of enkephalin on shock-motivated learning is weakened by removal of the adrenal medulla (Martinez and Rigter, 1982). Intracerebroventricular administration of Leu-enkephalin is ineffective in this particular setting (Martinez et al., 1984).

Much work needs to be done to further describe and clarify the behavioral activity of the enkephalins. To date, it is not clear whether the peptide affects memory processes. The available data are conflicting.

In one study, posttrial injection of Met-enkephalin, but not Leu-enkephalin, enhanced performance on a subsequent test for retrieval of an inhibitory avoidance response (Stein and Belluzzi, 1979). In contrast, another report mentions that posttrial treatment with this peptide impaired later performance in a shuttle-box task, i.e., produced amnesia (Lucion *et al.*, 1982).

B. Prolactin, Growth Hormone, Gonadotropins, and TSH

I. PROLACTIN

Prolactin has not been investigated extensively for its effect on learning. However, prolactin administration appears to result in various behavioral changes (Zarrow *et al.*, 1963; Moltz *et al.*, 1970; Bates *et al.*, 1964; Hartmann *et al.*, 1966; Svare *et al.*, 1979; Drago *et al.*, 1981a,b, 1983; Van Ree and de Wied, 1977). Some of these behavioral effects have been observed following administration of (exogenous) prolactin, whereas often behavioral changes have been seen in hyperprolactinemic animals bearing adenopituitary homografts under the kidney capsule.

Drago *et al.* (1983) examined the influence of prolactin on learning. It appeared that hyperprolactinemic rats displayed better acquisition of shuttle-box and pole-jump active avoidance responses. No difference was found in extinction of an active avoidance response. In contrast, no behavioral improvement was found when hyperprolactinemic animals were tested in an inhibitory avoidance paradigm. Thus, it appears that elevated levels of prolactin do not result in a general improvement of learning. Whether the effect of hyperprolactinemia on acquisition of the active avoidance response is due to the postulated hypersensitivity of postsynaptic dopamine receptor systems in hyperprolactinemic animals (Drago *et al.*, 1983) remains to be seen.

2. GROWTH HORMONE, GONADOTROPINS, AND TSH

Little, if anything, is known about the effect of growth hormone (GH) on behavior. In a few studies growth hormone or antisera to growth hormone have been used as a control treatment, and generally no effects of these treatments on the performance of animals in an avoidance situation have been observed. However, recently some effects of somatostatin have been reported (Vécsei *et al.*, 1987). This peptide, known for its GH release-inhibiting effects, has been shown to delay extinction of active avoidance behavior and to counteract ECS-induced amnesia in rats (Vécsei *et al.*, 1987).

As far as gonadotropins and TSH are concerned, we do not know of reports in the literature from which an influence of these pituitary hormones on learning can be inferred.

C. Neurohypophyseal Hormones

1. INTRODUCTION

The nonapeptides, vasopressin and oxytocin, are structurally similar but functionally different. Both hormones are of hypothalamic origin. Neurosecretory pathways transport them from the sites of synthesis to the posterior lobe of the pituitary, where they can be released into the peripheral circulation or to other areas of the nervous system, notably midbrain limbic structures.

Studies on the behavioral effects of these neuropeptides started with the observation that pitressin, a crude extract of the posterior lobe of the pituitary, inhibited extinction of an active avoidance response. Subsequent experiments rendered it likely that vasopressin was the active principle. Like ACTH, vasopressin retards extinction of a variety of responses. Nevertheless, the two hormones appear to have different behavioral functions. Relevant evidence will be discussed in the next section (for reviews, see van Wimersma Greidanus et al., 1983b; Kovács and Telegdy, 1983).

2. VASOPRESSINLIKE PEPTIDES

a. Active Avoidance As discussed above, hypophysectomy impairs both acquisition and extinction of active avoidance responses. Selective ablation of the pituitary lobes appears to differentially affect the behavior of rats in the shuttle-box test. Removal of the anterior lobe impaired acquisition, whereas removal of the combined posterior and intermediate lobes did not affect acquisition. Since the anterior lobe is the source of ACTH, it seems logical that injection of this hormone normalizes acquisition. However, vasopressin also normalized acquisition in adenohypophysectomized rats (de Wied, 1969). Ablation of the posterior and intermediate lobes impaired extinction of the shuttle-box response. Both vasopressin and ACTH ameliorated this deficit. The extinction deficit could be reversed by vasopressin given during either acquisition or extinction. In contrast, administration of ACTH during acquisition was not as effective as treatment during the extinction period (de Wied, 1969).

These data indicate that vasopressin may facilitate learning performance, but also that pituitary vasopressin is not essential for learning to take place. These conclusions are confirmed by the results of studies with rats lacking both pituitary and cerebral vasopressin. Some animals of the Brattleboro strain have hereditary hypothalamic diabetes insipidus (Valtin et al., 1965; Dogterom et al., 1978). The data on the ability of these rats to learn an active avoidance response are not consistent (see van Wimersma Greidanus, 1982b). In two studies only about 30% of the diabetes insipidus rats attained the learning criterion in a shuttle-box test compared to 50–80% of the control animals (Celestian et al., 1975; Miller et al., 1976). In a third experiment diabetes insipidus rats did learn the shuttle-box response, but more slowly than control animals. No deficit was seen in the pole-jump avoidance test (Bohus et al., 1975). Carey and Miller (1982) failed to find any effect of diabetes insipidus on acquisition of active avoidance behavior.

Extinction data are also contradictory. Celestian et al. (1975) reported that those diabetes insipidus rats that succeeded in mastering acquisition of the shuttle-box response showed retarded extinction. In contrast, accelerated extinction of a shuttle-box and a pole-jump response was seen in a study by Bohus et al. (1975).

It should be noted that the interpretation of the results of studies with Brattleboro rats is hampered by the fact that these animals also exhibit other endocrine abnormalities. The more or less conflicting results obtained with Brattleboro rats have been discussed extensively by various investigators (see Sokol and Valtin, 1982).

Bohus et al. (1973) administered vasopressin to hypophysectomized rats during the first week of training on the shuttle-box task. The hormone ameliorated the acquisition deficit. When treatment was discontinued, the effect of vasopressin persisted and performance of the animals did not break down. Such a long-lasting consequence of treatment with vasopressin suggests that the hormone modulates memory formation. This hypothesis is supported by the results of studies with intact rats. De Wied (1971) examined the effect of vasopressin on extinction of the pole-jump response. After successful learning of the avoidance response, rats were given an extinction session of 10 trials. Only animals that made 8 or more avoidance responses during this first session were selected for further extinction testing. Vasopressin or ACTH (4-10) were administered immediately after selection, and extinction was measured repeatedly between 2 and 72 hours after injection. The ACTH fragment had a short-term effect; it delayed extinction up until 4 hours after injection. On the other hand the effect of vasopressin persisted even up to 72 hours after injection. Two interpretations are possible to account for the effect of vasopressin. Either the hormone strengthened the originally learned avoidance response or it prevented learning of a new response, i.e., ignoring the warning signal. The data favor the first interpretation. Postponing treatment with vasopressin for longer than 1 hour after selection abolished the effect on extinction. This finding indicates that the hormone did not proactively interfere with new learning. Rather, vasopressin appeared to strengthen the original response. Note that the memory for this response already had been formed during acquisition. Possibly, vasopressin added quality or intensity to this preexisting memory.

King and de Wied (1974) attempted to elucidate the nature of the effect of vasopressin on memory. Rats were treated with vasopressin immediately after making their first pole-jump avoidance response. Training was then stopped for 6 hours in order to prevent any proactive effect of the hormone on subsequent avoidance responses. This early treatment with vasopressin was still able to delay later extinction. The hormone was even effective in delaying extinction, although to a lesser extent, when given at a time when the animals had not yet reacted to the warning signal by jumping onto the pole but had already developed a learned fear for this signal. Vasopressin had no effect when administered at a time when learning, in whatever form, had not yet taken place. These authors assumed that some minimum level of learning must be present before vasopressin could be effective in facilitating memory formation. This assumption is consistent with the above-mentioned hypothesis that vasopressin *adds* some feature to a preexisting memory.

b. Inhibitory Avoidance   The hypothesis that vasopressin modulates memory processes has been explored further in studies using inhibitory avoidance tasks. The data of these experiments meet the two criteria for the assessment of hormonal effects on memory formation, i.e., posttrial efficacy and time dependence of the posttrial effect. Thus, subcutaneous or intracerebroventricular injection of vasopressin immediately after the learning experience enhanced performance of rats during a later retrieval test (Ader and de Wied, 1972; Bohus *et al.*, 1978b; Krejci and Kupková, 1978). Vasopressin was still effective when given 3 hours after training, but not when injected 6 hours after training (Bohus *et al.*, 1978b).

If injection of vasopressin strengthens memory formation, inactivation of endogenous vasopressin should impair memory. This hypothesis is confirmed by the results of experiments using antiserum against vasopressin. Intracerebroventricular injection of antiserum in rats after the learning trial time-dependently impaired retrieval of an inhibitory avoidance response (van Wimersma Greidanus *et al.*, 1975; van Wimersma Greidanus and de Wied, 1976). Midbrain limbic structures are important in mediating this effect (van Wimersma Greidanus *et al.*, 1983a; van Wimersma Greidanus and Veldhuis, 1985). Thus, bilateral injection of vasopressin antiserum into the dentate gyrus of the dorsal hippocampus, into the ventral hippocampus, or into the septal area of rats also attenuated retrieval of the inhibitory avoidance response (Kovács *et al.*, 1982; van Wimersma Greidanus and Veldhuis, 1985). Intravenous injection of much larger quantities of antiserum removed the hormone from the circulation, as indicated by the absence of vasopressin in the urine and by increased urine production, but failed to influence inhibitory avoidance behavior (van Wimersma Greidanus *et al.*, 1975). This indicates that interference with central, but not peripheral, vasopressin impairs memory formation.

Recent publications have addressed the question of whether vasopressin acts on memory formation per se or rather on some supportive function, such as arousal. The effects of vasopressin on extinction of the pole-jump response and on acquisition of an inhibitory avoidance response could be attenuated by 1-deaminopenicillamine, 2-(O-methyl)tyrosine arginine vasopressin (Le Moal *et al.*, 1981; LeBrun *et al.*, 1984). The latter peptide is an antagonist of the pressor effect of vasopressin. Conceivably, the effectiveness of the pressor antagonist in attenuating the behavioral effects of vasopressin may indicate either that signals from peripheral visceral sources play an important role in the subsequent behavioral changes or that the receptors at which vasopressin elicits its pressor effect are similar to those leading to its behavioral action. However, the interpretation of these data is difficult due to the fact that the pressor antagonist has some behavioral activity of its own (LeBrun *et al.*, 1984). Furthermore, recent experiments performed by de Wied *et al.* (1984) indicate that AVP and related peptides, such as the very potent fragment [pGlu^4,Cyt^6]AVP-(4-8), affect inhibitory avoidance behavior by a direct central action and that the structural requirement for activation of central vasopressin receptors differs from that of the peripheral cardiovascular receptors, although both can be blocked by the same vasopressor antagonist (de Wied *et al.*, 1984).

There are data indicating that vasopressin may have arousing properties (Ettenberg *et al.*, 1983). If these properties are related to the ability of the hormone to facilitate memory formation, it still remains to be established how posttrial activation of arousal processes can contribute to memory formation. At any rate, it cannot be concluded that visceral effects of vasopressin exclude a role of central vasopressin in memory. Evidence for such a role has been cited earlier in this section. It should be stressed that this role is not an absolute one: vasopressin is not indispensable in memory formation. The effectiveness of vasopressin critically depends on the characteristics of the inhibitory avoidance task (Rigter, 1982).

Vasopressinlike peptides also promote the retrieval of an inhibitory avoidance response when given shortly before the test trial. This effect has been reported for lysine vasopressin (Ader and de Wied, 1972), arginine vasopressin (Bohus *et al.*, 1978a), and desglycinamide lysine vasopressin (de Wied *et al.*, 1974). The latter peptide has much-reduced or no pressor and antidiuretic activities, suggesting that the behavioral effects of vasopressinlike peptides are independent of, at least in part, the classical endocrine properties of the hormone. Furthermore, intracerebroventricular injection of anti-vasopressin serum 1 hour before the retrieval test resulted in an inhibitory avoidance deficit (van Wimersma Greidanus and de Wied, 1976). Also, in studies of prevention and reversal of experimentally induced amnesia in rats, vasopressin was effective both at the time of acquisition and at the time of the test trial (Rigter *et al.*, 1974).

It may be assumed that the postlearning and the pretest effects of vasopressin reflect one and the same action, e.g., on some arousal process. However, there are indications from structure–activity studies that the two effects are not identical. The ring structure of vasopressin may be most important for modulating memory formation, whereas the C-terminal appears to be more important for influencing retrieval performance (Van Ree *et al.*, 1978).

c. Nonshock Tasks   The ability of vasopressin injected posttrial to facilitate later test performance is not limited to avoidance learning. Bohus (1977) trained male rats to transverse a maze to gain access to a sexually receptive female. The male rats received desglycinamide lysine vasopressin after the last trial of each acquisition session. The peptide facilitated retrieval. Similar effects have been seen in a test requiring water-deprived rats to locate a water tube (Ettenberg *et al.*, 1983) and in a test in which mice were trained to avoid an aggressive fighter (Leshner and Roche, 1977). Vasopressin treatment also resulted in a delay of extinction of an appetitive discrimination task, in a delay of the postcastration decline in copulatory behavior of male rats, and in an enhanced development of tolerance to the analgesic effect of morphine. Finally, vasopressin prevented or reversed amnesia induced by electroconvulsive shock, CO_2^- inhalation, pentylenetetrazol, or puromycin. Accordingly, the effects of vasopressin on behavior are not restricted to aversively motivated behaviors (for review, see van Wimersma Greidanus *et al.*, 1983b, 1985a, 1986).

Neutralization or reduction of centrally available vasopression by intracerebroventricular administration of antivasopressin serum to rats also interfered with the development of tolerance to morphine (van Wimersma

Greidanus *et al.*, 1978). Time-gradient studies suggested that centrally available vasopressin may be physiologically involved in the development of tolerance to the analgesic effect of morphine in a manner comparable to its role in learning (and memory).

d. Effects in Man Vasopressin has been examined for its effectiveness in reversing amnesia in man. Recent reviews indicate that a majority of subjects described so far in the literature appear to benefit from peptide treatment (Jolles, 1983; Jolles *et al.*, 1983; Van Ree *et al.*, 1985; van Wimersma Greidanus *et al.*, 1985a, 1986). Interpretation of the data is difficult due to the heterogeneity of the cognitive disorders examined and the uncertain validity of the psychological tests used. Patients with similar memory complaints may suffer from various memory disorders. Vasopressin may have selective effects, depending on the nature of the disorder involved (Jolles, 1983, 1985; Jolles *et al.*, 1983; Van Ree *et al.*, 1985).

3. OXYTOCIN

Oxytocin may have behavioral effects opposite to those of vasopressin (for review, see Kovács and Telegdy, 1983). Schulz *et al.* (1974) reported that vasopressin and oxytocin delayed and accelerated, respectively, extinction of an active avoidance response. Small amounts of oxytocin administered intracerebroventricularly immediately after the learning experience affected retrieval of an inhibitory avoidance response in a manner opposite to that of vasopressin (Bohus *et al.*, 1978a,b). The intracerebroventricular route of injection appeared to be important in some studies (Bohus *et al.*, 1978a,b) but not in others (Kovács *et al.*, 1978). Antiserum against vasopressin reduced inhibitory avoidance behavior, whereas antiserum against oxytocin increased avoidance behavior (Bohus *et al.*, 1978a). It has been suggested that oxytocin may be a naturally occurring amnestic neuropeptide (Bohus *et al.*, 1978a; Kovács and Telegdy, 1983).

III. Gut and Brain Hormones

A. Introduction

During recent years the presence in the brain of hormones which were found originally in the gastrointestinal tract initiated research on the behavioral effects of these hormones. Although most of this research was not focused on learning processes, a few data from these studies are worthy of mention in this chapter.

B. Neurotensin

Numerous reports in the literature deal with central nervous system effects of this tridecapeptide (for reviews, see Nemeroff *et al.*, 1977, 1980; van Wimersma Greidanus *et al.*, 1982). Most of these reports deal with the effects of this neuropeptide on body temperature, locomotor activity, or responsiveness to a painful stimulus. Only a few studies have been performed on the influence of neurotension on learning, such as active and passive avoidance behavior. From these latter studies it appears that neurotensin affects extinction of a pole-jump avoidance response. When neurotensin was intracerebroventricularly administered during extinction of this response in a dose range of 3 ng–3 μg, a dose-dependent inhibition of extinction was found. In contrast, intracerebroventricular administration of an antiserum to neurotensin resulted in a facilitated extinction of the pole-jump avoidance response (van Wimersma Greidanus *et al.*, 1982).

Intracerebroventricular treatment of rats with 300 ng neurotensin immediately after the learning trial or prior to the first retrieval test in an inhibitory avoidance paradigm increased the median latency score during the first test only. A dose of 3 μg increased latency during the first and second tests. Thus, it appears that neurotensin enhances inhibitory avoidance behavior, although the doses used are relatively high. These data indicate that neurotensin may affect memory formation as well as retrieval processes (van Wimersma Greidanus *et al.*, 1982).

C. Bombesin

This peptide is known mainly for its hypothermic action and for its strong potency to induce excessive grooming behavior (van Wimersma Greidanus *et al.*, 1985b,c, 1987). A recent study on the influence of bombesin on inhibitory avoidance behavior revealed that this peptide attenuated passive avoidance behavior, when intracerebroventricularly administered in a dose of 300 ng either immediately after the learning trial or prior to the first retrieval test (M. Fekete and Tj. B. van Wimersma Greidanus, unpublished results). Thus, bombesin would appear to impair both memory formation and performance on the retrieval test. More studies are needed to further characterize this effect.

D. Cholecystokinin (CCK)

Recently various experiments have been performed on the effects of cholecystokinin octapeptides on active and inhibitory avoidance behavior following central and peripheral administration (for review, see Fekete *et al.*, 1987). Various studies have shown that CCK-8-related peptides, injected peripherally or intracerebroventricularly, facilitated the retrieval of an inhibitory avoidance response (Fekete *et al.*, 1981a,c; Kádár *et al.*, 1981; Telegdy *et al.*, 1982; Van Ree *et al.*, 1983) when administered 1 hour before the retrieval test. However, the same peptides injected into the nucleus accumbens (0.3 pg) attenuated passive avoidance behavior (Van Ree *et al.*, 1983; Fekete *et al.*, 1984). In addition, CCK-8-related peptides impaired acquisition and facilitated extinction of active avoidance behavior and of food-motivated conditioned behavior (Fekete *et al.*, 1981b, 1982; Cohen *et al.*, 1982; Telegdy *et al.*, 1982). It should be noted that an increased tendency to display inhibitory avoidance behavior would be expected to result in the observed changes in acquisition and extinction of active avoidance responses. CCK-8-related peptides were also able to attenuate electroconvulsive shock (ECS)-induced retrograde amnesia, as measured in a one-trial step-through passive avoidance paradigm. This effect, however, is possibly due to a reduction in the severity of ECS-induced seizures (Kádár *et al.*, 1984). The marked behavioral effects of CCK octapeptides may be due partly to neurolepticlike actions (mediated through the nucleus accumbens) and partly to an enhancement of arousal and/or fear motivation (Fekete *et al.*, 1984). At present, no definite conclusions about the nature of the influence of CCK on learning performance can be drawn.

E. Substance P

Since an extensive review appeared on substance P and its effects on learning and memory (Huston and Stäubli, 1981), we will discuss here only briefly the effects on learning of this interesting undecapeptide.

Intracranial injection of substance P can influence inhibitory avoidance learning and either facilitate or inhibit behavioral performance depending on the site of injection. Posttrial administration of substance P directly into the medial septal nucleus or the lateral hypothalamus improved the retrieval of an inhibitory avoidance response. The posttrial efficacy suggests that the peptide affected some feature of memory formation. In contrast, local application of substance P into the substantia nigra and amygdala resulted in amnesia. Interestingly, all these brain areas contain moderate to high levels of substance P located in nerve terminals.

It has been suggested (Huston and Stäubli, 1981) that the facilitatory effect of substance P on memory formation following its intraseptal administration may occur by influencing arousal or hippocampal theta activity.

Huston and Stäubli (1981) favor the hypothesis that substance P can have reinforcing as well as punishing effects, depending on the brain area involved. Facilitation of memory formation following posttrial administration of substance P in an inhibitory avoidance situation is obtained in cases where the peptide is applied to brain sites such as septum and lateral hypothalamus where electrical stimulation is predominantly reinforcing. Thus, substance P may have reinforcing or rewarding effects at these sites. In contrast, substance P may have aversive or mixed effects when administered directly into the amygdala or substantia nigra and consequently its local application in these brain areas may result in amnesia.

F. Vasoactive Intestinal Peptide (VIP)

Recently some studies have been undertaken on the behavioral actions of the 28-amino-acid peptide VIP (Cottrell *et al.*, 1984). The results of these experiments revealed that VIP facilitated extinction of a pole-jump avoidance response. Furthermore, the peptide attenuated inhibitory avoidance behavior, but only when injected prior to the retrieval test. These findings indicate that VIP does not influence memory formation, but rather some behavioral function important for optimal performance in the retrieval test.

Interestingly, the inhibitory effects of VIP on conditioned avoidance behavior are similar to or comparable with the ones observed after administration of corticosteroid (*vide infra*) and it has been suggested (Cottrell *et al.*, 1984) that VIP controls neural mechanisms underlying behavior that are also affected by corticosterone and 5-HT.

IV. Steroid Hormones

A. Adrenocorticosteroids

Adrenocortical steroids are known to affect performance of animals in tests of learning and memory. The limitation in the extensive literature available on the behavioral influences of corticosteroids is that the large majority originates from studies using avoidance behaviors (see Leshner *et al.*, 1981). From these studies it appears that corticosterone and related peptides facilitate the extinction of a conditioned avoidance response. This effect may be related to the negative feedback action of corticosterone on ACTH release, since ACTH has been shown to exert an opposite effect, i.e., it delays extinction of conditioned avoidance behavior. However, the observation that corticosterone also facilitates extinction in hypophysectomized rats indicates that this effect of the steroid is independent of its effects on the release of ACTH from the pituitary. Further evidence against the hypothesis that the corticosteroid effect on extinction of avoidance behavior solely depends on a blockade or inhibition of pituitary ACTH release was obtained from implantation studies. Bohus (1968) found that implantation of cortisol in the median eminence region of the hypothalamus, which inhibited ACTH release, facilitated extinction of shuttle-box avoidance behavior. The more the release of ACTH was suppressed, the stronger was the behavioral effect. However, cortisol implantation in the mesencephalic reticular formation, which hardly reduced ACTH release, markedly facilitated extinction of the avoidance response. Accordingly, corticosteroids have multiple effects on extinction of avoidance behavior, through inhibition of ACTH release or through direct action on the brain (de Wied *et al.*, 1970). Generally the locus of behavioral action of corticosteroids appears to be in the rostral mesencephalic/caudal diencephalic area at the level of the posterior thalamus, including the parafascicular nuclei (van Wimersma Greidanus and de Wied, 1969). It appears from studies of the structure–activity relationship that glucocorticoid rather than mineralocorticoid activity is responsible for the facilitatory effect on extinction of a conditioned avoidance response (de Wied, 1967). However, this behavioral action of the steroids is not directly correlated with their glucocorticoid activity, since progesterone and pregnenolone facilitated extinction of the avoidance response in a way similar to that of corticosterone (de Wied *et al.*, 1972; van Wimersma Greidanus, 1970; van Wimersma Greidanus *et al.*, 1973). Preretention treatment of rats with progesterone increases avoidance latencies in an inhibitory avoidance task (Tj. B. van Wimersma Greidanus, unpublished). It may be that these behavioral effects of progesterone are caused by a progesterone-induced reduction of arousal in fearful situations. Studies with progesterone using nonshock tasks are needed to clarify this issue. More recent studies indicate that corticosterone is physiologically involved in the modulation of adaptive behavior in the rat and that hippocampal receptors are involved in its mechanism of action (Bohus and de Kloet, 1977). It is likely that multiple corticosteroid binding systems exist in the brain, which are involved in the actions of steroid on brain functions related to behavioral responses.

B. Sex Steroids

As mentioned in the previous section, progesterone facilitates extinction of a conditioned avoidance response in a way similar to that of corticosteroids. Extensive experiments on the structure–activity relationship (van Wimersma Greidanus, 1970; de Wied *et al.*, 1972) revealed that pregnene- (or pregnadiene-) type steroids induced the same effect, whereas the cholestene, androstene, estratriene, and pregnane types were ineffective. This implies that the sex steroids estradiol and testosterone will not affect this type of behavior. Despite the well-known actions of sex steroids on various aspects of sexual behavior, there is no clear-cut evidence that these hormones are involved in brain processes related to learning.

Generally it can be stated that steroids affect behavior in which aspects of learning and memory are involved, but that these effects are generally restricted to certain types of corticosteroids. The behavioral effects of these steroids, to which belong cortisone, cortisol, corticosterone, and progesterone, are mainly related to adaptive behavior. Thus, steroids originating from the adrenal cortex fulfill their role in adaptation, a well-known physiological function of the pituitary adrenal axis, among others, in modulating behavior and in this way enable the organism to cope with environmental changes in an appropriate way.

V. Concluding Remarks

Many hormones influence behavior, but only some of them can be regarded as "learning modulatory hormones" (McGaugh and Martinez, 1981) due to their effects on arousal, vigilance, attention, motivation, or consolidation (see Section I). The most important hormones belonging to this latter group seem to be ACTH and related peptides, some opioids, the neurohypophyseal hormones vasopressin and oxytocin, CCK, substance P, and some steroids. As stated in Section I, learning involves a complex interaction between different neuronal functions. Optimal brain functioning depends in turn on a well-balanced hormonal climate in the brain. Thus, numerous neuropeptides and steroid hormones can affect learning performance and may in one way or another, directly or indirectly, influence brain processes mediating behavioral neuroendocrine responses in which elements of learning are involved.

References

Ader, R., and de Wied, D. (1972). Effects of lysine vasopressin on passive avoidance learning. *Psychon. Sci.* **29,** 46–48.

Anderson, D. C., Winn, W., and Tam, T. (1968). Adrenocorticotrophic hormone and acquisition of a passive avoidance response: A replication and extension. *J. Comp. Physiol. Psychol.* **66,** 497–499.

Applezweig, M. H., and Baudry, F. D. (1955). The pituitary-adreno-cortical system in avoidance learning. *Psychol. Rep.* **1,** 417–420.

Bates, R. W., Molkovic, S., and Garrison, M. M. (1964). Effects of prolactin, growth hormone and ACTH, alone and in combination, upon organ weights and adrenal function in normal rats. *Endocrinology (Baltimore)* **74,** 714–723.

Bélanger, D. (1958). Effets de l'hypophysectomie sur l'apprentissage d'un réaction échappement-évitement. *Can. J. Physiol.* **12,** 171–178.

Bohus, B. (1968). Pituitary ACTH release and avoidance behaviors of rats with cortisol implants in mesencephalic reticular formation and median eminence. *Neuroendocrinology* **3,** 355–365.

Bohus, B. (1977). Effects of desglycinamide-lysine vasopressin (DG-LVP) on sexually motivated T-maze behavior of the male rat. *Horm. Behav.* **8,** 52–61.

Bohus, B., and de Kloet, E. R. (1977). Behavioral effects of corticosterone related to putative glucocorticoid receptor properties in the rat brain. *J. Endocrinol.* **72,** 64P.

Bohus, B., and de Wied, D. (1966). Inhibitory and facilitatory effect of two related peptides on extinction of avoidance behavior. *Science* **153,** 318–320.

Bohus, B., and de Wied, D. (1981). Actions of ACTH- and MSH-like peptides on learning, performance and retention. *In* "Endogenous Peptides and Learning and Memory Processes" (J. L. Martinez, Jr., R. A. Jensen, R. B. Messing, H. Rigter, and J. L. McGaugh, eds.), pp. 59–77. Academic Press, New York.

Bohus, B., and Endröczi, E. (1965). The influence of pituitary-adrenocortical function on the avoiding conditioned reflex activity in rats. *Acta Physiol. Acad. Sci. Hung.* **26,** 183–189.

Bohus, B., Nyakas, C., and Endröczi, E. (1968). Effects of adrenocorticotropic hormone on avoidance behaviour of intact and adrenalectomized rats. *Int. J. Neuropharmacol.* **7,** 307–314.

Bohus, B., Gispen, W. H., and de Wied, D. (1973). Effect of lysine vasopressin and ACTH$_{4\text{-}10}$ on conditioned avoidance behavior of hypophysectomized rats. *Neuroendocrinology* **11,** 137–143.

Bohus, B., van Wimersma Greidanus, Tj. B., and de Wied, D. (1975). Behavioral and endocrine responses of rats with hereditary hypothalamic diabetes insipidus (Brattleboro strain). *Physiol. Behav.* **14,** 609–615.

Bohus, B., Urban, I., van Wimersma Greidanus, Tj. B., and de Wied, D. (1978a). Opposite effects of oxytocin and vasopressin on avoidance behavior and hippocampal theta rhythm in the rat. *Neuropharmacology* **17,** 239–247.

Bohus, B., Kovács, G. L., and de Wied, D. (1978b). Oxytocin, vasopressin and memory: Opposite effects on consolidation and retrieval processes. *Brain Res.* **157,** 414–417.

Bohus, B., de Boer, S., Zanotti, A., and Drago, F. (1983). Opiocortin peptides, opiate agonists and antagonists and avoidance behavior. *Dev. Neurosci.* **16,** 107–113.

Carey, R. J., and Miller, M. (1982). Absence of learning and memory deficits in the vasopressin-deficient rat (Brattleboro strain). *Behav. Brain Res.* **6,** 1–13.

Celestian, J. F., Carey, R. J., and Miller, M. (1975). Unimpaired maintenance of a conditioned avoidance response in the rat with diabetes insipidus. *Physiol. Behav.* **15,** 707–711.

Chang, J. K., Fong, B. T. W., Pert, A., and Pert, C. B. (1976). Opiate receptor affinities and behavioral effects of enkephalin: Structure-activity relationship of ten synthetic peptide analogues. *Life Sci.* **18,** 1473–1482.

Cohen, S. L., Knight, M., Tamminga, C. A., and Chase, T. N. (1982). Cholecystokinin-octapeptide effects on conditioned-avoidance behavior, stereotypy and catelepsy. *Eur. J. Pharmacol.* **83,** 213–222.

Cottrell, G. A., Veldhuis, H. D., Rostene, W. H., and de Kloet, E. R. (1984). Behavioural actions of vasoactive intestinal peptide (VIP). *Neuropeptides (Edinburgh)* **4,** 331–341.

de Wied, D. (1967). Opposite effects of ACTH and glucocorticosteroids on extinction of conditioned avoidance behavior. *Int. Congr. Ser.—Excerpta Med.* **132,** 945–951.

de Wied, D. (1968). Influence of vasopressin and of a crude CRF preparation on pituitary ACTH-release in posterior-lobectomized rats. *Neuroendocrinology* **3,** 129–135.

de Wied, D. (1969). Effects of peptide hormones on behavior. In "Frontiers in Neuroendocrinology" (W. F. Ganong and L. Martini, eds.), pp. 97–104. Oxford Univ. Press, London and New York.

de Wied, D. (1971). Long-term effect of vasopressin on the maintenance of a conditioned avoidance response in rats. *Nature (London)* **232,** 58–60.

de Wied, D. (1976). Behavioral effects of intraventricularly administered vasopressin and vasopressin fragments. *Life Sci.* **19,** 685–690.

de Wied, D., and Jolles, J. (1982). Neuropeptides derived from pro-opiocortin: Behavioral, physiological and neurochemical effects. *Physiol. Rev.* **62,** 976–1059.

de Wied, D., Witter, A., and Lande, S. (1970). Anterior pituitary peptides and avoidance acquisition of hypophysectomized rats. *Prog. Brain Res.* **32,** 213–220.

de Wied, D., Van Delft, A. M. L., Gispen, W. H., Weijnen, J. A. W. M., and van Wimersma Greidanus, Tj. B. (1972). The role of pituitary-adrenal system hormones in active avoidance conditioning. In "Hormones and Behavior" (S. Levine, ed.), pp. 135–171. Academic Press, New York.

de Wied, D., Bohus, B., and van Wimersma Greidanus, Tj. B. (1974). The hypothalamic-neurohypophyseal system and preservation of conditioned avoidance behavior in rats. *Prog. Brain Res.* **41,** 417–428.

de Wied, D., Bohus, B., Van Ree, J. M., and Urban, I. (1978). Behavioral and electrophysiological effects of peptides related to lipotropin (β-LPH). *J. Pharmacol. Exp. Ther.* **204,** 570–580.

de Wied, D., Gaffori, O., Van Ree, J. M., and De Jong, W. (1984). Central target for the behavioural effects of vasopressin neuropeptides. *Nature (London)* **308,** 276–278.

Dogterom, J., van Wimersma Greidanus, Tj. B., and de Wied, D. (1978). Vasopressin in cerebrospinal fluid and plasma of man, dog, and rat. *Am. J. Physiol.* **234,** E463–E467.

Drago, F., Pellegrini-Quarantotti, B., Scapagnini, U., and Gessa, G. (1981a). Short-term endogenous hyperprolactinaemia and sexual behavior of male rats. *Physiol. Behav.* **26,** 277–279.

Drago, F., Bohus, B., Canonico, P. L., and Scapagnini, U. (1981b). Prolactin induces grooming in the rat: Possible involvement of nigrostriatal dopaminergic system. *Pharmacol., Biochem. Behav.* **15,** 61–63.

Drago, F., Bohus, B., Gispen, W. H., Van Ree, J. M., Scapagnini, U., and de Wied, D. (1983). Behavioral changes in short-term and long-term hyperprolactinaemic rats. *Dev. Neurosci.* **16,** 417–427.

Duncan, C. P. (1949). The retroactive effect of electroshock on learning. *J. Comp. Physiol. Psychol.* **42,** 132–144.

Ettenberg, A., Van der Kooy, D., Le Moal, M., Koob, G. F., and Bloom, F. E. (1983). Can aversive properties of (peripherally-injected) vasopressin account for its putative role in memory? *Behav. Brain Res.* **7,** 331–350.

Fekete, M., and de Wied, D. (1982). Dose-related facilitation and inhibition of passive avoidance behavior by the ACTH 4-9 analog (Org 2766). *Pharmacol., Biochem. Behav.* **17,** 177–182.

Fekete, M., Kádár, T., Penke, B., and Telegdy, G. (1981a). Modulation of passive avoidance behaviour by cholecystokinin octapeptides in rats. *Neuropeptides (Edinburgh)* **1,** 301–307.

Fekete, M., Szábo, A., Balázs, M., Penke, B., and Telegdy, G. (1981b). Effect of intraventricular administration of cholecystokinin octapeptide sulfate ester and unsulfated cholecystokinin octapeptide on active avoidance and conditioned feeding behaviour of rats. *Acta Physiol. Acad. Sci. Hung.* **58,** 39–45.

Fekete, M., Penke, B., and Telegdy, G. (1981c). Structure-activity studies concerning effect of cholecystokinin octapeptide on passive avoidance behaviour of rats. *Acta Physiol. Acad. Sci. Hung.* **58,** 197–200.

Fekete, M., Bokor, M., Penke, B., and Telegdy, G. (1982). Effects of cholecystokinin octapeptide sulphate ester and unsulphated cholecystokinin octapeptide on active avoidance behavior in rats. *Acta Physiol. Acad. Sci. Hung.* **60,** 57–63.

Fekete, M., Lengyel, A., Hegedüs, B., Penke, B., Zarándy, M., Tóth, G. K., and Telegdy, G. (1984). Further analysis of the effects of cholecystokinin octapeptides on avoidance behaviour in rats. *Eur. J. Pharmacol.* **98,** 79–91.

Fekete, M., Szipöcs, I., Halmai, L., Fromczia, P., Kardos, A., Csonka, E., Szántó-Fekete, M., and Telegdy, G. (1987). Cholecystokinin and the central nervous system. *Front. Horm. Res.* **15,** 175–251.

Flood, J. F., Jarvik, M. E., Bennett, E. L., and Orme, A. E. (1976). Effects of ACTH peptide fragments on memory formation. *Pharmacol., Biochem. Behav.* **5,** 41–51.

Garrud, P., Gray, J. A., and de Wied, D. (1974). Pituitary-adrenal hormones and extinction of rewarded behaviour in the rat. *Physiol. Behav.* **12,** 109–119.

Gispen, W. H., van Wimersma Greidanus, Tj. B., and de Wied, D. (1970). Effects of hypophysectomy and ACTH 1-10 on responsiveness to electric shock in rats. *Physiol. Behav.* **5**, 143–146.

Gispen, W. H., van der Poel, A. M., van Wimersma Greidanus, Tj. B. (1973). Pituitary-adrenal influences on behavior: Responses to test situations with or without electric footshock. *Physiol. Behav.* **10**, 345–350.

Gold, P. E., and McGaugh, J. L. (1977). Hormones and memory. *In* "Neuropeptide Influences on the Brain and Behavior" (L. H. Miller, C. A. Sandman, and A. J. Kastin, eds.), pp. 127–143. Raven Press, New York.

Gold, P. E., and Van Buskirk, R. (1976). Enhancement and impairment of memory processes with post-trial injections of adrenocorticotrophic hormone. *Behav. Biol.* **16**, 387–400.

Gorelick, D. A., Catlin, D. H., George, R., and Li, C. H. (1978). Beta-endorphin is behaviorally active in rats after chronic intravenous administration. *Pharmacol., Biochem. Behav.* **9**, 385–386.

Greven, H. M., and de Wied, D. (1973). The influence of peptides derived from corticotropin (ACTH) on performance. Structure activity studies. *Prog. Brain Res.* **39**, 426–442.

Hartmann, G., Endröczi, E., and Lisak, K. (1966). The effects of hypothalamic implantation of 17-β-estradiol and systemic administration of prolactin (LTH) on sexual behaviour in male rabbits. *Acta Physiol. Acad. Sci. Hung.* **30**, 53–59.

Hennessy, J. W., Smotherman, W. P., and Levine, S. (1980). Investigations into the nature of the dexamethasone and ACTH effects upon learned taste aversion. *Physiol. Behav.* **24**, 645–649.

Huston, J. P., and Stäubli, U. (1981). Substance P and its effects on learning en memory. *In* "Endogenous Peptides and Learning and Memory Processes" (J. L. Martinez, Jr., R. A. Jensen, R. B. Messing, H. Rigter, and J. L. McGaugh, eds.), pp. 521–540. Academic Press, New York.

Jolles, J. (1983). Vasopressin-like peptides and the treatment of memory disorders in man. *Prog. Brain Res.* **60**, 169–182.

Jolles, J. (1985). Neuropeptides and cognitive disorders. *Prog. Brain Res.* **65**, 177–192.

Jolles, J., Gaillard, A. W. K., and Hijman, R. (1983). Memory disorders and vasopressin. *Dev. Neurosci.* **16**, 63–75.

Kádár, T., Fekete, M., and Telegdy, G. (1981). Modulation of passive avoidance behaviour of rats by intracerebroventricular administration of cholecystokinin octapeptide sulfate ester and nonsulfated cholecystokinin octapeptide. *Acta Physiol. Acad. Sci. Hung.* **58**, 269–274.

Kádár, T., Penke, B., Kovács, K., and Telegdy, G. (1984). The effects of sulfated and nonsulfated cholecystokinin octapeptides on electroconvulsive shock-induced retrograde amnesia after intracerebroventricular administration in rats. *Neuropeptides (Edinburgh)* **4**, 127–135.

Kastin, A. J., Dempsey, G. L., Le Blanc, B., Dyster-Aas, K., and Schally, A. V. (1974).

Extinction of an appetitive operant response after administration of MSH. *Horm. Behav.* **5,** 135–139.

Kendler, K., Hennessy, J. W., Smotherman, W. P., and Levine, S. (1976). An ACTH effect on recovery from conditioned taste aversion. *Behav. Biol.* **17,** 225–229.

King, A. R., and de Wied, D. 1974). Localized behavioral effects of vasopressin on maintenance of an active avoidance response in rats. *J. Comp. Physiol. Psychol.* **86,** 1008–1018.

Kovács, G. L., and de Wied, D. (1981). Endorphin influences on learning and memory. *In* "Endogenous Peptides and Learning and Memory Processes" (J. L. Martinez, Jr., R. A. Jensen, R. B. Messing, H. Rigter, and J. L. McGaugh, eds.), pp. 231–247. Academic Press, New York.

Kovács, G. L., and Telegdy, G. (1983). Role of oxytocin in memory and amnesia. *Pharmacol. Ther.* **18,** 375–395.

Kovács, G. L., Vécsei, L., and Telegdy, G. (1978). Opposite action of oxytocin to vasopressin in passive avoidance behavior in rats. *Physiol. Behav.* **20,** 801–802.

Kovács, G. L., Buijs, R. M., Bohus, B., and van Wimersma Greidanus, Tj. B. (1982). Microinjection of arginine[8]-vasopressin anti-serum into the dorsal hippocampus attenuates passive avoidance behavior in rats. *Physiol. Behav.* **28,** 45–48.

Krejci, I., and Kupková, B. (1978). Effects of vasopressin analogues on passive avoidance behavior. *Act. Nerv. Super.* **20,** 11–12.

LeBrun, C. J., Rigter, H., Martinez, J. L., Jr., Koob, G. F., Le Moal, M., and Bloom, F. E. (1984). Antagonism of effects of vasopressin (AVP) on inhibitory avoidance by a vasopressin antagonist peptide [dPtyr(ME)AVP]. *Life Sci.* **35,** 1505–1512.

Le Moal, M., Koob, G. F., and Bloom, F. E. (1979). Endorphins and extinction: Differential actions on appetitive and adversive tasks. *Life Sci.* **24,** 1631–1636.

Le Moal, M., Koob, G. F., Koda, L. Y., Bloom, F. E., Manning, M., Sawyer, W. H., and Rivier, J. (1981). Vasopressor receptor antagonist prevents behavioral effects of vasopressin. *Nature (London)* **291,** 491–493.

Leshner, A. I., and Roche, K. E. (1977). Comparison of the effects of ACTH and lysine vasopressin on avoidance-of-attack in mice. *Physiol. Behav.* **18,** 879–883.

Leshner, A. I., Merkle, D. A., and Mixon, J. F. (1981). Pituitary-adrenocortical effects on learning and memory in social situations. *In* "Endogenous Peptides and Learning and Memory Processes" (J. L. Martinez, Jr., R. A. Jensen, R. B. Messing, H. Rigter, and M. L. McGaugh, eds.), pp. 159–179. Academic Press, New York.

Ley, K. F., and Corson, J. A. (1971). ACTH: Differential effects on avoidance and discrimination. *Experientia* **27,** 958–959.

Leybin, L., Pinsky, C., LaBella, F. S., Havlicek, V., and Rezek, M. (1976). Intraventricular Met[5]-enkephalin causes unexpected lowering of pain threshold and narcotic withdrawal signs in rats. *Nature (London)* **264,** 458–459.

Lissák, K., and Bohus, B. (1972). Pituitary hormones and avoidance behavior of the rat. *Int. J. Psychobiol.* **2,** 103–115.

Lucion, A. B., Rosito, G., Sapper, D., Palimini, A. L., and Izquierdo, I. (1982). Intracerebroventricular administration of nanogram amounts of beta-endorphin and Met-enkephalin causes retrograde amnesia in rats. *Behav. Brain Res.* **4,** 111–115.

Martinez, J. L., Jr., and Rigter, H. (1980). Endorphins alter acquisition and consolidation of an inhibitory avoidance response in rats. *Neurosci. Lett.* **18,** 197–201.

Martinez, J. L., Jr., and Rigter, H. (1982). Enkephalin actions on avoidance conditioning may be related to adrenal medullary function. *Behav. Brain Res.* **6,** 289–299.

Martinez, J. L., Jr., Vasquez, B. J., Jensen, R. A., Soumireu-Mourat, B., and McGaugh, J. L. (1979). ACTH 4-9 analog (Org 2766) facilitates acquisition of an inhibitory avoidance response in rats. *Pharmacol., Biochem. Behav.* **10,** 145–147.

Martinez, J. L., Jr., Olson, K., and Hilston, C. (1984). Opposite effects of Met-enkephalin and Leu-enkephalin on a discriminated shock-escape task. *Behav. Neurosci.* **98,** 487–495.

McGaugh, J. L. (1961). Facilitative and disruptive effects of strychnine sulphate on maze learning. *Psychol. Rep.* **8,** 99–104.

McGaugh, J. L., and Martinez, J. L., Jr. (1981). Learning modulatory hormones: An introduction to Endogenous peptides and learning and memory processes. *In* "endogenous Peptides and Learning and Memory Processes" (J. L. Martinez, Jr., R. A. Jensen, R. B. Messing, H. Rigter, and J. L. McGaugh, eds.), pp. 1–3. Academic Press, New York.

Miller, M., Barranda, E. G., Dean, M. C., and Brush, F. R. (1976). Does the rat with hereditary hypothalamic diabetes insipidus have impaired avoidance learning and/or performance? *Pharmacol., Biochem. Behav.* **5,** 35–40.

Mirsky, I. A., Miller, R., and Stein, M. (1953). Relation of adrenocortical activity and adaptive behavior. *Psychosom. Med.* **15,** 574–588.

Moltz, H., Lubin, M., Leon, M., and Numan, M. (1970). Hormonal induction of maternal behavior in the ovariectomized nulliparous rat. *Physiol. Behav.* **5,** 1373–1377.

Nemeroff, C. B., Bissette, G., Prange, A. J., Jr., Loosen, P. T., Barlow, T. S., and Lipton, M. A. (1977). Neurotensin: Central nervous system effects of a hypothalamic peptide. *Brain Res.* **128,** 485–496.

Nemeroff, C. B., Luttinger, D., and Prange, A. J., Jr. (1980). Neurotensin: Central nervous system effects of a neuropeptides. *Trends NeuroSci.* **3,** 212–215.

Pigache, R. M., and Rigter, H. (1981). Effects of peptides related to ACTH on mood and vigilance in man. *Front. Horm. Res.* **8,** 193–207.

Rigter, H. (1978). Attenuation of amnesia in rats by systematically administered enkephalins. *Science* **200,** 83–85.

Rigter, H. (1982). Vasopressin and memory: The influence of prior experience with the training situation. *Behav. Neural Biol.* **34,** 337–351.

Rigter, H., and Crabbe, J. C. (1979). Modulation of memory by pituitary hormones and related peptides. *Vitam. Horm. (N.Y.)* **37,** 153–241.

Rigter, H., and Popping, A. (1976). Hormonal influences on the extinction of conditioned taste aversion. *Psychopharmacology* **46,** 255–261.

Rigter, H., van Riezen, H., and de Wied, D. (1974). The effect of ACTH- and vasopressin analogues on CO_2-induced retrograde amnesia in rats. *Physiol. Behav.* **13,** 381–388.

Rigter, H., Greven, H., and Riezen, H. (1977). Failure of naloxone to prevent reduction of amnesia by enkephalins. *Neuropharmacology* **16,** 545–547.

Rigter, H. H., Hannan, T. J., Messing, R. B., Martinez, J. L., Jr., Vasquez, B. J., Jensen, R. A., Veliquette, J., and McGaugh, J. L. (1980a). Enkephalins interfere with acquisition of an active avoidance response. *Life Sci.* **26,** 337–345.

Rigter, H., Jensen, R. A., Martinez, J. L., Jr., and McGaugh, J. L. (1980b). Enkephalin and fear-motivated behavior. *Proc. Natl. Acad. Sci. U.S.A.* **77,** 3729–2732.

Schulz, H., Kovács, G. L., and Telegdy, G. (1974). Effect of physiological doses of vasopressin and oxytocin on avoidance and exploratory behavior in rats. *Acta Physiol. Acad. Sci. Hung.* **45,** 211–215.

Sokol, H. W., and Valtin, H. (1982). *Ann. N.Y. Acad. Sci.* **394.**

Stein, L., and Belluzzi, J. D. (1979). Brain endorphins: Possible role in reward and memory formation. *Fed. Proc., Fed. Am. Soc. Exp. Biol.* **38,** 2468–2472.

Stratton, L. O., and Kastin, A. J. (1974). Avoidance learning at two levels of shock in rats receiving MSH. *Horm. Behav.* **5,** 149–155.

Svare, B., Martke, A., Boherty, P., Mason, I., Michael, S. D., and Smith, M. S. (1979). Hyperprolactinemia suppresses copulatory behavior in male rats and mice. *Biol. Reprod.* **21,** 529–535.

Telegdy, G. M., Fekete, M., and Kádár, T. (1982). Cholecystokinin in the brain. *Int. Med.* **2,** 2–5.

Valtin, H., Sawyer, W. H., and Sokol, H. W. (1965). Neurohypophysial principles in rats homozygous and heterozygous for hypothalamic diabetes insipidus (Brattleboro strain). *Endocrinology (Baltimore)* **77,** 701–706.

Van Ree, J. M., and de Wied, D. (1977). Heroin self-administration is under control of vasopressin. *Life Sci.* **21,** 315–320.

Van Ree, J. M., Bohus, B., Versteeg, D. H. G., and de Wied, D. (1978). Neurohypophyseal principles and memory proceses. *Biochem. Pharmacol.* **27,** 1793–1800.

Van Ree, J. M., Gaffori, O., and de Wied, D. (1983). In rats, the behavioral profile of CCK-8 related peptides resembles that of antipsychotic agents. *Eur. J. Pharmacol.* **93,** 63–78.

Van Ree, J. M., Hijman, R., Jolles, J., and de Wied, D. (1985). Vasopressin and related peptides: Animal and human studies. *Prog. Neuro-Psychopharmacol. Biol. Psychiatry* **9,** 551–559.

van Wimersma Greidanus, Tj. B. (1970). Effect of steroids on extinction of an avoidance response in rats. A structure-activity relationship study. *Prog. Brain Res.* **32,** 185–191.

van Wimersma Greidanus, Tj. B. (1977). Effects of MSH and related peptides on avoidance behavior in rats. *Front. Horm. Res.* **4**, 129–139.

van Wimersma Greidanus, Tj. B. (1982a). MSH/ACTH 4-10: A tool to differentiate between the role of vasopressin in memory consolidation or retrieval processes. *Peptides (N.Y.)* **3**, 7–11.

van Wimersma Greidanus, Tj. B. (1982b). Disturbed behavior and memory of the Brattleboro rat. *Ann. N.Y. Acad. Sci.* **394**, 655–662.

van Wimersma Greidanus, Tj. B., and de Wied, D. (1969). Effects of intracerebral implantation of corticosteroids on extinction of an avoidance response in rats. *Physiol. Behav.* **4**, 365–370.

van Wimersma Greidanus, Tj. B., and de Wied, D. (1976). Modulation of passive avoidance behavior of rats by intracerebroventricular administration of anti-vasopressin serum. *Behav. Biol.* **18**, 325–333.

van Wimersma Greidanus, Tj. B., and Veldhuis, H. D. (1985). Vasopressin: Site of behavioral action and role in human performance. *Peptides (N.Y.)* **6**, Suppl. 2, 177–180.

van Wimersma Greidanus, Tj. B., Wijnen, H., Deurloo, J., and de Wied, D. (1973). Analysis of the effect of progesterone on avoidance behavior. *Horm. Behav.* **4**, 19–30.

van Wimersma Greidanus, Tj. B., Dogterom, J., and de Wied, D. (1975). Intraventricular administration of anti-vasopressin serum inhibits memory consolidation in rats. *Life Sci.* **16**, 637–644.

van Wimersma Greidanus, Tj. B., Tjon Kon Fat-Bronstein, H., and Van Ree, J. M. (1978). Antisera to pituitary hormones modulate development of tolerance to morphine. *In* "Characteristics and Function of Opioids" (J. M. Van Ree and L. Terenius, eds.), pp. 73–74. Elsevier/North-Holland Biomedical Press, Amsterdam.

van Wimersma Greidanus, Tj. B., Van Praag, M. C. G., Kalmann, R., Rinkel, G. J. E., Croiset, G., Hoeke, E. C., van Egmond, M. A. H., and Fekete, M. (1982). Behavioral effects of neurotensin. *Ann. N.Y. Acad. Sci.* **400**, 319–329.

van Wimersma Greidanus, Tj. B., Bohus, B., Kovács, G. L., Versteeg, D. H. G., Burbach, J. P. H., and de Wied, D. (1983a). Sites of behavioral and neurochemical action of ACTH-like peptides and neurohypophyseal hormones. *Neurosci. Biobehav. Rev.* **7**, 453–463.

van Wimersma Greidanus, Tj. B., Van Ree, J. M., and de Weid, D. (1983b). Vasopressin and memory. *Pharmacol. Ther.* **20**, 437–458.

van Wimersma Greidanus, Tj. B., Jolles, J., and de Wied, D. (1985a). Hypothalamic neuropeptides and memory. *Acta Neurochir.* **75**, 99–105.

van Wimersma Greidanus, Tj. B., Donker, D. K., Van Zinnicq Bergmann, F. F. M., Bekenkamp, R., Maigret, C., and Spruijt, B. (1985b). Comparison between excessive grooming induced by bombesin or by ACTH: The differential elements of grooming and development of tolerance. *Peptides (N.Y.)* **6**, 369–372.

van Wimersma Greidanus, Tj. B., Donker, D. K., Walhof, R., Van Grafhorst, J. C.

A., De Vries, N., Van Schaik, S. J., Maigret, C., Spruijt, B. M., and Colbern, D. L. (1985c). The effects of neurotensin, naloxone and haloperidol on elements of excessive grooming behavior induced by bombesin. *Peptides (N.Y.)* **6,** 1179–1183.

van Wimersma Greidanus, Tj. B., Burbach, J. P. H., and Veldhuis, H. D. (1986). Vasopressin and oxytocin; their presence in the central nervous system and their functional significance in brain processes related to behaviour and memory. *Acta Endocrinol. (Copenhagen), Suppl.* **276,** 85–94.

van Wimersma Greidanus, Tj. B., Van de Brug, F., De Bruijckere, L. M., Pabst, P. H. M. A., Ruesink, R. W., Hulshof, R. L. E., Van Berckel, B. N. M., Arissen, S. M., De Koning, E. J. P., and Donker, D. K. (1987). Comparison of bombesin, ACTH and β-endorphin induced grooming: Antagonism by haloperidol, naloxone and neurotensin. *Ann. N.Y. Acad. Sci.* **525,** 519–527.

Vécsei, L., Balász, M., and Telegdy, G. (1987). Action of somatostatin on the central nervous system. *In* "Neuropeptides and Brain Function" (G. Telegdy, ed.), pp. 36–57. Karger, Basel.

Weijnen, J. A. W. M., and Slangen, J. L. (1970). Effects of ACTH-analogues on extinction of conditioned behavior. *Prog. Brain Res.* **32,** 221–235.

Zarrow, M. X., Farooq, A., Denenberg, V. H., Sawin, P. B., and Ross, S. (1963). Maternal behavior in the rabbit: Endocrine control of maternal nest building. *J. Reprod. Fertil.* **6,** 375–383.

CHAPTER SIX

Hormonal Modulation of Memory

James L. McGaugh and Paul E. Gold

I. Introduction

Hormones have a variety of influences on physiological systems involved in adaptation to environmental changes. For most if not all species of animals, the ability to learn and remember information based on experiences is essential for survival. The evidence summarized in the preceding chapter indicates that learning is markedly influenced by hormones (van Wimersma Greidanus and Rigter, this volume). It is clear that hormonal levels during learning can affect acquisition as well as retention and extinction. There is also extensive evidence, which is the focus of this chapter, that the retention of newly acquired information is influenced by hormonal levels following the learning experience. This evidence suggests that hormones can influence retention through effects on memory storage processes independently of effects the hormones may have on processes influencing acquisition (Gold and McGaugh, 1978; McGaugh, 1983; Versteeg and de Wied, 1985).

Most studies of the influence of hormones on memory have focused on hypothalamic-pituitary hormones (ACTH, vasopressin, oxytocin, and β-endorphin), hormones of the adrenal medulla (epinephrine, norepinephrine, and enkephalin) and, to a lesser extent, hormones of the adrenal cortex (glucocorticoid, corticosterone, in the rat). There are several reasons for this. First, there has been for many years considerable interest in stress-related hormones and, as a consequence, this area of neuroendocrinology and neu-

Psychoendocrinology, copyright © 1989 by Academic Press, Inc. All rights of reproduction in any form reserved.

ropharmacology has a long history (Dunn and Kramarcy, 1984). The development of new techniques and drugs has greatly accelerated research in these areas in recent years. Second, the fact that stress-related hormones are released by even relatively mildly stimulating experiences has suggested that these hormones may have an important adaptive role in the regulation of memory just as they do in the regulation of other adaptive physiological processes. It seems reasonable to suggest that experiences which are sufficiently exciting to elicit the release of glucose are likely to be worth remembering and that hormones which are involved in reacting to exciting events may also influence the storage of memory of the events (Gold and McGaugh, 1975, 1978; Kety, 1972).

The evidence suggesting that hormones affect memory is based primarily on experiments investigating the effects on retention of hormones or treatments that alter the functioning of endocrine systems. It is clear from such studies that such treatments can alter retention. And the findings of such studies offer strong support for the view that memory storage processes are modulated by endogenously released hormones (McGaugh, 1983). However, the fact that memory can be altered by treatments that affect hormonal functioning does not compel the conclusion that endogenous hormones are normally involved in memory storage. If endogenous hormones are involved in modulating memory storage, a number of findings are to be expected: (1) training experiences should release the hormones in question; (2) exogenously administered hormones should have greatest effects when administered shortly after training (at the time at which the training releases the hormones); (3) the effects of exogenously administered hormones should depend upon the levels of endogenously released hormones; (4) retention should be affected by treatments affecting the release of hormones or the activation of receptors; and (5) the effects of treatments interfering with the release of hormones should be attenuated by administration of the hormones. These implications have been extensively investigated in recent research. In subsequent sections we consider the extent to which the research findings are consistent with these implications.

Efforts to investigate the role of hormones in memory are complicated by the fact that many hormones which are released by endocrine glands in the periphery are also present in the brain, where they are considered to be putative transmitters or neuromodulators (Goldstein, 1976; Krieger and Liotta, 1979). As a consequence, many treatments that affect peripheral hormones no doubt also directly affect central transmission and neuromodulation. And, peripherally released hormones might well have direct central as well as peripheral effects. A further complication comes from the fact that there is, for many of the hormones of interest, no definitive evidence as to whether hormones of peripheral origin act directly on the brain. Evaluation of hypotheses concerning the mechanisms underlying the effects of hormones on memory will require identification of the locus of the receptors activated by the hormones. This is a particularly critical matter since, as we discuss below, there is extensive recent evidence that hormones interact with transmitter systems in their effects on memory. In the absence of knowledge of whether peripheral hormones pass the blood–brain barrier in amounts sufficient to produce central physiological effects, it is difficult to determine the locus of the interactions (see also Chapter 1, this volume).

It is now well established that hormones administered shortly after training can enhance or impair retention on tests given a day or longer following training. As is found with other posttraining treatments known to influence retention, the effects are time dependent: that is, the degree of the effect of hormones on retention decreases as the interval between the training and hormone administration is increased. Such findings are generally interpreted as indicating that the hormones affect retention by modulating memory storage processes (McGaugh, 1983). Since the hormones are administered after training, the effects on retention cannot, of course, be attributed to influences on factors influencing acquisition performance. Further, since the hormones are usually administered many hours or days before the retention test, the findings that the effects are time dependent indicate that it is unlikely that the effects on retention are due directly to influences acting at the time of the retention test.

However, hormones have been shown to affect retention performance when administered shortly before the retention test (Riccio and Concannon, 1981). Thus, it appears that hormones may affect both storage and retrieval processes. Recent evidence suggests that a stressful experience (Spear, 1978) or pretest injections of catecholamine, opioid agonists, or corticotrophin (ACTH) can enhance performance on a test trial (e.g., de Almeida and Izquierdo, 1984; Izquierdo, 1984; Izquierdo and Dias, 1983a; Quartermain, 1982; Riccio and Concannon, 1981; Sara and Deweer, 1982). Such findings are consistent with the view that some neuroendocrine systems may facilitate access to memories. It is interesting to note that treatments such as these can facilitate retention performance which is poor for any of a number of reasons, including forgetting, infantile amnesia, or retrograde amnesia (cf. Riccio and Concannon, 1981; Riccio and Ebner, 1981).

In the broadest terms, this research can be viewed as part of an attempt to learn the rules by which neuroendocrine states at the time of acquisition, storage, and retrieval interact to regulate all phases of learning and memory. Recent studies suggest that the effects of posttraining hormonal conditions on memory may be state dependent: Under some conditions the effects depend upon the congruence between the hormonal state immediately after learning and the state at the time of testing (e.g., Haroutunian and Riccio, 1977; Izquierdo, 1980, 1984; Izquierdo et al., 1981; Riccio and Concannon, 1981; Spear, 1978; Zornetzer, 1978). In particular, Izquierdo and his colleagues have found that pretest injections of epinephrine or ACTH can enhance retention performance.

Interestingly, there is some intercompatibility between these hormones; when animals are trained in an inhibitory avoidance task with weak footshock and given posttraining injections of ACTH or epinephrine, pretest injections of either ACTH or epinephrine will enhance retention performance (Izquierdo and Dias, 1983b). These results suggest that the relationship between hormonal states at the time of training and of testing will not be precisely those which have been obtained in classical state-dependent learning experiments (Overton, 1971). However, as we discuss below, one important consideration in evaluating these experiments is that none of the experiments, those that either support or do not support a state-dependency view, is quite complete; in no case is the necessary information provided regarding endogenous hormonal levels at the time of testing.

These findings are generally consistent with the view that the neuroendocrine status at the time of retention testing can affect retrieval and that similar neuroendocrine states following training and at the time of testing create optimal conditions for retrieval. However, it is important to note some of the problems raised by these views. First, if hormones can provide generally enhanced access to stored memories, as is often assumed, it seems unlikely that they would improve test performance since there would be little specificity to those memories enhanced; enhancing all memories would seem likely to cause confusion and impair retention performance. Therefore, it would seem that the specificity must come from cues specific at the test trial, i.e., the specific and contextual stimuli provided by the testing procedures and, of primary importance here, the internal stimuli provided by hormonal and neurotransmitter systems. Second, in contrast to the posttrial design found useful in memory storage experiments, there is no readily apparent way to examine drug effects on memory retrieval mechanisms without influencing other conditions, including motivation, motor abilities, and perceptual processes, which influence performance on a retention test.

Clearly, precise examination of these views of the roles of hormones in memory retrieval will require assessment of the status of endogenous systems following training as well as at the time of the test trials. To our knowledge, however, no studies have, as yet, directly measured hormonal states produced by retrieval-enhancing drugs and hormones administered to an animal prior to retention testing. Such measurements would alter the significance of the research in interesting and important ways. Once known, the specific neuroendocrine responses which are present at the time of testing become candidates for systems which may activate retrieval mechanisms. Moreover, if these systems are engaged during testing, any resulting changes in motivation, motor, or sensory systems need not be viewed as confounds of retrieval processes; such changes become part of the neurobiological processes which control performance of learned responses.

While this chapter emphasizes the effects of specific hormonal systems, it is clear that no system works alone. Each experience no doubt activates a myriad of hormonal and transmitter systems in the periphery and in different brain regions. The findings reviewed in this chapter suggest that we are beginning to understand the role of some hormones in memory as well as some of the ways that specific hormones interact with other hormonal and transmitter systems in modulating memory. We turn now to a review of the experimental evidence.

II. Peripheral Catecholamines

Of the hormones investigated, the adrenergic catecholamines have provided perhaps the most extensive evidence concerning the involvement of a peripheral hormonal system in memory modulation. The fact that there is information about catecholamines and memory is due in part to the availability of pharmacological tools, including drugs which affect catecholamines, and sensitive catecholamine assays. Peripheral catecholamines represent the only neuroendocrine system for which there are answers, albeit incomplete, relevant to each of the criteria listed above concerning the possible involvement of hormones in memory modulation.

Acute stress stimulation releases epinephrine and norepinephrine into plasma. The epinephrine found in plasma, at both basal and stress-related levels, is derived from the adrenal medulla. Epinephrine is not seen in the plasma of adrenalectomized rats either before or after stress. Although basal norepinephrine levels are due primarily to adrenal medullary release, the increased levels seen after stress appear to reflect overflow from sympathetic ganglia (Kvetnansky *et al.*, 1979). Levels of peripheral catecholamines, and epinephrine in particular, are highly sensitive to behavioral manipulations. For example, the plasma epinephrine levels of rats placed in a novel compartment are roughly double those of undisturbed rats (McCarty and Gold, 1981). Although this elevation is significant, the magnitude of the effect is small when compared to the 10-fold rise in plasma epinephrine found after administration of a footshock. Footshock at an intensity such as that used to produce good acquisition of inhibitory (passive) avoidance training results in a 10- to 15-fold rise in plasma catecholamine levels (McCarty and Gold, 1981). The rise in plasma catecholamine levels is maximal immediately after the footshock and levels return to normal within a few minutes. A mild footshock at an intensity which does not result in optimal retention performance adds little to the elevation in plasma epinephrine seen when animals are placed in the novel training apparatus. Thus, the first of the conditions listed above which implicates a hormone in memory modulation is met for peripheral catecholamines: the hormones are released by training procedures and the retention performance is correlated with the magnitude of the release.

The second implicating condition is that exogenous administration of the hormone should affect memory and, in particular, should be most effective when administered shortly after the training experience, i.e., when administered at a time at which endogenous levels would be high. This possibility has been investigated in several studies. In these studies, animals were usually trained with a mild footshock (which produced low endogenous plasma epinephrine levels) and were then given an injection of epinephrine immediately or at one of several delays posttraining. In the initial experiments of this type (Gold and van Buskirk, 1975, 1976b, 1978a,b; Gold *et al.*, 1977b; Izquierdo and Dias, 1983a), rats or mice which received a subcutaneous injection of epinephrine immediately after training on an inhibitory avoidance task had enhanced retention performance when tested the next day. Injections delayed by 30 minutes or more after training did not have an effect on retention performance. Furthermore, in those studies which examined a broad dose–response range, the epinephrine effects on memory exhibited an inverted-U dose–response curve. Intermediate doses facilitated retention performance while high doses produced retrograde amnesia. More recent findings indicate that epinephrine can enhance memory for appetitive tasks as well as avoidance tasks (Sternberg *et al.*, 1985b). In addition, peripheral injections of amphetamine or epinephrine can enhance the development of a neurophysiological analog of memory, long-term potentiation (Gold *et al.*, 1984). The effect on long-term potentiation follows the inverted-U dose–

response curve with maximal facilitation at doses comparable to those effective in behavioral studies of memory.

Epinephrine can also enhance memory of animals which have poor memory retention for any of several reasons. For example, epinephrine enhances memory in juvenile rats which otherwise exhibit rapid forgetting (infantile amnesia). In fact, the posttraining injection of epinephrine was more effective than an increase in the footshock intensity in enhancing 24-hour retention performance in 16-day-old rats (Gold et al., 1982). Such results suggest that the mechanisms for long-term memory storage are available to the juvenile rats, but the hormonal modulatory events which initiate the storage mechanisms may develop somewhat later. In addition, there is recent evidence that posttraining injections of epinephrine can enhance memory in aged rodents as well. Two-year-old rats have impaired release of adrenal medullary catecholamines following footshock (McCarty, 1981). These animals, as well as aged mice, also show very rapid forgetting of inhibitory avoidance and other learned responses (Bartus et al., 1979, 1980; Bartus and Dean, 1981; Gold and McGaugh, 1975; Gold et al., 1981; Kubanis and Zornetzer, 1981; Zornetzer et al., 1982). Young adult rodents demonstrate good retention for several weeks after inhibitory avoidance training. Aged rodents appear to forget the learned response within days or even hours. However, rats and mice which receive a posttrial injection of epinephrine have good retention performance for as long as a week after training (Sternberg et al., 1985a). These findings are consistent with the view that the impaired memory in aged rats and mice may be due, at least in part, to an impairment of stimulation-induced release of peripheral epinephrine.

The findings of another recent experiment indicate that epinephrine may even enable learning to occur during deep anesthesia (Weinberger et al., 1984). Rats in this study were anesthetized with nembutal and chloral hydrate. The animals then received a series of classical conditioning trials in which cutaneous shock was paired with white noise. Retention was assessed several days later by examining conditioned suppression of drinking during the presentation of white noise. Animals given a saline injection prior to conditioning showed no conditioned suppression. Remarkably, however, animals given epinephrine prior to training demonstrated significant suppression of drinking in the presence of the conditioned stimulus. These findings suggest that the anesthetized brain may retain the capacity for storing new information but that the mechanisms which underlie that storage cannot be initiated in the absence of an appropriate hormonal milieu. This experimental preparation may prove useful in identifying neuronal systems which must be activated for memory storage to occur and may thus facilitate our understanding of the way in which epinephrine acts on the central nervous system to promote information storage.

If exogenous epinephrine acts on memory in animals by adding to the endogenous hormonal response to training, changes in the footshock level which are known to affect plasma epinephrine levels should shift the dose–response curves for the effects of epinephrine on memory. The findings of several studies indicate that the dose of epinephrine which facilitates memory does change with footshock level. For example, we have already noted that the effects of epinephrine on memory vary with dose in an inverted-U manner: high doses produce retrograde amnesia and intermediate doses facilitate memory storage. Moreover, a dose which enhances memory after low footshock can produce amnesia if administered after high footshock (Gold and van Buskirk, 1976b, 1978a,b; Gold *et al.*, 1977b; Izquierdo and Dias, 1983b). Thus, these findings are consistent with the view that the inverted-U dose–response curve is based on the summation of exogenous and endogenous epinephrine.

There is now considerable evidence concerning the effects of adrenergic receptor antagonists on memory modulation. Gold and Sternberg (1978) reported that the α-receptor antagonist, phenoxybenzamine, attenuated the retrograde amnesia produced by several amnestic treatments. In this study, animals received an injection of phenoxybenzamine 30 minutes prior to training; the injection of the antagonist did not itself affect acquisition or retention performance. The animals were then trained in an inhibitory avoidance task and animals in different groups were given one of the following amnestic treatments: supraseizure electrical stimulation of frontal cortex, subseizure stimulation of the amygdala, cycloheximide (a protein synthesis inhibitor), diethyldithiocarbamate (DDC) (a norepinephrine synthesis inhibitor), or pentylenetetrazol (a convulsant drug). On retention tests the next day, each amnestic treatment was found to be effective in impairing memory. However, the pretraining injection of phenoxybenzamine attenuated the amnesia produced by all of the posttraining treatments. Additional studies demonstrated that peripheral, but not central, injections of a variety of α- and β-adrenergic receptor antagonists attenuated amnesia and memory facilitation produced by a wide range of memory modulation treatments (Sternberg and Gold, 1980, 1981; Sternberg *et al.*, 1983). Such results suggest that peripheral adrenergic influences may be involved in the effects on memory of a variety of treatments. These findings are consistent, for example, with the finding that posttrial frontal cortex stimulation, at an intensity which produces brain seizures and amnesia, adds to the plasma epinephrine levels reached after the training footshock (Gold and McCarty, 1981).

A number of studies have reported the retention performance is impaired by treatments that interfere with peripheral adrenergic mechanisms and that peripheral injections of adrenergic agonists can reverse the impairment. For example, Stein et al. (1975) found that injections of DDC administered prior to training impaired memory and that the impairment was attenuated by posttraining intracerebroventricular (i.c.v.) administration of norepinephrine. These results would appear to support the view that DDC interfered with memory storage by impairing brain norepinephrine synthesis and that the norepinephrine levels were restored by the i.c.v. injections. However, the effects of DDC on memory can also be reversed with peripheral injections of epinephrine or norepinephrine (McGaugh et al., 1979; Meligeni et al., 1978). Further, since epinephrine is known to enhance retention under conditions where retention would otherwise be poor it is difficult to interpret the basis of this effect. The apparent "reversal" of a deficit might be due to memory facilitation above the impaired level rather than specific reversal of the drug-induced neurobiological deficit.

In a somewhat similar study, Walsh and Palfai (1979) found that pretrial injections of reserpine, as well as the peripherally acting analog syringoserpine, impaired memory for inhibitory avoidance training. The amnesia produced by either drug was reversed with posttrial injections of epinephrine. However, comparable results were not obtained when similar procedures were used with animals trained in a discrimination task. The basis for these seemingly discrepant findings is unclear.

In view of the finding that retention is readily influenced by posttraining peripheral administration of epinephrine it is perhaps somewhat surprising that adrenalectomy and adrenal demedullation do not always interfere with memory storage or with memory modulation. Memory impairments in adrenalectomized animals have been reported (Borrell et al., 1983a), but adrenal demedullation does not always produce deficits in learning or retention (Bennett et al., 1985; Liang et al., 1985). It may be that following adrenalectomy or adrenal demedullation the norepinephrine overflow into plasma which is retained in adrenalectomized animals (Kvetnansky et al., 1979) partially compensates for the loss of epinephrine. Sternberg et al. (1982) found that bretylium, which blocks the release of norepinephrine from peripheral sympathetic nerve endings, did not affect acquisition of an inhibitory avoidance response and did not block the amnestic effects of cortical stimulation. However, there have, as yet, been no studies of the effects of bretylium on retention in adrenal demedullated animals.

The findings of a series of recent experiments indicate that adrenal demedullation has clear, albeit subtle, effects on memory modulation. These studies examined the effects of low-intensity stimulation of the amygdala on memory in adrenal demedullated rats. Of primary interest here is the finding that amygdala stimulation produced amnesia in intact animals but enhanced memory in demedullated rats (Bennett et al., 1985). However, the amygdala stimulation impaired retention in demedullated animals if they were given epinephrine intraperitoneally (i.p.) immediately after training but just prior to the amygdala stimulation. If the epinephrine was administered after the amygdala stimulation, retention was enhanced. Thus epinephrine appears to influence the effects of brain stimulation on memory (Liang et al., 1985).

Studies of the effects of epinephrine on memory have addressed most of the criteria, discussed above, which can be used to evaluate the involvement of a hormone in memory modulation: the hormone is released by training; memory is modulated by posttraining administration of epinephrine; the extent and direction of modulation (whether enhancement or impairment) varies with the endogenous epinephrine level at the time of injection; adrenergic antagonists affect memory modulation; and many treatments affecting memory appear to act through influences involving peripheral epinephrine.

Still unanswered is the mechanism by which epinephrine acts on the central nervous system. The amine does not readily enter the brain (Axelrod et al., 1959) but might affect neural function by such mechanisms as peripheral afferent monitoring of plasma epinephrine levels, modifications of cerebral blood flow (Berntman et al., 1978), or increasing glucose availability for central nervous system metabolism. Preliminary evidence for the latter possibility comes from three recent studies in which glucose injections were found to enhance memory (Gold, 1986; Hall and Gold, 1986; Messier et al., 1984). Furthermore, unlike other treatments which modulate memory, glucose facilitation of memory appears not to be attenuated by adrenergic antagonists (Gold et al., 1986). One interpretation of these findings is that epinephrine, and many other memory modulators, may act on memory by releasing glucose from liver stores via adrenergic receptors. According to this view, adrenergic antagonists would be expected to attenuate memory modulation produced by all treatments which act by this mechanism, but not alter the modulation produced by glucose itself.

III. ACTH and Corticosterone

Stress results not only in the release of adrenomedullary epinephrine, but also in the release of ACTH from the adenohypophysis and, as a consequence, the release of adrenal cortical steroids. The effects of ACTH on memory are in many ways similar to those described for epinephrine. First, the hormone is released after stress, presumably including footshock stress such as that used in training. Second, ACTH enhances or impairs memory in an inverted-U dose–response manner (Gold and van Buskirk, 1976a,b). Third, the effect of a given dose of ACTH on memory varies with the footshock level (Gold and van Buskirk, 1976b). A posttraining injection of ACTH either enhances or impairs retention depending on the intensity of the training footshock level. Impairments in learning and memory can be observed in hypophysectomized rats (de Wied, 1964, 1974) and these effects can be reduced by posttraining ACTH injections (Gold *et al.*, 1977a). As we have noted, however, there is a serious problem in interpreting the findings of such "replacement" effects when the hormone which is replaced is itself capable of enhancing memory in untreated rats as well.

The findings obtained in studies of the effects of posttraining administration of ACTH are much like those described previously for epinephrine. This evidence strongly supports the view that ACTH plays a role in endogenous modulation of memory storage. The most readily apparent mechanism by which ACTH might act on memory is by releasing adrenal cortical steroids. However, there is some evidence against this possibility. Much of the evidence comes from early studies of de Wied and colleagues (de Wied, 1969, 1974; de Wied *et al.*, 1972), who found that ACTH analogs which lacked steroidogenic action retained the effects of ACTH on extinction performance or when administered prior to retrieval tests in several behavioral tasks. Also, Gold and van Buskirk (1976a) found that posttraining injections of corticosterone did not affect retention performance for inhibitory avoidance training. Such findings are consistent with the view that ACTH may act on memory by direct actions on the brain, an interpretation supported by the finding that ACTH (4-10) can affect learning when directly injected into the brain.

However, more recent evidence suggests that adrenal steroids may be involved in the effects of ACTH on memory. Although there is extensive evidence that ACTH analogs affect performance on extinction and retrieval tests, the findings of the effects on memory of posttraining injections are, at best, inconclusive. Flood *et al.* (1976) reported that posttraining injections of

ACTH (4-10), a nonsteroidogenic ACTH analog, affected retention in mice trained in either an active or an inhibitory avoidance task. However, other studies which have tested the ACTH (4-9) and (4-10) analogs have not obtained comparable results (Fekete and de Wied, 1982; Martinez et al., 1979; Soumireu-Mourat et al., 1981). It is possible that injections prior to training and testing affect performance through influences on sensory, attentional, or motivational processes (McGaugh, 1983).

At the least, the differences in the behavioral actions of ACTH and its analogs suggest that the peptides may act through different biological mechanisms. Thus, the findings reopen the question of whether the effects of ACTH on memory might be mediated by corticosteroids. A number of recent studies have examined the question of whether adrenal steroids modulate memory. In one series of studies, Leshner and colleagues (Leshner and Potlitch, 1979; Leshner et al., 1981) examined the role of pituitary-adrenal hormones on retention of a learned change in social behavior in mice. Mice which have an encounter with a trained fighter exhibit submissive behavior on later tests. Postexperiential injections of ACTH resulted in increased submission on these subsequent tests. In this situation, postattack injections of corticosterone also enhanced retention of the submissive behavior. Thus, the effects of ACTH on this learned behavior may be mediated by ACTH-initiated release of steroids from the adrenal cortex.

A possible role for corticosterone in memory modulation is also indicated by the findings of another set of studies in which corticosterone was administered intracerebrally shortly after training in appetitive tasks. Micheau et al. (1981, unpublished findings) found that immediate posttrial intraventricular injections of the steroid facilitated retention of a discriminative operant response but did not influence retention of a continuously reinforced operant response. Because of the evidence indicating uptake of the glucosteroids by the hippocampus, these investigators also examined the effects on memory of posttraining intrahippocampal injections of corticosterone. As was found with intraventricular injections, the hippocampal injections did not enhance retention of animals trained on a continuously reinforced schedule, but did enhance retention when administered after successive discrimination training. Furthermore, the effects on memory were time dependent; memory enhancement was obtained with injections given immediately, 30 minutes, or 3 hours after training, but not when delayed by 6 hours (Micheau et al., 1985). Corticosterone injections into the hippocampus also facilitated later retention of extinction of operant conditioning (Micheau et al., 1985). The findings should produce a renewed interest in the possibility that the effect of ACTH on memory might be mediated at least in part by adrenal cortical steroids.

IV. Opioid Peptides

There is now extensive evidence that opioid peptides are released centrally as well as peripherally when animals are stressed (Amir *et al.*, 1980; Bodnar *et al.*, 1980; Izquierdo *et al.*, 1980b; Rossier *et al.*, 1977). β-Endorphin is released from the hypothalamus (Carrasco *et al.*, 1982a) and from the pituitary along with ACTH. Enkephalin is released from the adrenal medulla along with norepinephrine and epinephrine (Viveros *et al.*, 1980). Recent findings suggest that opioid peptides, particularly β-endorphin, released by stimulation may be involved in the endogenous modulation of memory storage.

An involvement of opioid peptides in memory storage was first suggested by findings of studies of the effects on memory of posttraining administration of opioid agonists and antagonists. Many studies have reported that retention is enhanced by posttraining administration of the opiate antagonist naloxone. Enhancing effects have been found in a variety of learning tasks including habituation, spatial maze learning, and active and inhibitory avoidance learning (Baratti *et al.*, 1984; Gallagher and Kapp, 1978; Gallagher *et al.*, 1981, 1983; Izquierdo, 1979; Messing *et al.*, 1979). The evidence from several studies indicating that the opiate agonist morphine blocks the effects of naloxone on memory suggests that the effects of naloxone are due selectively to antagonist action at opiate receptors (Izquierdo, 1979; Messing *et al.*, 1979). Memory-modulating effects have been obtained with central as well as peripheral administration of opioid agonists and antagonists. Gallagher and her colleagues reported that retention of an inhibitory avoidance response was enhanced by posttraining intraamygdala injections of naloxone. Further, intraamygdala injections of the opiate agonist levorphanol produced impairment of retention which was blocked by naloxone (Gallagher and Kapp, 1978; Gallagher *et al.*, 1981). The report that naloxone enhances spatial memory (Gallagher *et al.*, 1983) is of interest in view of evidence suggesting that spatial memory may be a special form of memory (O'Keefe and Nadel, 1978). However in contrast to the findings of Gallagher *et al.* (1983), Beatty (1983) reported that spatial memory was not influenced by either naloxone or morphine. In both studies rats were trained on an eight-arm maze, and the drugs were then administered after the animals completed the first four alley choices. Performance (i.e., retention) on the remaining four choices was tested after a delay of several hours. There was, however, a major difference in these two studies. In the Gallagher *et al.* (1983) study the maze was moved to a new room (i.e., a novel spatial environment) on the day that the drugs were first administered. Animals given naloxone (or diprenorphine) each day after the first four trials were, after a delay of several hours, superior to controls in maze performance in the new spatial environment. Thus the opiate antagonists appeared to affect retention performance by enhancing the storage of the memory of the new environment. Beatty's (1983) findings suggest, however, that opiate drugs do not influence spatial memory when the spatial environment is constant.

Findings of studies of the effects of morphine and other opiate agonists on memory are less consistent than studies of the effects of opiate antagonists. Most studies have reported that, when administered in low doses, opiate agonists impair retention and that the memory impairment is antagonized by opiate antagonists (Castellano, 1975; Castellano *et al.*, 1984; Introini and Baratti,1984; Izquierdo, 1979; Messing *et al.*, 1979). As we noted above, such effects have been obtained with central as well as peripheral administration. For example, Gallagher and her colleagues (1981) have shown that post-training intraamygdala administration of naloxone enhances retention of an inhibitory avoidance response and antagonizes the retention impairment produced by the opiate agonist levorphanol. When administered posttraining in high doses morphine has been found to enhance retention of inhibitory avoidance learning (Mondadori and Waser, 1979; Staubli and Huston, 1980). It may be that such effects are due to the aversive properties of morphine in high doses. This issue might be clarified by additional studies examining the effects of high doses of morphine in a variety of learning tasks.

In an interesting series of studies Castellano and his colleagues (Castellano and Puglisi-Allegra, 1983; Castellano *et al.*, 1984) have shown that, in mice, posttraining immobilization stress has impairing effects on retention of inhibitory avoidance comparable to that produced by low doses of morphine. In DBA mice the impairing effects of both treatments were blocked by naloxone. The effects of the treatments were time dependent and neither treatment affected performance in controls that did not receive footshock on the training trial. The effect of the immobilization stress was strain dependent: In C57BL/6 mice posttraining immobilization stress enhanced retention and the enhancement was antagonized by naloxone. Both morphine and immobilization stress were less effective in mice that were familiarized with the training apparatus prior to the training (Castellano *et al.*, 1984). These findings are consistent with the view that the effect of immobilization stress on memory involves the release of endogenous opioids. As we pointed out above, both enkephalins and endorphins are released in the brain and periphery following stressful stimulation. These findings also fit well with evidence from a number of studies that retention is influenced by posttraining administration of enkephalin and endorphin (Belluzzi and Stein, 1981; Izquierdo *et al.*, 1981; Kovács and de Wied, 1981). The findings of studies of β-endorphin are, however, conflicting. Kovács *et al.* (1981) and Kovács and de Wied (1981) reported that inhibitory avoidance retention is enhanced by α- and β-endorphin and impaired by γ-endorphin. Martinez and Rigter (1980), in contrast, found that β-endorphin impaired retention and that α-endorphin had no effect. Izquierdo *et al.* (1980a) also reported that β-endorphin impairs retention and that the effect was blocked by naloxone. These studies were comparable except for route of administration. Thus, the reason for the different results is not clear. There is, of course, extensive evidence indicating that the effects of drugs on memory vary widely in different strains of rats and under different experimental conditions; doses that enhance retention under some conditions are sometimes found to impair retention under other experimental conditions. As nonlinear dose–response effects are typically obtained in studies of the effects of drugs and hormones on retention of inhibitory avoidance tasks, additional dose–response studies examining a wide range of doses of β-endorphin might provide clarification of the reasons for these differing results.

In a series of studies using an active avoidance task and a habituation task, Izquierdo and his colleagues (1981) found that retention was impaired by posttraining endorphin or enkephalin as well as morphine. These effects were blocked by naloxone. These investigators also reported that brain β-endorphin immunoreactivity is decreased following training. The decrease presumably reflects a release and metabolism of β-endorphin. β-endorphin immunoreactivity was also decreased by electroconvulsive shock (ECS). This finding suggests that the release of β-endorphin might contribute to the amnesia produced by ECS. In support of this view, recent studies have found that naloxone blocks the amnestic effects of ECS (Carrasco *et al.*, 1982b) as well as electrical stimulation of the amygdala (Liang *et al.*, 1983).

On the basis of these findings, Izquierdo (Izquierdo *et al.*, 1982) has argued that endorphins and enkephalins may be endogenous amnestic peptides. According to this view the opioid peptides function to regulate information storage by preventing the storage of excessive amounts of information. But, if this is the case, why are these hormones released along with other hormones, including ACTH and epinephrine, that have been found to enhance retention? If endogenously released hormones have both amnestic and enhancing effects, what determines the ultimate effect of the combination of hormonal influences on memory? Recently, Izquierdo (1984) suggested that the amnesia induced by β-endorphin may reflect a retrieval deficit based on state-dependent effects. This view is supported by evidence that good retention is seen in animals given posttraining β-endorphin if the peptide is also administered prior to the retention test.

It is not yet clear, however, that opioid hormones are uniquely amnestic. Recent findings indicate that the effects depend upon the behavioral task and experimental conditions used to study learning and retention as well as the receptors influenced by the peptide. In a series of recent studies Jefferys and his colleagues (1983, 1984, 1985) have studied retention in a task which uses swimming behavior as the index of retention. In this task, rats are placed in a narrow cylinder of water at a depth of 15 cm for 15 minutes on the "training" day and given a 5-minute test 24 hours later. Over the first 15 minutes the rats become relatively immobile and on the retention test normal rats are immobile for about 70% of the test period. In animals given naloxone 15 minutes after the first session the immobility is reduced to approximately 30%. Thus the naloxone appears to impair the retention of the learned immobility. The naloxone effect appears to be due to central actions since retention was not influenced by a quaternary naloxone analog. Retention performance is also impaired in adrenalectomized (ADX) rats. The impairment seen in ADX rats was attenuated by posttraining administration of corticosterone and dexamethasone, as well as the opiate agonist, [D-Ala^2-Met^5]enkephalinamide (DMAE). Naloxone antagonized the effects of both corticosterone and DMAE in ADX rats. Naloxone and DMAE have relatively nonspecific influences on opiate receptors. Other findings of Jefferys and his colleagues suggest that the effects are mediated by κ-receptors. The retention impairment found in ADX rats was attenuated by several κ-selective agonists, including ketocyclazocine, dynorphin (1-17), and Met^5-enkephalin-Arg^6-Phe^7 administered 15 minutes after the first swimming session. Opioids with selectivity for μ-receptors (including morphine) and δ-receptors were ineffective. The effects of ketocyclazocine were blocked by a κ-selective antagonist, MR2266. Jefferys et al. (1985) interpret the findings as indicating that a central opioid pathway involving κ-receptors is involved in the consolidation of information in this learning task.

While the findings of these studies are internally consistent they contrast with those of a large number of studies reporting that, in other kinds of learning tasks, retention is enhanced by naloxone and impaired by morphine as well as endorphin and enkephalin. It may well be that different opioid peptides are released with the conditions used in different learning tasks. In particular, the degree of stress (and consequent release of endogenous hormones) may be a critical factor. Further, as Castellano and Puglisi-Allegra (1983) have shown, the effects of stress depend upon the strain of animals studied. It is clear from these studies that posttraining opioid hormones can influence retention and that both antagonists and agonists can produce impairment as well as enhancement. Additional work is needed to clarify the bases of the differential involvement to opioid hormones in different types of learning tasks.

V. Vasopressin and Oxytocin

The peptide hormones vasopressin and oxytocin are synthesized in the magnocellular neurons of the supraoptic and paraventricular nuclei of the hypothalamus. Fibers from these cells are known to project to a large number of brain regions (Sofroniew et al., 1981). These hormones are secreted into the circulation from terminals in the posterior pituitary. They also appear to be released into the cerebrospinal fluid. Vasopressin, which is also termed antidiuretic hormone (ADH), influences the regulation of a variety of physiological functions including osmolarity and volume of body fluids as well as blood pressure (Doris, 1984). In most mammals the vasopressin is arginine-8-vasopressin (AVP). Lysine vasopressin (LVP) is found in pigs and some other animals, and arginine vasotocin (AVT) is found in nonmammalian vertebrates. Oxytocin is known to regulate milk release and parturition.

The possibility that vasopressin may be involved in the regulation of learning and memory was first suggested by the findings (de Wied, 1964; de Wied and Bohus, 1966) that extinction of a learned active avoidance response was accelerated in rats with lesions of the posterior pituitary. The extinction was normalized, that is, retarded, by a crude extract of pituitary tissue containing vasopressin. There is now extensive evidence that vasopressin and analogs of vasopressin retard extinction (de Wied, 1976). The interpretation that the behavioral effects are due to influences on memory is supported by evidence that retention is enhanced by posttraining administration of vasopressin. The memory-enhancing effects can be produced by central as well as systemic administration (Bohus et al., 1978). Further, posttraining i.c.v. administration of vasopressin antiserum impairs retention and the impairment can be attenuated by systemic injections of vasopressin (van Wimersma

Greidanus and de Wied, 1976). De Wied and his colleagues interpret these findings as indicating that vasopressin facilitates memory through influences on brain processes underlying memory consolidation (van Wimersma Greidanus *et al.*, 1981; see also chapter 5, this volume).

While the findings of several laboratories support the view that vasopressin enhances memory in a variety of training tasks (Davis *et al.*, 1982; Ettenberg *et al.*, 1983b; Hagan, 1983), the hypothesis has recently been criticized on two general grounds (Gash and Thomas, 1983; Sahgal, 1984). First, a number of studies have failed to find enhancing effects of vasopressin on learning or memory in various learning tasks (e.g., Alliot and Alexinsky, 1982; Buresova and Skopkova, 1980, 1982; Sahgal *et al.*, 1982). Second, the results of several studies suggest that the effects of vasopressin may be due to effects such as influences on peripheral blood pressure or arousal rather than to direct influences of the peptide on the brain (Ettenberg *et al.*, 1983a; Le Moal *et al.*, 1981; Sahgal, 1984). For example, Le Moal *et al.* (1981) reported that the effect of systemically administered AVP on extinction was blocked by a peptide antagonist of AVP that prevented the pressor effect of AVP. Thus, they suggested that the effects of vasopressin may be mediated by peripheral responses including alterations in blood pressure. Recent findings from de Wied's laboratory, however, argue against this interpretation (Burbach *et al.*, 1983; de Wied *et al.*, 1984). Posttraining i.c.v. administration of AVP, as well as two peptide metabolites of AVP that have no effect on blood pressure or heart rate, was found to enhance retention of an inhibitory avoidance response. The metabolites were effective in very low doses, in comparison with effective doses of AVP when administered either s.c. or i.c.v. Further, they found that central (i.c.v.) administration of a vasopressin antagonist blocked the behavioral effect of systemically administered AVP without blocking the effect of AVP on blood pressure. These findings suggest that the vasopressin receptors mediating the memory-enhancing effects of AVP are located in the brain, while those mediating the pressor effect are located in the periphery. The evidence that these potent AVP metabolites are found in the brain suggests that they may be involved in the endogenous modulation of memory storage processes (Versteeg and de Wied, 1985).

It is clear from findings of Ettenberg et al. (1983b) that AVP can produce aversive effects. Rats developed an aversion to saccharin if they were injected (s.c.) with AVP after each of four training sessions. Significant aversion was not obtained with only two pairings. Rats also learned to avoid a place in an apparatus where they are placed each day for 8 days following an injection of AVP. However, it is not clear that these effects are relevant to studies in which the animals are given only a single training trial and a single injection of AVP. Several studies have shown that AVP does not affect retention latencies of unshocked controls in inhibitory avoidance tasks. (Burbach et al., 1983; Sahgal et al., 1982). The latencies should be increased if AVP effects are due to aversive properties. Further, Ettenberg et al. (1983b) have found that posttraining AVP enhanced retention of a one-trial appetitive learning task. In this study satiated rats were given AVP (s.c.) after exploring a box with an alcove containing a drinking tube. They were then deprived of water and returned to the box 48 hours later. The AVP-treated rats approached the drinking tube with latencies significantly shorter than those of controls. Further, posttraining AVP did not affect retention latencies unless the drinking tube was present in the box during the exposure trial. Thus, a single injection of AVP appears to have neither punishing nor rewarding effects. These findings are difficult to reconcile with a view that posttraining AVP effects on retention are due simply to aversive properties. They are, however, as Ettenberg et al. conclude, ". . . consistent with the notion that AVP can act to facilitate the processing and/or storage of learned information" (Ettenberg et al., 1983b, p. 345).

Ettenberg et al. (1983b) and Sahgal (1984) suggest that AVP may enhance retention through influences on arousal. However, it is difficult to argue that the arousal effect is due to the aversive effects of AVP since AVP does not appear to have aversive effects in one-trial learning tasks. It seems more likely that, if AVP affects arousal, it does so by modulating brain systems involved in arousal. The central issue is whether AVP affects memory. The evidence rather strongly suggests that it does. The recent evidence also suggests that the memory effects are not due to peripheral effects such as increases in blood pressure. The hypothesis that AVP influences memory by increasing arousal has not yet been systematically addressed. If subsequent evidence should indicate that AVP affects memory by modulating arousal, such findings would suggest that such effects may be normally involved in the endogenous regulation of memory storage. There is extensive evidence that memory is influenced by activation of brain systems involved in arousal (Bloch and Laroche, 1984). As we discussed above, there is evidence that adrenergic and opioid hormones affect memory through central modulating systems (McGaugh, 1985; McGaugh et al., 1984). And, as we discuss below, vasopressin effects on memory appear to be influenced by adrenal epinephrine (Borrell et al., 1983b). Whether AVP affects memory through influences on central modulating systems or by directly affecting neuronal processes at the site of memory storage remains to be determined.

The findings that AVP did not affect retention performance in a spatial maze (Buresova and Skipkova, 1982) are of some interest. As we noted earlier, Beatty (1983) reported that spatial memory was not affected by opiate agonists and antagonists, whereas Gallagher *et al.* (1983) found that opiate antagonists enhanced performance when animals that had learned a spatial maze were then trained and tested in a novel environment. It would be of interest to know whether AVP has comparable effects on spatial learning and retention. Additional studies are needed to resolve the conflicting evidence from studies that have failed to find effects of AVP on retention.

As we noted above, a number of studies have reported that retention is enhanced by intraventricular administration of AVP as well as vasopressin metabolites. Retention is also enhanced by minute doses of AVP injected posttraining into a number of specific brain regions including the dorsal septal area, the dentate gyrus of the hippocampus, and the dorsal raphe area. Microinjections of AVP into the central amygdala or the locus coeruleus were ineffective (Kovács *et al.*, 1979). However, administration of AVP into the central amygdala nucleus or dentate gyrus prior to retention attenuated amnesia produced by posttraining pentylenetetrazol (Bohus *et al.*, 1982). These findings are consistent with the view that AVP influences memory processes by direct effects on brain systems. Evidence from lesion studies provides additional support for this view. AVP effects on consolidation are blocked by 6-OHDA lesions of the medial forebrain bundle. Lesions of the fornix are ineffective. However, fornix lesions attenuated the facilitating effects of AVP administered prior to a retention test (van Wimersma Greidanus *et al.*, 1979).

Posttraining i.c.v. administration of oxytocin and oxytocin metabolites has been found to impair retention (de Wied *et al.*, 1984). Microinjections of oxytocin also impaired retention when administered into the dentate gyrus or dorsal raphe area. Retention was enhanced by injections into the dorsal septum. Amygdala injections were ineffective (Kovács *et al.*, 1979). Since oxytocin has generally been found to impair retention it has been suggested (Bohus *et al.*, 1982; van Wimersma Greidanus *et al.*, 1981) that oxytocin may be an endogenous amnestic peptide. However, Davis *et al.* (1982) have reported that posttraining i.c.v. injections of oxytocin enhance retention in chicks. Enhancement of retention in chicks was also obtained with either i.c.v. or s.c. administration of AVP. Since both enhancement and impairment of memory have been found with both AVP and oxytocin it is premature to conclude that oxytocin has an amnestic role. Careful dose–response studies are needed to determine the effects of these peptides on memory. Since the dose–response effect found for hormones is generally U-shaped it seems likely that the direction of the effect of vasopressin and oxytocin depends upon the doses used as well as the experimental conditions used.

Other attempts to investigate the role of vasopressin have involved studies of learning and retention in rats of the Brattleboro strain. These rats are virtually unable to synthesize AVP. While de Wied and his colleagues have reported that Brattleboro rats are deficient in learning (van Wimersma Greidanus *et al.*, 1981), other laboratories have obtained conflicting findings (Bailey and Weiss, 1979; Carey and Miller, 1982). Gash and Thomas (1983) have argued that the negative findings weaken the argument that vasopressin affects memory. In order to draw this conclusion it would be necessary to know that the Brattleboro rats are normal in all other systems that might influence learning and retention. It might be that other hormonal systems are altered in Brattleboro rats and that such alterations produce effects that compensate for the deficiency in vasopressin. Without such information the contribution of these studies to understanding the role of vasopressin in memory is unclear.

The reports of memory enhancement with vasopressin in animals have stimulated studies of the effects of vasopressin and vasopressin analogs on memory tasks in humans. Several studies have reported finding positive effects (e.g., Beckwith *et al.*, 1982; P. W. Gold *et al.*, 1979; LeBoeuf *et al.*, 1978; Legros *et al.*, 1978; Oliveros *et al.*, 1978; Weingartner *et al.*, 1981). A number of studies have, however, reported negative findings (e.g., Blake *et al.*, 1978; Jenkins *et al.*, 1981). There are many differences in the types of patients and subjects as well as the procedures used in these investigations. In view of the mixed results of these studies it is too early to draw any conclusions concerning the effects of vasopressin on human memory.

VI. Hormonal and Transmitter Interactions

Studies of the involvement of hormones in memory have each tended to focus on a particular hormonal system. It is clear, however, that the conditions used in such studies are highly artificial. Hormones act in the presence of other hormones. Stimulation of the kind used in laboratory studies of learning and memory releases a variety of hormones including epinephrine, ACTH, corticosteroids, and endorphin (Dunn and Kramarcy, 1984). Each of these hormones has been reported to have a modulating influence on memory storage. Further, it seems likely that many newly discovered as well as many yet-to-be discovered peptide hormones may also be released by experiences and play roles in memory storage. Our efforts to understand the polyhormonal regulation of memory are complicated by the fact that there are many known and, no doubt, even more as-yet unknown interactions among hormones as well as interactions between hormonal and transmitter systems (Gold and Zornetzer, 1983; Zornetzer, 1978). It is, of course, well known that corticosterone, which is released from the adrenal cortex by ACTH, is involved in feedback regulation of pituitary release of ACTH. Recent studies have shown that adrenergic agonists (epinephrine and isoproterenol) increase plasma levels of immunoreactive ACTH and corticosterone as well as β-endorphin and α-MSH. These effects are blocked by adrenergic antagonists (Berkenbosch et al., 1981; Pettibone and Mueller, 1982). Other recent findings suggest that central adrenergic activity modulates the negative feedback effects of corticosterone. Feldman et al. (1983) found that the inhibitory effect of glucocorticoids on adrenocortical secretory response was potentiated in rats with 6-OHDA lesions which block noradrenergic innervation of the hypothalamus. Thus, it would seem that it is impossible to influence one particular hormonal system without influencing several others. As these recent findings indicate, alterations in either central or peripheral adrenergic activity influence the release of several hormones including ACTH and corticosterone. In view of the extensive (albeit controversial) evidence that vasopressin influences memory it is perhaps surprising that stress does not elevate plasma vasopressin in normal rats. However, this finding is consistent with evidence that opiates and opioid hormones inhibit the release of vasopressin (Iversen et al., 1980; Knepel and Reimann, 1982) and that plasma vasopressin is elevated in animals pretreated either with naloxone or with dexamethasone, which inhibits the release of pituitary ACTH and β-endorphin (Knepel et al., 1982). In view of these findings it seems unlikely that peripherally released vasopressin is normally involved in the modulation of memory. Thus, these findings are consistent with the view (Versteeg and de Wied, 1985) that vasopressin effects on memory are due to activation of central receptors. However, peripheral vasopressin might well be involved in the effects on memory of drugs such as naloxone and dexamethasone, which influence the release of vasopressin.

A number of recent studies have examined interactions among hormones found to affect memory. The studies by Jefferys and his colleagues (1983, 1984, 1985) discussed above, reported that, in adrenalectomized rats, the learning impairment found in a swimming task was attenuated by either opioid agonists or corticosterone. Further, naloxone but not quaternary naloxone impaired retention in normal rats. These findings suggest that both the adrenal cortex and the adrenal medulla are involved in endogenous modulation of memory storage in this learning task. They interpret these findings as suggesting that corticosterone and opioid peptides (κ-active) activate different central neurons which converge on a common opioid system. In contrast, Borrell *et al.* (1983a) found that corticosterone did not attenuate the impaired retention performance on an inhibitory avoidance task seen in rats adrenalectomized 5 days before training. Retention was not impaired if rats were adrenalectomized 10 days before training. They suggested that the attenuation of the impairment was due to high levels of ACTH 10 days after the adrenalectomy. The finding that retention was impaired in rats given corticosterone following the adrenal surgery is consistent with this interpretation. Borrell *et al.* (1983a) also found that the retention impairment seen 2 days after either adrenalectomy or adrenal demedullation was attenuated by posttraining epinephrine or norepinephrine. Other recent findings by Borrell and his colleagues (1984a) indicate that the effects of posttraining epinephrine on retention are markedly influenced by corticosterone levels. The enhancing effects of posttrial epinephrine on retention in ADX rats were attenuated in rats given corticosterone replacement: the dose required for enhancement in ADX rats given corticosterone was 10,000 times as high as that required in ADX rats not given corticosterone (5 ng/kg in ADX rats versus 50 mg/kg in ADX rats given corticosterone). Borrell *et al.* suggest that the corticosterone attenuates the effects of epinephrine through influences on receptors in hippocampal, septal, and amygdaloid neurons. Understanding the basis of the interaction between corticosterone and epinephrine should help provide clues to the basis (or bases) of the effects of epinephrine on memory.

Other recent findings point to an interaction between epinephrine and vasopressin. Borrell *et al.* (1984b) reported that posttraining AVP (s.c.) did not enhance retention in rats which were adrenalectomized 2 days before training. However, AVP potentiated the effect of a low dose of epinephrine. Thus it seems that the effects of AVP on memory may require the presence of epinephrine, and the epinephrine effects are influenced by AVP. This conclusion is supported by results of studies of the effects of epinephrine in heterozygous (HE) and homozygous (HO) Brattleboro rats. As we discussed above, the HO rats are unable to synthesize vasopressin. Borrell *et al.* (1984b) reported that AVP had no effects on retention in either HO or HE rats adrenalectomized 2 days before training. Such effects would be expected if epinephrine is required for AVP effects on memory. In addition, they found that posttraining epinephrine enhanced retention in ADX HE rats but had no effect on retention in either ADX or intact HO rats. These results suggest that epinephrine effects on memory may require the presence of AVP. Since the evidence, on balance, argues that AVP modulates memory through activation of central receptors, these findings suggest that the interaction occurs centrally. It would be of interest to know whether central administration of AVP also potentiates the effects of epinephrine on memory. It seems clear that the effects of AVP on memory are not due to influences on the release of epinephrine from the adrenal medulla since AVP potentiates the effects of peripherally administered epinephrine. These findings are also of interest in view of the finding that AVP effects on learning are blocked by 6-OHDA lesions of the dorsal adrenergic bundle (Bohus *et al.*, 1982). Bohus *et al.* suggest that AVP effects on memory involve modulation of neurotransmission in limbic-midbrain structures. Since the effects of epinephrine on memory are blocked by lesions of the stria terminalis (ST) (Liang and McGaugh, 1983) it would be of interest to know whether ST lesions also block the effects of peripherally administered AVP. Research to date has provided few clues concerning the locus of the interaction between AVP and epinephrine.

Evidence from recent investigations suggests that effects of opioid hormones on memory involve interactions with central catecholaminergic and cholinergic systems. The findings suggest that opioid hormones may impair hormones by inhibiting the release of transmitters. Izquierdo and Graudenz (1980) reported that the effects of naloxone on memory are antagonized by the β-antagonist propranolol and the dopamine antagonist haloperidol, but not by the α-antagonist phenoxybenzamine. Introini and Baratti (1984) reported that the impairing effects of posttraining β-endorphin on memory were attenuated by the muscarinic agonist oxotremorine and the acetylcholinesterase inhibitor physostigmine. β-Endorphin effects on retention were not blocked by the peripherally acting anticholinesterase neostigmine. In other recent studies Baratti *et al.* (1984) found that the enhancing effects of posttrial naloxone were blocked by atropine but not methylatropine. Thus the effects appear to involve central cholinergic neurons. These findings are of interest in view of evidence that the enhancing effect of a variety of central nervous system stimulants, including strychnine and D-amphetamine (McGaugh, 1973), is blocked by atropine and scopolamine (Yonkov, 1983; Yonkov and Roussinov, 1981). It would be interesting to know whether epinephrine and AVP effects on retention are also blocked by interference with central cholinergic activity.

These recent studies have begun to reveal some of the many and complex interactions among hormonal and transmitter systems that are thought to be involved in the modulation of memory storage. But, clearly, the work of understanding the nature and sites of the interaction has only just begun. But, it now seems that understanding the physiological processes involved in the regulation of memory storage will require detailed information of the interactions among hormonal and transmitter systems.

Acknowledgments

Preparation of this chapter was supported by research grants MH 12526, AG 00538, and Contract N00014-84-k-0391 from the Office of Naval Research (JLM) and research grants MH31141, NSF BNS 83-17940, and a James McKeen Cattell Fellowship (PEG).

References

Alliot, J., and Alexinsky, T. (1982). Effects of posttrial vasopressin injections on appetitively motivated learning in rats. *Physiol. Behav.* **28**, 525–530.

Amir, S., Brown, Z. W., and Amit, Z. (1980). The role of endorphin in stress: Evidence and speculation. *Neurosci. Biobehav. Rev.* **4**, 77–86.

Axelrod, J., Weil-Malherbe, H., and Tomchick, R. (1959). The physiological disposition of 3H-epinephrine and its metabolite metanephrine. *J. Pharmacol. Exp. Ther.* **127**, 251–256.

Bailey, W. H., and Weiss, J. M. (1979). Evaluation of 'memory deficit' in vasopressin-deficient rats. *Brain Res.* **162**, 174–178.

Baratti, C. M., Introini, I. B., and Huygens, P. (1984). Possible interaction between central cholinergic muscarinic and opioid peptidergic systems during memory consolidation in mice. *Behav. Neural Biol.* **40**, 155–169.

Bartus, R. T., and Dean, R. L. (1981). Age-related memory loss and drug therapy: Possible directions based on animal models. *In* "Brain Neurotransmitters and Receptors in Aging and Age-related Disorders" (S. J. Enna, T. Samorajski, and B. Beer, eds.), pp. 209–224. Raven Press, New York.

Bartus, R. T., Dean, R. L., and Fleming, D. L. (1979). Aging in the rhesus monkey: Effects on visual discrimination learning and reversal learning. *J. Gerontol.* **34**, 209–219.

Bartus, R. T., Dean, R. L., Goss, J. A., and Lippa, A. S. (1980). Age-related changes in passive avoidance retention: Modulation with dietary choline. *Science* **209**, 301–313.

Beatty, W. (1983). Opiate antagonists, morphine and spatial memory in rats. *Pharmacol., Biochem. Behav.* **19**, 397–401.

Beckwith, B. E., Petros, T., Kanan-Beckwith, S., Couk, D. I., Haug, R. J., and Ryan, C. (1982). Vasopressin analog (DDAVP) facilitates concept learning in human males. *Peptides (N.Y.)* **3**, 627–630.

Belluzzi, J. D., and Stein, L. (1981). Facilitation of long-term memory by brain endorphins. *In* "Endogenous Peptides and Learning and Memory Processes" (J. L. Martinez, Jr., R. A. Jensen, R. B. Messing, J. L. McGaugh, and H. Rigter, eds.), pp. 291–303. Academic Press, New York.

Bennett, C., Liang, K. C., and McGaugh, J. L. (1985). Depletion of adrenal catecholamines alters the amnestic effect of amygdala stimulation. *Behav. Brain Res.* **15**, 83–91.

Berkenbosch, F., Vermes, I., Binnekade, R., and Tilders, F. J. H. (1981). Beta-adrenergic stimulation induces an increase of the plasma levels of immunoreactive alpha-MSH, beta-endorphin, ACTH and of corticosterone. *Life Sci.* **29**, 2249–2256.

Berntman, L., Dahlgren, N., and Siesjuo, B. K. (1978). Influence of intravenously administered catecholamines on cerebral oxygen consumption and blood flow in the rat. *Acta Physiol. Scand.* **104**, 101–108.

Blake, D. R., Dodd, M. J., and Evans, J. G. (1978). Vasopressin in amnesia. *Lancet* **1**, 608.

Bloch, V., and Laroche, S. (1984). Facts and hypotheses related to the search for the engram. *In* "Neurobiology of Learning and Memory" (G. Lynch, J. L. McGaugh, and N. M. Weinberger, eds.), pp. 249–260. Guilford Press, New York.

Bodnar, R. J., Kelly, D. D., Brutus, M., and Glusman, M. (1980). Stress-induced analgesia: Neural and hormonal determinants. *Neurosci. Biobehav. Rev.* **4**, 87–100.

Bohus, B., Kovács, G. L., and de Wied, D. (1978). Oxytocin, vasopressin and memory: Opposite effects on consolidation and retrieval processes. *Brain Res.* **157**, 414–417.

Bohus, B., Conti, L., Kovács, G. L., and Versteeg, D. H. G. (1982). Modulation of memory processes by neuropeptides: Interaction with neurotransmitter systems. *In* "Neuronal Plasticity and Memory Formation" (C. Amjone Marsan and H. Matthies, eds.), pp. 75–87. Raven Press, New York.

Borrell, J., de Kloet, E. R., Versteeg, D. H. G., and Bohus, B. (1983a). Inhibitory avoidance deficit following short-term adrenalectomy in the rat: The role of adrenal catecholamines. *Behav. Neural Biol.* **39**, 241–258.

Borrell, J., de Kloet, E. R., Versteeg, D. H. G., and Bohus, B. (1983b). The role of adrenomedullary catecholamines in the modulation of memory by vasopressin. *In* "Integrative Neurohumoral Mechanisms: Developments in Neuroscience" (E. Endröczi, D. de Wied, L. Angelucci, and V. Scapagnini, eds.), pp. 85–90. Elsevier/North-Holland, Amsterdam.

Borrell, J., de Kloet, E. R., and Bohus, B. (1984a). Corticosterone decreases the efficacy of adrenaline to affect passive avoidance retention of adrenalectomized rats. *Life Sci.* **34**, 99–105.

Borrell, J., de Kloet, E. R., Versteeg, D. H. G., Bohus, B., and de Wied, D. (1984b). Neuropeptides and memory: Interactions with peripheral catecholamines. *In* "The Role of Catecholamines and Other Neurotransmitters" (E. Usdin, R. Kvetnansky, and A. Axelrod, eds.), pp. 391–402. Gordon & Breach, New York.

Burbach, J. P. H., Kovács, G. L., de Wied, D., van Nispen, J. W., and Greven, H. M. (1983). A major metabolite of arginine vasopressin in the brain is a highly potent neuropeptide. *Science* **221**, 1310–1312.

Buresova, O., and Skopkova, J. (1980). Vasopressin analogues and spatial short-term memory in rats. *Peptides (N.Y.)* **1**, 261–263.

Buresova, O., and Skopkova, J. (1982). Vasopressin analogues and spatial working memory in the 24-arm radial maze. *Peptides (N.Y.)* **3**, 725–728.

Carey, R. J., and Miller, M. (1982). Absence of learning and memory deficits in the vasopressin-deficient rat (Brattleboro strain). *Behav. Brain Res.* **6**, 1–3.

Carrasco, M. A., Perry, M. L., Dias, R. D., Wofchuk, S., and Izquierdo, I. (1982a). Effect of tones, footshocks, shuttle avoidance, and electroconvulsive shock on met-enkephalin immunoreactivity of rat brain. *Behav. Neural Biol.* **34**, 1–4.

Carrasco, M. A., Dias, R. D., and Izquierdo, I. (1982b). Naloxone reverses retrograde amnesia induced by electroconvulsive shock. *Behav. Neural Biol.* **34**, 352–357.

Castellano, C. (1975). Effects of morphine and heroin on discrimination learning and consolidation in mice. *Psychopharmacology* **42**, 235–242.

Castellano, C., and Puglisi-Allegra, S. (1983). Strain-dependent modulation of memory by stress in mice. *Behav. Neural Biol.* **38**, 133–138.

Castellano, C., Pavone, F., and Puglisi-Allegra, S. (1984). Morphine and memory in DBA/2 mice: Effects of stress and of prior experience. *Behav. Brain Res.* **11**, 3–10.

Davis, J. L., Pico, R. M., and Cherkin, A. (1982). Arginine vasopressin enhances memory retroactively in chicks. *Behav. Neural Biol.* **35**, 242–250.

de Almeida, M., and Izquierdo, I. (1984). Effect of the intraperitoneal and intracerebroventricular administration of ACTH, epinephrine, of beta-endorphin on retrieval of an inhibitory avoidance task in rats. *Behav. Neural Biol.* **40**, 119–122.

de Wied, D. (1964). Influence of anterior pituitary on avoidance learning and escape behavior. *Am. J. Physiol.* **207**, 255–259.

de Wied, D. (1969). Effects of peptide hormones on behavior. *In* "Frontiers in Neuroendocrinology" (W. F. Ganong and L. Martini, eds.), pp. 97–140. Oxford Univ. Press, London and New York.

de Wied, D. (1974). Pituitary-adrenal system hormones and behavior. *In* "The Neurosciences: Third Study Program" (F. O. Schmitt and F. G. Worden, eds.), 3rd ed., pp. 653–666. MIT Press, Cambridge, Massachusetts.

de Wied, D. (1976). Behavioral effects of intraventricularly administered vasopressin and vasopressin fragments. *Life Sci.* **19**, 685–690.

de Wied, D., and Bohus, B. (1966). Long-term and short term effects on retention of a conditioned avoidance response in rats by treatment with long acting pitressin and alpha-MSH. *Nature (London)* **212**, 1484–1486.

de Wied, D., van Delft, A. M. L., Gispen, W. H., Weijnen, J. A. W. M., and van Wimersma Greidanus, T. (1972). The role of pituitary-adrenal system hormones in active avoidance conditioning. *In* "Hormones and Behavior" (S. Levine, ed.), pp. 135–171. Academic Press, New York.

de Wied, D., Gaffori, O., Van Ree, J. M., and de Jong, W. (1984). Central target for the behavioural effects of vasopressin neuropeptides. *Nature (London)* **308**, 276–278.

Doris, P. A. (1984). Vasopressin and central integrative processes. *Neuroendocrinology* **38**, 75–85.

Dunn, A. J., and Kramarcy, N. R. (1984). Neurochemical responses in stress: Relationships between the hypothalamic pituitary-adrenal and catecholamine systems. *In* "Handbook of Psychopharmacology" (L. L. Iverson, S. D. Iversen, and S. H. Snyder, eds.), pp. 455–515. Plenum, New York.

Ettenberg, A., van der Kooy, D., Le Moal, M., Koob, G. F., and Bloom, F. E. (1983a). Can aversive properties of (peripherally-injected) vasopressin account for its putative role in memory? *Behav. Brain Res.* **7**, 331–350.

Ettenberg, A., van der Kooy, D., Le Moal, M., Koob, G. F., and Bloom, F. E. (1983b). Vasopressin potentiation in the performance of a learned appetitive task: Reversal by a pressor antagonist analog of vasopressin. *Pharmacol., Biochem. Behav.* **18**, 645–647.

Fekete, M., and de Wied, D. (1982). Potency and duration of action of ACTH (4-9) on active and passive avoidance behavior of rats. *Pharmacol., Biochem. Behav.* **16**, 387–392.

Feldman, S., Siegel, R. A., Weidenfeld, J., Conforti, N., and Melamed, E. (1983). Role of medial forebrain bundle catecholaminergic fibers in the modulation of glucocorticoid negative feedback effects. *Brain Res.* **260**, 297–300.

Flood, J. F., Jarvik, M. E., Bennett, E. L., and Orme, A. A. (1976). Effects of ACTH peptide fragments on memory formation. *Pharmacol., Biochem. Behav.* **5**, Suppl. 1, 41–51.

Gallagher, M., and Kapp, B. S. (1978). Manipulation of opiate activity in the amygdala alters memory processes. *Life Sci.* **23**, 1973–1978.

Gallagher, M., Kapp, B. S., Pascoe, J. P., and Rapp, P. R. (1981). A neuropharmacology of amygdaloid systems which contribute to learning and memory. *In* "The Amygdaloid Complex" (Y. Ben-Ari, ed.), pp. 343–354. Elsevier/North-Holland, Amsterdam.

Gallagher, M., King, R. A., and Young, N. B. (1983). Opiate antagonists improve spatial memory. *Science* **221**, 975–976.

Gash, D. M., and Thomas, G. J. (1983). What is the importance of vasopressin in memory processes?. *Trends NeuroSci.* **6**, 197–198.

Gold, P. E. (1986). Glucose modulation of memory storage processing. *Behav. Neural Biol.* **45**, 342–349.

Gold, P. E., and McCarty, R. (1981). Plasma catecholamines: Changes after footshock and seizure-producing frontal cortex stimulation. *Behav. Neural Biol.* **31**, 247–260.

Gold, P. E., and McGaugh, J. L. (1975). A single-trace, two process view of memory storage processes. *In* "Short Term Memory" (D. Deutsch and J. A. Deutsch, eds.), pp. 355–378. Academic Press, New York.

Gold, P. E., and McGaugh, J. L. (1978). Neurobiology and memory: Modulators, correlates, and assumptions. *In* "Brain and Learning" (T. Teyler, ed.), pp. 93–104. Greylock Publishers, Stamford, Connecticut.

Gold, P. E., and Sternberg, D. B. (1978). Retrograde amnesia produced by several treatments: Evidence for a common neurobiological mechanism. *Science* 201, 367–369.

Gold, P. E., and van Buskirk, R. (1975). Facilitation of time-dependent memory processes with posttrial epinephrine injections. *Behav. Biol.* 13, 145–153.

Gold, P. E., and van Buskirk, R. (1976a). Enhancement and impairment of memory processes with posttrial injections of adrenocorticotrophic hormone. *Behav. Biol.* 16, 387–400.

Gold, P. E., and van Buskirk, R. (1976b). Effects of post-trial hormone injections on memory process. *Horm. Behav.* 7, 509–517.

Gold, P. E., and van Buskirk, R. (1978a). Posttraining brain norepinephrine concentrations: Correlation with retention performance of avoidance training with peripheral epinephrine modulation of memory processing. *Behav. Biol.* 23, 509–520.

Gold, P. E., and van Buskirk, R. (1978b). Effects of alpha and beta adrenergic receptor antagonists on post-trial epinephrine modulation of memory: Relationship to posttraining brain norepinephrine concentrations. *Behav. Biol.* 24, 168–184.

Gold, P. E., and Zornetzer, S. F. (1983). The mnemon and its juices: Neuromodulation of memory processes. *Behav. Neural Biol.* 38, 151–189.

Gold, P. E., Rose, R. P., Spanis, C. W., and Hankins, L. L. (1977a). Retention deficit for avoidance training in hypophysectomized rats: Time dependent enhancement with posttraining ACTH injection. *Horm. Behav.* 8, 363–371.

Gold, P. E., van Buskirk, R., and Haycock, J. (1977b). Effects of posttraining epinephrine injections on retention of avoidance training in mice. *Behav. Biol.* 20, 197–207.

Gold, P. E., McGaugh, J. L., Hankins, L. L., Rose, R. P., and Vasquez, B. J. (1981). Age-dependent changes in retention in rats. *Exp. Aging Res.* 8, 53–58.

Gold, P. E., Murphy, M. J., and Cooley, S. (1982). Neuroendocrine modulation of memory during development. *Behav. Neural Biol.* 35, 277–293.

Gold, P. E., Delanoy, R. L., and Merrin, J. (1984). Modulation of long-term potentiation by peripherally administered amphetamine and epinephrine. *Brain Res.* 305, 103–107.

Gold, P. E., Vogt, J., and Hall, J. L. (1986). Glucose effects on memory: Behavioral and pharmacological characteristics. *Behav. Neural Biol.* 46, 145–155.

Gold, P. W., Weingartner, H., Ballenger, J. C., Goodwin, F. K., and Post, R. M. (1979). Effects of 1-desamo-8-D-arginine vasopressin on behaviour and cognition in primary affective disorder. *Lancet* 2, 992–994.

Goldstein, A. (1976). Opioid peptides (endorphins) in pituitary and brain. *Science* 193, 1081–1086.

Hagan, J. J. (1983). Post-training lysine-vasopressin injections increase conditioned suppression in rats. *Physiol. Behav.* 31, 765–769.

Hall, J. L., and Gold, P. E. (1986). The effects of training, epinephrine, and glucose injections on plasma glucose levels in rats. *Behav. Neural Biol.* 46, 156–167.

Haroutunian, V., and Riccio, D. C. (1977). Effect of arousal conditions during reinstatement treatment upon learned fear in young rats. *Dev. Psychobiol.* 10, 24–32.

Introini, I., and Baratti, C. M. (1984). The impairment of retention induced by beta endorphin in mice may be mediated by reduction of central cholinergic activity. *Behav. Neural Biol.* 41, 152–163.

Iversen, L. L., Iversen, S. D., and Bloom, F. E. (1980). Opiate receptors influence vasopressin release from nerve terminals in rat neurohypothesis. *Nature (London)* 284, 350–351.

Izquierdo, I. (1979). Effect of naloxone and morphine on various forms of memory in the rat: Possible role of endogenous opiate mechanisms in memory consolidation. *Psychopharmacology* 66, 199–203.

Izquierdo, I. (1980). Effects of a low and a high dose of b-endorphin on acquisition and retention in the rat. *Behav. Neural Biol.* 30, 460–464.

Izquierdo, I. (1984). Endogenous state dependency: Memory depends on the relation between the neurohumoral and hormonal states present after training and at the time of testing. *In* "Neurobiology of Learning and Memory" (G. Lynch, J. L. McGaugh, and N. M. Weinberger, eds.), pp. 333–350. Guilford Press, New York.

Izquierdo, I., and Dias, R. D. (1983a). Memory as a state dependent phenomenon: Role of ACTH and epinephrine. *Behav. Neural Biol.* 38, 144–149.

Izquierdo, I., and Dias, R. D. (1983b). Memory modulation by the administration of ACTH, adrenaline or beta-endorphin after training and prior to testing in an inhibitory avoidance task in rats. *Braz. J. Med. Biol. Res.* 16, 54–64.

Izquierdo, I., and Graudenz, M. (1980). Memory facilitation by naloxone is due to release of dopaminergic and beta-adrenergic systems from tonic inhibition. *Psychopharmacology* 67, 265–268.

Izquierdo, I., Paiva, A. C. M., and Elisabetsky, E. (1980a). Post-training intraperitoneal administration of leu-enkephalin and beta-endorphin causes retrograde amnesia for two different tasks in rats. *Behav. Neural Biol.* 28, 246–250.

Izquierdo, I., Souza, D. O., Carrasco, M. A., Dias, R. D., Perry, M. L., Eisinger, S., Elisabetsky, E., and Vendite, D. A. (1980b). Beta-endorphin causes retrograde amnesia and is released from the rat brain by various forms of training and stimulation. *Psychopharmacology* 70, 173–177.

Izquierdo, I., Perry, M. L., Dias, R. D., Souza, D. O., and Elisabetsky, E. (1981). Endogenous opioids, memory modulation, and state dependency. *In* Endogenous Peptides and Learning and Memory Processes" (J. L. Martinez, Jr., R. A. Jensen, R. B. Messing, H. Rigter, and J. L. McGaugh, eds.), pp. 269–290. Academic Press, New York.

Izquierdo, I., Dias, R. D., Perry, M. L., Souza, D. O., Elisabetsky, E., and Carrasco, M. A. (1982). A physiological amnestic mechanism mediated by endogenous opioid peptides, and its possible role in learning. *In* "Neuronal Plasticity and

Memory Formation" (A. Marsan and H. Matthies, eds.), pp. 89–113. Raven Press, New York.

Jefferys, D., Copolov, D., Irby, D., and Funder, J. (1983). Behavioural effect of adrenalectomy: Reversal by glucocorticoids or [D-ALA, MET] enkephalinamide. *Eur. J. Pharmacol.* **92**, 99–103.

Jefferys, D., Copolov, D., and Funder, J. W. (1984). Naloxone inhibits both glucocorticoid and [D-ALA, MET] enkephalinamide reversal of behavioural effect of adrenalectomy. *Eur. J. Pharmacol.* **102**, 205–210.

Jefferys, D., Boublik, J., and Funder, J. W. (1985). A k-selective opioidergic pathway is involved in the reversal of behavioural effect of adrenalectomy. *Eur. J. Pharmacol.* **107**, 331–335.

Jenkins, J. S., Mather, H. M., Coughlan, A. K., and Jenkins, D. G. (1981). Desmopressin and desglycinamide vasopressin in post-traumatic amnesia. *Lancet* **1**, 39.

Kety, S. (1972). Brain catecholamines, affective states and memory. *In* "The Chemistry of Mood, Motivation and Memory" (J. L. McGaugh, ed.), pp. 65–80. Plenum, New York.

Knepel, W., and Reimann, W. (1982). Inhibition of morphine and beta-endorphin of vasopressin release evoked by electrical stimulation of the rat medial basal hypothalamus in vitro. *Brain Res.* **238**, 484–488.

Knepel, W., Nutto, D., and Hertting, G. (1982). Evidence for inhibition by beta-endorphin of vasopressin release during footshock-induced stress in the rat. *Neuroendocrinology* **34**, 353–356.

Kovács, G. L., and de Wied, D. (1981). Endorphin influences on learning and memory. *In* "Endogenous Peptides and Learning and Memory Processes" (J. L. Martinez, Jr., R. A. Jensen, R. B. Messing, H. Rigter, and J. L. McGaugh, eds.), pp. 231–247. Academic Press, New York.

Kovács, G. L., Bohus, B., Versteeg, D. H. G., de Kloet, E. R., and de Wied, D. (1979). Effect of oxytocin and vasopressin on memory consolidation: Sites of action and catecholaminergic correlates after local microinjection into limbic-midbrain structures. *Brain Res.* **175**, 303–314.

Kovács, G. L., Versteeg, D. H. G., de Kloet, E. R., and Bohus, B. (1981). Passive avoidance performance correlates with catecholamine turnover in discrete limbic regions. *Life Sci.* **28**, 1109–1116.

Krieger, D. T., and Liotta, A. S. (1979). Pituitary hormones in brain. Where, why and how?. *Science* **205**, 366–372.

Kubanis, P., and Zornetzer, S. F. (1981). Age-related behavioral and neurobiological changes: A review with an emphasis on memory. *Behav. Neural Biol.* **32**, 241–247.

Kvetnansky, R., Weise, V. K., Thoa, N. B., and Kopin, I. J. (1979). Effects of chronic guanethidine treatment and adrenal medullectomy of plasma levels of catecholamines and corticosterone in forcibly immobilized rats. *J. Pharmacol. Exp. Ther.* **209**, 287–291.

LeBoeuf, A., Lodge, J., and Eames, P. G. (1978). Vasopressin and memory in Korsakoff syndrome. *Lancet* **2**, 1370.

Legros, J. J., Gilot, P., Seron, X., Claessens, J., Adam, A., Moeglen, J. M., Audibert, A., and Berchier, P. (1978). Influence of vasopressin on learning and memory. *Lancet* **1**, 41–42.

Le Moal, M., Koob, G. F., Koda, L. Y., Bloom, F. E., Manning, M., Sawyer, W. H., and River, J. (1981). Vasopressor receptor antagonist prevents behavioural effects of vasopressin. *Nature (London)* **291**, 491–493.

Leshner, A. I., and Potlitch, J. A. (1979). Hormonal control of submissiveness in mice: Irrelevance of the androgens and relevance of the pituitary adrenal hormones. *Physiol. Behav.* **22**, 531–534.

Leshner, A. I., Merkle, D. A., and Mixon, J. F. (1981). Pituitary-adrenocortical effects on learning and memory in social situations. *In* "Endogenous Peptides and Learning and Memory Processes" (J. L. Martinez, Jr., R. A. Jensen, R. B. Messing, H. Rigter, and J. L. McGaugh, eds.), pp. 159–179. Academic Press, New York.

Liang, K. C., and McGaugh, J. L. (1983). Lesions of the stria terminalis attenuate the enhancing effect of posttraining epinephrine on retention of an inhibitory avoidance response. *Behav. Brain Res.* **9**, 49–58.

Liang, K. C., Messing, R. B., and McGaugh, J. L. (1983). Naloxone attenuates amnesia caused by amygdaloid stimulation: The involvement of a central opioid system. *Brain Res.* **271**, 41–49.

Liang, K. C., Bennett, C., and McGaugh, J. L. (1985). Peripheral epinephrine modulates the effects of posttraining amygdala stimulation on memory. *Behav. Brain Res.* **15**, 93–100.

Martinez, J. L., Jr., and Rigter, H. (1980). Endorphins alter acquisition and consolidation of an inhibitory avoidance response in rats. *Neurosci. Lett.* **19**, 197–201.

Martinez, J. L., Jr., Vasquez, B. J., Jensen, R. A., Soumireu-Mourat, B., and McGaugh, J. L. (1979). ACTH4-9 analog (ORG 2766) facilitates acquisition of an inhibitory avoidance response in rats. *Pharmacol., Biochem. Behav.* **10**, 145–147.

McCarty, R. (1981). Aged rats: Diminished sympathetic-adrenal medullary responses to acute stress. *Behav. Neural Biol.* **33**, 204–212.

McCarty, R., and Gold, P. E. (1981). Plasma catecholamines: Effects of footshock level and hormonal modulators of memory storage. *Horm. Behav.* **15**, 168–182.

McGaugh, J. L. (1973). Drug facilitation of learning and memory. *Annu. Rev. Pharmacol.* **13**, 229–241.

McGaugh, J. L. (1983). Preserving the presence of the past: Hormonal influences on memory storage. *Am. Psychol.* **38**, 161–174.

McGaugh, J. L. (1985). Peripheral and central adrenergic influences on brain systems involved in the modulation of memory storage. *In* "Memory Dysfunctions: An Integration of Animal and Human Research from Preclinical and Clinical

Perspectives" (D. S. Olton, E. Gamzu, and S. Corkin, eds.), pp. 150–161. N.Y. Acad. Sci., New York.

McGaugh, J. L., Gold, P. E., Handwerker, M. J., Jensen, R. A., and Martinez, J. L., Jr. (1979). Altering memory by electrical and chemical stimulation of the brain. *In* "Brain Mechanisms in Memory and Learning: From the Single Neuron to Man" (M. A. B. Brazier, ed.), pp. 154–164. Raven Press, New York.

McGaugh, J. L., Liang, K. C., Bennett, C., and Sternberg, D. B. (1984). Adrenergic influences on memory storage: Interaction of peripheral and central systems. *In* "Neurobiology of Learning and Memory" (G. Lynch, J. L. McGaugh, and N. M. Weinberger, eds.), pp. 313–333. Guilford Press, New York.

Meligeni, J. A., Ledergerber, S. A., and McGaugh, J. L. (1978). Norepinephrine attenuation of amnesia produced by diethyldithiocarbamate. *Brain Res.* **149**, 155–164.

Messier, C., Blackburn, J., and White, N. M. (1984). Effect of glucose and insulin on memory and stereotype. *Neurosci. Abstr.* **10**, 54.

Messing, R. B., Jensen, R. A., Martinez, J. L., Jr., Spiehler, V. R., Vasquez, B. J., Soumireu-Mourat, B., Liang, K. C., and McGaugh, J. L. (1979). Naloxone enhancement of memory. *Behav. Neural Biol.* **27**, 266–275.

Micheau, J., Destrade, C., and Soumireu-Mourat, B. (1981). Intraventricular corticosterone injection facilitates memory of an appetitive discriminative task in mice. *Behav. Neural Biol.* **31**, 100–104.

Micheau, J., Destrade, C., and Soumireu-Mourat, B. (unpublished findings). Dissociation of effects of intrahippocampally administered corticosterone and dexamethasone on extinction of operant conditioning in BALB/c mice.

Micheau, J., Destrade, C., and Soumireu-Mourat, B. (1985). Time-dependent effects of posttraining intrahippocampal injections of corticosterone in retention of appetitive learning tasks in mice. *Eur. J. Pharmacol.* **106**, 39–46.

Mondadori, C., and Waser, P. G. (1979). Facilitation of memory processing by post-trial morphine: Possible involvement of reinforcement mechanisms?. *Psychopharmacology* **63**, 297–300.

O'Keefe, J., and Nadel, L. (1978). "The Hippocampus as a Cognitive Map." Oxford Univ. Press (Clarendon).

Oliveros, J. C., Jandali, M. K., Timsit-Berthier, M., Remy, R., Benghezal, A., Audibert, A., and Moeglen, J. M. (1978). Vasopressin in amnesia. *Lancet* **1**, 42.

Overton, D. A. (1971). Discriminative control of behavior by drug states. *In* "Stimulus Properties of Drugs" (G. Thompson and R. Dicken, eds.), pp. 87–110. Appleton-Century-Crofts, New York.

Pettibone, D. J., and Mueller, G. P. (1982). Adrenergic control of immunoactive beta-endorphin release from the pituitary of the rat: In vitro and in vivo studies. *J. Pharmacol. Exp. Ther.* **222**, 103–108.

Quartermain, D. (1982). The role of catecholamines in memory processing. *In* "The Physiological Bases of Memory" (J. A. Deutsch, ed.), 2nd ed., pp. 387–423. Academic Press, New York.

Riccio, D. C., and Concannon, J. T. (1981). ACTH and the reminder phenomena. *In* "Endogenous Peptides and Learning and Memory Processes" (J. L. Martinez, Jr., R. A. Jensen, R. B. Messing, H. Rigter, and J. L. McGaugh, eds.), pp. 117–142. Academic Press, New York.

Riccio, D. C., and Ebner, D. L. (1981). Postacquisition modifications of memory. *In* "Information Processing in Animals: Memory Mechanisms" (N. E. Spear and R. R. Miller, eds.), pp. 291–315. Erlbaum, Hillsdale, New Jersey.

Rossier, J., French, E. D., Rivier, C., Ling, N., Guillemin, K. R., and Bloom, F. E. (1977). Foot-shock induced stress increases beta-endorphin levels in blood but not brain. *Nature (London)* **270**, 618–620.

Sahgal, A. (1984). A critique of the vasopressin-memory hypothesis. *Psychopharmacology* **83**, 215–228.

Sahgal, A., Keith, A. B., Wright, C., and Edwardson, J. A. (1982). Failure of vasopressin to enhance memory in a passive avoidance task in rats. *Neurosci. Lett.* **28**, 87–92.

Sara, S. J., and Deweer, B. (1982). Memory retrieval enhanced by amphetamine after a long retention interval. *Behav. Neural Biol.* **36**, 146–160.

Sofroniew, M. V., Weindl, A., Schrell, U., and Wetzstein, R. (1981). Immunohistochemistry of vasopressin, oxytocin and neurophysin in the hypothalamus and extrahypothalamic regions of the human and primate brain. *Acta Histochem.* **24**, 79–95.

Soumireu-Mourat, B., Micheau, J., and Franc, C. (1981). ACTH(4-9) analog (ORG 2766) and memory processes in mice. *In* "Endogenous Peptides and Learning and Memory Processes" (J. L. Martinez, Jr., R. A. Jensen, R. B. Messing, H. Rigter, and J. L. McGaugh, eds.), pp. 143–158. Academic Press, New York.

Spear, N. E. (1978). "The Processing of Memories, Forgetting and Retention." Erlbaum, Hillsdale, New Jersey.

Staubli, U., and Huston, J. P. (1980). Avoidance learning enhanced by post-trial morphine injection. *Behav. Neural Biol.* **28**, 487–490.

Stein, L., Belluzzi, J. D., and Wise, C. D. (1975). Memory enhancement by central administration of norepinephrine. *Brain Res.* **84**, 329–335.

Sternberg, D. B., and Gold, P. E. (1980). Effects of alpha and beta-adrenergic receptor antagonists on retrograde amnesia produced by frontal cortex stimulation. *Behav. Neural Biol.* **29**, 289–302.

Sternberg, D. B., and Gold, P. E. (1981). Retrograde amnesia produced by electrical stimulation of the amygdala: Attenuation with adrenergic antagonists. *Brain Res.* **211**, 59–65.

Sternberg, D. B., Gold, P. E., and McGaugh, J. L. (1982). Noradrenergic sympathetic blockade: Lack of effect on memory or retrograde amnesia. *Eur. J. Pharmacol.* **81**, 133–136.

Sternberg, D. B., Gold, P. E., and McGaugh, J. L. (1983). Memory facilitation and impairment with supraseizure electrical brain stimulation: Attenuation with pretrial propranolol injections. *Behav. Neural Biol.* **38**, 261–268.

Sternberg, D. B., Martinez, J. L., Jr., Gold, P. E., and McGaugh, J. L. (1985a). Age-related memory deficits in rats and mice: Enhancement with peripheral injections of epinephrine. *Behav. Neural Biol.* **44**, 213–220.

Sternberg, D. B., Isaacs, K., Gold, P. E., and McGaugh, J. L. (1985b). Epinephrine facilitation of appetitive learning: Attenuation with adrenergic receptor antagonists. *Behav. Neural Biol.* **44**, 447–453.

van Wimersma Greidanus, T., and de Wied, D. (1976). Modulation of passive avoidance behavior of rats by intracerebroventricular administration of antivasopressin serum. *Behav. Biol.* **18**, 325–333.

van Wimersma Greidanus, T., Croiset, G., Bakker, E., and Bouman, H. (1979). Amygdaloid lesions block the effect of neuropeptides (vasopressin, ACTH4-10) on avoidance behavior. *Physiol. Behav.* **22**, 291–295.

van Wimersma Greidanus, T., Bohus, B., and de Wied, D. (1981). Vasopressin and oxytocin in learning and memory. *In* "Endogenous Peptides and Learning and Memory Processes" (J. L. Martinez, Jr., R. A. Jensen, R. B. Messing, H. Rigter, and J. L. McGaugh, eds.), pp. 413–427. Academic Press, New York.

Versteeg, D. H. G., and de Wied, D. (1985). Central modulation of memory processes by neuropeptides of the vasopressin family: Recent findings. *Prog. Neuroendocrinol.* **1**, 45–62.

Viveros, O. H., Diliberto, J. R., Hazum, E., and Chang, K. J. (1980). Enkephalins as possible adrenomedullary hormones: Storage secretions, and regulation of synthesis. *In* "Neural Peptides and Neuronal Communications" (E. Costa and M. Trabucchi, eds.), pp. 191–204. Raven Press, New York.

Walsh, T. J., and Palfai, T. (1979). Peripheral catecholamines and memory characteristics of syrosingopine-induced amnesia. *Pharmacol., Biochem. Behav.* **11**, 449–452.

Weinberger, N. M., Gold, P. E., and Sternberg, D. B. (1984). Epinephrine enables Pavlovian fear conditioning under anesthesia. *Science* **223**, 605–607.

Weingartner, H., Gold, P., Ballenger, J. C., Smallberg, S. A., Summers, R., Rubinow, D. R., Post, R. M., and Goodwin, F. K. (1981). Effects of vasopressin on human memory functions. *Science* **211**(6), 601–603.

Yonkov, D. I. (1983). Correlations between the effects of the central stimulants, cholinolytic and their combinations on the behaviour in open field and on the memory processes of albino rats in maze. *C. R. Acad. Bulg. Sci.* **36**, 1245–1248.

Yonkov, D. I., and Roussinov, K. S. (1981). Importance of application time of central stimulants and cholinoblockers to their effect on memory. *C. R. Acad. Bulg. Sci.* **34**, 441–443.

Zornetzer, S. F. (1978). Neurotransmitter modulation and memory: A new neuropharmacological phrenology. *In* "Psychopharmacology: A Generation of

Progress" (M. S. Lipton, A. DiMascio, and K. F. Killam, eds.), pp. 637–649. Raven Press, New York.

Zornetzer, S. F., Thompson, R., and Rogers, J. (1982). Rapid forgetting in aged rats. *Behav. Neural Biol.* **36,** 49–60.

CHAPTER SEVEN

Psychoneuroendocrinology of Stress: A Psychobiological Perspective

Seymour Levine, Christopher Coe, and
Sandra G. Wiener

I. Introduction

Despite numerous and valiant attempts to define stress, a clear and universally accepted definition still remains, to say the least, elusive. It is not the purpose of this chapter, therefore, to dwell extensively on the problems related to the definition of stress. This would be an exercise in futility. It is important to note, however, that historically the concept of stress has always been associated with changes in the endocrine system. Initially, these changes were specifically related to either increased secretion of catecholamines or activation of the pituitary–adrenal (P–A) system. It has now been clearly demonstrated that many endocrine changes occur following those environmental events which are typically called stressful. The focus of a great deal of research related to stress physiology is still on the P–A system. It is, of course, impossible to discuss any facet of the P–A system and its relationship to environmental factors without acknowledging the landmark contributions of Hans Selye. His formulation of the General Adaptation Syndrome (GAS) highlighted the importance of this hormonal system and the diversive physiological effects of glucocorticoids following adverse stimulation (Selye, 1950).

Selye's emphasis on the role of the glucocorticoids in biological adapta-

Psychoendocrinology, copyright © 1989 by Academic Press, Inc. All rights of reproduction in any form reserved.

tion was indeed prophetic. With the advent of modern molecular biology there has been increasing evidence of the fundamental role that the adrenal glucocorticoids play in a wide variety of physiological processes. Glucocorticoids have an effect on many aspects of development, on metabolism, on the expression of genes in essentially all tissues of the body, and on basic immune processes, and there is increasing evidence that they have important influences on many functions of the central nervous system (CNS), including learning and memory (Munck *et al.*, 1984; De Kloet and Reul, 1987). In view of the essential function that the glucocorticoids have in the maintenance of numerous physiological functions, it is not surprising that the regulation of the secretion of glucocorticoids has been studied extensively. The regulatory mechanisms which govern the secretory rates of glucocorticoids have been examined at every level, including the adrenal, pituitary, hypothalamus, midbrain, and limbic system.

In Selye's original theory, which elaborated the elements of the GAS, one of the major assumptions was that the stimuli eliciting the stress response were nonspecific. This hypothesis was based largely on the results of early research which demonstrated that a variety of physical and chemical challenges seemed to elicit the same pattern of endocrine responses. Mason (1975a,b) in his critique of Selye's theory pointed out that there were important psychological variables embedded in many of the experimental situations in which P–A activity was observed to increase. Moreover, in some cases the psychological reaction appeared to be the primary stimulus for initiating the cascade of neuroendocrine responses observed following stressful stimulation. Mason cited several experiments in which seemingly equivalent physical stimuli had different effects on the P–A system, depending upon temporal aspects of the challenge. He hypothesized that the psychological aspects of the situation, related to novelty, had to be considered and controlled for by altering the rate of presentation. One of the best examples of graded adrenocortical responses, which are dependent upon the rate of presentation, was reported by Gann (1969). In this experiment, dogs were first hemorrhaged 10 ml/kg at the rate of 6.6 ml/kg per minute. This rate of hemorrhage actively stimulated the adrenal cortex of the dog. In contrast, if the same volume of blood loss was achieved at a slower rate of hemorrhage (i.e., 3.3 ml/kg per minute), the P–A system did not activate. That rapid hemorrhaging induced adrenal cortical activity, while a slow rate of hemorrhaging did not, indicates that the rate of stimulus change is one important parameter for the induction of P–A activity. The fact that dexamethasone blocked the P–A response at the high rate of hemorrhaging indicated that the neuroendocrine system was involved and the effect was not mediated at the adrenal level. These data along with the studies presented by Mason indicate that there are many appraisal processes located in the CNS which determine whether or not the P–A system will respond to a specific set of stimuli.

In this chapter we shall attempt to accomplish two goals. First, we will describe the specific psychological variables which are involved in the regulation of P–A activity and, second, we will demonstrate that specific aspects of the psychological variables can selectively affect and regulate the secretion of gonadal hormones as well.

II. Pituitary–Adrenal System

As we have just stated, a psychobiological approach to understanding the participation of psychological factors in the regulation of the P–A system cannot escape making reference to cognitive processes (Berlyne, 1960, 1967). Berlyne's arousal theory provides a framework for the description of the processes by which stimulators of arousal operate. Novelty, uncertainty, and conflict are considered arousing. These have been labeled by Berlyne as collative factors, because in order to evaluate them it is necessary to compare similarities and differences between stimulus elements (novelty) or between stimulus-evoked expectations (uncertainty). Hennessy and Levine (1979) have assembled evidence in support of the hypothesis that the responses of the P–A system are a reflection of changes in the level of the hypothetical construct of arousal. The approach that was followed was to demonstrate that the psychological arousal and the P–A response were both subject to the same parametric relationships between stimulus and training variables. It thus follows that the basic cognitive process involved in regulation of the P–A system is one of comparison. The cognitive processes of comparison can best be understood by considering the concept of uncertainty.

A. Uncertainty

Neuroendocrine activation in response to uncertainty is best explained by Sokolov's model (1960) accounting for the general process of habituation. The pattern of habituation has been well described. A subject is presented with an unexpected stimulus and initially shows an orienting reaction. Physiological components of the orienting reaction include general activation of the brain, decreased blood flow into the extremities, changes in electrical resistance of the skin, and increases in both adrenal hormones. If the stimulus is frequently repeated, most of these reactions gradually diminish and eventually disappear, and the subject is said to be habituated. It does appear, however, that some physiological responses may habituate more slowly than the overt behavioral reactions (Hennessy and Levine, 1979). Sokolov's model, in essence, is based on a matching system in which new stimuli or situations are compared with a representation in the CNS of prior events. This matching process results in the development of expectancies whereby the organism is either habituated or gives an alerting reaction to a mismatch (Pribram and Melges, 1969). If the environment does not contain any new stimuli or contingencies, the habituated organism no longer responds with the physiological responses related to the alerting reaction. Activation of the P–A system to novel stimuli can also be accounted for by invoking the powerful explanatory principles of the Sokolov model.

B. Novelty and Uncertainty

Exposure of an animal to novelty is one of the most potent experimental conditions leading to an increase in adrenal activity (Hennessy and Levine, 1979). Novelty can be classified as a collative variable, since the recognition of any stimulus situation as being novel requires a comparison between present stimulus events and those experienced in the past. Increases in adrenal activity in response to novelty have been demonstrated in humans as well as animals. Increased adrenal activity, as evidenced by elevated levels of circulating cortisol, were observed in individuals during their first exposure to procedures involved in drawing blood at a blood bank (Mendoza and Barchas, 1982). However, if they have had prior blood-bank experience, there were no such increases. Further, in an experiment to be discussed in detail later, young adults experiencing their first jump off a tower during parachute training also showed a dramatic elevation of adrenal activity which is not observed on subsequent jumps from the tower (Levine, 1978).

Studies on animals also indicate another important characteristic of the cognitive process which results in P–A activation, that is, the ability of the animal to discriminate familiar versus unfamiliar stimulus elements (Fig. 1). In a series of experiments on rats and mice it was demonstrated that, if novelty was varied along a continuum with increasing changes in the stimulus elements, there was a graded adrenocortical response according to the degree to which the environment represented a discrepancy from the normal living environment of the organism (Hennessy and Levine, 1977; Hennessy et al., 1979). Thus, minor changes, such as placing the animal in a different cage, but one identical with its home cage, resulted in a moderate elevation of plasma corticosterone, but the response was significantly less than when the animal

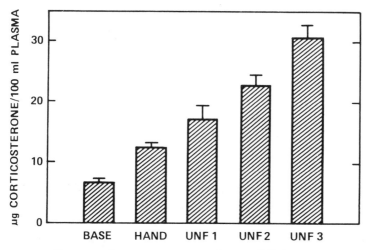

F I G U R E 1. Plasma corticosterone response of rats to differing degrees of novelty. HAND, handling and return to familiar home cage; UNF 1, handling and placement in an unfamiliar metal cage with new bedding; UNF 2, handling and placement in an unfamiliar cage without bedding; UNF 3, handling and placement in a plastic container (Hennessy *et al.*, 1979).

was placed in a totally novel cage that was distinctly different from its familiar home cage (Fig. 1). This capacity for graded elevations of P–A activity clearly demonstrates the remarkable capacity of the CNS to regulate the output of the adrenal glands.

Novelty, according to the theory presented by Sokolov, should be one of the most potent variables that elicits increases in P–A activity. Insofar as an organism has no familiarity with a novel environment, that environment should stimulate a degree of uncertainty that maximally elicits increases in neuroendocrine activity. Although novelty can be subsumed under the general heading of uncertainty, not all conditions which create uncertainty are novel. Uncertainty can also be evoked by insufficient information concerning the nature of upcoming events. Uncertainty can be seen to vary along the continuum from highly certain and predictable events to highly uncertain, unpredictable events. The presentation of a novel stimulus is likely to lead to an increase in uncertainty, by definition, because there is little information the organism can use to predict forthcoming events. Uncertainty can also be defined in terms of contingencies between environmental events. Experimentally, the dimension of uncertainty can be controlled by limiting the amount of information available to the organism to predict the occurrence of a specific event. Thus, one would hypothesize that if an organism is given information about the occurrence of either an appetitive or an aversive stimulus, such predictability should lead to a reduction in the P–A response. Further, situations in which there is an absence of predictability should lead to a marked increase.

There are many experiments that illustrate the value of predictability in modifying the P–A response to a variety of stimuli (Weinberg and Levine, 1980). One illustration of the effects of reducing uncertainty by providing predictability can be seen in a study by Dess-Beech et al. (1983). Dogs were subjected to a series of electric shocks which were either predictable or unpredictable. The predictable condition involved presenting the animal with a tone prior to the onset of shock. In the unpredictable condition, no tone was presented. The adrenocortical response observed on subsequent testing of these animals clearly indicated the importance of reducing uncertainty by predictability. Animals that did not have the signal preceding the shock showed an adrenocortical response which was two to three times that observed in animals which had previous predictable shock experiences. It should be noted that the procedures used in this experiment are typical of those utilized in experiments examining learned helplessness (Seligman, 1975). Learned helplessness refers to the protracted effects resulting from prolonged exposure to unpredictable and uncontrollable stimuli of an aversive nature. It has been observed that organisms exposed to this type of an experimental regime show long-term deficits in their ability to perform appropriately under subsequent testing conditions. Further, these animals show a much greater increase in adrenocortical response when exposed to novelty (Levine et al., 1973) than do control animals. Thus, an organism exposed to uncontrollable and unpredictable aversive stimuli not only shows increases in adrenocortical activity while exposed to these conditions, but there are also long-term consequences observed in other test conditions. These effects may be mediated in part by a chronic change in brain norepinephrine (NE) activity. Nursery-reared monkeys show reduced brain NE, behavioral pathology, and an increased responsiveness to stressful situations (Kraemer, 1985).

There is yet another series of experiments related to the issue of uncertainty. These experiments do not utilize aversive stimuli usually typical of stress research, but are more directly related to a psychological process commonly described as frustration. Frustration can be evoked when the organism fails to achieve a desired goal following a history of successful fulfillment of these goals. Experimentally, the operations utilized to produce frustration involve either preventing an animal from making the appropriate response to

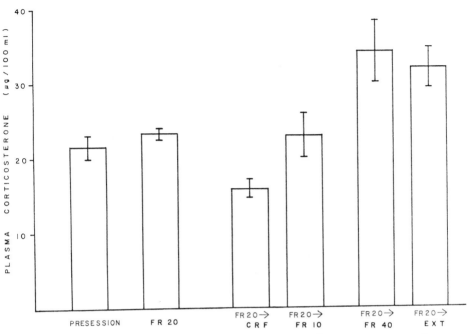

F I G U R E 2. The first two bars represent the mean corticosterone concentrations for all animals sampled before (presession) and after (FR 20) a normal session of stabilized FR-20 responding. The last four bars represent the mean value for each of four subgroups after a shift from stabilized FR 20 to one session of either more frequent reward (CRF, FR 10) or less frequent or no reward (FR 40, EXT). Vertical lines indicate standard error of the mean (Levine *et al.*, 1972).

achieve a desired object or not reinforcing the animal for a response that has had a prior history of reinforcement. In a broader sense, frustration involves the failure of an animal to fulfill expectancies developed through previous experiences and, thus, can be subsumed under the larger heading of uncertainty. For example, rats trained to press a lever for water, in which each lever-press delivered a small amount of water, showed an elevation of plasma corticosterone when the water was no longer available following the lever-press response (Coover, 1983; Coe *et al.*, 1983b; Levine *et al.*, 1972). Elevations of plasma corticosterone have been shown to be a robust and reliable phenomenon occurring under many experimental conditions in which reinforcement contingencies are altered (Fig. 2). Thus, not only is an elevation of plasma corticosterone observed when reinforcement is eliminated, but if the animal receives less reinforcement than it has previously become accustomed to, then elevations of plasma corticosterone also occur.

A similar phenomenon can be observed when using aversive stimuli. If an animal has learned to make an appropriate avoidance response that eliminates the occurrence of an electric shock, and this animal is then prevented from making the response, an increase in P–A activity occurs even when no electric shock is delivered (Coover *et al.*, 1973). These experiments have led to the hypothesis that one of the primary conditions for activation of neuroendocrine responses is a change in expectancies concerning well-established behaviors. In the case of the appetitive learning situation when reinforcement is eliminated, as well as in avoidance conditions, activation of the P–A system is a common occurrence following disruption of previously predictable behavior and outcomes.

C. Inhibition of Pituitary–Adrenal Activity

To a large extent, most of the research examining the influence of psychological factors on P–A hormones has been concerned with activation. However, considerable evidence has been accumulating which indicates that the P–A system is bidirectional. That is, psychological stimuli not only participate in activating this system, but can also effectively inhibit it. This inhibition is manifested either by reduced elevations of plasma corticoids during aversive stimulation or by an actual decrease in circulating levels of corticoids (Goldman *et al.*, 1973; Hennessy *et al.*, 1977; Levine *et al.*, 1979). The rationale for introducing this concept at this point is that if the absence of reinforcement leads to uncertainty and activation of the adrenal, then it might be expected that the presence of predictable and reinforcing events could result in lowering of adrenal activity. This effect is particularly pronounced when the predictable or reinforcing stimulus involves consummatory events, although we have also observed it using "contact comfort" in infant monkeys and rodents (Stanton *et al.*, 1987).

To summarize these experiments, it has been demonstrated that if access to food and water is restricted to the morning hours, the normal daily rhythm of adrenal hormones is reoriented to the period of consummatory behavior (i.e., the levels of plasma corticosteroids are elevated in the morning, whereas they are usually low during that period in a nocturnal rodent) (Johnson and Levine, 1973). Within 5 minutes of consuming food or water, however, there is a marked decline in plasma corticoids accompanied by a concomitant drop in ACTH (Gray *et al.*, 1978; Heybach and Vernikos-Danellis, 1979). These studies have shown that consummatory behavior initiates an inhibitory process at the level of the CNS which reduces both ACTH and corticoid secretion. There is also evidence indicating that when reinforcement occurs in an operant conditioning experiment, there is a comparable drop in glucocorticoids following consummatory behavior (Coe et al., 1983b; Goldman et al., 1973). It has further been demonstrated that the availability of a consummatory response can reduce an organism's P–A response to novelty (Levine et al., 1979). Thus, when exposed to novelty, rats usually show an elevation of plasma corticosterone, as has been discussed previously. However, if a consummatory response such as drinking is available in the novel environment, the response is reduced or absent.

There is yet another situation in which consummatory behavior has been shown to inhibit the elevation of plasma corticosterone. An unusual behavior pattern has frequently been observed in hungry rats on intermittent feeding schedules. If water is freely available during the experimental session, usually excessive amounts are drunk during intervals between rewards. The extent of this "abnormal" water consumption or schedule-induced polydipsia (SIP) (Falk, 1961) appears to be primarily determined by two experimental variables: the degree of intermittence of the schedule (i.e., the length of the intertrial intervals) and the food deprivation state of the animal (Falk, 1969). Rats in SIP studies are typically maintained at about 80% of their free-feeding body weight by restricting their daily food intake. However, no restrictions are imposed on water intake. There are two classes of explanation for SIP: those that are exclusively related to drinking behavior and those that view the phenomena in a broader perspective of schedule-induced behaviors. The explanations restricted only to drinking have been generally discredited. One can induce many different types of schedule-induced behaviors, including wheel running (Levitsky and Collier, 1968), aggression (Azrin et al., 1966), and chain pulling (Dantzer and Moremede, 1981), etc.

The other explanations rely on activation models, suggesting that intermittent delivery of a reinforcer causes a change in state (Roper, 1980). Whether these models focus on frustration (Denny and Ratner, 1970; Kissileff, 1973), general activation (Wayner, 1970, 1974; Wayner et al., 1981), arousal (Killeen, 1975, 1979; Killeen et al., 1978), or arousal reduction (Delius, 1967; Brett and Levine, 1979), they are general activation models that share the premise that an increased level of arousal is associated with the pattern of reinforcement. The induction of schedule-induced behaviors is primarily a function of increased arousal levels. Manipulations that are associated with increases in arousal are also associated with an enhancement of schedule-induced behavior. For example, mild shock enhances the expression of SIP

(Galantowicz and King, 1975; Segal and Oden, 1969). Frustration, as produced by lesser amounts of reinforcement than expected, also produces greater amounts of schedule-induced drinking (Thomka and Rosellini, 1975). Conversely, rats trained to tolerate frustration reduce levels of SIP (Amsel, 1971; Thomka and Rosellini, 1975).

In a series of experiments (Levine *et al.*, 1979) it was determined that this excess drinking results in a reduction of plasma corticoids which are normally high during intermittent food reinforcement if the animal is not permitted to drink. Since it has been postulated that elevated levels of corticoids are associated with high levels of arousal, then it can be argued that engaging in schedule-induced behaviors functionally reduces arousal. There are several interesting features of this phenomenon. The time course of the corticosterone decline following SIP has been shown to be rapid, occurring within 10 minutes of the onset of the induction of SIP. Further, plasma corticosterone levels decline even in rats which are allowed to consume only a small volume of liquid. Thus, although these animals consumed less than 4 ml of the fluid a decline in plasma corticosterone was still observed. It appears that the ability to engage in schedule-induced behaviors is more important than consumption. Further, Dantzer and Moremede (1981) reported that the induction of schedule-induced chain pulling in pigs also results in a significant decline in circulating levels of glucocorticoid hormones.

There is further evidence that the phenomenon of consummatory-induced inhibition of the P–A system does not require an ingestion. It has been demonstrated that once an animal has been exposed to consummatory behavior for a period of time, presentation of the stimuli associated with consummatory behavior can also serve to inhibit the P–A system (Coover *et al.*, 1977; Levine and Coover, 1976). That is, the phenomenon of adrenocorticoid inhibition is conditionable. It appears that whereas uncertain events can induce activation of the P–A system, events that are associated with reinforcement and which are highly predictable are capable of serving the opposite function.

D. Control and Feedback

The consequences of inappropriate adrenal secretion are evident from the effects of both hyper- and hypoadrenal output. Organisms deprived of adrenocorticoids are clearly in jeopardy and unable to deal effectively with even minor stresses, such as water or salt restriction, that are of little consequence to the intact organism. Conversely, chronic excessive secretion of glucocorticoids is also maladaptive. Prolonged elevations of these hormones can have a high biological cost, leading to a number of immunological changes as well as effects on digestive and cardiovascular physiology (Coe *et al.*, 1987; Munck *et al.*, 1984). It would follow, therefore, that there must have evolved a set of mechanisms available to the organism whereby it can regulate and modulate excessive output of glucocorticoids. We believe that many of these mechanisms are predominantly psychological. Perhaps the most important single psychological factor involved in modulating hormonal responses to aversive stimuli is the dimension of control. Control can best be defined as the capacity to make active responses during the presence of an aversive stimulus. These responses are frequently effective in allowing the animal to avoid or escape from the stimulus, but they may also function by providing the animal with the opportunity to change from one set of stimulus conditions to another, rather than to escape the aversive stimulus entirely. Control can reduce an organism's physiological response to such noxious stimuli as electric shock. It has been observed that rats able to press a lever to terminate shock show less severe physiological disturbances (e.g., weight loss and gastric lesions) than yoked controls which cannot respond, even though both animals receive identical amounts of shock (Weiss, 1984). Similarly, animals able to escape from shock show a reduction in plasma corticosterone following repeated exposures to shock (Davis *et al.*, 1977). The effects of control have been demonstrated clearly in an experiment by Hanson *et al.* (1976). These investigators studied rhesus monkeys that were exposed to the noxious stimulus of loud noise. One group of monkeys was permitted to control the duration of the noxious stimulus by making a lever-press response to terminate the noise. A comparable group of monkeys was given the identical amount of noise, but was not permitted to regulate the duration. The animals that were allowed to control the stimulus showed plasma cortisol levels like those of undisturbed subjects, whereas their yoked counterparts showed extremely high levels of plasma cortisol.

The effect of control over stimuli on adrenal responses was also demonstrated very clearly in an experiment using dogs (Dess-Beech *et al.*, 1983). These animals were subjected to a standard procedure used to produce learned helplessness. They were placed in a hammock and given uncontrollable and unpredictable threshold-level shocks. Other dogs were allowed to control the shock and terminate it by making a panel-press response with their heads. We have previously discussed the role of predictability in this experiment. The results further indicated that controllability also affected the magnitude of the cortisol response to the shock. Having neither control nor predictability elicited the maximum cortisol response, while having both minimized the impact of the shock. In addition, the capacity to control the stimuli, even in the absence of predictability, resulted in reducing the cortisol response to shock.

The concept of control in reducing adrenocortical activity in humans has been observed in the work environment. Rose and associates (1982a,b,c) investigated a variety of endocrine parameters, including growth hormone and cortisol, in a large group of air-traffic controllers during and after the work day. The job demands placed upon air-traffic controllers have been considered to be extremely stressful. Blood samples were obtained automatically and noninvasively via catheter at 20-minute intervals over a 5-hour period from a large group of working air-traffic controllers in their routine occupational environment. The data indicated that cortisol and growth-hormone levels were not appreciably elevated during the average work day. There appeared to be little in the way of increased endocrine activity under these working conditions which are usually presumed to be stressful. It is important to note, however, that the population selected for study was composed of highly experienced individuals who had been on the job an average of 11 years. One might conclude, therefore, that control over the environment resulting from their extensive experience could enable them to minimize the physiological consequences of their stressful occupation.

Further evidence that control can reduce cortisol in a task that would be considered to be predominantly stressful has been obtained in a study on drivers of heavy-goods vehicles. Cullen (1980) has reported that a task involving highway truck driving for 11 hours, on each of four successive days, immediately preceded by a nondriving day, produced significant effects on cortisol levels. Thus drivers with high initial levels on the nondriving day showed a drop in cortisol levels over days. These data were interpreted as indicating that, for experienced drivers, engaging in a task that contains a high degree of control and predictability once again resulted in a reduction of levels of circulating cortisol.

E. Feedback

Although it is clear that control is a major factor involved in modulating endocrine responses induced by stress, there is yet another factor which is also involved in this process: feedback. Feedback refers to stimuli or information occurring after a behavioral response has been made in reaction to an event. These stimuli may be used to convey information to the responding organism indicating that it has made the correct response to a noxious event, for example, or that the aversive event is terminated for at least some interval of time. According to Weiss (1971a,b,c), the amount of stress an animal actually experiences when exposed to noxious stimuli depends upon the number of coping attempts the animal makes (control) and the amount of relevant information the coping response produces (feedback). As the required number of coping responses increases and/or the amount of relevant feedback decreases, the amount of stress experienced increases. He has also argued that high operant responders, as an individual trait characteristic, will show more stress pathophysiology than low responders. In an extensive series of studies, Weiss demonstrated that if two groups of rats were subjected to the same amount of electric shock, the severity of gastric ulceration was reduced if the animal could respond—avoid and escape—and if the situation had some feedback information, i.e., a signal following the termination of shock. Although feedback information usually occurs in the context of control, namely, information about the efficacy of a response, it has been reported that feedback information per se, even in the absence of control, can reduce the P–A response to noxious stimuli. Hennessy et al. (1977) reported that the presence of a signal following the delivery of shock resulted in a reduced adrenocortical response even in the absence of control. In contrast, the P–A response of animals given a random signal was not significantly different from those animals that had no signal at all.

Human data concerning the role of feedback are not abundant, but there is one large study (Ursin et al., 1978) that investigated the cortisol response in humans following repeated experience with parachute training. In this study, the hormonal and behavioral responses of a group of Norwegian paratroop trainees were examined following repeated exposure to jumping off a 10-meter tower on a guide wire. After the first jump experience, there was a dramatic elevation of plasma cortisol, but as early as the second jump, there was a significant drop to basal levels. Thereafter, basal levels were observed on subsequent jumps (Fig. 3). It is also important to note that the fear ratings changed following the first and second jumps, so that there was very little fear expressed after the second jump, even though there had been a very high rating of fear prior to the first jump. We believe that both aspects of the "coping" model presented by Weiss can be applied to this situation. The individuals were able to make appropriate responses; that is, after the first experience they had already improved their performance of the task. However, since the individuals were jumping on a guide wire, performance was probably not the critical factor. The second aspect of the "coping" model, feedback, may be more important in this context. Although the situation was potentially dangerous and threatening, the trainees had gone through the

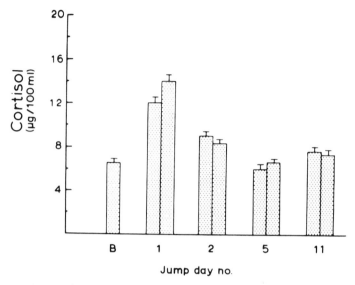

F I G U R E 3. Plasma cortisol levels of Norwegian paratroopers immediately after and 20 minutes after successive training jumps compared to base (B) (Levine, 1978).

experience and suffered no bad consequences. Thus, the feedback about the absence of danger in a potentially threatening situation was immediately obvious.

When one closely examines the three factors—control, predictability, and feedback—they all have a common element which can be viewed within the framework of a determinant of P–A activation proposed earlier, namely, uncertainty. Each of these factors, either acting alone or more probably in concert, has the capacity to reduce uncertainty. Control provides the organism with the capacity to eliminate, or at least to regulate, the duration of the aversive stimuli. Thus, the uncertainty involved in an unpredictable and uncontrollable situation is reduced. Predictability, by definition, serves to reduce uncertainty. Feedback can also be viewed in terms of reducing uncertainty, since feedback provides information to the organism about the efficacy and success of the response being emitted. We can therefore speculate that any cognitive or behavioral process that reduces uncertainty can result in a reduction or elimination of one endocrine response to a stressful situation.

F. Social Factors Influencing Pituitary–Adrenal Activity

There has been a great deal of speculation concerning the importance of social relationships in determining an individual's response to stress (Hamburg and Adams, 1967; Cohen and Wills, 1985). The available information emphasizes the great need for a continuity of personal relationships when individuals are involved in crises and highlights the critical role of social support as a major determinant in a number of health-related issues. For example, Nuckolls *et al.* (1972) reported that there was a higher incidence of pregnancy complications in women who experienced stressful life events before and during pregnancy and had weak social support system. Similarly, Cobb (1976) has shown that social support affects the length of hospitalization and the rate of recovery from illness and minimizes the effects of bereavement and retirement. Cobb has also reported that social support affects the physiological consequences of unemployment. Men whose jobs were terminated were evaluated periodically by public health nurses for 2 years after the onset of unemployment. The data revealed a 10-fold increase in arthritic joints in men who came from families with low social support as compared to men from highly supportive families. More recently, Kiecolt-Glaser *et al.* (1984) have demonstrated that social support can have an important effect on immune competence during stressful experiences. In one study on the effects of college examinations, medical students showed changes in a number of immunological parameters, including decreased natural killer cell activity, as compared to 2 months before. Moreover, greater immunosuppression, both before and after exams, occurred in those medical students who had reported low social support based on assessment by the UCLA Loneliness Scale.

The effects of social variables on P–A activity have been studied in an extensive series of studies using nonhuman primates. These studies have shown that, at almost every stage in the development of the organism, social variables can modulate the level of endocrine activity, both in terms of increased hormone release and/or the inhibition of endocrine responses to environmental perturbations. Disruption of social relationships is a potent psychogenic stressor leading to increased P–A activity in nonhuman primates. Conversely, the presence of social partners can reduce, and at times completely eliminate, the normal P–A response which would occur when social companions were not available. The neuroendocrine response to the disturbance of social relationships emerges early in life. This is best exemplified by the mother–infant relationship, which for the infant is the basis of its first experience with any kind of social interaction. Beginning with the work of Harry Harlow (Harlow and Harlow, 1969), numerous studies have shown that the mother serves not only to provide sustenance, but also to function as a source of emotional security. This latter role is evident in behavioral observations of even older, independent infants who will immedi-

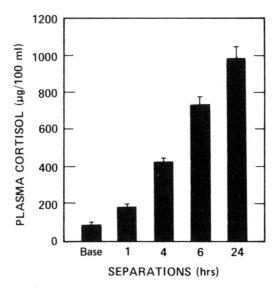

F I G U R E 4. Plasma cortisol response of infant squirrel monkeys to maternal separations of various duration (Coe *et al.*, 1985a).

ately seek proximity and make contact with the mother when a novel or fear-eliciting situation occurs. Soon after contact has been achieved, the infants begin to show signs of reduced behavioral agitation.

Psychoendocrine responses to disturbance of the mother–infant relationship also occur in even older, independent infants. The infant will show pronounced behavioral and hormonal responses to involuntary separation from the mother (Coe *et al.*, 1985a). As shown in Fig. 4, removing a squirrel monkey infant from its mother and placing it in a novel environment leads to an elevation of plasma cortisol levels over the first 24 hours. Although the unusually high levels begin to decline, with time, we have now ascertained that plasma cortisol levels remain elevated for up to 3 weeks after separation (Wiener *et al.*, 1989b). In similar studies on rhesus macaque infants (Gunnar *et al.*, 1981), we have also observed increased adrenal responses to maternal separation during the first day, but, in keeping with their lower corticoid levels in general, the infants typically return to basal values within 1–3 days.

It is important to note, however, that in these experiments not only was the infant removed from its mother, disrupting the social relationship, but perhaps as important, it was placed in a novel environment. Thus, the infant's level of uncertainty was increased in a number of ways. First, we have placed an organism in an unfamiliar environment, one of the conditions which definitely increases uncertainty. Second, by removing the infant from its mother, we introduced a loss of control (that is, removing the most important behavioral option that it normally uses to cope in stressful situations). To delineate the contribution of the novel separation environment versus the social disturbance, the following experiment was conducted on subadult squirrel monkeys (S. Levine, unpublished). From previous work we knew that peer separation also evoked an adrenal response in both prepubescent and adult monkeys. The subjects were male monkeys that had lived for over 2 years in a stable heterosexual group. They were housed in a unique double-pen environment which could be used to isolate an animal on one side of a connecting tunnel. Thus, it was possible to restrict access to the social group by closing a guillotine door, but still allow the subject to remain in a familiar environment. Comparison of the adrenal response to a 1-hour separation of this type with that to a separation in a novel environment demonstrated that the loss of social partners was the predominant factor causing the increase in plasma cortisol levels. Thus, these data support our hypotheses from studies on infants that the novelty of the unfamiliar separation environment aggravates the infant's response, but that the aversiveness is primarily due to the loss of the mother or other familiar social partners.

In experiments on infants, it was demonstrated that the magnitude of the adrenal response to maternal loss is dependent in part upon its previous housing conditions during rearing (Wiener et al., 1987). Infants were reared either in small maternal groups consisting of 3–4 dyads or with only their mothers in individual cages. When the infants reared in social groups were removed from the mother and placed in novel environments for 1 or 6 hours, there was once again the reliable elevation in plasma cortisol level. When the same infants were allowed to remain in the social home environment during 1- and 6-hour separations, there was a significant diminution in the plasma cortisol response (see Fig. 5). This finding is in keeping with our previous reports on the beneficial effects of providing a familiar physical and social environment for separated infants (Coe et al., 1983c). In contrast, when the same experimental conditions were imposed on the infants reared alone with their mothers, the resulting adrenal responses were different. No decrease in the adrenal response to separation was observed in the infants left alone in the home cage at either 1 or 6 hours. Therefore, for the home environment to be of equivalent comfort to the separated infant, the critical aspect of the home environment appears to be the presence of familiar conspecifics.

Yet another example of the influence of social variables on the response to separation was demonstrated in the following experiment. This experi-

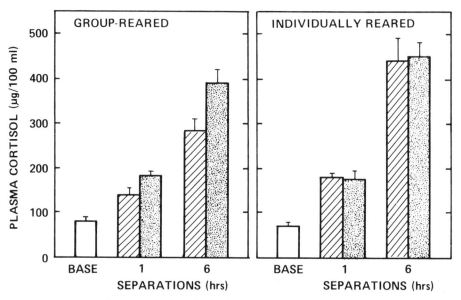

FIGURE 5. Plasma cortisol level of infants allowed to remain in home cage (cross-hatched bars) after mother was removed, or placed in a novel cage (stippled bars) alone, compared to nondisturbed base (Wiener et al., 1987).

ment utilized two paradigms previously employed in our laboratory. The first paradigm involved either removing the infant from the social group and placing it in a novel environment, or removing the mother and permitting the infant to remain with its social group. The second paradigm utilized the response following reunion with the mother as a means of evaluating the impact of the two types of separation. Typically, when the infant is not reunited with the mother, there is a continuing elevation in plasma cortisol levels over the first day; but when reunion is permitted, the infants begin to show a rapid return to normal levels of adrenal output. The following experiment examined the rate of recovery over a 6-hour period in 10 infants that were either separated for 30 minutes in a novel cage or separated for 30 minutes in the home cage with other mothers and infants present. Figure 6 shows that the adrenal response to separation was significantly lower when the infants were allowed to remain in the social group and also throughout the 6 hours after reunion with the mother. In addition, behavioral observations of the infants during the 6 hours after reunion indicated that they showed a more rapid adjustment in terms of their willingness to depart from the mother and resume play behavior following separations in the home cage.

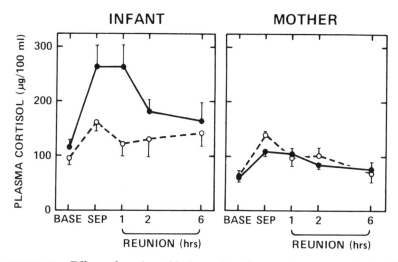

F I G U R E 6.   Effects of reunion with the mother after 30-minute separation in either the home cage (mother removed, ○---○) or novel cage (infant removed, ●——●) (Coe *et al.*, 1985c).

Thus, these data again show how the familiarity of the physical and social environment facilitates the infant's ability to modify the P–A response induced by separation, and how reunion with the mother initiates an active process that ameliorates the endocrine activation.

There is another set of data which illustrates the importance of social factors in the regulation of the P–A response. In studies on infant rhesus macaques we had found that the adrenal response to separation was greater in the totally isolated infants as compared to infants that remained within visual contact of the mother (Gunnar *et al.*, 1981). Recently (Wiener *et al.*, 1989a), an experiment was conducted with squirrel monkeys in which the infants were separated under three different conditions: (1) mother absent but infant in the presence of familiar social partners, (2) infant separated with the mother visually present, and (3) infant totally isolated. The data presented in Fig. 7 indicated a graded cortisol response to these separation conditions. Thus, the adrenal activation was the smallest in the presence of familiar conspecifics—a condition containing the least uncertainty—whereas a dramatic elevation was induced by the uncertainty of total isolation.

The early studies on separation in primates interpreted the infant's response to separation primarily in behavioral and emotional terms (Rosenblum and Kaufman, 1967; Hinde and Spencer-Booth, 1971; Mineka and

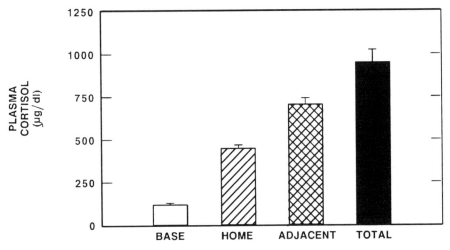

F I G U R E 7.   Plasma cortisol response of infant squirrel monkeys placed under different environmental conditions following maternal separation compared to base.

Suomi, 1978). Based on the theoretical formulations of Bowlby (1973), the infant's agitated activity and vocalizations were described as indicative of a "protest" response. As the infant's agitated behavior subsided, it was described as entering a stage of "despair." The emphasis on the emotional basis of the reaction can be thought of as a ballistic model, in that the infant's behavior follows a fixed curve following maternal loss (Rosenblum and Plimpton, 1981). From this perspective, the infant is viewed as a passive respondent. However, the infant's behavior following maternal loss can also be described in functional and cognitive terms as active "coping" attempts to resolve the aversive aspects of the situation. Thus, the infant's calls are not simply a reflection of affective disturbance, but actually represent a concerted effort to reestablish contact with the mother. The agitated activity would normally manifest itself as active searching behavior to locate the mother, and the high-intensity calls would elicit retrieval efforts by the mother. In this view, the infant is actively attempting to exert control over the situation. By resuming contact with the mother, the infant is attempting to use one of the most important behavioral responses for lowering its arousal levels (i.e., interaction with its mother). Thus, control over the reunion process is extremely important for the infant, and one of the aversive aspects of maternal separation appears to be loss of this behavioral control. We have argued previously that one of the most potent psychological variables modulating the P–A response to stress is control; and conversely, the loss of control, with the resulting increase in uncertainty, would activate the P–A axis.

G. Social Buffering

Probably as an outgrowth of the primate infant's reliance on its mother and other conspecifics, social factors also significantly influence the P–A response to stressful stimuli in adult monkeys. In the following studies, it has been demonstrated that the presence of a social group can ameliorate the neuroendocrine response to aversive stimuli that normally evoke a marked adrenal response in individuals housed alone. Nonhuman primates typically show strong behavioral reactions indicative of fear when exposed to a snake or a snakelike object, and these reactions do not readily extinguish following repeated exposure (Mineka et al., 1980). In order to determine whether a social group can serve as an effective modulator of physical responses to this type of fear stimulus, squirrel monkeys were exposed to a live boa constrictor which was presented above their cage in a wire-mesh box (Vogt et al., 1981). Monkeys living in four social groups, each consisting of two males and four females, were exposed to a snake for 30 minutes both while in the group and also after being removed from the group and placed in an individual cage. To control for the effects of general disturbance and handling, an empty stimulus box was placed on top of the troop and individual cages on different test days. All of the monkeys showed increased vigilance, agitated activity, and total avoidance of the snake in both the group and the individual conditions. However, while the behavioral response was consistent in both housing conditions, the snake did not elicit an adrenal response when the monkeys were tested as a group. This study, therefore, indicates that a social group could prevent a physiological response from occurring in a potentially threatening situation. In further experiments we have tried many other stimuli, including a loud, mobile robot with flashing lights, and have not observed an adrenal response in group-housed subjects tested in their home cage.

One interesting dimension of this social buffering is that it appeared to be dependent upon the number of available partners. In a subsequent study, we evaluated the response of adult males tested either alone or as a pair (Coe et al., 1982). The results indicated that a single partner was not sufficient to ameliorate the adrenal response to snake presentation and that, in some cases, the reaction was even aggravated. In an extension of these studies, we have carried out a classical conditioning study which more clearly delineates the effect of two or more partners in ameliorating stress responses (Stanton et al., 1985). In this study, the stressor and behavioral reactions were also more

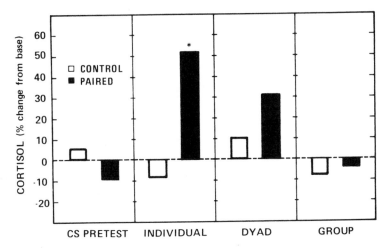

F I G U R E 8. Cortisol responses to 10 reinforced CS presentations in the home cage before training (CS Pretest) and after training as a function of social housing condition. The CS pretest occurred under individual housing. (*Differs from control and from 0% baseline, $p < 0.05$) (Stanton et al., 1985).

clearly delineated by associating a previously neutral light stimulus with an aversive event (electric shock). Thus, animals were individually presented in a test chamber with multiple pairings of the light [conditioned stimulus, (CS)] and the shock [unconditioned stimulus, (UCS)]. All subjects were later tested in the home cage and their adrenal responses monitored when they were housed individually, as pairs, or in a group of six animals (Fig. 8). When the monkeys were housed either alone or in pairs, the presentation of the flashing light led to a significant elevation in plasma cortisol levels. In contrast, when the light was presented to the animals in a social group with six subjects, there was no elevation in plasma cortisol levels even though increased behavioral activity was observed.

While there were several possible interpretations for the initial experiment on responses to snakes, including that the group test provided more behavioral options (i.e., avoidance) and control, the conditioning experiment unequivocally demonstrates that social factors can modulate the P–A response to fear-eliciting stimuli. These data concur with the view held by ethologists that one of the most important adaptations developed by higher primates is a sustained mother–infant relationship and group living. As Hans Kummer has stated, ". . . primates seem to have only one unusual asset in coping with their environments: a type of society which, through constant association of young and old and through a long life duration, exploits their large brains to produce adults of great experience. One may, therefore, expect to find some specific primate adaptations in the way primates do things as groups" (Kummer, 1971, pp. 37–38).

As we indicated earlier, we have hypothesized that one of the primary psychological stimuli for eliciting P–A activation is uncertainty. How then is it possible to fit the data on social support within the general context of the proposition that uncertainty leads to an activation of the P–A response and that the reduction of uncertainty diminishes or eliminates this response? Cobb (1976) discusses social support as a moderator of life stress and sees it as providing information that falls into three major categories. The first leads the individual to believe that he is loved or cared for. The second leads the individual to have higher self-esteem as a public expression of approval. The third is a perception of social congruity derived from a shared network of information and mutual obligation within which each member participates and a common knowledge is shared and accepted by all. It is possible to speculate that all three types of information derived from social support serve to minimize an individual's level of uncertainty about aversive situations. Although Cobb's propositions are expressed in humanistic terms, we can propose that the availability of stable and familiar social relationships also provides a set of predictable outcomes, due in part to the long history of previous interactions and experiences. Thus, the predictability of the social interactions in a group of familiar conspecifics tends, by definition, to reduce uncertainty. This hypothesis would lead to the prediction that an unfamiliar social group would provide none of these beneficial features and, in fact, should evoke a state of high uncertainty leading to an elevation of P–A activity.

In one experiment, an infant was separated from its mother and placed in an unfamiliar social group. Not only did we see no amelioration of the stress response, but the data indicated that being placed with nonfamiliar conspecifics was far more aggravating than simply being isolated (Levine and Wiener, 1989). In adult organisms, we have also found that one of the most reliable elicitors of increased P–A activation is the formation of new social groups. In several experiments we have shown that there was a marked elevation of plasma cortisol when animals were placed in a new social group, and that this elevation continued for several months (Gonzalez et al., 1981; Coe et al., 1983a). However, the formation of a social group not only has profound effects on the P–A system, it also affects other endocrine systems, in particular, the pituitary–gonadal system.

III. Pituitary–Gonadal System and Stress

Whereas we have emphasized the importance of uncertainty and reduction of uncertainty as modifiers of the P–A system, the pituitary–gonadal (P–G) system appears to be influenced not only by stressful events, but also by events related to reproductive activities and social interactions. The suppressive effect of stress on testicular and ovarian functions is probably the most well-known phenomenon (Levine, 1985), but it is apparent that gonadal hormone secretion can also be activated by certain types of external stimulation (Allen and Adler, this volume). Much of the latter research stems from studies on reproduction in animals that have shown the importance of environmental and physiological processes in stimulating sexual behavior (Desjardins, 1978). In particular, experiments on the hormone cycles of seasonally breeding species have demonstrated that these animals rely heavily on environmental cues for synchronizing the timing of the mating and birth periods (Michael and Zumpe, 1976). The seasonal recrudescence of testicular and ovarian activity in many of these species is also affected by psychological processes, such as territorial behavior and the presence of receptive mates (Rose et al., 1978; Rosenblatt, 1978; Vandenbergh and Post, 1976). These findings have served to reshape our earlier view of gonadal hormone secretion as invariant and have provided us with a large body of research on how psychological processes influence gonadal function.

The effect of physical trauma on gonadal function can be best exemplified by the testosterone decline observed during surgery (Aono et al., 1972). Typically, following moderate surgery involving a blood loss of less than 300 ml and a duration of 1 to 2 hours, there are decreased levels of testosterone for 3 to 4 days with a rebound to normal or slightly above normal levels at the end of a week. Following major surgery, such as pulmonary lobectomy, heart surgery, or total gastrectomy, the decreased testosterone output persists for 2 to 3 weeks. While these testosterone decreases are associated with concomitant drops in leutinizing hormone (LH) levels, it is important to note that the gonadal decrease actually precedes the change at the pituitary level. This is probably mediated in part by a decrease in blood flow to the testes—less blood to the periphery during stress. Moreover, attempts to stimulate testicular secretion directly with human chronic gonadotropin (HCG) did not return the circulating levels of testosterone to normal, revealing that the inhibition occurred at both the pituitary and the gonadal levels.

The evidence for a suppressive effect of physical trauma concurs with experimental studies on the effects of psychologically induced environmental disturbances. Some of the earliest studies on this issue were conducted on the effects of overcrowding in both natural and captive animal populations (Christian, 1963). In addition to causing adrenal hypertrophy and higher mortality, overcrowding had a number of inhibitory effects on reproductive physiology and behavior. Subsequent studies have refined this analysis and have shown that overcrowding has differential effects depending upon the aggressivity and social rank of the animals. That is, only those animals that receive the most aggression and are low in the social hierarchy show the inhibitory effects. In fact, aggressive animals of higher social rank may actually show increases in testosterone levels.

Similar effects have also been observed in studies on the correlation between dominance, aggression, and testosterone levels in nonhuman primates. While the correlation between social rank and testosterone is often low in stable groups under normal conditions, during periods of aggression there is a higher correlation. One way of studying this phenomenon is to examine hormone levels in newly formed social groups, since there is usually a period of aggression before a social hierarchy is established between the previously unfamiliar animals. An example of the effect of establishing dominance relationships is shown in Fig. 9, which illustrates the adrenal and gonadal responses observed in seven pairs of male squirrel monkeys (Coe *et al.*, 1985b; Coe and Levine, 1981). As can be seen, both males showed increased adrenal secretion in response to the behavioral agitation, but the gonadal response was more selective. Males that became dominant showed a striking increase in testosterone levels at 24 hours after pair formation, whereas subordinate males tended to show a decline. In this study, the effect on hormone levels was transient, but in other experiments conducted during the mating period, when hormone levels are higher, we have observed sustained effects for several weeks. Similar findings have been obtained in other primate species, such as the talapoin monkey and the rhesus macaque (Eberhart *et al.*, 1980; Rose *et al.*, 1975). In general, researchers have found that winners of fights show testosterone elevations, whereas losers of fights show lower levels of gonadal hormones. In time, as the aggression level subsides and the group becomes stable, the correlation decreases and may be entirely absent unless wounding occurs, in which case the testosterone suppressions may be prolonged.

A legacy of this animal heritage can be detected in humans. Even in moderately aversive environments, such as in military training school or during periods of extended school examinations, lower-than-normal levels of

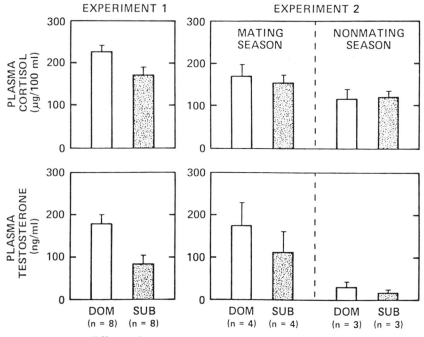

F I G U R E 9.   Effects of dominance rank on hormone levels in pairs of Bolivian males. Experiment 1 was conducted during the mating season; experiment 2 was a replication at two different times of the year (Coe and Levine, 1981).

testosterone may occur (Kreuz *et al.*, 1972). Moreover, the transient effect of aggression on testosterone titers has also been observed in men. Elevations in testosterone have been observed in winners of fights or contests in several situations including mock combat, collegiate wrestling, and highly competitive tennis matches (Mazur and Lamb, 1980).

Just as with the adrenal glucocorticoids, then, we must be cognizant of the potential for bidirectional changes in testosterone secretion. While the effect of severe and prolonged stress is usually suppressive, acute stress can actually result in transient increases. Thus, moderate exercise on a treadmill can increase testosterone in man, but sustained running, such as in a marathon, will depress testosterone levels (Nieschlag, 1979). An example of transient increases and decreases following psychogenic stress was also obtained in the study on Norwegian paratroopers (Davidson *et al.*, 1978). As shown in Fig. 10, testosterone levels were suppressed immediately before and 20 minutes after the first jump. However, on successive jumps the decline was not

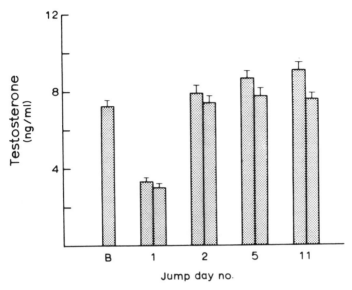

F I G U R E 10. Plasma testosterone levels of Norwegian paratroopers immediately after and 20 minutes after successive training jumps compared to base (B) (Davidson *et al.*, 1978).

observed, and following the fifth and eleventh jumps, significant increases occurred immediately after jumping. This finding of a transient elevation in testosterone immediately after acute stress concurs with a number of studies on LH and testosterone in other stressful situations (Coe *et al.*, 1978; Gray *et al.*, 1977; Nieschlag, 1979). That is, the initial response of the pituitary to a stressful situation is to release LH, and the testes respond with increased testosterone secretion for 15–30 minutes before the suppression occurs.

Up to now we have emphasized the general effects of stress on adrenal and gonadal physiology, but the selective response of the gonadal hormones to social and sexual stimuli is equally important. It should not be too surprising that psychological processes related to reproduction have been shown to influence hormone secretion and other aspects of reproductive physiology. For example, in a wide variety of species, ranging from rodents to boars to primates, the presence of opposite sex partners has been shown to raise testosterone levels in males. Conversely, the presence of same-sexed animals has been shown to have an inhibitory effect in a number of species. In the case of female mice, the presence of only other female conspecifics will cause the females to become anestrus, while the introduction of a male will initiate synchronized estrus (Rosenblatt, 1978). Another example of this type of effect can be found in a number of monogamous primate species from South America. Only the dominant male and female in these family groups will reproduce, and it has been found that the inhibition in females is due to a gonadotropin-releasing-factor responsiveness at the hypothalamic level (Hearn, 1983).

To provide an example of this kind of psychological stimulation, we have included a figure from a study on the response of male squirrel monkeys to females (Mendoza *et al.*, 1979). The study had three phases: males living alone, three all-male groups, and three male–female groups. As described before, when the male triads were housed together as a group, there was a differential effect of the resulting dominance ranks on both adrenal and gonadal hormones. All males showed increases in adrenocortical output following group formation, but dominant males showed the greatest increases. Moreover, there was a selective gonadal response (Fig. 11). Dominant males had the highest levels of testosterone, and this effect became more pronounced after females were introduced. During the weeks with females, tes-

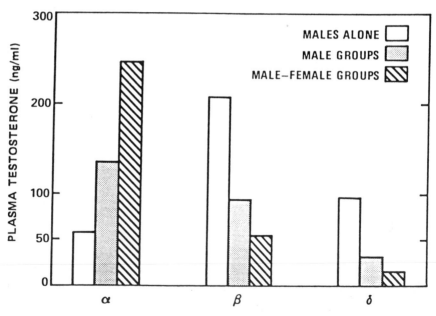

FIGURE 11. Mean levels of plasma testosterone for males at each dominance status during each of the three phases of group formation (Mendoza *et al.*, 1979).

tosterone levels in dominant males increased further, whereas subordinate males showed decreased levels. Although some of these effects could be attributed to the occurrence of reproductive behavior, since it is known that sexual activity can increase testosterone levels, several studies have now shown that mating is not a prerequisite for these hormone changes to occur. For example, it has been demonstrated in rats that LH increases can be conditioned to cues previously associated with the presentation of females (Graham and Desjardins, 1980).

IV. Conclusion

It is apparent from this discussion of psychological factors and the regulation of endocrine activity that we must modify the original concept of nonspecificity and incorporate a view which focuses on how specific psychological events affect the various hormonal systems. In the case of the P–A system we have hypothesized a set of cognitive factors that can be incorporated under the label of uncertainty. Activation or inhibition of the P–G system seems to occur mostly in the context of social interactions related to dominance, aggression, and reproductive behavior. When aspects of social relationships or a situation involve elements of uncertainty, we predict that they would elicit both P–A activation and changes in the P–G system. However, in experimental settings, at least, we can purposely design conditions that selectively alter either the P–A or the P–G systems (Coe *et al.*, 1985b). That is not to say that physical trauma does not directly affect adrenal or gonadal activity, but that most normal environmental challenges involve a large psychological component as well. This point was emphasized in John Mason's studies (1968) when he also argued for a revision of the nonspecificity concept and pointed out the importance of the psychological aspects of the stimulus.

In 1933, the eminent British neurophysiologist, Lord Sherrington, wrote, "The question who turns the key, to use that metaphor, is soon answered, the outside world." This statement was used by Sherrington in a discussion attempting to describe the relationship between environmental factors and internal biological rhythms. What has become increasingly apparent since the time of Sherrington is that organisms live in concert with their environment and that many factors which occur in what is seemingly the outside world do indeed have profound consequences on endogenous processes. The secretion of hormones from many glands in the body is sensitive to a whole variety of environmental changes. In this chapter, we have focused primarily on two hormonal systems, the pituitary–adrenal and pituitary–

gonadal axes. This is not to imply that these are the only hormonal systems in which environmental factors have a regulatory role. It is clear that virtually all peptide hormones emanating from the pituitary, including growth hormone, prolactin, and β-endorphins, are also responsive to environmental and psychological processes.

The field of neuroendocrinology has made tremendous advances in the past few decades in identifying many chemical substances which qualify as hormones, not only in the pituitary and peripheral organs, but in the central nervous system. Increasingly, the distinction between hormones and neurotransmitters has blurred as we have come to appreciate the neuromodulatory role of many hormones. The way in which these new CNS hormones respond to experiential factors has not yet been thoroughly investigated. This is perhaps due to the greater effort spent in attempting to isolate and identify these hormones. However, when the methods of measuring these particular substances become routine and readily available, there is no question that many of the issues dealt with in this chapter will be investigated with regard to these new and potentially important substances.

Acknowledgments

This research was supported by a grant from the National Institute of Mental Health (NIMH) HD-02881 and by a Research Scientist Award MH-19963 to Seymour Levine.

References

Amsel, A. (1971). Frustration, persistence, and regression. *In* "Experimental Psychopathology: Recent Research and Theory" (H. D. Kimmel, ed.). Academic Press, New York.

Aono, T., Kurachi, K., Mizutani, S., Hamanaka, Y., Uozumi, T., Nakasima, A., Koshiyama, K., and Matsumoto, K. (1972). Influence of major surgical stress on plasma levels of testosterone, luteinizing hormone and follicle-stimulating hormone in male patients. *J. Clin. Endocrinol. Metab.* **35,** 535–542.

Azrin, N. H., Hutchinson, R. R., and Hake, D. F. (1966). Extinction induced aggression. *J. Exp. Anal. Behav.* **9,** 191–204.

Berlyne, D. E. (1960). "Conflict, Arousal and Curiosity." McGraw-Hill, New York.

Berlyne, D. E. (1967). Arousal and reinforcement. *In* "Nebraska Symposium on Motivation" (D. Levine, ed.). Univ. Nebraska Press, Lincoln.

Bowlby, J. (1973). "Attachment and Loss," Vol. 2. Basic Books, New York.

Brett, L. P., and Levine, S. (1979). Schedule-induced polydipsia suppresses pituitary-adrenal activity in rats. *J. Comp. Physiol. Psychol.* **93,** 946–956.

Christian, J. J. (1963). Endocrine adaptive mechanisms and the physiological regulation of population growth. In "Physiological Mammalogy" (W. V. Mayer and R. G. Van Gelder, eds.), Vol. 1. Academic Press, New York.

Cobb, S. (1976). Social support as a moderator of life stress (presidential address). Psychosom. Med. 38, 300–314.

Coe, C. L., and Levine, S. (1981). Psychoneuroendocrine relationships underlying reproductive behavior in the squirrel monkey. Int. J. Mental Health 10, 22–42.

Coe, C. L., Mendoza, S. P., Davidson, J. M., Smith, E. R., Dallman, M. F., and Levine, S. (1978). Hormonal response to stress in the squirrel monkey (Saimiri sciureus). Neuroendocrinology 26, 367–377.

Coe, C. L., Franklin, D., Smith, E. R., and Levine, S. (1982). Hormonal responses accompanying fear and agitation in the squirrel monkey. Physiol. Behav. 29, 1051–1057.

Coe, C. L., Smith, E. R., Mendoza, S. P., and Levine, S. (1983a). Varying influence of social status on hormone levels in male squirrel monkeys. In "Hormones, Drugs and Social Behavior in Primates" (H. D. Steklis and A. S. Kling, eds.), pp. 7–32. Spectrum, New York.

Coe, C. L., Stanton, M. E., and Levine, S. (1983b). Adrenal responses to reinforcement and extinction: Role of expectancy vs. instrumental responding. Behav. Neurosci. 97, 654–657.

Coe, C. L., Wiener, S. G., and Levine, S. (1983c). Psychoendocrine responses of mother and infant monkeys to disturbance and separation. In "Symbiosis in Parent-Offspring Interactions" (L. A. Rosenblum and H. Moltz, eds.), Plenum, New York.

Coe, C. L., Wiener, S. G., Rosenberg, L. T., and Levine, S. (1985a). Physiological consequences of maternal separation and loss in the squirrel monkey. In "Handbook of Squirrel Monkey Research" (L. A. Rosenblum and C. L. Coe, eds.), pp. 127–148. Plenum, New York.

Coe, C. L., Smith, E. R., and Levine, S. (1985b). The endocrine system of the squirrel monkey. In "Handbook of Squirrel Monkey Research" (L. A. Rosenblum and C. L. Coe, eds.), pp. 191–218. Plenum, New York.

Coe, C. L., Wiener, S. G., Rosenberg, L. T., and Levine, S. (1985c). Endocrine and immune responses to separation and maternal loss in nonhuman primates. In "The Psychobiology of Attachment and Separation" (M. Reite and T. Field, eds.), pp. 163–199. Academic Press, New York.

Coe, C. L., Rosenberg, L. T., Fischer, M., and Levine, S. (1987). Psychological factors capable of preventing the inhibition of antibody responses in separated infant monkeys. Child Dev. 58, 1420–1430.

Cohen, S., and Wills, T. A. (1985). Stress, social support and the buffering hypothesis. Psychol. Bull. 98(2), 310–357.

Coover, G. D. (1983). The rat's reward environment and pituitary-adrenal activity. In "Biological and Psychological Basis of Psychosomatic Disease" (H. Ursin and R. Murison, eds.). Pergamon, Oxford.

Coover, C. D., Ursin, H., and Levine, S. (1973). Plasma corticosterone levels during active avoidance learning in rats. *J. Comp. Physiol. Psychol.* **82**, 170–174.

Coover, C. D., Sutton, B. R., and Heybach, J. P. (1977). Conditioning decreases in plasma corticosterone level in rats by pairing stimuli with daily feedings. *J. Comp. Physiol. Psychol.* **91**, 716–726.

Cullen, J. (1980). Coping and health—a clinician's perspective. In "Coping and Health" (S. Levine and H. Ursin, edsd.). Plenum, New York.

Dantzer, R., and Mormede, P. (1981). Pituitary-adrenal consequences of adjunctive activities in pigs. *Horm. Behav.* **15**, 386–395.

Davidson, J. M., Smith, E. R., and Levine, S. (1978). Testosterone. In "Psychobiology of Stress: A Study of Coping Men" (H. Ursin, E. Baade, and S. Levine, eds.). Academic Press, New York.

Davis, H., Porter, J. W., Livingstone, J., Herrmann, T., MacFadden, L., and Levine, S. (1977). Pituitary-adrenal activity and lever press shock escape behavior. *Physiol. Psychol.* **5**, 280–284.

De Kloet, E. R., and Reul, J. M. H. M. (1987). Feedback action and tonic influence of corticosteroids on brain function: A concept arising from heterogeneity of brain receptor system. *Psychoneuroendocrinology* **12**, 83–105.

Delius, J. D. (1967). Displacement activities and arousal. *Nature (London)* **214**, 1259–1260.

Denny, M. R., and Ratner, S. C. (1970). "Comparative Psychology: Research in Animal Behavior." Dorsey Press, Homewood, Illinois.

Desjardins, C. (1978). Potential sources of variation affecting studies on pituitary-gonadal function. In "Symposium on Animal Models for Research on Contraception" (N. J. Alexander, ed.). Harper & Row, Hagerstown, Maryland.

Dess-Beech, N. K., Linwick, D., Patterson, J., and Overmier, J. B. (1983). Immediate and proactive effects of controllability and predictability on plasma cortisol responses to shocks in dogs. *Behav. Neurosci.* **97**, 1005–1016.

Eberhart, J. A., Keverne, E. B., and Meller, R. E. (1980). Social influences on plasma testosterone levels in male talapoin monkeys. *Horm. Behav.* **14**, 247–265.

Falk, J. L. (1961). Production of polydipsia in normal rats by an intermittent food schedule. *Science* **133**, 195–196.

Falk, J. L. (1969). Conditions producing psychogenic polydipsia in animals. *Ann. N. Y. Acad. Sci.* **157**, 569–593.

Galantowicz, E. P., and King, G. D. (1975). The effects of three levels of lick-contingent footshock on schedule-induced polydipsia. *Bull. Psychon. Soc.* **5**, 113–116.

Gann, D. C. (1969). Parameters of the stimulus initiating the adrenocortical response to hemorrhage. *Ann. N. Y. Acad. Sci.* **156**, 740–755.

Goldman, L., Coover, G. D., and Levine, S. (1973). Bidirectional effects of reinforcement shifts on pituitary-adrenal activity. *Physiol. Behav.* **10**, 209–214.

Gonzalez, C. A., Hennessy, M. B., and Levine, S. (1981). Subspecies differences in hormonal and behavioral responses after group formation in squirrel monkeys. *Am. J. Primatol.* **1**, 439–452.

Graham, J. M., and Desjardins, C. (1980). Classical conditioning: Induction of luteinizing hormone and testosterone secretion in anticipation of sexual activity. *Science* **210,** 1039–1041.

Gray, G. D., Smith, E. R., Damassa, C. A., Ehrenkranz, J. R. L., and Davidson, J. M. (1977). Chronic suppression of pituitary-testicular function by stress in rats. *Fed. Proc., Fed. Am. Soc. Exp. Biol.* **36,** 322.

Gray, G. D., Bergfors, A. M., Levine, R., and Levine, S. (1978). Comparison of the effects of restricted morning or evening water intake on adrenocortical activity in female rats. *Neuroendocrinology* **25,** 236–246.

Gunnar, M. R., Gonzalez, C. A., Godlin, B. L., and Levine, S. (1981). Behavioral and pituitary-adrenal responses during a prolonged separation period in infant rhesus macaques. *Psychoneuroendocrinology* **6,** 65–75.

Hamburg, D. A., and Adams, J. E. (1967). A perspective on coping behavior: Seeking and utilizing information in major transitions. *Arch. Gen. Psychiatry* **17,** 277–284.

Hanson, J. D., Larson, M. E., and Snowdon, C. T. (1976). The effects of control over high intensity noise on plasma cortisol levels in rhesus monkeys. *Behav. Biol.* **16,** 333–340.

Harlow, H. F., and Harlow, M. K. (1969). Effects of various mother-infant relationships on rhesus monkey behaviors. *In* "Determinants of Infant Behaviour" (B. M. Foss, ed.). Methuen, London.

Hearn, J. (1983). "Reproduction in New World Primates." MTP Press, Lancaster.

Hennessy, J. W., and Levine, S. (1979). Stress, arousal and the pituitary-adrenal system: A psychoendocrine model. *Prog. Psychobiol. Physiol. Psychol.* **8.**

Hennessy, J. W., King, M. G., McClure, T. A., and Levine, S. (1977). Uncertainty, as defined by the contingency between environmental events, and the adrenocortical response of the rat to electric shock. *J. Comp. Physiol. Psychol.* **91,** 1447–1460.

Hennessy, M. B., and Levine, S. (1977). Effects of various habituation procedures on pituitary-adrenal responsiveness in the mouse. *Physiol. Behav.* **18**(5), 799–802.

Hennessy, M. B., Heybach, J. P., Vernikos, J., and Levine, S. (1979). Plasma corticosterone concentrations sensitively reflect levels of stimulus intensity in the rat. *Physiol. Behav.* **22,** 821–825.

Heybach, J. P., and Vernikos-Danellis, J. (1979). Inhibition of adrenocorticotrophin secretion during deprivation-induced eating and drinking in rats. *Neuroendocrinology* **28,** 329–338.

Hinde, R. A., and Spencer-Booth, Y. (1971). Effects on brief separations from mothers on rhesus monkeys. *Science* **173,** 111–118.

Johnson, J. T., and Levine, S. (1973). Influence of water deprivation of adrenocortical rhythms. *Neuroendocrinology* **11,** 268–273.

Kiecolt-Glaser, J. K., Garner, W., Speicher, C., Penn, G. M., Holliday, J., and Glaser, R. (1984). Psychosocial modifiers of immunocompetence in medical students. *Psychosom. Med.* **46,** 7–14.

Killeen, P. R. (1975). On the temporal control of behavior. *Psychol. Rev.* **85**, 89–115.

Killeen, P. R. (1979). Arousal: Its genesis, modulation and extinction. *In* "Reinforcement and the Organization of Behavior" (M. D. Zeiler and P. Harzem, eds.). Wiley, New York.

Killeen, P. R., Hanson, S. J., and Osborne, S. R. (1978). Arousal: Its genesis and manifestation of response rate. *Psychol. Rev.* **85**, 571–581.

Kissileff, H. R. (1973). Nonhomeostatic controls of drinking. *In* "The Neuropsychology of Thirst: New Findings and Advances in Concepts" (A. N. Epstein, H. R. Kissileff, and E. Stellar, eds.), V. H. Winston, Washington, D.C.

Kraemer, G. W. (1985). Effects of differences in early social experience on primate neurobiological-behavioral development. *In* "The Psychobiology of Attachment and Separation" (M. Reite and T. Field, eds.). Academic Press, New York.

Kreuz, L. E., Rose, R. M., and Jennings, J. R. (1972). Suppression of plasma testosterone levels and psychological stress. *Arch. Gen. Psychiatry* **26**, 479–482.

Kummer, H. (1971). "Primate Societies: Group Techniques of Ecological Adaptation." Aldine-Atherton, Chicago, Illinois.

Levine, S. (1985). A definition of stress? *In* "Animal Stress" (G. P. Moberg, ed.). Am. Physiol. Soc., Bethesda, Maryland.

Levine, S. (1978). Cortisol changes following repeated experiences with parachute training. *In* "Psychobiology of Stress: A Study of Coping Men" (H. Ursin, E. Baade, and S. Levine, eds.). Academic Press, New York.

Levine, S., and Coover, G. D. (1976). Environmental control of suppression of the pituitary-adrenal system. *Physiol. Behav.* **17**, 35–37.

Levine, S., and Wiener, S. G. (1989). Psychoendocrine aspects of mother-infant relationships in nonhuman primates. *Psychoneuroendocrinology* (in press).

Levine, S., Goldman, L., and Coover, G. D. (1972). Expectancy and the pituitary-adrenal system. *Ciba Found. Symp.* **8**, 281–296.

Levine, S., Madden, J., IV, Conner, R. L., Moskal, J. R., and Anderson, D. C. (1973). Physiological and behavioral effects of prior aversive stimulation (preshock) in the rat. *Physiol. Behav.* **10**, 467–471.

Levine, S., Weinberg, J., and Brett, L. P. (1979). Inhibition of pituitary-adrenal activity as a consequence of consummatory behavior. *Psychoneuroendocrinology* **4**, 275–286.

Levine, S., Wiener, S. G., Coe, C. L., Bayart, F. E. S., and Hayashi, K. T. (1987). Primate vocalization: A psychobiological approach. *Child Dev.* **58**, 1408–1419.

Levitsky, D., and Collier, G. (1968). Schedule-induced wheel running. *Physiol. Behav.* **3**, 571–573.

Mason, J. W. (1968). A review of psychoendocrine research on the pituitary-adrenal cortical system. *Psychosom. Med.* **30**, 576–607.

Mason, J. W. (1975a). A historical view of the stress field. *J. Hum. Stress* **1**, 6–12.

Mason, J. W. (1975b). A historical view of the stress field. *J. Hum. Stress* **1**, 22–36.

Mazur, A., and Lamb, T. A. (1980). Testosterone, status and mood in human males. *Horm. Behav.* **14,** 236–246.

Mendoza, S. P., and Barchas, P. (1982). Social mediation of stress: Human applications of a nonhuman primate model. *In* "Symposium on Hormones and Behavior." Western Psychological Association, Sacramento, California.

Mendoza, S. P., Coe, C. L., Lowe, E. L., and Levine, S. (1979). The physiological response to group formation in adult male squirrel monkeys. *Psychoneuroendocrinology* **3,** 221–229.

Michael, R. P., and Zumpe, D. (1976). Environmental and endocrine factors influencing annual changes in sexual potency in primates. *Psychoneuroendocrinology* **1,** 303–313.

Mineka, S., and Suomi, S. J. (1978). Social separation in monkeys. *Psychol. Bull.* **85,** 1376–1400.

Mineka, S., Keir, R., and Price, V. (1980). Fear of snakes in wild- and lab-reared rhesus monkeys. *Anim. Learn. Behav.* **8,** 653–663.

Munck, A., Guyre, P. M., and Holbrook, N. J. (1984). Physiological functions of glucocorticoids in stress and their relation to pharmacological actions. *Endocrinol. Rev.* **5,** 25–44.

Nieschlag, E. (1979). The endocrine function of the human testis in regard to sexuality. *Ciba Found. Symp.* **62,** 1978.

Nuckolls, K. B., Cassel, J., and Kaplan, B. H. (1972). Psychosocial assets, life crisis and the prognosis of pregnancy. *Am. J. Epidemiol.* **95,** 431–441.

Pribram, K. H., and Melges, F. T. (1969). Psychophysiological basis of emotion. *In* "Handbook of Clinical Neurology" (P. J. Vinken and G. W. Bruyn, eds.), North-Holland Publ., Amsterdam.

Roper, T. J. (1980). Behaviour of rats durinhg self-initiated pauses in feeding and drinking, and during periodic response-independent delivery of food and water. *Q. J. Exp. Psychol.* **32,** 459–472.

Rose, R. M., Bernstein, I. S., and Gordon, T. P. (1975). Consequences of social conflict on plasma testosterone levels in rhesus monkeys. *Psychosom. Med.* **37,** 50–61.

Rose, R. M., Bernstein, I. S., Gordon, T. P., and Lindsley, J. G. (1978). Changes in testosterone and behavior during adolescence in the male rhesus monkey. *Psychosom. Med.* **40,** 60–70.

Rose, R. M., Jenkins, C. D., Hurst, H., Livingston, L., and Hall, R. P. (1982a). Endocrine activity in air traffic controllers at work. I. Characterization of cortisol and growth hormone levels during the day. *Psychoneuroendocrinology* **7,** 101–111.

Rose, R. M., Jenkins, C. D., Hurst, M., Herd, J. A., and Hall, R. P. (1982b). Endocrine activity in air traffic controllers at work. II. Biological, psychological and work correlates. *Psychoneuroendocrinology* **7,** 113–123.

Rose, R. M., Jenkins, C. D., Hurst, B. E., Kreger, J., Barrett, J., and Hall, R. P.

(1982c). Endocrine activity in air traffic controllers at work. III. Relationship to physical and psychiatric morbidity. *Psychoneuroendocrinology* 7, 125–134.

Rosenblatt, J. S. (1978). Behvioral regulation of reproductive physiology: A selected review. *In* "Comparative Endocrinology" (P. J. Galliard and H. H. Boer, eds.). Elsevier, Amsterdam.

Rosenblum, L. A., and Kaufman, I. C. (1967). Laboratory observations of early mother-infant relations in pigtail and bonnet macaques. *In* "Social Communication among Primates" (S. A. Altmann, ed.), pp. 33–41. Univ. of Chicago Press, Chicago, Illinois.

Rosenblum, L. A., and Plimpton, E. H. (1981). Adaptation to separation: The infant's effort to cope with an altered environment. *In* "The Uncommon Child: Genesis of Behavior" (M. Lewis and L. A. Rosenblum, eds.), pp. 225–257. Plenum, New York.

Segal, E. F., and Oden, D. L. (1969). Effects of drinometer current and of foot shock on psychogenic polydipsia. *Psychon. Sci.* 14, 14–15.

Seligman, M. E. P. (1975). "Learned Helplessness: On Depression, Development and Death." Freeman, San Francisco, California.

Selye, H. (1950). "Stress." Acta, Montreal.

Sherrington, C. S. (1933). "The Brain and Its Mechanisms." Cambridge Univ. Press, London and New York.

Sokolov, E. N. (1960). Neuronal models and the orienting reflex. *In* "The Central Nervous System and Behavior" (M. A. B. Brazier, ed.). Josiah Macy, Jr. Found., New York.

Stanton, M. E., Patterson, J. M., and Levine, S. (1985). Social influences on conditioned cortisol secretion in the squirrel monkey. *Psychoneuroendocrinology* 10, 125–134.

Stanton, M. E., Walstrom, J., and Levine, S. (1987). Maternal contact inhibits pituitary-adrenal stress responses in preweanling rats. *Dev. Psychobiol.* 20, 131–145.

Thomka, M. L., and Rosellini, R. A. (1975). Frustration and production of schedule-induced polydipsia. *Anim. Learn. Behav.* 3, 380–384.

Ursin, H., Baade, E., and Levine, S., eds. (1978). "Psychobiology of Stress: A Study of Coping Men." Academic Press, New York.

Vandenbergh, J. G., and Post, W. (1976). Endocrine coordination in rhesus monkeys: Female responses to the male. *Physiol. Behav.* 17, 979–984.

Vogt, J. L., Coe, C. L., and Levine, S. (1981). Behavioral and adrenocorticoid responsiveness of squirrel monkeys to a live snake: Is flight necessarily stressful? *Behav. Neural Biol.* 32, 391–405.

Wayner, M. J. (1970). Motor control functions of the lateral hypothalamus and adjunctive behavior. *Physiol. Behav.* 5, 1319–1325.

Wayner, M. J. (1974). Specificity of behavioral regulation. *Physiol. Behav.* 12, 851–869.

Wayner, M. J., Barone, F. C., and Loullis, C. C. (1981). The lateral hypothalamus and adjunctive behavior. *In* "Behavioral Studies of the Hypothalamus" (B. J. Morgane and J. Panksepp, eds.), Part B. New York. Dekker.

Weinberg, J., and Levine, S. (1980). Psychobiology of coping in animals: The effects of predictability. *In* "Coping and Health" (S. Levine and H. Ursin, eds.). Plenum, New York.

Weiss, J. M. (1971a). Effects of coping behavior in different warning signal conditions on stress pathology in rats. *J. Comp. Physiol. Psychol.* **77,** 1–13.

Weiss, J. M. (1971b). Effects of punishing the coping response (conflict) on stress pathology in rats. *J. Comp. Physiol. Psychol.* **77,** 14–21.

Weiss, J. M. (1971c). Effects of coping behavior with and without a feedback signal on stress pathology in rats. *J. Comp. Physiol. Psychol.* **77,** 22–30.

Weiss, J. M. (1984). Behavioral and psychological influences on gastrointestinal pathology. *In* "Handbook of Behavioral Medicine" (W. D. Gentry, ed.). Guilford Press, New York.

Wiener, S. G., Johnson, D. F., and Levine, S. (1987). Influence of postnatal rearing conditions on the response of squirrel monkey infants to brief perturbations in mother-infant relationships. *Physiol. Behav.* **39,** 21–26.

Wiener, S. G., Bayart, F., Faull, K. F., and Levine, S. (1989a). Behavioral and physiological responses to maternal separation in the squirrel monkey. *Behav. Neurosci.* (submitted for publication).

Wiener, S. G., Lowe, E., and Levine, S. (1989b). Influence of weaning on the pituitary-adrenal system in infant squirrel monkeys. In preparation.

CHAPTER EIGHT

Endogenous Opioids and Behavior

F. Robert Brush and
Carolyn Nagase Shain

I. Introduction

Human use of the prototypical opiate, opium, is first mentioned in the third century B.C. by Theophratus (Seiden, 1983), but most people probably associate its use with the opium dens of nineteenth-century China. They supported a lucrative opium trade controlled by the British who naturally wished to maintain it. However, China attempted to regulate and reduce the use of opium, which resulted in two wars with the British in 1839 and 1865. Regulation of opiates remains a significant social and legal problem today, although social attitudes about the use of opiates have changed considerably over the years (Siegel, 1986).

In this chapter, we will distinguish between *opiates,* such as morphine and related alkaloid pharmaceutical agents, and *opioids,* a family of compounds that includes the alkaloid opiates as well as a subclass of neuropeptides that have opiatelike activity and that are endogenously produced by the brain, pituitary gland, and other peripheral organs. In order to help make sense of what has become a voluminous and complex literature, we will first identify the major brain structures that bind opiates and review some of the pharmacological effects of alkaloid opiates. Then we shall identify the molecular structures, sources, and brain distributions of the endogenous opioids. The interactions of these opioids with the stress-related neuroen-

Psychoendocrinology, copyright © 1989 by Academic Press, Inc. All rights of reproduction in any form reserved.

docrine systems will be described in an effort to provide a basis for unraveling the *direct* behavioral effects of the opioids from possible *indirect* effects that may be mediated by their neuroendocrine influences. The opioids have been shown to have two principal effects, analgesia and euphoria, but we will focus only on the analgesic effects, particularly stress-induced analgesias. Smith and Lane (1983) provide a useful review of the rewarding (euphoric) effects of the opioids.

II. Distribution of Opiate Receptors in the Brain

A. Structures of Morphine and Its Agonists and Antagonists

Morphine was first isolated in pure form from opium in 1806 by Serturner, who named it after Morpheus, the Greek god of dreams (see Fig. 1). Codeine, a methylated form of morphine, was isolated in 1832 (see Fig. 1). Heroin, which is also derived from opium, and codeine are both quickly converted to morphine *in vivo*. Needless to say, following the invention of the hypodermic needle extensive abuse of the opiates developed. This, in turn, led to a search for effective antagonists of the opiates, and for agents to counteract the withdrawal symptoms in addicts who stopped taking the drugs. Nalorphine (see Fig. 1) was found to counteract the actions of morphine, and it was successfully used to treat morphine poisoning as early as 1951. However, it also produced anxiety in doses that were effectively antagonistic, and it was subsequently found to produce analgesia in even larger doses, with the likelihood of addiction with repeated use. So the search for an effective antagonist continued. Naloxone and a closely related compound, naltrexone (see Fig. 1), were found to be relatively pure antagonists of morphine but were relatively ineffective in antagonizing the effects of cyclazocine (see Fig. 1), which is a potent analgesic with addicting properties; but the withdrawal syndrome following addiction to cyclazocine is different from that following withdrawal from morphine. Other compounds were found to have mixed agonistic and antagonistic effects. Methadone, which is structurally so different from heroin, was found to effectively counter withdrawal symptoms in heroin addicts and is in wide use today in our major metropolitan areas.

B. Early Analyses of Opiate Receptor Subtypes

These findings from the 1950s and 1960s led Martin (1967) to postulate the existence of multiple receptors for the opiates (Martin *et al.*, 1976; Gilbert and Martin, 1976). This idea was based primarily on a careful analysis of the

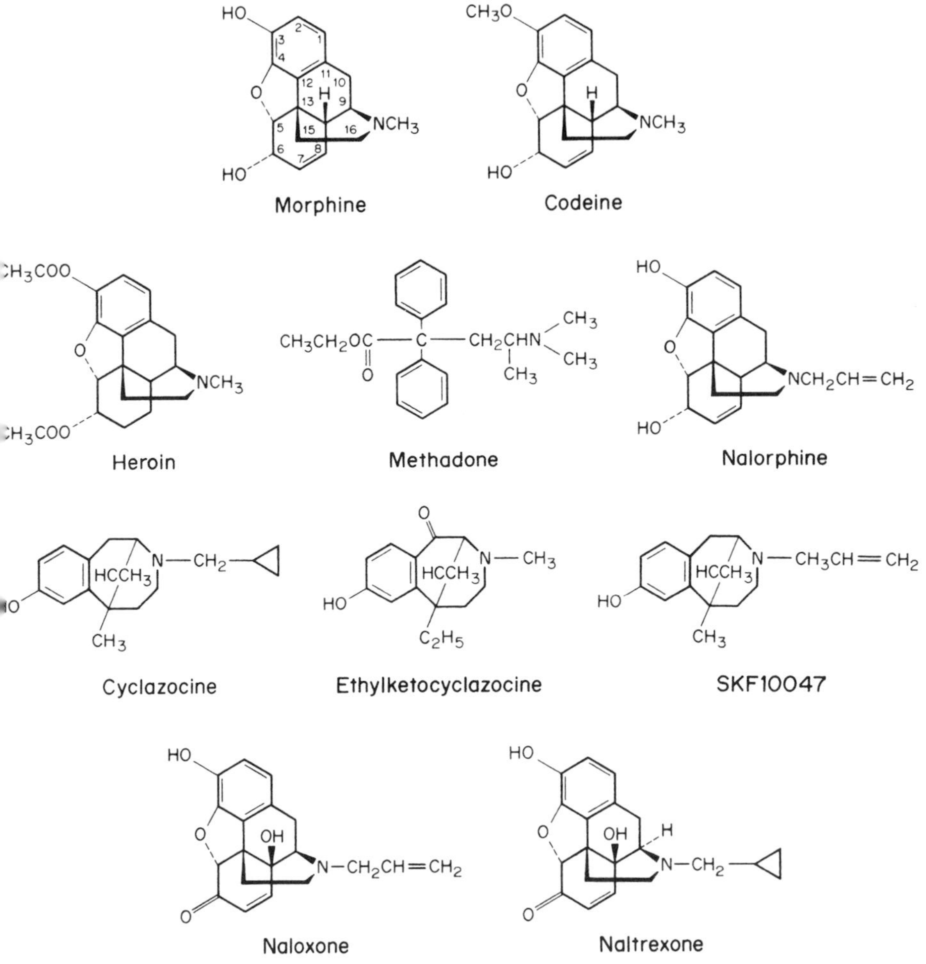

FIGURE I. Structures of some opiates and their antagonists.

varying pharmacological effects of opiate compounds (see Martin, 1983, for a recent review of that work). In brief, it was postulated that morphinelike drugs acted mostly on the μ-receptor, and their actions were antagonized by naloxone. Benzomorphans, e.g., ethylketocyclazocine (see Fig. 1), were postulated to have a high affinity for κ-receptors, and N-allylnormetazocine, also known as SKF10047 (see Fig. 1), was thought to be a selective agonist for the σ-receptor; it also antagonized the μ-receptor but with low affinity (Jaffe and Martin, 1985; Miller, 1986). However, subsequent work has suggested that the σ-receptor may not be a true opiate receptor (Wood, 1982). Cyclazocine (see Fig. 1) was more catholic in its taste and showed affinity for both κ- and δ-receptors with only weak affinity for the μ-receptors (Atweh, 1983). Pasternak and Snyder (1975) had previously found *in vitro* evidence for two binding sites having high and low affinity for opiates. Zhang and Pasternak (1980) subsequently suggested that the high- and low-affinity binding sites might correspond to Martin's μ- and δ-receptors, respectively. They further speculated that the μ-receptor may mediate analgesia, whereas the δ-receptor may mediate respiratory depression. As a precautionary note, the reader should be aware that all of the above findings, as well as subsequent ones, regarding selective binding and agonistic versus antagonistic effects are demonstrations of *relative*, not *absolute* effects, many of which remain controversial today. Nevertheless, it should be apparent from Fig. 1 that relatively minor variations in structure determine the receptor-binding properties of these compounds and, as we will see later, their pharmacological actions. Conversely, compounds differing markedly in structure can have similar pharmacological and/or receptor-binding characteristics. See Jaffe and Martin (1985) and Martin (1983) for a discussion of structural variations and characteristics of binding sites.

C. Identification of Opiate Receptors in the Brain

Using methods based on an approach initiated by Goldstein (Goldstein *et al.*, 1971), three groups, working independently, found *in vitro* evidence that opiates do indeed bind stereospecifically to brain cell membranes (Pert and Snyder, 1973; Simon *et al.*, 1973; Terenius, 1973). This binding was saturable, could be blocked with known opiate antagonists, and the binding affinity of various compounds corresponded roughly to their clinical effectiveness in relieving pain (Atweh, 1983). Furthermore, Pert *et al.* (1974) localized opiate receptors in synaptosomal fractions of brain homogenates and on synaptic membranes. The regional density of opiate receptors in the brain was found to be highly variable, with the greatest density of binding sites being found in limbic structures. The amygdala, especially the anterior region, had the highest concentration, and all regions of the hypothalamus and the medial thalamus were also rich in opiate receptors. The basal ganglia and central (periaqueductal) gray area of the midbrain also contained many opiate receptors (Kuhar *et al.*, 1973).

Tritiated ligands (diprenorphine and etorphine) with higher affinity and lower dissociation constants were found and used to identify more precisely many small regions of the brain and spinal cord, involving sensory as well as limbic structures, that contain opiate receptors (Atweh, 1983). Not surprisingly, opiates were found to bind in the substantia gelatinosa of the spinal cord and in the spinal nucleus of the trigeminal nerve, two structures, of course, that are involved in mediating sensory pain (Thomas, 1974). The sensory vagal nuclei were also found to be rich in opiate receptors, as were sensory nuclei of the visual system. Many medullary nuclei having autonomic cardiovascular and respiratory functions also contain rich concentrations of opiate receptors. It should be noted that age and sex also influence the distribution of opiate receptors. For example, Messing et al. (1980, 1981) reported that binding of tritiated dihydromorphine decreased in various brain regions as a function of age; the affected regions include the frontal poles, anterior cortex, and striatum. Similar effects of age were found for binding of tritiated etorphine in these same regions as well as in hypothalamus, hippocampus, thalamus, midbrain, and pons-medulla. In addition, the changing patterns of opiate binding as a function of age may be different in males and females. Greater complexity was introduced by the report by Codd and Byrne (1980) that the number of μ-receptors in mouse brain, as indexed by binding of tritiated naloxone, varies by nearly a factor of two as a function of season, being low in the summer and high in the winter.

III. Some Pharmacological Effects of Morphine

Morphine has many properties, the best known being relief of pain and induction of feelings of well-being. Although most people report feelings of lethargy and pleasant relaxation, some individuals, on initial exposure, experience hedonically negative feelings (Seiden, 1983). Morphine, of course, is highly addicting as indexed by the development of tolerance and the presence of withdrawal symptoms during abstinence following development of tolerance. Some of the varied actions of morphine include suppression of coughing by direct action on the hypothalamus (hence the use of codeine in cough remedies), suppression of gut motility, which results in its effective use in the treatment of diarrhea (probably this was the earliest and most frequent use of morphine), an increase in the frequency and depth of sleep, depression of respiratory rate, and constriction of the pupil. It also has effects on the cardiovascular system, but these effects are complex and vary among species. In humans, morphine generally causes only modest depression of heart rate and blood pressure but significantly decreases systemic and pulmonary resistance to blood flow (Martin, 1983). Effects on body temperature are also complex, species specific, and dependent on ambient room temperature and prior exposure to morphine. In rats, for example, morphine initially produces hypothermia followed by hyperthermia, and at analgesic doses the hyperthermic response predominates. In cats morphine induces hyperthermia, whereas in the dog it results in hypothermia. In humans, morphine causes peripheral vasodilation, as noted above, and induces the subjective experience of warmth, especially of the face, neck, and upper thorax (Jaffe and Martin, 1985). In addition, morphine inhibits α- and increases δ-wave activity in the electroencephalogram (EEG). It also is generally reported to block the activation pattern of the EEG induced by nociceptive (painful) and other peripheral stimulation (Martin, 1983). Intraventricular administration of morphine results in cortical seizure activity and increased activity of neurons in the mesencephalic central gray (Frenk *et al.*, 1978; Urca *et al.*, 1977). Large doses of morphine can result in tonic-clonic convulsions. In acute opiate poisoning the symptoms of coma, pinpoint pupils, and depressed respiration may be reversed by naloxone, which is the treatment of choice for overdoses of heroin (Jaffe and Martin, 1985).

IV. Precursors and Structures of Endogenous Opioids

A. The Search for the Endogenous Ligands

Because opiates are not normally present in the body, it was argued early in the 1970s that there must be endogenously produced ligands for the newly discovered opiate receptors of the brain. Kosterlitz's group in Aberdeen pioneered the guinea-pig ileum preparation as a sensitive *in vitro* method for investigating the actions of various opiate molecules (e.g., Kosterlitz *et al.*, 1972). Hughes (1975), using the Kosterlitz assay, found material in extracts of pig brain that behaved much like morphine. In the same year Hughes *et al.* (1975) identified the structures of two such substances and conservatively named them Met- and Leu-enkephalin, from the Greek *en kephalos* meaning "in the head" (Hughes, 1984).[1] The enkephalins were identified as pentapeptides having the amino acid sequence Tyr-Gly-Gly-Phe-Met (or Leu). The ratio of Met- to Leu-enkephalin varies among species. For instance, there is three to four times more Met- than Leu-enkephalin in pig brain, in contrast to the reverse relation found in bovine brain (Simantov and Snyder, 1976).

---

[1] This reference also provides some amusing anecdotes about the working conditions in Aberdeen. In Great Britain it often appears that the quality of science is inversely related to the quality of the working environment; the work of the late Mortyn T. Jones is a case in point (e.g., Jones and Hillhouse, 1976, 1977; Jones *et al.*, 1977, 1987).

At about the same time another hormone, considerably larger than the enkephalins but also having opiatelike activity, was isolated from pituitary extracts; it is called β-endorphin (Teschmaker et al., 1975; Cox et al., 1975). It is composed of 31 amino acids including the Met-enkephalin sequence at its N-terminal, which may be critical for its opiatelike activity (Goldstein, 1976). However, recent analyses of the normal enzymatic cleavage of β-endorphin suggest that the Met-enkephalin sequence is not separated from the larger molecule and hence does not, by itself, induce the analgesic and other effects of β-endorphin (Burbach, 1986). β-Endorphin and the enkephalins represent two independent neural opioid systems (Rossier et al., 1977a; Bloom et al., 1978a; Watson et al., 1978).

Yet another peptide with opiatelike activity was isolated from porcine pituitary by Goldstein et al. (1979) and was named dynorphin; it is composed of 13 amino acids and is currently known as dynorphin B. Its first five amino acids are identical to Leu-enkephalin, but it is many times more potent than Leu-enkephalin.

B. The Three Families of Endogenous Opioids

1. β-ENDORPHIN

It is now known that the above three opioid peptides are each the product of a larger precursor peptide or prohormone. Proopiomelanocortin (POMC) was isolated from pituitary by three groups in 1979, proenkephalin was characterized from adrenal medulla in 1982, and prodynorphin from hypothalamic tissue in 1982 (see the excellent reviews by Akil et al., 1984, and de Kloet et al., 1986, for references). Table I lists the amino acid sequences of the various opioid peptides that have been identified as deriving from each prohormone. Note that in different species the prohormone and the derived opioid peptides may differ in one or more specific amino acids and even in the number of amino-acid residues (Burbach, 1986).

POMC (265 amino acids, Nakanishi et al., 1979) is cleaved to form β-lipotropin (β-LPH, 91 amino acids) and adrenocorticotropin (ACTH, 39 amino acids), the latter, of course, having direct behavioral effects (see, for example, Urban, 1986; Strand and Smith, 1986; Gispen and Isaacson, 1986; Bohus, 1986; van Nispen and Greven, 1986, for recent reviews; de Wied, 1964, for an early report). β-LPH, in turn, is cleaved to produce γ-lipotropin (58 amino acids) and the 31-amino-acid β-endorphin, the only opioid peptide derived from POMC. Although slightly smaller fractions of β-endorphin, such as β-endorphin (1-26) and β-endorphin (1-27) are found in some regions of the brain, they have less opioid activity than β-endorphin (Millan and Herz, 1985) and may not be of physiological or behavioral significance (Bur-

TABLE I The Three Families of Opioid Peptides

Precursors		
Proopiomelanocortin (rat)	Proenkephalin (human)	Prodynorphin (porcine)
β-Endorphin	Met-enkephalin[a][b]	α-Neoendorphin
H-Tyr-Gly-Gly-Phe-Met-	H-Tyr-Gly-Gly-Phe-Met-OH	H-Tyr-Gly-Gly-Phe-Leu-
Thr-Ser-Glu-Lys-Ser-		Arg-Lys-Tyr-Pro-Lys-OH
Gln-Thr-Pro-Leu-Val-		
Thr-Leu-Phe-Lys-Asn-	Leu-enkephalin	β-Neoendorphin
Ala-Ile-Ile-Lys-Asn-		
Ala-His-Lys-Lys-Gly-	H-Tyr-Gly-Gly-Phe-Leu-OH	H-Tyr-Gly-Gly-Phe-Leu-
Glu-OH		Arg-Lys-Tyr-Pro-OH
	Met-enkephalin-8	Dynorphin A (1-8)
	H-Tyr-Gly-Gly-Phe-Met	H-Tyr-Gly-Gly-Phe-Leu-
	Arg-Gly-Leu-OH	Arg-Arg-Ile-OH
	Met-enkephalin-Arg⁶-Phe⁷	Dynorphin A (1-17)
	H-Tyr-Gly-Gly-Phe-Met-	H-Tyr-Gly-Gly-Phe-Leu-
	Arg-Phe-OH	Arg-Arg-Ile-Arg-Pro-
		Lys-Leu-Lys-Trp-Asp-
		Asn-Gln-OH
	Peptide E	
	H-Tyr-Gly-Gly-Phe-Met	Dynorphin B (1-13)
	Arg-Arg-Val-Gly-Arg-	
	Pro-Glu-Trp-Trp-Met-	H-Tyr-Gly-Gly-Phe-Leu-
	Asp-Tyr-Gln-Lys-Arg-	Arg-Arg-Gln-Phe-Lys-
	Try-Gly-Gly-Phe-Leu-OH	Val-Val-Thr-OH

[a]The common amino acid sequences of Met- and Leu-enkephalin are underscored.
[b]Ala, Alanine; Arg, Arginine; Asn, Asparagine; Asp, Aspartate; Cys, Cysteine; Gln, Glutamine; Glu, Glutamate; Gly, Glycine; His, Histidine; Ile, Isoleucine; Leu, Leucine; Lys, Lysine; Met, Methionine; Phe, Phenylalanine; Pro, Proline; Ser, Serine; Thr, Threonine; Trp, Tryptophan; Tyr, Tyrosine; Val, Valine.

bach, 1986). ACTH can also be cleaved to produce α-melanocyte-stimulating hormone [α-MSH = $ACTH_{(1-13)}$] and corticotropinlike intermediate lobe peptide [CLIP = $ACTH_{(18-39)}$], the former having direct behavioral effects (de Wied and Bohus, 1966). These cleavage processes are schematically illustrated in Fig. 2.

2. THE ENKEPHALINS

Proenkephalin (263 amino acids, Noda *et al.*, 1982) is cleaved to produce four copies of Met-enkephalin, one copy of Leu-enkephalin, and one

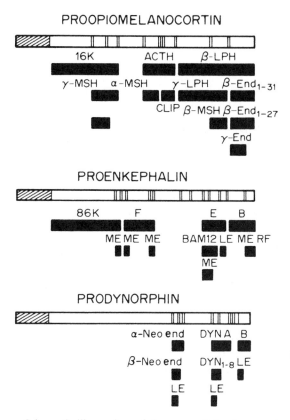

F I G U R E 2. Schematic illustration of cleavages of the three prohormone precursors of the endogenous opioids. Reproduced from de Kloet *et al.* (1986), with permission of the authors and publisher.

each of Met-enkephalin-8 (Met-enkephalin-Arg^6-Gly^7-Leu^8) and Met-enkephalin-Arg^6-Phe^7. An additional longer fragment, having opioid properties, has been identified. It is known as peptide E, but it is found mainly in the adrenal medulla and may be without behavioral effects. These cleavage processes are also illustrated schematically in Fig. 2.

3. THE DYNORPHINS

Prodynorphin (256 amino acids, Kakidani *et al.*, 1982) is cleaved into five Leu-enkephalin-containing peptides, some of which may be further cleaved to produce the Leu-enkephalin pentapeptide. The major products having opioid activity are α-neoendorphin, β-neoendorphin, dynorphin A (1-8), dynorphin A (1-17), and dynorphin B (1-13). The two neoendorphins and dynorphin A (1-8) and dynorphin B can all be cleaved further to produce Leu-enkephalin, as illustrated in Fig. 2. Whether such cleavages occur in normal physiological states is open to question, but *in vitro* receptor binding does not depend upon such cleavage (Bloom, 1983).

C. Current Models of Opioid Receptor Classes

The identification of these three families of endogenous opioid peptides stimulated additional work aimed at identifying receptor subtypes for the various opioid peptides. An additional receptor for β-endorphin was found and named η (Blanc *et al.*, 1983; Wood, 1982). The η-receptor was identified using an *in vitro* rat vas deferens assay which is quite insensitive to morphine and related opiates (Lemaire *et al.*, 1978); the action of β-endorphin in that assay is virtually unaffected by the prior addition of morphine, dynorphin, or the enkephalins, suggesting the existence of a rather specific binding site for β-endorphin (Schultz *et al.*, 1978). Nonetheless, many have argued that β-endorphin binds selectively to the μ-receptor, as evidenced by the fact that the analgesic and cataleptic effects of β-endorphin can be blocked by naloxone, long thought to be a selective μ-receptor antagonist (van Ree and de Wied, 1983). However, it is now known that the selective antagonism of specific receptors by naloxone or naltrexone is a function of dose (Martin, 1983). The enkephalins rather selectively bind to the δ-receptor in a mouse vas deferens assay, and dynorphin-derived peptides show selective affinity for the κ-receptors in both the mouse vas deferens and the guinea-pig ileum assays. Table II provides a list of these preparations and the receptors present in each. Table III provides a rough indication of the relative affinities of representatives of the three opioid families for the receptors in those assays. The previous caveat regarding the relativity of the binding and agonistic versus antagonistic characteristics of the opiates applies equally well to the members of these three families of endogenous opioids. Truly *selective* antagonists of opioid receptors have yet to be identified, although a putative

TABLE II   Receptors in *in Vitro* Assay Systems

Assay	Receptors
Guinea-pig ileum	μ, ϰ
Mouse vas deferens	μ, δ, ϰ
Rabbit vas deferens	ϰ
Rat vas deferens	η

TABLE III Relative Receptor Specificity of
Opioid Peptides

Peptide system	Relative specificity
β-Endorphin	$\eta > \mu, \delta$
The enkephalins	$\delta >> \mu$
Dynorphins	
Dynorphin A and B	$\varkappa > \mu, \delta$
β-Neoendorphin	$\varkappa = \mu = \delta$

selective δ-antagonist may have been found (Cotton *et al.*, 1984; Dray and Nunan, 1984), and claims have been made for a selective κ-antagonist (Bechara and van der Kooy, 1987). The degree of selectivity varies as a function of the particular assay used to measure binding affinities. Quirion and Pert (1981), for example, using rat striatal tissue, found ratios of affinity constants for μ-, δ-, and κ-receptors to be 3.6, 1.9, and 1, respectively, for dynorphin A (1-17) and 2.7, 1.5, and 1, respectively, for dynorphin B (1-13), indicating relatively poor selectivity of the dynorphins for the κ-receptors in that part of the brain. On the other hand, β-endorphin is highly active in the rat vas deferens assay (η-receptors) but is relatively inactive in the rabbit vas deferens assay (κ-receptors).

V. Sources and Brain Distributions of the Opioid Peptides

A. β-Endorphin

I. SOURCES

The pituitary gland is a major source of this endogenous opioid. The anterior and intermediate lobes of the pituitary contain β-endorphin-secreting cells and this POMC-derived opioid is released into systemic circulation from those cells. The anterior lobe also contains the other POMC-derived peptide hormones: β-LPH, ACTH, and α-MSH, with ACTH predominating over α-MSH. β-Endorphin and ACTH are stored and released in roughly equimolar amounts (Millan and Herz, 1985). In contrast, in the intermediate lobe, ACTH and β-LPH are cleaved to the smaller α-MSH and β-endorphin peptides, respectively. Both peptides are N-acetylated, which potentiates the physiological activity of α-MSH but destroys the opioid activity of β-endorphin (Millan and Herz, 1985). It is not clear what the physiological effect of N-acetylated β-endorphin actually is, but it appears not to be opioid in nature.

2. BRAIN DISTRIBUTIONS

In the brain, cell bodies that synthesize and/or contain β-endorphin are found predominantly in the arcuate nucleus of the hypothalamus, which innervates other hypothalamic and extrahypothalamic regions as well as the median eminence. In these areas almost none of the β-endorphin is acetylated and presumably it is physiologically active either as a neurotransmitter or as a neuromodulator involved in regulating the action of "classical" neurotransmitters (Miller, 1986; North and Williams, 1983). Neurons in the arcuate nucleus send dense projections to midline diencephalic nuclei which are particularly prominent around the anterior commissure, the stria terminalis, the median eminence, and the paraventricular, supraoptic, and suprachiasmatic nuclei. In addition, these neurons send projections to the ventral septum and nucleus accumbens, where they turn caudally and traverse the dorsal central gray area to reach the locus coeruleus, the parabrachial nucleus in the pons, the nuclei of the solitary tract, and the reticular formation in the medulla (Bloom *et al.*, 1978b; Kosterlitz and McKnight, 1981; Watson and Barchas, 1979). These regions of the brain contain cells that synthesize, transport, and release β-endorphin. However, it should be noted that the pituitary may also be a source of at least some of the β-endorphin found in these brain regions. Although sources outside the blood–brain barrier have not been thought to contribute to brain content of β-endorphin, evidence is growing that retrograde transport of anterior pituitary peptides to the brain does occur and might be an additional source of such brain peptides. This work has been reviewed by de Kloet *et al.* (1986).

B. The Enkephalins

I. SOURCES

The anterior and intermediate lobes of the pituitary contain Met- and Leu-enkephalin-synthesizing cells. In addition, enkephalinergic cells are found in the arcuate, paraventricular, and supraoptic nuclei and these cells richly innervate the median eminence, possibly providing an additional source of the immunoreactive enkephalins found in the anterior and intermediate lobes of the pituitary.

2. BRAIN DISTRIBUTIONS

In the central nervous system, the enkephalins are found in laminas I, II, V, and VII of the spinal cord, in several of the medullary nuclei of the cranial nerves, in the central gray, and in the zona compacta of the substantia nigra. In the limbic system, the highest concentrations of the enkephalins are found in the amygdala, especially in the central amygdaloid nucleus. In the hypothalamus, the enkephalins are widely distributed throughout the medial nuclei with somewhat lower concentrations in lateral nuclei. The bed nucleus of the stria terminalis, a pathway interconnecting amygdaloid and hypothalamic nuclei, also contains a high concentration of the enkephalins. The septal region also contains enkephalinergic terminals, especially the lateral septal nucleus. The highest concentration of enkephalins, however, is found in the striatum, particularly in the globus pallidus and, to a lesser extent, in the caudate putamen. In contrast, the enkephalins are barely detectable in the cortex (Atweh, 1983; de Kloet *et al.*, 1986; Kosterlitz and McKnight, 1981).

C. The Dynorphins

1. SOURCES

The pituitary origins of the dynorphin family of opioids are less clear than is the case for β-endorphin and the enkephalins. Dynorphin-staining cells in the supraoptic and paraventricular nuclei of the hypothalamus send axons to the posterior lobe of the pituitary, and these axons probably account for most of the immunoreactive dynorphin found there. In the anterior lobe of the pituitary, it is inferred that dynorphin-secreting cells are present, because immunocytochemical studies indicate that the dynorphins, especially the larger molecular forms of the dynorphin family, are present. This may represent a somewhat parallel process to that of POMC-derived peptides in the anterior lobe, although the specific cells containing the dynorphins have not yet been visualized in the anterior lobe (Millan and Herz, 1985). Apparently the intermediate lobe is devoid of peptides of the dynorphin family.

2. BRAIN DISTRIBUTIONS

The major opioid peptides of the dynorphin family that are found in the central nervous system are dynorphin A (1-8) and α-neoendorphin (de Kloet *et al.*, 1986). In the spinal cord neurons staining positively for these peptides are found in laminas I, II, V, and X of the dorsal horn. In the pons and medulla, dynorphin-containing cells and fibers are diffusely scattered in the reticular formation. They have also been identified in various nuclei such as the mesencephalic and main sensory nuclei of the trigeminal nerve and their spinal projections, the motor nucleus of the facial nerve, the vestibular and cochlear nuclei, the olivary complex, and the nucleus of the solitary tract. Dynorphin-containing neurons are also found scattered throughout the mesencephalic reticular formation, in the lateral tegmental nucleus, and in ventral and lateral regions of the central gray. The dynorphin-containing neurons in the central gray are located more ventrally than the enkephalin-containing neurons, which represent a separate opioid system from that of the dynorphin family (Khachaturian *et al.*, 1982). In the diencephalon, dynorphin-containing cells are located in supraoptic, paraventricular, and lateral hypothalamic nuclei. As noted above, the fibers from the supraoptic and paraventricular nuclei terminate in the posterior pituitary gland and appear to be the primary source of the dynorphins in the neurohypophysis (Millan and Herz, 1985; Palkovits *et al.*, 1983). It is interesting that these posterior pituitary cells exhibit colocalization of dynorphin and vasopressin [vasopressin is also known as antidiuretic hormone (ADH) because of its urine-concentrating effect on the kidney]. In addition, dynorphin-containing cells are found also in the arcuate and suprachiasmatic nuclei of the hypothalamus. In the telencephalon, dynorphin-containing cells are found in the dentate gyrus and mossy fibers of the dorsal and ventral hippocampus (McGinty *et al.*, 1983) and again, these appear to be separate from the enkephalin-containing cells in the hippocampus; they are also found in the nucleus accumbens, in the central amygdaloid nucleus, and in large numbers in the rostral head of the caudate nucleus. These cells project via the medial forebrain bundle to the lateral and medial portions of the substantia nigra (Vincent *et al.*, 1982).

VI. Interactions between Opioids and Stress-Related Endocrine Systems

It is not possible in this chapter to review in detail how the brain regulates the endocrine systems, but Fig. 3 illustrates some of the ways in which this regulation is achieved. We will focus on the role of the hypothalamo–pituitary–adrenal systems in regulating CNS and systemic opioid activity and on the effects of opioids on the stress-related endocrine systems. Details of the nonopioid mechanisms of endocrine regulation can be obtained elsewhere, but some aspects will be touched on in the context of this brief summary.

A. Anterior and Intermediate Lobes of the Pituitary

I. POMC-DERIVED HORMONES

Because of their anatomical proximity and similarity of function, the anterior and intermediate lobes and their POMC-derived hormones will be treated together. Corticotropic-releasing hormone (CRH) was first struc-

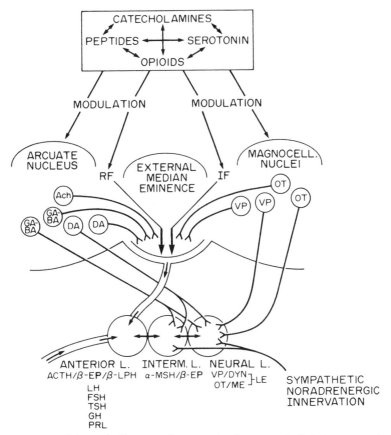

FIGURE 3.   Schematic diagram of neuroendocrine control of pituitary secretion of traditional hormones and endogenous opioids. Abbreviations: Ach, acetylcholine; ACTH, adrenocorticotropic hormone; DA, dopamine; DYN, dynorphin; β-EP, β-endorphin; FSH, follicle-stimulating hormone; GABA, γ-aminobutyric acid; GH, growth hormone; IF, inhibiting factors; LE, Leu-enkephalin; LH, luteinizing hormone; β-LPH, β-lipotropin; magnocell., magnocellular; ME, Met-enkephalin; α-MSH, α-melanocyte-stimulating hormone; OT, oxytocin; PRL, prolactin; RF, releasing factors; TSH, thyroid-stimulating hormone; VP, vasopressin (ADH). Adapted from Millan and Herz (1985), with permission of authors and publisher.

turally identified by Vale *et al.* (1981). It is a 41-amino-acid peptide which is synthesized in a number of basomedial hypothalamic nuclei. Upon delivery to the anterior pituitary, via the portal vein system, it stimulates synthesis and release of β-LPH which is cleaved to produce equimolar amounts of ACTH and β-endorphin, which, of course, can induce analgesia (Hargreaves *et al.*,

1987; see Figs. 2 and 3). In the hypothalamus CRH-containing neurons are stimulated by inputs from cells containing acetylcholine (Ach) and serotonin (5-HT). Initially, based on bioassays for ACTH, CRH release *in vitro* was found to be inhibited by γ-aminobutyric acid (GABA) and norepinephrine (NE, α subtype) (Jones and Hillhouse, 1977). With the development of radioimmunoassays for CRH, however, NE was found to stimulate CRH release *in vitro*, which leaves only GABA as an inhibitory neurotransmitter for CRH release (Jones *et al.*, 1987). In addition, glucocorticoids provide negative fast and slow feedback to the CRH cells (Jones *et al.*, 1972, 1974; Jones and Hillhouse, 1976), and Hauger *et al.* (1987) have recently shown that the negative feedback in the hypothalamus depends, in part, upon down-regulation of CRH receptors, an effect which apparently does not occur in the anterior pituitary following administration of corticosterone. Receptor modulation by steroid feedback in sites remote from the hypothalamus, e.g., hippocampus, has also been implicated in feedback regulation by the glucocorticoids, and de Kloet's group has identified two receptor types that are sensitive to experiential factors and has begun to map the distribution of the two types of receptors in the CNS (de Kloet *et al.*, 1987). The action of CRH on the anterior pituitary is strongly potentiated by the presence of vasopressin from the posterior pituitary or hypothalamus (Buckingham and Leach, 1979; Carlson *et al.*, 1982; Jones and Hillhouse, 1977; Rivier and Vale, 1983).

Additional interactions between the neurotransmitters and the POMC-derived peptides have been found. Serotonin has been found to bind selectively to the amino-acid sequences common to α-MSH and ACTH (see Root-Bernstein, 1987), which may be a factor controlling their physiological action, and Van Loon and De Souza (1978) found evidence that β-endorphin increases 5-HT turnover in hypothalamus and brain, while exerting the opposite effects in hippocampus. Thus, there are many levels of interaction among the excitatory and inhibitory neurotransmitters and the neuropeptides, including the opioid peptides, in the brain (see Sections VI,B and VI,C which follow).

In the intermediate lobe of the pituitary, release of α-MSH and β-endorphin is primarily under the inhibitory control of hypothalamic cells containing dopamine (DA) and GABA (see Fig. 3). This control may be neural or humoral (via the portal vessels) or both (Millan and Herz, 1985). In addition, there is some suggestion that CRH may, by neural delivery, stimulate the release of α-MSH from the intermediate lobe (Millan and Herz, 1985) (Fig. 4 illustrates the comparison of anterior and intermediate lobe function).

ANTERIOR LOBE   INTERMEDIATE LOBE

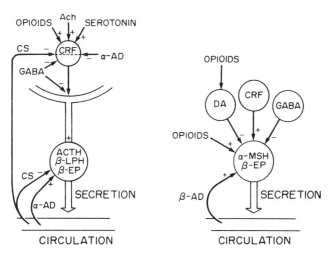

F I G U R E 4.   Schematic diagram illustrating differences between anterior and intermediate lobe control and release of POMC-derived opioids and peptides. See Fig. 3 legend for most abbreviations. α-AD, α-adrenergic; β-AD, β-adrenergic; CRF corticotropin (ACTH)-releasing factor; CS, corticosteroids. Adapted from Millan and Herz (1985), with permission of the authors and publisher.

Systemic α-adrenergic input *directly* stimulates the anterior pituitary to release its subset of POMC-derived peptides, whereas systemic β-adrenergic input stimulates the intermediate lobe to release its subset. It is presumed that these adrenergic inputs are of adrenal medullary origin (Kvetnansky *et al.*, 1987; Millan and Herz, 1985).

2. EFFECTS OF ACUTE ADMINISTRATION OF MORPHINE

An additional control feature of particular importance here is demonstrated by the finding that acute administration of opioids and a variety of opiate agonists stimulates the release of the POMC-derived peptides. In the case of the anterior lobe, the effects of morphine and both Met- and Leu-enkephalin on peptide release are central, mediated by stimulation of CRH release, and are naloxone blockable (Buckingham 1980, 1982; Buckingham and Cooper, 1984). However, morphine can acutely reduce stimulation-induced release of vasopressin from the *in vitro* hypothalamus (Bradbury *et al.*, 1974; Knepel and Reimann, 1982). Thus, the effect of morphine on the secretion of ACTH and β-endorphin represents an integration of a direct stimulatory effect on CRH release and a reduction of the potentiation of CRH's action by vasopressin. In addition, β-endorphin exerts an antagonistic effect on morphine- and enkephalin-induced release of CRH. Thus, it appears that β-endorphin functions in a short negative feedback loop to inhibit further secretion of CRH (Buckingham, 1980), which is analogous to the short-loop negative feedback effect of ACTH (Motta *et al.*, 1965). In the case of the intermediate lobe, the stimulatory effects of acute administration of opioids on the release of α-MSH and β-endorphin are thought to be indirectly mediated by a reduction in the inhibitory signal from dopaminergic cells (de Rotte *et al.*, 1981), although others have suggested a direct action (Celis, 1980).

3. EFFECTS OF CHRONIC ADMINISTRATION OF MORPHINE

Chronic administration of morphine results in the development of tolerance, as evidenced by a gradual diminution of the stimulatory effect on CRH release. For example, Buckingham and Cooper (1984) reported that daily administration of 20 mg/kg of morphine for 8 days results in the progressive loss of the stimulatory effect on ACTH and corticosterone release and a reduction in the magnitude of the stress-induced (ether and laparotomy) release of these hormones. Moreover, tolerance was also seen in the *in vitro* hypothalamus preparation of Jones and Hillhouse (1976, see above), i.e., the CRH release by hypothalamic tissue of untreated rats normally seen in response to morphine, Met-enkephalin, Ach, or 5-HT stimulation was no longer found in hypothalamic tissue taken from rats chronically pretreated with morphine. In addition, naloxone binding in the hypothalamic tissue of the morphine-tolerant rats was normal, and the *in vitro* response of the anterior pituitary to CRH was only slightly depressed by naloxone. Thus decreased release of POMC-derived peptides from the anterior lobe seen after repeated opiate administration appears to be largely due to induction of hypothalamic tolerance which results in decreased release of CRH, a process which does not appear to involve down-regulation of morphine (μ-) receptors. Because of the inhibitory effects of GABA on the *in vitro* hypothalamus, one might speculate that hypothalamic tolerance to chronic opiate administration might involve enhanced inhibition by this neurotransmitter, but this possibility has not yet been evaluated.

Chronic administration of morphine over a 4-week period has been reported to produce a decrease in β-endorphin content of the intermediate lobe, a reduction of *in vitro* release of β-endorphin, and a decrease in plasma levels of β-endorphin (Hollt *et al.*, 1980; Millan *et al.*, 1981; Przewlocki *et al.*, 1979; Wuster *et al.*, 1980). Thus, although differing in detail, roughly parallel effects of acute and chronic morphine administration on POMC-derived peptides are found in both anterior and intermediate lobes of the pituitary (see Section VI,A,2).

B. *Posterior Lobe of the Pituitary*

VASOPRESSIN AND NON-POMC OPIOID PEPTIDES

As illustrated in Fig. 3, neurons in the paraventricular and supraoptic nuclei of the hypothalamus project to and terminate in the neurohypophysis. They synthesize, transport, and release vasopressin and oxytocin and opioids of the enkephalin and dynorphin families (see also Fig. 5). As noted above, vasopressin powerfully potentiates the action of CRH on the anterior and probably also the intermediate lobe of the pituitary. Vasopressin can reach the anterior lobe via the portal veins which not only interconnect the pituitary with the median eminence but also interconnect the various lobes of the pituitary itself (Bergland and Page, 1979; Page, 1982). The neural lobe also receives direct dopaminergic and GABAergic inputs from the hypothalamus, as well as sympathetic noradrenergic innervation, as illustrated in Fig. 3 (Millan and Herz, 1985). The noradrenergic inputs are apparently primarily to the vasculature of the lobe, which could, therefore, influence endocrine synthesis and release by vasodilation and constriction, whereas the dopaminergic and GABAergic influences are presumably more direct, acting as neurotransmitters to regulate release of the hormones of the neural or other lobes of the pituitary.

There has been considerable controversy regarding the co-occurrence of the two opioids of the neural lobe with vasopressin and oxytocin neurons and terminals. In general, the evidence suggests that Met- and Leu-enkephalin coexist in neurons and terminals that also secrete oxytocin, whereas Leu-enkephalin and dynorphins coexist in vasopressin-secreting neurons and terminals. This evidence finds support from work with Brattleboro rats that are genetically unable to synthesize vasopressin and which, therefore, suffer diabetes insipidus and drink as much as 100% of their body weight in water each day. These rats show depressed, but not zero, levels of peptides from the dynorphin family in the neural lobe. The continued presence of the dynorphins in the neural lobe of Brattleboro rats suggests that not only

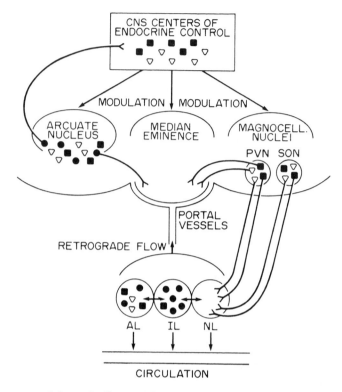

FIGURE 5. Schematic diagram illustrating the organization and distribution of opioids involved in the hypothalamic–pituitary axis. Abbreviations: AL, anterior lobe; IL, intermediate lobe; NL, neural lobe; PVN, paraventricular nucleus; SON, supraoptic nucleus. ●, β-endorphin cell bodies; ▽, dynorphin cell bodies; ■, enkephalin cell bodies. Adapted from Millan and Herz (1985), with the permission of the authors and publisher.

different precursors, but also different transport mechanisms, exist for the dynorphins and vasopressin (Millan and Herz, 1985).

In addition to the fact that under basal conditions there is a strong positive correlation between neural-lobe content of vasopressin and the dynorphin peptides (Millan et al., 1983), other physiological evidence suggests a common regulatory mechanism for these neuropeptides. For example, dehydration, long known to be the adequate stimulus for vasopressin release, also causes posterior lobe depletion of Leu-enkephalin and the dynorphin peptides (Hollt et al., 1981). Pharmacological manipulations, such as chronic administration of haloperidol, a dopamine antagonist, elevate neural lobe content of vasopressin and the dynorphin peptides (Hollt, 1981).

There also is recent evidence to suggest that catecholamines such as dopamine, epinephrine, and norepinephrine are not only costored and co-released with enkephalins, but also selectively bind to morphine and the enkephalins. Indeed, it has been suggested that the capacity to bind with the catecholamines may determine the opiatelike activity of the endogenous opioids (Root-Bernstein, 1987).

C. Adrenal Medulla and Cortex

Acute administration of morphine causes a partial depletion of adrenal medullary catecholamines, and this effect shows tolerance with repeated administrations, i.e., analogously to its effects in the anterior and intermediate lobes of the pituitary, the stimulatory effect of morphine on catecholamine release declines with repeated administration. Acute morphine increases the rate of synthesis of catecholamines and the rate of vesicle formation and induces the content of these vesicles to be expelled in an all-or-none fashion (Anderson and Slotkin, 1975). Because enkephalins are stored in the same granules with catecholamines, it is probable that both are released upon morphine stimulation. It appears that these effects are mediated by direct transsynaptic stimulation of the splanchnic nerve by morphine (Anderson and Slotkin, 1976). It has also been shown that chromaffin cells and splanchnic-nerve terminals both contain μ- and ACh-receptors, so that these effects of morphine may be mediated both neurally and by direct action on the cells themselves (Yang et al., 1980).

Given the presence of μ-receptors on these cells and terminals, it is possible that endorphins released from the pituitary could act in parallel with ACTH and exhibit tropic effects at the level of the adrenal gland. Similarly, Amir et al. (1980) speculated that enkephalins released from the medulla could interact with the cells in the cortex, since both β-endorphin and ACTH stimulate in vitro synthesis of corticosterone. In addition, Lymangrover et al. (1981) have shown that Met-enkephalin potentiates the in vitro steroidogenic response to ACTH in superfused adrenocortical tissue. Thus it appears that the endogenous opioids directly stimulate both the adrenal medulla to release catecholamines and the adrenal cortex to release glucocorticoids, the latter being enhanced by the medullary release of the catecholamines. Thus, there are a number of levels at which the opioids and the hypothalamo–pituitary–adrenal axis interact.

VII. Pain and Analgesia

A. Neural Pathways

The neural pathways which transmit painful sensations from the periphery to central structures are composed of A-δ and C fibers. These first-order neuronal fibers differ in size as well in their degree of myelinization and, thus, differ in the rate at which they conduct impulses: A-δ fibers are myelinated and conduct impulses rapidly at speeds of 5–30 m/sec, whereas C fibers are unmyelinated and conduct impulses more slowly, transmitting at rates of 0.5–2 m/sec.

Axons of these first-order neurons enter the spinal cord and form synapses with second-order neurons located in the dorsal horn. Fibers of the second-order neurons then ascend via the anterior portion of the lateral spinothalamic tract and form synapses in the thalamus with third-order neurons whose axons ascend to the cortex. Cells within the posterior thalamus readily respond to noxious stimulation occurring within a large cutaneous receptive field, and this thalamic area is organized in a crude topographic manner (Carpenter, 1976).

According to the Melzack and Wall (1965) gate-control theory of pain, the primary gating mechanism is located in the dorsal horn of the spinal cord, where fibers from the periphery form synapses with (1) interneurons in the substantia gelatinosa located in laminas II and III, and (2) the central transmission or "T" cells located in lamina V of the dorsal horn (Melzack and Wall, 1975). Axons of these interneurons in the substantia gelatinosa synapse on the primary afferent fibers and, by presynaptic inhibition, alter the firing rate of those primary afferents. Thus, impulses generated by afferent fibers are regulated presynaptically by either opening or closing of the gate (controlled by the interneurons in the substantia gelatinosa), which subsequently determines the amount of input to the T cells in the dorsal horn. In addition, Melzack (1973) suggested that there may be postsynaptic modulation of T-cell firing, and descending fibers from the brain could either increase or decrease the excitability of the T cells.

Thus, central pain processing involves both presynaptic and postsynaptic modifications of stimulus input. Impulses from the periphery can be modulated presynaptically at the level of the dorsal horn before they reach the brain. Also, descending inputs from higher brain structures, which reflect the psychological state of the animal, can modify nociception (pain perception) via postsynaptic modulation. Therefore, noxious input could be modulated by both ascending and descending influences which can serve either to enhance or to diminish nociception.

B. Tests of Pain Sensitivity

Because animal experimentation requires objective measurement of subjective pain, several tests have been devised which measure thresholds for reflexes or behaviors which are typically elicited by noxious stimuli. The hot-plate test was first devised by Woolfe and MacDonald (1944) and later modified by Eddy and Leimbach (1953). In this test the animal is placed on a heated grid or metal surface which is maintained at a temperature generally within the range of 48–52°C, although temperatures as low as 44.5°C and as high as 80°C have been used (Bolles and Fanselow, 1982). It has been suggested that lower temperatures offer finer discrimination of nociception than higher ones because of the slower rate of rise of painful input (O'Callaghan and Holtzman, 1975). Latency to lick either a fore or a hind paw is the most commonly used measure of pain threshold, although some investigators use a "jump" response as the dependent variable.

The tail-flick test for measuring pain threshold also uses a thermal stimulus. The general procedure consists of focusing radiant heat from a lamp onto the tail of a restrained animal and measuring the latency to move the tail away from the heat. The tail-flick response is a spinal reflex (Vyklicky, 1978) and is considered a more sensitive measure of analgesia than the hot-plate test because of the uniformity of application of the stimulus and the absence of contact stimuli (Beecher, 1959). The original test, designed by D'Amour and Smith (1941), was used to compare the analgesic effect of morphine to other compounds on the response to pain in mice. However, it should be noted that a variety of procedural differences, which include where or how heat is applied to the tail, may influence the estimates of pain threshold (Yoburn et al., 1984). Although other tests of pain sensitivity have been used, the hot-plate and tail-flick tests appear to be the most reliable and the most frequently used. Our discussion of the role of endogenous opioids in pain sensitivity will focus on studies which use these two tests. The vast majority of the studies rely on the spinal reflex of the tail-flick test. Generalization of the observed effects of opioids in these tests to other measures of pain sensitivity or to supraspinal mechanisms of coping with pain must be made cautiously, if at all.

C. Opioid Modulation of Pain at the Spinal Level

Morphine, administered i.v., inhibits the response of the T cells in lamina V of the dorsal horn, and this inhibition is reversed by naloxone (see

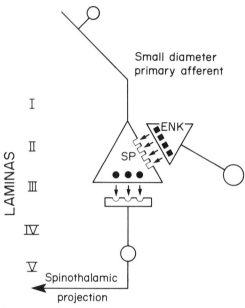

FIGURE 6. Schematic diagram of presynaptic inhibition by an enkephalinergic (ENK) interneuron of release of substance P (SP) by primary afferent neuron. Adapted from Jessel and Iversen (1977), with the permission of the authors and publisher.

Kosterlitz and McKnight, 1981). In addition, iontophoretic injection of morphine or enkephalins near the interneurons in the substantia gelatinosa, but not near the T cells, suppresses T-cell firing to noxious but not to mild tactile stimulation (Duggan et al., 1977a,b). Again, the suppression is reversed by naloxone. A current model (illustrated in Fig. 6) proposes that enkephalinergic interneurons in the substantia gelatinosa may presynaptically inhibit the release of substance P from the first-order afferents; substance P is the excitatory neurotransmitter for the T cells (Jessell and Iversen, 1977). Consistent with this model is the finding that opiate binding sites and enkephalin-containing cells are abundant in the substantia gelatinosa (Atweh and Kuhar, 1977; Hokfelt et al., 1977a,b). It may well be that the effects of morphine at the spinal level are the result of morphine stimulation of the enkephalinergic interneuron, an effect which is dose dependent.

D. Opioid Modulation of Pain at the Supraspinal Level

1. STIMULATION-INDUCED ANALGESIA

Reynolds (1969) was the first to report analgesia produced by electrical stimulation of the brain. In his study, rats which were stimulated in the dorsolateral perimeter of the central gray area (often also called the periaqueductal gray) showed no reactions to painful pinches to the paws and tail. The rats exposed to central gray stimulation were sufficiently analgesic to permit surgical laparotomy without additional anesthetic. Normal responses to noxious stimulation returned roughly 5 minutes after termination of the central gray stimulation. Since this original report, many other investigators have confirmed Reynolds' findings (e.g., Akil and Liebeskind, 1975; Akil and Mayer, 1972; Akil *et al.*, 1976b; Mayer and Liebeskind, 1974; Mayer *et al.*, 1971; Soper, 1976).

In examining areas both within the central gray and those lying more rostrally in the brainstem, Rhodes and Liebeskind (1978) found that responses to a variety of noxious stimuli were suppressed during and after stimulation of areas located in the posterior hypothalamus adjacent to the caudal portion of the third ventricle. All active sites were localized within the periventricular system and were found to have reciprocal connections with the caudal central gray, an area which appears to be critically involved in the mediation of analgesia.

2. CENTRAL ADMINISTRATION OF OPIATES

Further evidence of involvement of the central gray in analgesia is provided by studies using microinjections of opiates directly into the central gray. Injection of morphine, for example, yields an initial (10 to 15-minute) period of hyperreactivity to previously neutral stimuli, combined with explosive motor behavior. This is followed by hyporeactivity to previously painful stimuli, and maximal analgesia appears about 1 hour after morphine injection, with (Jacquet and Lajtha, 1974) or without (Yaksh *et al.*, 1976) catatonia. The extensive study by Yaksh *et al.* (1976) indicates that morphine injections into rostral regions of the central gray result in analgesia restricted to rostral regions of the body, whereas injections into caudal portions of the central gray render the whole body analgesic.

3. EFFECTS OF CENTRAL LESIONS

In contrast, lesions of the central gray decreased basal tail-flick latencies (Rhodes, 1979) and enhanced responding in an approach-avoidance conflict situation (Liebman *et al.*, 1970).

E. Descending Pathways

The fact that electrical or morphine stimulation of the central gray induces profound analgesia implies that the central gray exerts some inhibitory control over nociceptive responding, and the data suggest that stimulation of caudal areas is more effective than stimulation of rostral areas. The lesion data are consistent with the notion of inhibitory control, because such lesions appear to attenuate analgesia. Taken together, these data suggest the existence of a descending pathway which subserves this inhibitory control, and two such systems have been identified: a descending serotonergic system arising from the raphe nuclei and a descending noradrenergic system arising primarily from nucleus reticularis lateralis (Yaksh, 1979). The former has received much attention, appears to be essential in supraspinal modulation of pain, and will be the focus of our discussion.

I. THE RAPHE COMPLEX

The raphe complex is made up of several diffuse nuclei located within the mesencephalon: nucleus raphe magnus, nucleus raphe dorsalis, and nucleus raphe medianus. The major ascending efferents from the nucleus raphe magnus (NRM) project to more rostral areas such as the septum and limbic structures (Hamilton, 1976). Descending efferent fibers from the NRM are less numerous than descending efferents from other raphe nuclei, but these fibers comprise the largest proportion of efferents to the spinal cord via the dorsal half of the lateral funiculus (Taber et al., 1960). The nucleus raphe dorsalis (dorsal raphe) lies within the lower portion of the central gray (Hamilton, 1976) and has major ascending projections which course through and terminate in the central gray (Conrad et al., 1974; Pierce et al., 1976).

Stimulation of the NRM is effective in elevating the response threshold of animals subjected to tail pinch, but only during stimulation. Satch et al. (1980) found that prior treatment with methysergide, a serotonin blocking agent, inhibits the appearance of analgesia and that naloxone could not block this effect. Similarly, stimulation of the dorsal raphe induces analgesia in the tail-flick test, but to induce analgesia which outlasts the duration of brain stimulation, higher current levels are required in the raphe than in the central gray. However, the analgesias induced by dorsal raphe stimulation are more sensitive to naloxone blockade than those produced by stimulation of the central gray (Cannon et al., 1982). Consistent with these findings, Oleson et al. (1978) found that multiple unit recordings from the three raphe nuclei in the awake rat showed increased activity during noxious stimulation (foot-shock or paw pinch) and exhibited decreased activity following systemic morphine injection or during stimulation of the central gray or medial thalamus. However, the NRM is not a structure with a unitary function, because single-unit activity in the NRM can be either increased (87% of neurons) or decreased (13% of neurons) by noxious stimuli such as skin pinch

(Labatz *et al.*, 1977). Systemic morphine increased activity in 57% and decreased activity in 43% of neurons tested (Labatz *et al.*, 1977), although others find different proportions of cells that increase, decrease, or show no change in firing rate (Deakin *et al.*, 1977). Not surprisingly, lesions of the raphe nuclei, which greatly reduce 5-HT levels in the forebrain, eliminate the analgesic effect of morphine in hot-plate (Samanin *et al.*, 1973) and tail-flick (Abbot and Melzack, 1982) tests, and in one study NRM lesions not only blocked morphine-induced analgesia but, in fact, rendered the animals hyperalgesic in the tail-flick test; lesions more rostral to the NRM had no effect on pain sensitivity (Proudfit and Anderson, 1975).

It is well known that the raphe complex is a rich source of 5-HT, and many have proposed that 5-HT mediates the analgesia induced by morphine and by chemical or electrical stimulation of the raphe (e.g., Akil and Mayer, 1972). This premise is based on the findings that electrolytic or chemical lesions of the NRM, which depletes 5-HT stores, eliminate morphine- and stimulation-induced analgesia. Indeed, raphe-lesioned animals given replacement injections of 5-HT or its precursors displayed analgesia (Akil and Mayer, 1972; Akil and Liebeskind, 1975; Dennis and Melzack, 1980; Proudfit and Anderson, 1975; Samanin and Benasconi, 1972). These findings appear to present a strong case for the involvement of this serotonergic system in the production of analgesia.

2. THE DORSOLATERAL FUNICULUS

As noted previously, the majority of efferents from the raphe are ascending and terminate in limbic structures and in the central gray where opioid receptors are numerous; descending serotonergic fibers from the NRM travel in the dorsolateral funiculus (DLF) in what appears to be a monosynaptic pathway (Dahlström and Fuxe, 1965) to terminate in the cord. Proof of this direct pathway is offered by a study in which tritiated leucine was injected in the NRM and labeled terminals were found in laminas I, II, V, and VI of the dorsal horn (Basbaum *et al.*, 1976). In addition, activity of dorsal-horn cells, induced by noxious peripheral stimulation, was inhibited by electrical stimulation of the NRM. Furthermore, lesioning of the DLF eliminates the analgesias produced by morphine or electrical stimulation of the raphe, in regions caudal to the DLF lesion. However, this effect could be overcome by increasing the dose of morphine (Basbaum *et al.*, 1976, 1977). These findings indicate that although the DLF is one pathway which mediates morphine- and stimulation-induced analgesia, other mechanisms must also contribute to these analgesias. In addition to these supraspinal inhibitory mechanisms it is likely that intraspinal mechanisms caudal to the DLF lesion also inhibit nociceptive responding, presumably by the hypothesized presynaptic inhibition of release of substance P noted above (see Fig. 6).

VIII. Stress-Induced Analgesia

Stress-induced analgesia has been of considerable interest because this form of analgesia involves the integrated responses of an animal's opioid and nonopioid systems for coping with stress. To establish the opioid nature of a stress-induced analgesia researchers have attempted to (1) correlate the degree of analgesia with endogenous opioid activity, (2) demonstrate cross-tolerance to morphine in analgesic animals, and (3) block the analgesia with opioid antagonists such as naloxone or naltrexone. Rarely are all three criteria used in a single experiment and, because of its ease, the last criterion is the most popular.

Peripherally administered electric shock has often been used as a stressor, because it is relatively easy to control. However, generalization across studies is difficult because of methodological differences. For example, effectiveness of shock as a stressor depends upon the parameters of the stimulus, e.g., intensity, duration, and frequency of application. Intensity varies widely across studies but is generally in the range of 0.8 to 3 mA. Duration and frequency range from a single exposure to 20 seconds or 3 minutes of shock to 6 days of exposure to 90 minutes of intermittent shock. Other features of the shock stimulus which differ across studies are where on the body the shock is delivered, e.g., to the forefeet or hindfeet or both, or to the tail, whether the shock is delivered via surface electrodes or through a grid flood, and whether the shock is escapable or inescapable.

A. Spinal Mechanisms of Stress-Induced Analgesia

Hayes et al. (1978b), replicating the findings of Basbaum et al. (1976, 1977), demonstrated that lesions of the DLF attenuated morphine-induced analgesia but did not affect analgesia induced by a 20-second footshock. These results contradict the findings of another study conducted by members of this group of investigators which indicated that lesions of the DLF dramatically reduced (by 89%) the analgesia induced by the 20-second footshock (Hayes et al., 1978a). Although there are problems with this study because of decreased basal pain responsivity in lesioned animals and because of the lack of some control groups, support for these results has been offered by Snow and Dewey (1983), who found that transection of the spinal cord at $T_{11}-T_{13}$

eliminated the analgesia normally seen after 20 seconds of footshock but that this surgical procedure did not directly interfere with the tail-flick response.

Another group of investigators found that the region of the body receiving shock is an important factor in determining not only the type of analgesia but also the effect of DLF lesions on that analgesia. For example, DLF lesions at C3 or T2 abolished the analgesia from 90 seconds of shock to the forepaws but only attenuated the analgesia from shock to the hindpaws (Watkins and Mayer, 1982b). Furthermore, naloxone administered either systemically (Watkins et al., 1982) or intrathecally (Watkins and Mayer, 1982a) prevented the development of analgesia, but did not relieve an existing analgesia from forepaw shock and was ineffective in altering the analgesia from hindpaw shock. However, there are methodological problems with these studies. The shock intensity was not controlled: 1.6 mA for forepaws, 1.2 mA for hindpaws. Because of variation in surface area of forepaws and hindpaws and the greater tactile sensitivity of forepaws than hindpaws, the difference in subjective intensity could be much greater than the actual difference in current would suggest. In addition, to administer shock to either the forepaws or the hindpaws, the rats were suspended in a loop which forced contact between the paws and the grid floor. It is conceivable that the "head-down" position is more traumatic than the more normal "head-up" body posture that rats spontaneously exhibit.

Despite these difficulties and despite the variation in the effect of DLF lesions, it appears that at the spinal level, there are both opioid and nonopioid mechanisms involved in mediating stress-induced analgesia. Given that DLF lesions attenuate some analgesias, it would appear that supraspinal mechanisms, be they opioid or nonopioid, are involved as well.

B. Nonopioid Stress-Induced Analgesia

The mechanisms underlying nonopioid analgesias remain largely unknown. However, several neurotransmitter systems have been implicated. For example, administration of reserpine, which depletes central and peripheral stores of monoamines, attenuated the analgesia from brief footshock stress, whereas naltrexone was without effect (Lewis et al., 1982). This finding spawned a series of studies which attempted to identify the critical monoamine(s) mediating this form of stress-induced analgesia [see also Nagase Shain (1985) for evidence of significant genetic determinants of the effects of reserpine on nonopioid forms of stress-induced analgesias and for genetic determinants of the lethal effects of reserpine and stress]. Terman et al. (1982) hypothesized that histamine was a likely candidate, because they found that α-fluoromethylhistidine, which disrupts synthesis and depletes neuronal stores of histamine, rendered animals significantly less analgesic than saline-treated controls following brief footshock. Further support for this hypothesis was provided when it was found that this analgesia could be reduced by administration of a histamine antagonist, diphenhydramine. Neither 5-HT, nor dopamine, nor norepinephrine had any effect on the stress-induced analgesia from this procedure (Terman et al., 1983b). It should be noted, however, that prior research by these investigators indicated that the nature of the analgesia is influenced by the intensity and duration of shock (see preceding discussion and Sections VIII,D,3 and VIII,D,4 below) and that the shock parameters used in these studies were not within the restricted ranges that would ensure (see Section VIII,D,4) the induction of either an opioid or a nonopioid analgesia. Thus, the stimulus parameters used were such as to produce a potentially labile form of analgesia, and tests with opioid antagonists were not run, so it is not clear which form of analgesia occurred in these experiments.

Another possible substrate of nonopioid analgesia is 5-HT which, as we have noted above, is involved in mediating the analgesia from electrical stimulation of the raphe nuclei. Although Terman et al. (1983b) found no effect of 5-HT on the brief footshock analgesia, others have found that the stress-induced analgesia from 30 seconds of footshock was enhanced by prior administration of p-chlorophenylalanine (PCPA) in a dose sufficient to decrease brain 5-HT content by 81%. Paradoxically, PCPA also attenuated the analgesia induced by a prolonged stress procedure (30 minutes of footshock). Naloxone blocked the analgesia from this procedure but had no effect on the analgesia produced by the brief procedure (Tricklebank et al., 1982, 1984).

These data seem to suggest that the release of neuronal 5-HT enhances opioid forms of analgesia and inhibits nonopioid analgesia. However, two out of the four 5-HT agonists that were tested by Tricklebank et al. (1982) and none of the 5-HT antagonists they had used had any effect on the analgesia induced by the brief footshock procedure, presumably a nonopioid analgesia. Thus, this pharmacological manipulation of 5-HT results in conflicting findings and hard conclusions are not possible at this time. Histamine and 5-HT are both possible mediators of nonopioid analgesias, but they may also be involved in modulating opioid analgesias as well.

C. Opioid Stress-Induced Analgesia

ENDOGENOUS OPIOID ACTIVITY AND STRESS-INDUCED ANALGESIA

Receptor displacement assays have been used to determine opioid concentrations in brain and to correlate these concentrations with tail-flick latencies following either 1 or 12 daily exposures to 30 or 60 minutes of intermittent footshock (Akil *et al.*, 1976a). Central opioid levels and tail-flick latencies were significantly elevated after 1 shock session, but after 12 days of repeated shock exposure brain receptor binding of opiates and tail-flick latencies did not differ between shocked and nonshocked animals, suggesting that the animals which were repeatedly shocked had habituated their behavioral and opioid responses to the stressor. These findings were replicated in another experiment which demonstrated increases in both opiate receptor binding in brain tissue and tail-flick latencies following 30 minutes of footshock stress as well as development of tolerance during additional shock sessions (Madden *et al.*, 1977).

In contrast to these results are the findings of Rossier *et al.* (1977b), who used the same 30-minute footshock procedure and found that β-endorphin content in the hypothalamus was reduced, while plasma content was elevated six-fold over that of controls. As noted by these investigators, different assays were used in the two sets of experiments, and interassay differences in binding specificity may account for the discrepant findings. In general, this problem of opioid specificity remains a difficulty today; the common amino-acid sequence at the N-terminal of all of the endogenous opioids makes selective measurement of individual opioids difficult. As noted previously, inactivation of β-endorphin by N-acetylation creates additional measurement problems in either tissue or plasma. In addition, Rossier *et al.* (1977a) have reported that peripheral administration of β-endorphin, which elevated plasma concentration to three times the level induced by shock stress, failed to produce analgesia. Thus, efforts to correlate degree of analgesia with peripheral concentrations of the opioids seem doomed to failure, and we may conclude that the opioid stress-induced analgesias are the result of their central rather than peripheral action.

D. Opioid Versus Nonopioid Stress-Induced Analgesias

1. USE OF ANTAGONISTS TO DIFFERENTIATE THE ANALGESIAS

Rather than attempting to establish close correlations between analgesia and the endogenous opioid concentrations in various brain regions, many investigators have turned to the use of broad-spectrum opioid antagonists such as naloxone and naltrexone to determine the opioid or nonopioid nature of stress-induced analgesias. One may question the validity of the inference that an analgesia is opioid in nature if it can be abolished or significantly attenuated by prior treatment with an antagonist, but this has become the standard test used in many experiments.

2. DIRECT ACTIONS OF THE ANTAGONISTS

One question of interest, of course, is whether these antagonists directly affect measures of pain sensitivity. A difficulty here is that most tests of pain sensitivity also activate the endogenous opioid systems. Nonetheless, acute effects are rarely found in the doses customarily used in these experiments, but Amir and Amit (1979) found effects of chronic administration. They administered 10 mg/kg of naltrexone per day for 21 days, and beginning 3 days after the last injection the rats were tested every other day on a hot plate. Chronic naltrexone administration prior to induction of analgesia by repeated stress (10-minute footshock) prevented the development of tolerance seen in saline-treated controls. Chronic naltrexone also potentiated the stress-induced analgesia from footshock, but had no effect on the latencies of nonshocked animals. These results suggest that prolonged administration of opioid antagonists may later produce a "supersensitivity" of the opioid system, which in turn may affect the regulation of the hyopothalamo–pituitary–adrenocortical system, as noted previously in Section VI.

3. PROCEDURAL DIFFERENTIATION OF THE ANALGESIAS

Some stress-induced analgesias are blocked by opioid antagonists, but others are not. For example, the analgesias produced by prolonged (20-, 30-, or 60-minute) intermittent footshock were partially reversed by 3 or 10 mg/kg of naloxone (Akil et al., 1976a; Lim et al., 1982), but the analgesia induced by a brief 15-second footshock was unaffected by naloxone in dosages up to 50 mg/kg (Chance, 1980). Repeated exposure to prolonged shock for 14 days resulted in the development of tolerance, i.e., tail-flick latencies decreased over days to levels at or below those of controls, whereas comparable repeated exposure to brief shock did not show the tolerance effect, i.e., they were as analgesic on day 14 as they were on day 1 (Lewis et al., 1981). In another related study, Terman et al. (1983a) found that rats exposed to either the brief or the prolonged footshock for 14 days showed "robust" analgesia on day 15 when they were switched to the opposite stressor, indicating that no cross-tolerance developed between the two analgesias. These results also suggest that it was not sensory receptor adaptation, tissue damage, or habituation to stress that mediated these analgesias.

Various efforts have been made to integrate and understand the often contradictory evidence about the determinants of opioid versus nonopioid analgesias. For example, Watkins and Mayer (1982b) suggested that all analgesias could be classified as either opioid versus nonopioid or neural versus hormonal. The former dichotomy was to be decided on the basis of the effects of opioid antagonists, whereas the latter was to be decided by the effects of hypophysectomy (see discussion below). Terman *et al.* (1984) studied the opioid versus nonopioid nature of analgesias induced by continuous shock of various intensities and durations. The hypothesis generated by those experiments suggested that the total coulometric severity of the shock [intensity (mA) × time (min)] was the critical value, such that footshocks of less than 7.5 mA/min resulted in opioid analgesias, whereas footshocks of greater coulometric value produced nonopioid analgesias. They indicated that although this conclusion followed from their own experiments with continuous footshock, it might not apply to analgesias produced by intermittent footshock. Members of the same research group (Lewis *et al.*, 1980, 1982) suggested the converse relationship, i.e., that prolonged shock induced an opioid analgesia in contrast to brief shock, which induced a nonopioid analgesia. They also speculated that analgesias induced by intermittent footshock were naloxone reversible, whereas analgesias induced by continuous shock were not. Difficulties for the coulometric idea are provided in recent work by Grau (1987), who found evidence that brief (20-second) tail shock induced a transient nonopioid analgesia followed by an opioid one, whereas long (80-second) tail shock resulted in a rapidly decaying opioid analgesia. In keeping with the cognitive times in which we live, Grau interprets these and the results of related experiments in terms of a memory model (Wagner, 1981). Clearly, these varied approaches have not yet resolved or integrated the existing evidence. In this context, Rodgers and Randall (1987) have cogently noted that "a close functional relationship may exist between specific defensive strategies (e.g., active vs passive) and particular forms of pain inhibition (i.e., non-opioid vs opioid)" (p. 185). They suggest that noxious stimulation elicits defensive behaviors in the animal's repertoire and that the specific experimental conditions in which the noxious stimulation is administered determine the form of defensive reactions and hence the nature of the analgesia. This is a refreshing and new approach to the problem of identifying the factors that induce the specific forms of analgesia, but it has not yet been extensively employed.

Adding further complexity to the problem is the elegant work by Fanselow (1984). He uses the frequency of recuperative behavior (paw licking, paw elevation, etc.) after dorsal hindpaw subcutaneous injection of .05 ml of 15% formalin to index analgesia. Control animals lick and protect the injected paw with great frequency, whereas after footshock they lick and protect the injected paw hardly at all, indicating analgesia to the painful formalin injection. In a series of studies Fanselow demonstrated that the analgesias induced by intermittent shocks of 1 or 4 mA (2–3 shocks of 0.75-second duration at 20-second intervals) could be blocked by prior administration of naloxone. These analgesias were unaffected by hypophysectomy and, hence, were "nonhormonal" using the Watkins and Mayer (1982b) criteria. They were also controlled by the contextual stimuli associated with the footshock, i.e., were a product of Pavlovian conditioning, showed no deficits over a 48-hour retention interval, and could be eliminated by Pavlovian extinction of the contextual cues. Additional research indicated that these effects were centrally mediated (Calcagnetti *et al.*, 1987; Fanselow *et al.*, 1988a) and most probably involved δ-receptors (Fanselow *et al.*, 1988b). Clearly, integration of this more clinically relevant form of analgesia with the tail-flick literature is needed.

4. A "META-ANALYSIS" POINTS TO OPIOID RECRUITMENT TIME

Rather than review the many experiments on the opioid versus non-opioid nature of the stress-induced analgesias, Nagase-Shain (1984) conducted a "meta-analysis" of the tail-flick experiments up to that time, and Fig. 7 illustrates the resulting relationship between shock intensity and the duration of the stress session in determining the opioid or nonopioid nature of the induced analgesia. In the figure, filled symbols represent reports of stress-induced analgesias which apparently are nonopioid in nature, i.e., the are unaffected by opioid antagonists and/or show no cross-tolerance with morphine; open symbols represent reports of analgesias that apparently do involve the opioid systems of the brain. It is clear from the figure that the major factor determining the opioid or nonopioid nature of the induced analgesia is not shock intensity, but rather session duration. In the formalin test Fanselow (1984) noted that even at 4 mA, a single 0.75-second shock is insufficient to induce analgesia, whereas two or three shocks 20 seconds apart are sufficient to do so, suggesting that some recruitment time is necessary.

In Fig. 7 there are only a few exceptions to the general rule that brief shock sessions produce nonopioid analgesias whereas prolonged shock sessions rather consistently induce opioid analgesia, which is consistent with the analysis of Lewis *et al.* (1982) noted above. The cutoff appears to be in the neighborhood of 10 minutes or so. However, it must be acknowledged that the presence in Fig. 7 of the few exceptions to this general rule indicates that there are factors other than recruitment time that determine the opioid versus nonopioid nature of stress-induced analgesias. These are as yet not identified.

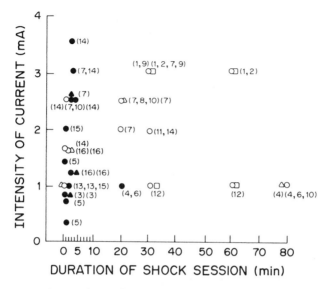

F I G U R E 7.   Meta-analysis of reports of opioid versus nonopioid stress-induced analgesia as a function of shock intensity and duration of shock session. ●, stress-induced analgesia unaffected by naloxone or naltrexone; ○, stress-induced analgesia attenuated by naloxone or naltrexone; △, cross-tolerance with morphine; ▲, no cross-tolerance with morphine; □, opioid involvement determined by assay; references are cited numerically in parentheses. References: (1) Akil *et al.*, 1976a; (2) Baizman *et al.*, 1979; (3) Chance, 1980; (4) Grau *et al.*, 1981; (5)Hayes *et al.*, 1978a; (6) Hyson *et al.*, 1982; (7) Lewis *et al.*, 1980, 1981, 1982; (8) Lim and Funder, 1983; Lim *et al.*, 1982; (9) Madden *et al.*, 1977; (10) Maier *et al.*, 1980, 1983; (11) Maixner and Randich, 1984; (12) Rossier *et al.*, 1977b; (13) Snow and Dewey, 1983; (14) Terman *et al.*, 1983b, 1984; (15) Tricklebank *et al.*, 1982, 1984; (16) Watkins and Mayer, 1982b.

5. PAVLOVIAN CONDITIONING MAY BE INVOLVED

Some evidence suggests that Pavlovian conditioning of contextual cues may be an important factor. For example, Matzel and Miller (1987) showed that brief exposure to contextual stimuli that were previously associated with shock could not induce an opioid analgesia if the animals were tested right after exposure to the conditioned stimuli but could do so if a delay of approximately 1.5 minutes followed exposure to the contextual cues. It appears that recruitment of the opioid systems of the brain to induce analgesia takes time, whether the eliciting stimuli are conditioned or unconditioned. In the case of well-established conditioned stimuli (Matzel and Miller, 1987), which are based on multiple days of paired association between the contextual cues and the unsignaled shocks, the required recruitment time appears to be considerably shorter than that required in most of the studies illustrated in Fig. 7, in which the opioid analgesias were induced in a single session with the unconditioned stimuli (shocks).

6. BOTH ANALGESIAS CAN OCCUR SEQUENTIALLY IN THE SAME SUBJECT

The studies reviewed in the above meta-analysis differentiate opioid from nonopioid stress-induced analgesias in experiments which use between-subjects comparisons. Maier's group, however, was able to induce sequentially both a nonopioid and an opioid analgesia with inescapable tail shock in a within-subjects design (Grau et al., 1981). This striking finding is illustrated in Fig. 8, which indicates that two peaks of analgesia were found in rats receiving a series of 80 shocks with measurement of tail-flick latencies after every twentieth shock; tail-flick latencies were also obtained before the first shock to establish an unstressed baseline. The first peak of analgesia occurred after 20 shocks and could not be attenuated by naltrexone in doses up to 28 mg/kg. Full recovery of baseline latencies was obtained after 40 shocks, and this was followed by a second, longer-lasting analgesia after 60 and 80 shocks. This second peak could be reduced or eliminated by pretreatment with naltrexone (see Fig. 8) in a dose-dependent fashion: tail-flick latencies of animals treated with 14 or 28 mg/kg of naltrexone did not differ from those of nonshocked controls (Hyson et al., 1982). Thus, it appears that these data are consistent with the general rule that relatively brief shock sessions result in nonopioid analgesia, whereas more prolonged sessions induce opioid analgesia. Again, it is apparent that recruitment of the central opioid systems is a relatively slow process. However, Terman et al. (1984) reported that increasing the severity of brief footshock stress, as defined by coulometric value (intensity × time), reduced the effectiveness of 5 mg/kg of naltrexone in attenuating the induced analgesia. These data suggest that subtle differences within brief-shock stress procedures can determine whether the induced analgesia is opioid or nonopioid.

An interesting feature of the 80-shock stress procedure is that analgesia could be reinstated 24 hours after exposure to 80 but not 20 inescapable tail shocks if the animals were given 5 brief footshocks in a shuttle box prior to the testing session (Jackson et al., 1979). Similarly, the analgesia from 20 minutes of intermittent, but not from 3 minutes of continuous, footshock could be reinstated 24 hours later if the animals were presented with reminder shocks (Maier et al., 1983). These reinstated analgesias were eliminated by the administration of either 14 mg/kg of naltrexone or 50 mg/kg of naloxone

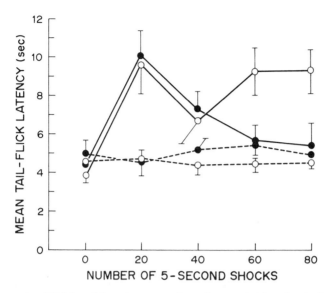

F I G U R E 8.   Within-subject demonstration of nonopioid analgesia followed by opioid analgesia using an 80-tail-shock procedure. ○---○, saline and restraint; ○——○, saline and shock; ●---●, naltrexone and restraint; ●——●, naltrexone and shock. Reproduced from Grau *et al.* (1981), with permission of the authors and the publisher.

before reexposure to the brief "reminder" foot shocks, or they were reduced in a dose-dependent manner by naltrexone (1–14 mg/kg) administered before the initial exposure to inescapable shock (Maier *et al.*, 1980). Animals injected with morphine either 24 hours before (Grau *et al.*, 1981) or after 80 inescapable shocks were hyperresponsive to the analgesic effects of morphine (Hyson *et al.*, 1982), suggesting the development of cross-tolerance between this stressor and morphine.

7. GENETIC FACTORS AND GENE–ENVIRONMENT INTERACTIONS HAVE BEEN OVERLOOKED

The above set of experiments from Maier's laboratory using the 80-shock procedure presents a rather consistent and dramatic picture. However, the above findings may be specific to the rat strain used in those experiments. Struck by the power of the within-subject design, we attempted to replicate those results using the Syracuse rat strains (Long-Evans derived) that have been selectively bred for high (SHA) and low (SLA) levels of avoidance performance in a shuttle box (Brush *et al.*, 1979, 1985). We carefully replicated

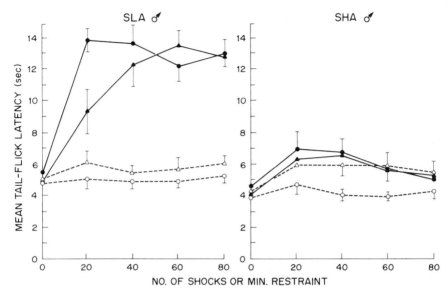

FIGURE 9. Mean tail-flick latencies as a function of number of shocks or duration of restraint for SHA and SLA rats previously injected with saline or 7 mg/kg naltrexone. ●——●, saline/shock; ○---○, saline/restraint; ▲——▲, naltrexone/shock; △---△, naltrexone/restraint.

Maier's 80-shock procedure with three tail-flick tests before the first tail shock and after every twentieth shock. The results are illustrated in Fig. 9. Animals of both strains showed statistically significant stress-induced analgesia, but only SLA animals showed anything like the profound analgesia reported by Grau *et al.* (1981). In neither case was the time course of analgesia biphasic, and in neither case could the analgesia be blocked by prior administration of 7 or 14 mg/kg of naltrexone (see Fig. 9 for the lower dose and Nagase Shain and Brush, 1986, for the higher dose).

It seems to us, therefore, that genetic factors are very important in determining whether stress (specifically the 80-shock procedure of Maier's laboratory) will induce an analgesia at all, whether it will induce an analgesia with a unimodal or biphasic time course, and whether the analgesia(s), if present, is(are) opioid in nature. The voluminous literature on this topic, although appearing to be relatively consistent (see Fig. 7), has ignored important sources of individual differences, especially differences between strains and genetic differences within strains.

Recent reports by Helmstetter and Fanselow (1987) and Panocka *et al.* (1986a,b) nicely underscore this point. Panocka *et al.* selectively bred mice for large (HA) or small (LA) stress-induced analgesias and showed that the stress-induced analgesia in HA animals could be blocked with naloxone, whereas that in LA animals could not. In addition, the LA animals required 12 times the dose of morphine to produce a level of analgesia equal to that of the HA animals. Clearly, the opioid systems in mice are under genetic control. Working with rats, Helmstetter and Fanselow examined the effects of 7 mg/kg of naloxone and naltrexone on the analgesia induced by exposure to contextual cues that had been paired with footshock the previous day. Analgesia was indexed by the frequency of recuperative behavior in response to dorsal hindpaw subcutaneous injection of .05 ml of 15% formalin. They compared these effects in Long-Evans hooded rats and in Sprague-Dawley albino rats. The Long-Evans animals were derived from stock obtained from Blue Spruce Farms, and the albinos came from two different commercial breeders, Holtzman and Charles River. Holtzman and Long-Evans rats exhibited the same relatively low amount of freezing during a 6-minute test in the conditioning environment, whereas Charles River animals showed twice as much freezing as the others. All saline-treated animals showed the expected suppression of recuperative behavior in the conditioning environment. However, the Charles River animals differed from the others in their response to the opiate antagonists: their recuperative behavior was significantly increased by naltrexone but not by naloxone. In contrast, both Long-Evans and Holtzman Sprague-Dawleys showed increased recuperative behavior (less analgesia) in response to both opiate antagonists, and the effect of naltrexone tended to be greater than that of naloxone, although the difference between the two antagonists was not reliable. Taken together, the data from these experiments clearly indicate that important genetic factors and gene–environment interactions play a major role in determining the magnitude of, and the opioid versus nonopioid nature of, stress-induced analgesias. Unfortunately, these factors have been largely ignored until recently.

E. Stress-Induced Analgesia and the Pituitary–Adrenal Axis

Given the complex interactions between opioids and the stress-related endocrine systems (see Section VI above), it is not surprising that the effects of manipulations of components of these systems on basal pain sensitivity and on stress-induced analgesia have been investigated. Some experiments have investigated the effects of hypophysectomy, which leaves brain concentrations of β-endorphin intact (Bloom *et al.*, 1978a) but reduces plasma concentrations of β-endorphin, the enkephalins, ACTH, and corticosterone. Other studies have used bilateral adrenalectomy, which has no effect on brain concentration of β-endorphin but removes the negative feedback of the glucocorticoids and thus increases the peripheral concentrations of β-endorphin and ACTH; in addition, of course, enkephalins and catecholamines of adrenal medullary origin are reduced. To separate the adrenal cortical and medullary effects some experiments have employed adrenal demedullation, which depletes peripheral enkephalins and catecholamines, or denervation, which in the rat results in increased synthesis and release of Met- and Leu-enkephalin, apparently a result of decreased inhibitory control by the splanchnic nerves. Finally, some studies have administered dexamethasone, a powerful synthetic glucocorticoid which suppresses hypothalamic release of CRH and pituitary release of β-endorphin and ACTH. It should be noted, however, that some stressors are capable of inducing a normal release of the POMC-derived peptides from the pituitary by breaking through the feedback inhibition by dexamethasone.

I. BASAL PAIN SENSITIVITY

With respect to basal pain sensitivity, the literature indicates that hypophysectomy and adrenalectomy have no acute effect, but induce increased pain sensitivity if sufficient postoperative time is allowed. Adrenal demedullation over these same time periods appears to be without effect, suggesting that the hyperalgesia of adrenalectomy is not mediated by the reduction in peripheral enkephalins. Because hypophysectomy reduces peripheral β-endorphin and enkephalin concentrations, either or both could be involved in the induced increase in basal pain sensitivity. However, the endocrine events following adrenalectomy are incompatible with hyperalgesia, so multiple mechanisms must be involved, and more research will be needed to unravel this confusing picture.

2. "META-ANALYSIS" OF STRESS-INDUCED ANALGESIAS

Nagase Shain (1984) also conducted a meta-analysis of the effects of such manipulations on the opioid versus nonopioid nature of stress-induced analgesia, and the results are illustrated in Fig. 10. As in Fig. 7, the axes locate experiments in terms of shock intensity and duration of the shock session. Solid symbols represent studies in which the pituitary–adrenal manipulation did not affect the induced analgesia, whereas open symbols represent studies where the manipulation was effective in reducing the stress-induced analgesia. It is clear from this analysis that short-duration shock sessions, re-

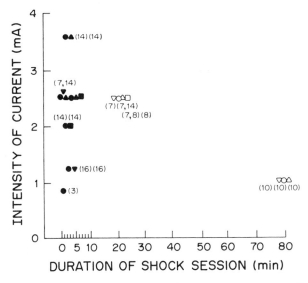

F I G U R E 10. Meta-analysis of reports of attenuation of, or no effect on, stress-induced analgesia by manipulations of the pituitary–adrenal axis. ●, stress-induced analgesia (SIA) unaffected by hypophysectomy; ○, SIA attenuated by hypophysectomy; ▲, SIA unaffected by adrenalectomy; △, SIA attenuated by adrenalectomy; ■, SIA unaffected by demedullation; □, SIA attenuated by demedullation; ▼, SIA unaffected by dexamethasone; ▽, SIA attenuated by dexamethasone; references are cited numerically in parentheses. References: (3) Chance, 1980; (7) Lewis *et al.*, 1980, 1981, 1982; (8) Lim and Funder, 1983; Lim *et al.*, 1982; (10) Maier *et al.*, 1980, 1983; (14) Terman *et al.*, 1983b, 1984; (16) Watkins and Mayer, 1982b.

gardless of shock intensity, induced analgesias that were not attenuated by these manipulations of the pituitary–adrenal axis, whereas long-duration shock sessions induced analgesias that were attenuated or abolished by these pituitary–adrenal manipulations. Recall that short-duration shock sessions typically induce nonopioid analgesias, whereas long-duration sessions usually induce opioid analgesias, although there are a few exceptions. Thus, the analysis presented in Fig. 10 suggests that the stress-related hormones of the pituitary–adrenal axis interact only with the opioid forms of stress-induced analgesia. Presumably, the pituitary and adrenal stores of β-endorphin and the enkephalins are the active agents in opioid forms of the stress-induced analgesias, because procedures which eliminate those stores or inhibit their release eliminate or attenuate the analgesia. However, demedullation also reduces a presumably opioid analgesia, which suggests a possible role for adrenocortical hormones as well.

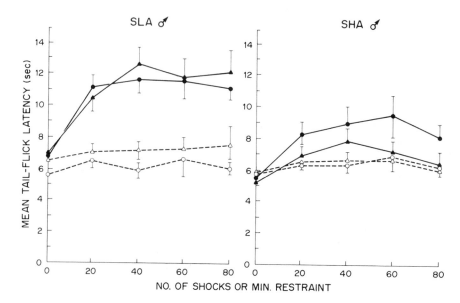

FIGURE 11. Mean tail-flick latencies as a function of number of shocks or duration of restraint for SHA and SLA rats previously adrenalectomized or sham adrenalectomized. ●——●, sham/shock; ○---○, sham/restraint; ▲——▲, adrenalectomy/shock; △---△, adrenalectomy/restraint.

Nagase Shain (1985) also examined the effects of bilateral adrenalectomy on the analgesias induced by the 80-shock procedure in SHA or SLA rats (see Fig. 11). Consistent with the results of our previous experiment (see Fig. 9), the stress-induced analgesia was significantly greater in SLA than SHA animals, and adrenalectomy was without effect in either strain. These two sets of results indicate that in neither strain is the stress-induced analgesia opioid in nature, despite the fact that Nagase Shain (1985) was able to demonstrate normal opiate functioning in these strains. Specifically, she was able to induce normal analgesia in both SHA and SLA animals with 2 mg/kg of morphine and to block the action of morphine with 7 mg/kg of naltrexone administered 20 minutes before the morphine. Thus, the opioid system in these animals appears normal, and once again the influence of genetic variables is shown to be importantly involved in stress-induced analgesia, be it opioid or nonopioid in nature.

IX. Summary and Conclusions

In the last 15 years enormous progress has been made toward increasing our understanding of the roles of central and peripheral exogenous opiates and endogenous opioids in modulating behavior, and in particular, nociceptively elicited reflexes and aversively motivated behaviors. The development of relatively specific radioimmunoassays and radiocytochemicalassays has established that there are three families of endogenous opioids, and this has been confirmed by the identification of the genetic codes for their precursor hormones. Our understanding of the sources and distributions of the endogenous opioids, both peripherally and centrally, has increased, and it has been established rather unambiguously that the opioids coexist intracellularly with various neurotransmitters and neuro- or tropic hormones. Some of these findings are still fraught with ambiguity because of possible nonspecificity of the assays. However, the generally accepted picture outlined in this review has emerged, subject to presumably minor modifications. Despite methodological difficulties, the general sources and distributions of the endogenous opioids have been established, and the findings are congruent with what is known about the role of various structures, mostly subcortical, in mediating nociceptively and/or aversively motivated behaviors.

The neuroanatomy and neurophysiology of the afferent and efferent mechanisms involved in nociception have advanced to the point that we now have a rather good understanding of the hard- and soft-wiring of pain perception mechanisms, including the spinal and supraspinal components. Indeed, the opioids are only one of several systems that may potentially modulate nociception.

Finally, the role of the hypothalamo–pituitary–adrenal axis in stress-induced analgesia has been the subject of intensive investigation, but unfortunately, little attention has been paid, until quite recently, to the influence of genetic and gene–environment interactive influences on the opioid versus nonopioid nature of the stress-induced analgesias. Much work has been done to identify the procedural variables that determine the opioid versus non-opioid character of stress-induced analgesias, but the work seems futile in view of the field's disregard of major genetic and genetic–environmental interactive factors which seem to determine the nature of the stress-induced analgesias. The limitation of the research, almost exclusively to measures dependent on one spinal reflex, tail-flick latency, seems inappropriate, given the enormous importance research in this field has for human application. Much has been learned, but much remains to be done in this terribly complex field of the behavioral effects of opioids.

Acknowledgments

Preparation of this chapter was supported in part by NIMH Grant MH39230 to FRB and by the Department of Psychology of the University of Hawaii at Manoa at which the first author was a visiting professor in 1987. We thank Dr. J. C. Froehlich for her thoughtful and helpful comments on an earlier draft of this chapter.

References

Abbott, F. V., and Melzack, R. (1982). Brainstem lesions dissociate neural mechanisms of morphine analgesia in different kinds of pain. *Brain Res.* **251,** 149–155.

Akil, H., and Liebeskind, J. C. (1975). Monoaminergic mechanisms of stimulation-produced analgesia. *Brain Res.* **94,** 279–296.

Akil, H., and Mayer, D. J. (1972). Antagonism of stimulation-produced analgesia by PCPA, a serotonin synthesis inhibitor. *Brain Res.* **44,** 692–697.

Akil, H., Madden, J., IV, Patrick, R. L., and Barchas, J. D. (1976a). Stress-induced increases in endogenous opiate peptides: Concurrent analgesia and its partial reversal by naloxone. *In* "Opiates and Endogenous Opioid Peptides" (H. W. Kosterlitz, ed.), pp. 63–78. North-Holland Publ., Amsterdam.

Akil, H., Mayer, D. J., and Liebeskind, J. C. (1976b). Antagonism of stimulation-produced analgesia by naloxone, a narcotic antagonist. *Science* **191,** 961–962.

Akil, H., Watson, S. J., Young, E., Lewis, M.C., Khachaturian, H., and Walker, J. M. (1984). Endogenous opioids: Biology and function. *Annu. Rev. Neurosci.* **7,** 223–255.

Amir, S., and Amit, Z. (1979). Enhanced analgesic effects of stress following chronic administration of naltrexone in rats. *Eur. J. Pharmacol.* **59,** 137–140.

Amir, S., Brown, Z. W., and Amit, Z. (1980). The role of endorphins in stress: Evidence and speculations. *Neurosci. Biobehav. Rev.* **4,** 77–86.

Anderson, T. R., and Slotkin, T. A. (1975). Effects of morphine on the rat adrenal medulla. *Biochem. Pharmacol.* **24,** 671–679.

Anderson, T. R., and Slotkin, T. A. (1976). The role of neural input in the effects of morphine on the rat adrenal medulla. *Biochem. Pharmacol.* **25,** 1071–1074.

Atweh, S. F. (1983). Characterization and distribution of brain opiate receptors and endogenous opioid peptides. *In* "The Neurobiology of Opiate Reward Processes" (J. E. Smith and J. D. Lane, eds.), pp. 59–88. Elsevier, Amsterdam.

Atweh, S. F., and Kuhar, S. J. (1977). Autoradiographic localization of opiate receptors in rat brain. *Brain Res.* **124,** 53–67.

Baizman, E. R., Cox, B. M., Osman, H., and Goldstein, A. (1979). Experimental alterations of endorphin levels in rat pituitary. *Neuroendocrinology* **28,** 402–424.

Barraclough, C. A., and Sawyer, C. H. (1955). Inhibition of the release of pituitary ovulatory hormone in the rat by morphine. *Endocrinology (Baltimore)* **57,** 329–337.

Basbaum, A. I., Clanton, C. H., and Fields, H. L. (1976). Opiate and stimulation-produced analgesia: Functional anatomy of a medullospinal pathway. *Proc. Natl. Acad. Sci. U.S.A.* **73,** 4685–4688.

Basbaum, A. I., Morley, N. J. E., O'Keefe, J., and Clanton, C. H. (1977). Reversal of morphine and stimulation-produced analgesia by subtotal spinal cord lesions. *Pain* **3**, 43–56.

Bechara, A., and van der Kooy, D. (1987). Kappa receptors mediate the peripheral aversive effects of opiates. *Pharmacol., Biochem. Behav.* **28**, 227–233.

Beecher, H. K. (1959). "Measurement of Subjective Responses." Oxford Univ. Press, London and New York.

Bergland, R. M., and Page, R. B. (1979). Pituitary-brain vascular relations: A new paradigm. *Science* **204**, 18–24.

Blanc, J. P., Taylor, J. W., Miller, R. J., and Kaiser, E. T. (1983). Examination of the requirement for an amphiphilic helical structure in beta-endorphin through the design, synthesis and study of model peptides. *J. Biol. Chem.* **258**, 8277–8284.

Bloom, F. E. (1983). The endorphins: A growing family of pharmacologically pertinent peptides. *Annu. Rev. Pharmacol. Toxicol.* **23**, 151–170.

Bloom, F. E., Battenberg, E., Rossier, J., Ling, N., and Guillemin, R. (1978a). Neurons containing beta-endorphin in rat brain exist separately from those containing enkephalin: Immunocytochemical studies. *Proc. Natl. Acad. Sci. U.S.A.* **75**, 1591–1595.

Bloom, F. E., Rossier, J., Battenberg, L. F., Bayon, A., French, E., Henrickson, S. J., Siggins, G. R., Segal, D., Browne, R., Ling, N., and Guillemin, R. (1978b). Beta-endorphin: Cellular localization, electrophysiological and behavioral effects. *Adv. Biochem. Psychopharmacol.* **19**, 89–109.

Bohus, B. (1986). Opiomelanocortins and behavioral adaptation. *In* "Neuropeptides and Behavior, Volume 1: CNS Effects of ACTH, MSH and Opioid Peptides" (D. de Wied, W. H. Gispen, and Tj. B. van Wimersma Greidanus, eds.), pp. 313–348. Pergamon, Elmsford, New York.

Bolles, R. C., and Fanselow, M. S. (1982). Endorphins and behavior. *Annu. Rev. Psychol.* **33**, 87–101.

Bradbury, M. W. B., Burden, J. L., Hillhouse, E. W., and Jones, M. T. (1974). Stimulation electrically and by acetylcholine of the hypophysiotrophic area of the rat hypothalamus *in vitro*. *J. Physiol. (London)* **239**, 269–283.

Bruni, J. F., Van Vugt, D. A., Marshall, S., and Meites, J. (1977). Effects of naloxone, morphine and methionine enkephalin on serum prolactin, luteinizing hormone, follicle stimulating hormone, thyroid stimulating hormone, and growth hormone. *Life Sci.* **21**, 461–466.

Brush, F. R., Froehlich, J. C., and Sakellaris, P. C. (1979). Genetic selection for avoidance behavior in the rat. *Behav. Genet.* **9**, 309–316.

Brush, F. R., Baron, S., Froehlich, J. C., Ison, J. R., Pellegrino, L. J., Phillips, D. S., Sakellaris, P. C., and Williams, V. N. (1985). Genetic differences in avoidance learning by *Rattus norvegicus:* Escape/avoidance responding, sensitivity to electric shock, discrimination learning and open-field behaviors. *J. Comp. Psychol.* **99**, 60–73.

Buckingham, J. C. (1980). Corticotrophin releasing factor. *Pharmacol. Rev.* **31**, 253–273.

Buckingham, J. C. (1982). Secretion of corticotrophin and its hypothalamic releasing factor in response to morphine and opioid peptides. *Neuroendocrinology* **35,** 111–116.

Buckingham, J. C., and Cooper, T. A. (1984). Influence of opioid substances on hypothalamo-pituitary-adrenocortical activity in the rat. *In* "Opioid Modulation of Endocrine Function" (G. Delitala, M. Motta, and M. Serio, eds.), pp. 81–87. Raven Press, New York.

Buckingham, J. C., and Leach, J. H. (1979). Vasopressin and hypothalamo-pituitary-adrenocorticotrophic activity. *J. Physiol. (London)* **296,** 87P.

Burbach, J. P. H. (1986). Action of proteolytic enzymes on lipotropins and endorphins: Biosynthesis, biotransformation and fate. *In* "Neuropeptides and Behavior, Volume 1: CNS Effects of ACTH, MSH and Opioid Peptides" (D. de Wied, W. H. Gispen, and Tj. B. van Wimersma Greidanus, eds.), pp. 43–76. Pergamon, Elmsford, New York.

Calcagnetti, D. J., Helmstetter, F. J., and Fanselow, M. S. (1987). Quaternary naltrexone reveals the central mediation of conditional opioid analgesia. *Pharmacol., Biochem. Behav.* **27,** 529–531.

Cannon, J. T., Prieto, G. J., Lee, A., and Liebeskind, J. C. (1982). Evidence for opioid and non-opioid forms of stimulation-produced analgesia in the rat. *Brain Res.* **243,** 315–321.

Carlson, D. E., Dornhurst, A., Seif, S. M., Robinson, A. G., and Gann, D. S. (1982). Vasopressin-dependent and -independent control of the release of adrenocorticotropin. *Endocrinology (Baltimore)* **110,** 680–682.

Carpenter, M. B. (1976). "Human Neuroanatomy." Williams & Wilkins, Baltimore, Maryland.

Celis, M. E. (1980). Effects of different opiate agonists on melanocyt-stimulating hormone release: In vivo and in vitro studies. *Can. J. Physiol. Pharmacol.* **58,** 326–329.

Chance, W. T. (1980). Autoanalgesia: Opiate and non-opiate mechanisms. *Neurosci. Biobehav. Rev.* **4,** 55–67.

Cicero, T. J. (1980). Effects of exogenous and endogenous opiates on the hypothalamic-pituitary-gonadal axis in the male. *Fed. Proc., Fed. Am. Soc. Exp. Biol.* **39,** 2551–2554.

Codd, E. E., and Byrne, W. L. (1980). Seasonal variation in the apparent number of ³H-Naloxone binding sites. *In* "Endogenous and Exogenous Opiate Agonists and Antagonists" (E. L. Way, ed.), pp. 63–66. Pergamon, New York.

Conrad, L. C. A., Leonard, C. M., and Pfaff, D. W. (1974). Connections of the median and dorsal raphe nuclei in the rat: An autoradiographic and degeneration study. *J. Comp. Neurol.* **156,** 179–206.

Cotton, R., Giles, M. G., Miller, L., Shaw, J. S., and Timms, D. (1984). ICI 174864: A highly selective antagonist for the opioid delta-receptor. *Eur. J. Pharmacol.* **97,** 331–332.

Cox, B. M., Opheim, K. E., Teschmaker, H., and Goldstein, A. (1975). A peptide-like substance from pituitary that acts like morphine. *Life Sci.* **16,** 1777–1782.

Dahlström, A., and Fuxe, K. (1965). Experimentally induced changes in the intra-neuronal amine levels of bulbospinal neuron systems. *Acta Physiol. Scand.,* **72,** Suppl. 247, 1–36.

D'Amour, F. E., and Smith, D. L. (1941). A method for determining loss of pain sensation. *J. Pharmacol. Exp. Ther.* **72,** 74–79.

Deakin, J. F. W., Dickenson, A. H., and Dostrovsky, J. O. (1977). Morphine effects on rat raphe magnus neurons. *J. Physiol. (London)* **267,** 43–45.

de Kloet, E. R., Palkovits, M., and Mezey, E. (1986). Opioid peptides: Localization, source and avenues of transport. *In* "Neuropeptides and Behavior, Volume 1: CNS Effects of ACTH, MSH and Opioid Peptides" (D. de Wied, W. H. Gispen, and Tj. B. van Wimersma Greidanus, eds.), pp. 1–41. Pergamon, Elmsford, New York.

de Kloet, E. R. Ratka, A., Reul, J. M. H. M., Sutano, W., and van Eekelen, J. A. M. (1987). Corticosteroid receptor types in brain: Regulation and putative function. *Ann. N. Y. Acad. Sci.* **512,** 351–361.

Dennis, S. G., and Melzack, R. (1980). Pain modulation by 5-hydroxytryptaminic agents and morphine as measured by three pain tests. *Exp. Neurol.* **69,** 260–270.

de Rotte, A. A., van Wimersma Greidanus, Tj. B., Van Ree, J. M., Andringa-Bakker, E. A. D., and de Wied, D. (1981). The influence of beta-LPH fragments on alpha-MSH release: The involvement of a dopaminergic system. *Life Sci.* **29,** 825–832.

de Wied, D. (1964). Influence of anterior pituitary on avoidance learning and escape behavior. *Am. J. Physiol.* **207,** 255–259.

de Wied, D., and Bohus, B. (1966). Long term and short term effects on retention of a conditioned avoidance response in rats by treatment with long acting pitressin and alpha-MSH. *Nature (London)* **212,** 1484–1486.

Dray, A., and Nunan, L. (1984). Selective delta-opioid receptor antagonism by ICI 174,864 in the central nervous system. *Peptides* **5,** 1015–1016.

Duggan, A. W., Hall, J. G., and Headley, P. M. (1977a). Suppression of transmission of nociceptive impulses by morphine; selective effects of morphine administered in the region of the substantia gelatinosa. *Br. J. Pharmacol.* **61,** 65–76.

Duggan, A. W., Hall, J. G., and Headly, P. M. (1977b). Enkephalins and dorsal horn neurones of the cat: Effects on responses to noxious and innocuous skin stimuli. *Br. J. Pharmacol.* **61,** 399–408.

Eddy, N. B., and Leimbach, D. (1953). Synthetic analgesics II: Dithienylbutenyl- and dithienylbutyl-amines. *J. Pharmacol. Exp. Ther.* **107,** 385–393.

Fanselow, M. S. (1984). Shock-induced analgesia on the formalin test: Effects of shock severity, naloxone, hypophysectomy, and associative variables. *Behav. Neurosci.* **98,** 79–95.

Fanselow, M. S., Calcagnetti, D. J., and Helmstetter, F. J. (1988a). Peripheral versus intracerebroventricular administration of quaternary naltrexone and the enhancement of Pavlovian conditioning. *Brain Res.* **444**, 147–152.

Fanselow, M. S., Calcagnetti, D. J., and Helmstetter, F. J. (1988b). Delta opioid receptor involvement in conditional fear-induced analgesia: Antagonism by 16-Me Cyprenorphine and [D-Ala², Leu⁵, Cys⁶] Enkephalin. *Pap., Meet. East. Psychol. Assoc., 1988.*

Forman, L. J., Sonntag, W. E., and Meites, J. (1983). Elevation of plasma LH in response to systemic injection of beta-endorphin antiserum in adult male rats. *Proc. Soc. Exp. Biol. Med.* **173**, 14–16.

Frenk, H., McCarty, B., and Liebeskind, J. C. (1978). Different brain areas mediate the analgesic and epileptic properties of enkephalin. *Science* **200**, 335–337.

Gilbert, P. E., and Martin, W. R. (1976). The effects of morphine- and naloprphine-like drugs in the non-dependent, morphine-dependent and cyclazocine-dependent chronic spinal dog. *J. Pharmacol. Exp. Ther.* **198**, 66–82.

Gispen, W. H., and Isaacson, R. L. (1986). Excessive grooming in response to ACTH. *In* "Neuropeptides and Behavior, Volume 1: CNS Effects of ACTH, MSH and Opioid Peptides" (D. de Wied, W. H. Gispen, and Tj. B. van Wimersma Greidanus, eds.), pp. 273–312. Pergamon, Elmsford, New York.

Goldstein, A. (1976). Opioid peptides (endorphins) in pituitary and brain. *Science* **193**, 1081–1086.

Goldstein, A., Lowney, L. I., and Pal, B. K. (1971). Sterospecific and non-specific interactions of the morphine congener levorphanol in subcellular fractions of mouse brain. *Proc. Natl. Acad. Sci. U.S.A.* **68**, 1742–1747.

Goldstein, A., Tachibana, S., Lowney, L. I., Hunkapiller, M., and Hood, L. (1979). Dynophin-(1-13), an extraordinaly potent peptide. *Proc. Natl. Acad. Sci. U.S.A.* **76**, 6666–6670.

Grau, J. W. (1987). Activation of the opioid and nonopioid analgesic systems: Evidence for a memory hypothesis and against the coulometric hypothesis. *J. Exp. Psychol., Anim. Behav. Processes* **13**, 215–225.

Grau, J. W., Hyson, R. L., Maier, S. F., Madden, J., and Barchas, J. D. (1981). Long term stress-induced analgesia and activation of the opiate system. *Science* **213**, 1409–1411.

Hamilton, L. W. (1976). "Basic Limbic System Anatomy of the Rat." Plenum, New York.

Hargreaves, K. M., Mueller, G. P., Dubner, R., Goldstein, D., and Dionne, R. A. (1987). Corticotropin-releasing factor (CRF) produces analgesia in humans and rats. *Brain Res.* **422**, 154–157.

Hauger, R. L., Millan, M. A., Catt, K. J., and Aguilera, G. (1987). Differential regulation of brain and pituitary corticotropin-releasing factor receptors by corticosterone. *Endocrinology (Baltimore)* **120**, 1527–1533.

Hayes, R. L., Bennett, G. J., Newlon, P. G., and Mayer, D. J. (1978a). Behavioral and physiological studies of non-narcotic analgesia in the rat eleicited by certain environmental stimuli. *Brain Res.* **155**, 69–90.

Hayes, R. L., Price, D. D., Bennett, G. J., Wilcox, G. L., and Mayer, D. J. (1978b). Differential effects of spinal cord lesions on narcotic and non-narcotic suppression of nociceptive reflexes: Further evidence for the physiologic multiplicity of pain modulation. *Brain Res.* 155, 91–101.

Helmstetter, F. J., and Fanselow, M. S. (1987). Strain differences in reversal of conditioned analgesia by opioid antagonists. *Behav. Neurosci.* 101, 735–737.

Hokfelt, T., Elde, R., Johansson, O., Terenius, L., and Stein, L. (1977a). The distribution of enkephalin-immunoreactive cell bodies in the rat central nervous system. *Neurosci. Lett.* 5, 25–31.

Hokfelt, T., Ljungdahl, A., Terenius, L., Elde, R., and Nilsson, G. (1977b). Immunohistochemical analysis of peptide pathways possibly related to pain and analgesia: Enkephalin and substance P. *Proc. Natl. Acad. Sci. U.S.A.* 74, 3081–3085.

Hollt, V. (1981). Effects of neuroleptic drugs on endogenous opioid peptides in the rat. *In* "The Role of Endorphins in Neuropsychiatry" (M. M. Emrich, ed.), pp. 1–18. Karger, Basel.

Hollt, V., Haarmann, I., Przewlocki, R., and Jerlicz, M. (1980). *In* "Neural Peptides and Neuronal Communication" (E. Costa and M. Trabucchi, eds.), pp. 399–407. Raven Press, New York.

Hollt, V., Haarman, I., Seizinger, B. R., and Herz, A. (1981). Levels of dynorphin-(1-13) immunoreactivity in rat neurointermediate pituitaries are concomitantly altered with those of leucine enkephalin and vasopressin in response to various endocrine manipuations. *Neuroendocrinology* 33, 333–339.

Hughes, J. (1975). Isolation of an endogenous compound from the brain with pharmacological properties similar to morphine. *Brain Res.* 88, 295–308.

Hughes, J. (1984). Reflections on opioid peptides. *In* "Opioids Past, Present and Future" (J. Hughes, H. O. J. Collier, M. J. Rance, and M. B. Tyers, eds.), pp. 9–19. Taylor & Francis, London.

Hughes, J., Smith, T., Kosterlitz, H. W., Fothergill, L. A., Morgan, B., and Morris, H. R. (1975). Identification of two related pentapeptides from the brain with potent opiate agonist activity. *Nature (London)* 258, 577–579.

Hyson, R. L., Ashcraft, L. J., Drugan, R. C., Grau, J. W., and Maier, S. F. (1982). Extent and control of shock affects naltrexone sensitivity of stress-induced analgesia and reactivity to morphine. *Pharmacol., Biochem. Behav.* 17, 1019–1025.

Jackson, R. L., Maier, S. F., and Coon, D. J. (1979). Long term analgesic effects of inescapable shock and learned helplessness. *Science* 206, 91–93.

Jacquet, Y. F., and Lajtha, A. (1974). Paradoxical effects after microinjection of morphine in the periaqueductal gray matter in the rat. *Science* 185, 1055–1057.

Jaffe, J. H., and Martin, W. R. (1985). Opioid analgesics and antagonists. *In* "The Pharmacological Basis of Therapeutics" (A. G. Gilman, L. S. Goodman, T. W. Rall, and F. Murad, eds.), 7th ed., pp. 491–531. Macmillan, New York.

Jessell, T. M., and Iversen, L. L. (1977). Opiate alagesics inhibit substance P release from rat trigeminal nucleus. *Nature (London)* 268, 549–551.

Jones, M. T., and Hillhouse, E. W. (1976). Structure-activity relationship and the mode of action of corticosteroid feedback on the secretion of corticotrophin-releasing factor (corticoliberin). *J. Steroid Biochem.* **7**, 1189–1202.

Jones, M. T., and Hillhouse, E. W. (1977). Neurotransmitter regulation of corticotropin-releasing factor *in vitro. Ann. N. Y. Acad. Sci.* **297**, 536–558.

Jones, M. T., Brush, F. R., and Neame, R. L. B. (1972). Characteristics of fast feedback of cortictrophin release by corticosteroids. *J. Endocrinol.* **55**, 489–497.

Jones, M. T., Tiptaft, E. M., Brush, F. R., Fergusson, D. A. N., and Neame, R. L. B. (1974). Evidence for dual corticosteroid receptor mechanisms in the feedback control of adrenocorticotrophin secretion. *J. Endocrinol.* **60**, 223–233.

Jones, M. T., Hillhouse, E. M., and Burden, J. L. (1977). Effect of various putative neurotransmitters on the secretion of corticotrophin-releasing hormone from the rat hypothalamus *in vitro. J. Endocrinol.* **69**, 1–10.

Jones, M. T., Gillham, B., Campbell, E. A., Al-Taher, A. R. H., Chuang, T. T., and Di Sciullo, A. (1987). Pharmacology of neural pathways affecting CRH secretion. *Ann. N. Y. Acad. Sci.* **512**, 162–175.

Kakidani, H., Furutani, Y., Takahashi, H., Noda, M., Morimoto, Y., Hirose, T., Asai, M., Inayami, S., Nakanishi, S., and Numa, S. (1982). Cloning and sequence analysis of cDNA for porcine beta-neo-endorphin/dynorphin precursor. *Nature (London)* **298**, 245–249.

Khachaturian, H., Watson, S. J., Lewis, M. E., Coy, D., Goldstein, A., and Akil, H. (1982). Dynorphin immunocytochemistry in the rat central nervous system. *Peptides (N.Y.)* **3**, 941–954.

Knepel, W., and Reimann, W. (1982). Inhibition by morphine and beta-endorphin of vasopressin release evoked by electrical stimulation of the rat medial basal hypothalamus in vitro. *Brain Res.* **238**, 484–488.

Kosterlitz, H. W., and McKnight, A. T. (1981). Opioid peptides and sensory function. *In* "Progress in Sensory Physiology" (D. Ottoson, ed.), pp. 31–95. Springer-Verlag, Berlin.

Kosterlitz, H. W., Lord, J. A. H., and Watt, A. J. (1972). Morphine receptor in the myenteric plexus of the guinea-pig ileum. *In* "Agonist and Antagonist Actions of Narcotic Analgesic Drugs" (H. W. Kosterlitz, H. O. J. Collier, and J. E. Villarreal, eds.), pp. 45–61. Macmillan, London.

Kuhar, M. J., Pert, C. B., and Snyder, S. H. (1973). Regional distribution of opiate receptor binding in monkey and human brain. *Nature (London)* **245**, 447–450.

Kvetnansky, R., Tilders, F. J. H., van Zoest, I. D., Dobrakovova, M., Berkenbosch, F., Culman, J., Zeman, P., and Smelik, P. G. (1987). Sympathoadrenal activity facilitates beta-endorphin and alpha-MSH secretion but does not potentiate ACTH secretion during immobilization stress. *Neuroendocrinology* **45**, 318–324.

Labatz, J., Proudfit, H. K., and Anderson, E. G. (1977). Effects of pain, and the iontophoresis of morphine and acetylcholine on neurons in the rat nucleus raphe magnus. *Proc. Soc. Neurosci.* p. 946.

Lemaire, S., Magnan, J., and Regoli, D. (1978). Rat vas deferens: A specific bioassay for endogenous opioid peptides. *Br. J. Pharmacol.* **64,** 327–329.

Lewis, J. W., Cannon, J. T., and Liebeskind, J. C. (1980). Opioid and nonopioid mechanisms of stress analgesia. *Science* **208,** 623–625.

Lewis, J. W., Sherman, J. E., and Liebeskind, J. C. (1981). Opioid and non-opioid stress analgesia: Assessment of tolerance and cross-tolerance with morphine. *J. Neurosci.* **1,** 358–363.

Lewis, J. W., Tordoff, M. G., Sherman, J. E., and Liebeskind, J. C. (1982). Adrenal medullary enkephalin-like peptides may mediate opioid stress analgesia. *Science* **217,** 557–559.

Liebman, J. M., Mayer, D. J., and Liebeskind, J. C. (1970). Mesencephalic central gray lesions and fear-motivated behavior in rats. *Brain Res.* **23,** 353–370.

Lim, A. T., and Funder, J. W. (1983). Stress-induced changes in plasma, pituitary and hypothalamic immunoreactive beta-endorphin: Effects of dirunal variation, adrenalectomy, corticosteroids and opiate agonists and antagonists. *Neuroendocrinology* **36,** 225–234.

Lim, A. T., Wallace, M., Oei, T. P., Gibson, S., Romas, N., Pappas, W., Clements, J., and Funder, J. W. (1982). Foot shock analgesia: Lack of correlation with pituitary and plasma immunoreactive beta-endorphin. *Neuroendocrinology* **35,** 235–241.

Lymangrover, J. R., Dokas, L. A., Konig, A., Martin, R., and Saffran, M. (1981). Naloxone has a direct effect on the adrenal cortex. *Endocrinology (Baltimore)* **109,** 1132–1137.

Madden, J., Akil, H., Patrick, R. L., and Barchas, J. D. (1977). Stress-induced parallel changes in central opioid levels and pain responsiveness in the rat. *Nature (London)* **265,** 358–360.

Maier, S. F., Davies, S., Grau, J. W., Jackson, R. L., Morrison, D. H., Moye, T., Madden, J., and Barchas, J. D. (1980). Opiate antagonists and long-term analgesic reaction induced by inescapable shock in rats. *J. Comp. Physiol. Psychol.* **94,** 1172–1183.

Maier, S. F., Sherman, J. E., Lewis, J. W., Terman, G. W., and Liebeskind, J. C. (1983). The opioid/nonopioid nature of stress-induced analgesia and learned helplessness. *J. Exp. Psychol., Anim. Behav. Processes* **9,** 80–90.

Maixner, W., and Randich, A. (1984). Role of the right vagal nerve trunk in antinociception. *Brain Res.* **298,** 374–377.

Martin, W. R. (1967). Opioid antagonists. *Pharmacol. Rev.* **19,** 463–521.

Martin, W. R. (1983). Pharmacology of opioids. *Pharmacol. Rev.* **35,** 283–323.

Martin, W. R., Eades, C. G., Thompson, J. A., Huppler, R. E., and Gilbert, P. E. (1976). The effects of morphine- and nalorphine-like drugs in the non-dependent and morphine-dependent chronic spinal dog. *J. Pharmacol. Exp. Ther.* **197,** 517–532.

Matzel, L. D., and Miller, R. R. (1987). Recruitment time of conditioned opioid analgesia. *Physiol. Behav.* **39,** 135–140.

Mayer, D. J., and Liebeskind, J. C. (1974). Pain reduction by focal electrical stimulation of the brain: An anatomical and behavioral analysis. *Brain Res.* 68, 73–93.

Mayer, D. J., Wolfle, T. L., Akil, H., Carder, B., and Liebeskind, J. C. (1971). Analgesia from electrical stimulation in the brainstem of the rat. *Science* 174, 1351.

McGinty, J. F., Henriksen, S. J., Goldstein, A., Terenius, L., and Bloom, F. E. (1983). Dynorphin is contained within hippocampal mossy fibers: Immunochemical alterations after kainic acid administration and cochicine-induced neurotoxicity. *Proc. Natl. Acad. Sci. U.S.A.* 80, 589–593.

Melzack, R. (1973). "The Puzzle of Pain." Basic Books, New York.

Melzack, R., and Wall, P. D. (1965). Pain mechanisms: A new theory. *Science* 150, 971–978.

Melzack, R., and Wall, P. D. (1975). Psychophysiology of pain. *In* "Pain: Clinical and Experimental Perspectives" (M. Weisenberg, ed.), pp. 8–23. Mosby, St. Louis, Missouri.

Messing, R. B., Vasquez, B. J., Spiehler, V. R., Martinez, J. L., Jr., Jensen, R. A., Rigter, H., and McGaugh, J. L. (1980). ^{3}H-dihydromorphine binding in brain regions of young and aged rats. *Life Sci.* 26, 921–927.

Messing, R. B., Vasquez, B. J., Samaniego, B., Jensen, R. A., Martinez, J. L., Jr., and McGaugh, J. L. (1981). Alterations in dihydromorphine binding in cerebral hemispheres of aged male rats. *J. Neurochem.* 36, 784–787.

Millan, M. J., and Herz, A. (1985). The endocrinology of the opioids. *Int. Rev. Neurobiol.* 26, 1–83.

Millan, M. J., Przewlocki, R., Jerlicz, M., Gramsch, C., and Herz, A. (1981). Stress-induced release of brain and pituitary beta-endorphin: Major role of endorphins in generation of hyperthermia, not analgesia. *Brain Res.* 208, 325–338.

Millan, M. J., Millan, M. H., and Herz, A. (1983). Contribution of the supra-optic nucleus to brain and pituitary pools of immunoractive vasopressin and particular opioid peptides, and the interrelationships between these, in the rat. *Neuroendocrinology* 36, 310–319.

Miller, R. J. (1986). Peptides as neurotransmitters: Focus on the enkephalins. *In* "Neuropeptides and Behavior, Volume 1: CNS Effects of ACTH, MSH and Opioid Peptides" (D. de Wied, W. H. Gispen, and Tj. B. van Wimersma Greidanus, eds.), pp. 95–136. Pergamon, Elmsford, New York.

Motta, M., Mangili, G., and Martini, L. (1965). A "short" feedback loop in the control of ACTH secretion. *Endocrinology (Baltimore)* 77, 392–395.

Nagase Shain, C. (1984). Shock-induced analgesia: A review and synthesis. Unpublished manuscript, Department of Psychological Sciences, Purdue University, West Lafayette, Indiana.

Nagase Shain, C. (1985). Genetic differences in stress-induced analgesia: Covariation with avoidance learning. Unpublished Ph.D. Dissertation, Department of Psychological Sciences, Purdue University, West Lafayette, Indiana.

Nagase Shain, C., and Brush, F. R. (1986). Genetic differences in avoidance learning covary with non-opioid stress-induced analgesia. *Ann. N. Y. Acad. Sci.* **467**, 416–418.

Nakanishi, S., Inour, A., Kita, T., Nakamura, M., Chang, A. C. Y., Sohen, S. N., and Numa, S. (1979). Nucleotide sequence of cloned cDNA for bovine corticotropin-B-LPH precursor. *Nature (London)* **278**, 423–427.

Noda, M., Furutani, Y., Takahashi, H., Toyosato, M., Hirose, T., Inayama, S., Nakanishi, S., and Numa, S. (1982). Cloning and sequence analysis of cDNA for bovine adrenal pre-proenkephalin. *Nature (London)* **295**, 202–206.

North, R. A., and Williams, J. T. (1983). Neurophysiology of opiates and opioid peptides. *In* "The Neurobiology of Opiate Reward Processes" (J. E. Smith and J. D. Lane, eds.), pp. 89–105. Elsevier, Amsterdam.

O'Callaghan, J. P., and Holtzman, S. G. (1975). Quantification of the analgesic activity of narcotic antagonists by a modified hot-plate procedure. *J. Pharmacol. Exp. Ther.* **192**, 497–505.

Oleson, T. D., Twombly, D. A., and Liebeskind, J. C. (1978). Effects of pain-attenuating brain stimulation and morphine on electrical activity in the raphe nuclei of the awake rat. *Pain* **4**, 211–230.

Page, R. B. (1982). Pituitary blood flow. *Am. J. Physiol.* **243**, E427–E442.

Palkovits, M., Brownstein, M., and Zamir, N. (1983). Immunoreactive dynorphin and alpha-neo-endorphin in rat hypothamo-neurohypophyseal system. *Brain Res.* **278**, 258–261.

Panocka, I., Marek, P., and Sadowski, B. (1986a). Inheritance of stress-induced analgesia in mice. Selective breeding study. *Brain Res.* **397**, 152–155.

Panocka, I., Marek, P., and Sadowski, B. (1986b). Differentiation of neurochemical basis of stress-induced analgesia in mice by selective breeding. *Brain Res.* **397**, 156–160.

Pasternak, G. W., and Snyder, S. H. (1975). Identification of novel high affinity opiate receptor binding in rat brain. *Nature (London)* **253**, 563–565.

Pert, C. B., and Snyder, S. H. (1973). Opiate receptor: Demonstration in nervous tissue. *Science* **179**, 1011–1014.

Pert, C. B., Snowman, A. M., and Snyder, S. H. (1974). Localization of opiate receptors binding in synaptic membranes of rat brain. *Brain Res.* **70**, 184–188.

Pierce, E. T., Foote, W. E., and Hobson, J. A. (1976). The efferent connections of the nucleus raphe dorsalis. *Brain Res.* **107**, 137–144.

Proudfit, H. K., and Anderson, E. G. (1975). Morphine analgesia: Blockade by raphe magnus lesions. *Brain Res.* **98**, 612–618.

Przewlocki, R., Hollt, V., Duka, T., Kleber, G., Haarmann, I., and Herz, A. (1979). Long-term morphine treatment decreases endorphin levels in rat brain and pituitary. *Brain Res.* **174**, 357–361.

Quirion, R., and Pert, C. B. (1981). Dynorphins: Similar relative potencies on mu-, delta-, and kappa-opiate receptors. *Eur. J. Pharmacol.* **76**, 467–468.

Reynolds, D. V. (1969). Surgery in the rat during electrical analgesia induced by focal brain stimulation. *Science* **164,** 444–445.

Rhodes, D. L. (1979). Periventricular system lesions and stimulation-produced analgesia. *Pain* **7,** 51–63.

Rhodes, D. L., and Liebeskind, J. C. (1978). Analgesia from rostral brain stem stimulation in the rat. *Brain Res.* **143,** 521–532.

Rivier, C., and Vale, W. (1983). Modulation of stress-induced ACTH release by corticotropin-releasing factor, catecholamines and vasopressin. *Nature (London)* **305,** 325–327.

Rodgers, R. J., and Randall, J. I. (1987). On the mechanisms and adaptive significance of intrinsic analgesia systems. *Rev. Neurosci.* **1,** 185–200.

Root-Bernstein, R. S. (1987). Catecholamines bind to enkephalins, morphiceptin, and morphine. *Brain Res. Bull.* **18,** 509–532.

Rossier, J., Vargo, T. M., Minick, S., Ling, N., Bloom, F. E., and Guillemin, R. (1977a). Regional dissociation of beta-endorphin and enkephalin contents in rat brain and pituitary. *Proc. Natl. Acad. Sci. U.S.A.* **74,** 5162–5165.

Rossier, J., French, E. D., Rivier, C., Ling, N., Guillemin, R., and Bloom, F. E. (1977b). Foot-shock induced stress increases beta-endorphin levels in blood but not brain. *Nature (London)* **270,** 618–620.

Samanin, R., and Benasconi, S. (1972). Effects of intraventricularly injected 6-OH dopamine or midbrain raphe lesions on morphine analgesia in rats. *Psychopharmacology* **25,** 175–182.

Samanin, R., Ghezzi, D., Nauron, C., and Valzelli, L. (1973). Effect of midbrain raphe lesions on the antinociceptive action of morphine and other analgesics in rats. *Psychopharmacologia* **33,** 365–368.

Satch, M., Akaike, A., Nakazawa, T., and Takagi, H. (1980). Evidence for involvement of separate mechanisms in the production of analgesia by electrical stimulation of the nucleus reticularis paragigantocellularis and nucleus raphe magnus in the rat. *Brain Res.* **104,** 525–529.

Schultz, R., Faase, T. E., Wuster, M., and Herz, A. (1978). Selective receptors for beta-endorphin in the rat vas deferens. *Life Sci.* **24,** 843–850.

Seiden, L. S. (1983). Foreword. *In* "The Neurobiology of Opiate Reward Processes" (J. E. Smith and J. D. Lane, eds.), pp. xii–xix. Elsevier, Amsterdam.

Siegel, S. (1986). Alcohol and opiate dependence: Reevaluation of the Victorian perspective. *Pap., Meet. East. Psychol. Assoc., 1986.*

Simantov, R., and Snyder, S. H. (1976). Morphine-like peptides in mammalian brain: Isolation, structure, elucidation, and interactions with the opiate receptor. *Proc. Natl. Acad. Sci. U.S.A.* **73,** 2515–2519.

Simon, E. J., Hiller, J. M., and Edelman, I. (1973). Stereospecific binding of the potent narcotic analgesic [^{3}H]etorphine to rat brain homogenate. *Proc. Natl. Acad. Sci. U.S.A.* **70,** 1947–1949.

Smith, J. E., and Lane, J. D., eds. (1983). "The Neurobiology of Opiate Reward Procecesses." Elsevier, Amsterdam.

Snow, A. E., and Dewey, W. L. (1983). A comparison of antinociception induced by foot shock and morphine. *J. Pharmacol. Exp. Ther.* **227**, 42–50.

Soper, W. Y. (1976). Effects of analgesic midbrain stimulation on reflex withdrawal and thermal escape in the rat. *J. Comp. Physiol. Psychol.* **90**, 91–101.

Strand, F. L., and Smith, C. M. (1986). LPH, ACTH, MSH and motor systems. In "Neuropeptides and Behavior, Volume 1: CNS Effects of ACTH, MSH and Opioid Peptides" (D. de Wied, W. H. Gispen, and Tj. B. van Wimersma Greidanus, eds.), pp. 245–272. Pergamon, Elmsford, New York.

Taber, E., Brodal, A., and Walberg, F. (1960). The raphe nuclei of the brainstem of the cat. I. Normal typography and cytoarchitecture and general discussion. *J. Comp. Neurol.* **114**, 239–260.

Terenius, L. (1973). Characteristics of the 'receptor' for narcotic analgesics in synaptic membrane fractions from rat brain. *Acta Pharmacol. Toxicol.* **33**, 377–384.

Terman, G. W., Lewis, J. W., and Liebeskind, J. C. (1982). Evidence for the involvement of histamine in stress analgesia. *Soc. Neurosci. Abstr.* **8**, 619.

Terman, G. W., Lewis, J. W., and Liebeskind, J. C. (1983a). Opioid and non-opioid mechanisms of stress analgesia: Lack of cross-tolerance between stressors. *Brain Res.* **260**, 147–150.

Terman, G. W., Lewis, J. W., and Liebeskind, J. C. (1983b). Multiple endogenous analgesia systems activated by stress in the rat. In "Current Topics in Pain Research and Therapy" (T. Yokota and R. Dubner, eds.), pp. 81–94. Am. Elsevier, New York.

Terman, G. W., Shavit, Y., Lewis, J. W., Cannon, J. T., and Liebeskind, J. C. (1984). Intrinsic mechanisms of pain inhibition: Activation by stress. *Science* **226**, 1270–1277.

Teschmaker, H., Opheim, K. E., Cox, B. M., and Goldstein, A. (1975). A peptide-like substance from pituitary that acts like morphine. *Life Sci.* **15**, 1771–1776.

Thomas, P. K. (1974). The anatomical substratum of pain. Evidence derived from morphometric studies on peripheral nerve. *Can. J. Neurol. Sci.* **1**, 92–97.

Tricklebank, M. D., Hutson, P. H., and Curzon, G. (1982). Analgesia induced by brief footshock is inhibited by 5-hydroxytryptamine but unaffected by antagonists of 5-hydroxytryptamine or by naloxone. *Neuropharmacology* **21**, 52–56.

Tricklebank, M. D., Hutson, P. H., and Curzon, G. (1984). Analgesia induced by brief or more prolonged stress differs in its dependency on naloxone, 5-hydroxytryptamine and previous testing of analgesia. *Neuropharmacology* **23**, 417–421.

Urban, I. J. A. (1986). Electrophysiological effects of peptides derived from pro-opiomelanocortin. In "Neuropeptides and Behavior, Volume 1: CNS Effects of ACTH, MSH and Opioid Peptides" (D. de Wied, W. H. Gispen, and Tj. B. van Wimersma Greidanus, eds.), pp. 211–244. Pergamon, Elmsford, New York.

Urca, G., Frenk, H., Liebeskind, J. C., and Taylor, A. N. (1977). Morphine and enkephalin: Analgesic and epileptic properties. *Science* **197**, 83–86.

Vale, W., Spiess, J., Rivier, C., and Rivier, J. (1981). Characterization of a 41-residue ovine hypothalamic peptide that stimulates secretion of corticotropin and beta-endorphin. *Science* **213**, 1394–1397.

Van Loon, G. R., and De Souza, E. B. (1978). Effects of beta-endorphin on brain serotonin metabolism. *Life Sci.* **23**, 971–978.

van Nispen, J. W., and Greven, H. M. (1986). Structure-activity relationships of peptides derived from ACTH, beta-LPH and MSH with regard to avoidance behavior in rats. *In* "Neuropeptides and Behavior, Volume 1: CNS Effects of ACTH, MSH and Opioid Peptides" (D. de Wied, W. H. Gispen, and Tj. B. van Wimersma Greidanus, eds.), pp. 349–384. Pergamon, Elmsford, New York.

van Ree, J. M., and de Wied, D. (1983). Behavioral effects of endorphins—Modulation of opiate reward by neuropeptides related to pro-opiocortin and neurohypophyseal hormones. *In* "The Neurobiology of Opiate Reward Processes" (J. E. Smith and J. D. Lane, eds.), pp. 109–145. Elsevier, Amsterdam.

Van Vugt, D. A., and Meites, J. (1980). Influence of endogenous opiates on anterior pituitary function. *Fed. Proc., Fed. Am. Soc. Exp. Biol.* **39**, 2533–2538.

Vincent, S., Hokfelt, T., Christensson, C., and Terenius, L. (1982). Immunohistochemical evidence for a dynorphin immunoreactive striato-nigral pathway. *Eur. J. Pharmacol.* **85**, 251–252.

Vyklicky, L. (1978). Techniques for the study of pain in animals. *In* "Advances in Pain Research and Therapy" (J. J. Bonica, J. C. Liebeskind, and D. G. Albe-Fessard, eds.), pp. 727–745. Raven Press, New York.

Wagner, A. R. (1981). SOP: A model of automatic memory processes in animal behavior. *In* "Information Processing in Animals: Memory Mechanisms" (N. E. Spear and R. R. Miller, eds.), pp. 5–47. Erlbaum, Hillsdale, New Jersey.

Watkins, L. R., and Mayer, D. J. (1982a). Involvement of spinal opioid systems in footshock-induced analgesia: Antagonism by naloxone is possible only before induction of analgesia. *Brain Res.* **242**, 309–316.

Watkins, L. R., and Mayer, D. J. (1982b). Organization of endogenous opiate and nonopiate pain control systems. *Science* **216**, 1185–1192.

Watkins, L. R., Cobelli, D. A., Farris, P., Aceto, M. D., and Mayer, D. J. (1982). Opiate vs. non-opiate footshock-induced analgesia (FSIA): The body region shocked is a critical factor. *Brain Res.* **242**, 299–308.

Watson, S. J., and Barchas, J. D. (1979). Anatomy of the endogenous opioid peptides and related substances: The enkephalins, beta-endorphin, beta-lipotropin and ACTH. *In* "Mechanisms of Pain and Analgesic Compounds" (R. F. Beers, Jr. and E. G. Bennett, eds.), pp. 227–237. Raven Press, New York.

Watson, S. J., Akil, H., Richard, C. W., III, and Barchas, J. D. (1978). Evidence for two separate opiate peptide neuronal systems. *Nature (London)* **275**, 226–228.

Wood, P. L. (1982). Multiple opiate receptors: Support for unique mu, delta and kappa sites. *Neuropharmacology* **21**, 487–492.

Woolfe, H. G., and MacDonald, A. D. (1944). The evaluation of the analgesic action of pethidine hydrochloride (Demerol). *J. Pharmacol. Exp. Ther.* **80,** 300–307.

Wuster, M., Schulz, R., and Herz, A. (1980). Inquiry into endorphinergic feedback mechanisms during the development of opiate tolerance/dependence. *Brain Res.* **189,** 403–411.

Yaksh, T. L. (1979). Direct evidence that spinal serotonin and noradrenaline terminals mediate the spinal antinociceptive effect of morphine in the periaqueductal gray. *Brain Res.* **160,** 180–185.

Yaksh, T. L., Yeung, J. C., and Rudy, T. A. (1976). Systematic examination of brain sites sensitive to direct application of morphine: Observation of differential effects within the periaqueductal gray. *Brain Res.* **114,** 83–103.

Yang, H.-Y. T., Hexum, T., and Costa, E. (1980). Opioid peptides in adrenal gland. *Life Sci.* **27,** 119–125.

Yoburn, B. C., Morales, R., Kelly, D. D., and Inturrisi, C. E. (1984). Constraints on the tailflick assay: Morphine analgesia and tolerance are dependent upon locus of tail stimulation. *Life Sci.* **34,** 1755–1762.

Zhang, A., and Pasternak, G. W. (1980). mu- and delta-opiate receptors: Correlation with high and low affinity opiate binding sites. *Eur. J. Pharmacol.* **67,** 323–324.

CHAPTER NINE

Temperature Regulation

Lawrence C. H. Wang and T. F. Lee

I. Introduction

The optimization of an internal thermal state for maximizing biochemical efficiency is of obvious survival value to organisms. In ectotherms (poikilotherms), the selection of a preferred thermal environment is accomplished mainly by behavioral means, whereas in endotherms (homeotherms), the maintenance of a relatively constant internal thermal state requires both behavioral and autonomic measures. Although they appear superficially divergent, both ectotherms and endotherms share many common neural elements governing thermoregulation; for instance, temperature transducers for the detection of ambient and internal thermal states, integrators or comparators for deciphering whether or not the internal thermal state is optimal, controllers for the activation or the inhibition of specific behavioral or autonomic measures, and effectors to carry out specific thermoregulatory tasks. To appreciate fully the neuroendocrine modulation of thermoregulatory functions, especially among the endotherms, it is perhaps beneficial to first have a firm grasp on the functional aspects of the individual neural elements followed by an understanding of their regulation by the neuroendocrine influences. Due to space constraint, only the most relevant points will be included, but recent reviews on specific topics will be listed wherever appropriate. Thermoregulation under modified physiological states, e.g., sleep, fever, and hibernation, will also be included to reflect species adaptations. An emphasis has been placed on neuroendocrine regulation of hibernation, not only because it represents an important adaptive strategy for energy conservation under adverse conditions in many birds and mammals, but also because it epitomizes the biochemical and physiological versatilities evolution is capable of amplifying to an existing thermoregulatory function.

A. Thermoreceptors

Nearly all excitable membranes are thermally sensitive. Changes in electrophysiological properties elicited by temperature are typically consequent of ionic or metabolic alterations, including changes in electrogenic sodium-pump activity and passive ionic permeability (Carpenter, 1981). Despite extensive studies into various systems, the molecular mechanisms for thermoreception remain unknown; however, changes in membrane proteins and/or the physical state of their lipid environments have been proposed as possible causes (see Hennessey *et al.*, 1983). In the most primitive unicellular organism, e.g., the paramecium, heat avoidance is thought to be initiated by a temperature-induced increase in graded generator potential, leading to a calcium-mediated action potential and corrective ciliary contractions (Hennessey *et al.*, 1983). In mammals, specialized thermosensory transducers exist which may be classified by their anatomical locations.

I. CUTANEOUS THERMORECEPTORS

The cutaneous thermoreceptors can be divided into two groups: cold receptors and warm receptors. Electron microscopy reveals that beneath each cold spot (receptor) there is a dermal papilla which contains a single, small, myelinated fiber with divided unmyelinated terminals penetrating into the basal layer of the epidermis (Hensel *et al.*, 1974). No similar structures can be identified beneath the warm spot (Hensel *et al.*, 1974); it is assumed that warm receptors are located in deeper layers of the skin. Thus, the identification of cold and warm receptors has mainly been based on afferent neural activity of single fibers following cold or warm stimulation to localized skin surfaces.

Cutaneous thermoreceptors are characterized by having both spontaneous static discharge and phasic discharge. The rates of static discharge depend on the skin temperature: cold receptors typically respond between 5 and 43°C with maximal rates between 25 and 30°C; warm receptors respond between 30 and 48°C with maxima between 40 and 46°C (Hensel, 1974, 1983). The rate of phasic discharge is positively, linearly related to stimulus intensity for both types of thermoreceptors at low intensities; at high intensities, the response is plateaued. The sensitivity of phasic response to temperature change seems better at lower rates (e.g., 0.4°C/sec) than it is at higher rates (e.g., 2°C/sec; Necker, 1981). The "cold" fibers belong to two major groups; the C-fiber, with a conduction velocity of less than 1 m/sec, and the thinly myelinated Aδ-fiber, with a conduction velocity of up to 40 m/sec (the "warm" fibers belong to the C-fiber group) (Necker, 1981). In cats, intravenous infusion of calcium decreases the static discharges of cold fibers (Schafer et al., 1982) but increases those of warm fibers (Hensel and Schafer, 1974). This differential effect is in contrast with the general depressing effect of calcium on other receptors, but the ionic mechanisms remain unknown. This, however, seems to provide a neurophysiological basis for the warm sensation experienced by humans following intravenous calcium administration.

2. DEEP-BODY THERMORECEPTORS (EXCLUDING CNS)

The characterization of deep-body thermoreceptors is based on changes of either afferent neural activity or thermoregulatory response consequent to heating or cooling of specific structures. Based on neural activities, the following structures are shown to be thermally sensitive: the carotid bodies (Gallego et al., 1979), the muscle spindle (Mense, 1978), the abdominal wall (Riedel, 1976), and the femoral vein (Thompson and Barnes, 1969). Based on thermoregulatory responses, the ruminal and intestinal wall of sheep (Rawson and Quick, 1972), the dorsal wall of abdominal cavity of rabbits (Riedel et al., 1973), the "inner body" of goats (Mercer and Jessen, 1978) including limb bones (Jessen et al., 1983), and the deep body of guinea pigs (Mercer and Simon, 1983) and ducks (Simon and Simon-Oppermann, 1979) are known to be thermally sensitive, although in many cases the specific structures have not been identified.

3. CENTRAL NERVOUS SYSTEM (CNS) THERMORECEPTORS

Using local heating or cooling of circumscribed brain structures and observing changes in (1) single-unit activities, (2) autonomic responses for heat loss or heat production, and (3) behavioral responses for heat loss or heat conservation/production, there is ample evidence to show that thermal-sensitive structures exist in the CNS in all vertebrate classes (see reviews by Hensel, 1974; Simon, 1974; Necker, 1981; Jessen, 1985). In mammals and birds, single-unit recordings have shown that thermal-sensitive neurons are most abundant in the preoptic–anterior hypothalamic area (POAH) and the spinal cord, but they also exist in the posterior hypothalamus, the midbrain and medulla, and the cortex. The impulse frequency versus temperature curve may be bell shaped, linear, or linear but with a threshold response and may be either facilitory or inhibitory (Necker, 1981). The slope may be either positive or negative, depending on whether the neurons are warm sensitive or cold sensitive. Neurons with linear responses are thought to be detective in function, similar to the cutaneous thermoreceptors, whereas neurons with threshold responses are thought to be integrative in function (interneurons) since thermoregulatory responses are characterized by a threshold (Eisenman, 1972; Necker, 1981). Modulations of firing activities of these thermosensitive neurons in the POAH by extrahypothalamic inputs (e.g., skin, spinal cord) have also been observed (Necker, 1981).

4. AFFERENT PROJECTIONS OF THERMORECEPTORS

The thermoafferent systems and their monoaminergic nature from seven main body areas, skin, spinal cord, lower brainstem, midbrain, hypothalamus, thalamus, and sensory cortex, have been extensively reviewed recently (Bruck and Hinckel, 1982). By recording along the sensory pathway of cutaneous thermoreceptors at different anatomical levels, it is possible to decipher the degree of signal processing during transmission. Using the scrotal inputs in the rat and recordings at the entry zone of the scrotal nerve in the dorsal horn of the spinal cord, the ventrobasal complex of the thalamus, and the SI somatosensory area of the cortex (Hellon and Mitchell, 1975), it was determined that thermal inputs synapse in the dorsal horn, project rostrally to the raphe nuclei (in particular, the nucleus raphe magnus) in the brainstem, and ascend to the ventrobasal thalamus, hypothalamus, and cerebral cortex. Similar pathways have also been found for thermal inputs from the trunk, limb, and abdomen (Hellon, 1983; Schingnitz and Werner, 1983; Hinckel and Schroder-Rosenstock, 1981; Nakayama et al., 1979). There is also evidence showing a convergence of thermal inputs from scrotal, back, and abdominal origins onto a single thalamic unit: the threshold for "switching response" of the thalamic unit to scrotal heating and cooling can be modulated by thermal inputs from the back and abdomen (Schingnitz and Werner, 1983). Since thalamic units are characterized by "switching responses," i.e., a steep increase in firing rate within a narrow temperature range (Hellon and Misra, 1973), the dynamic components of thermal inputs from the thermoreceptors are not reproduced. This indicates that there are certain degrees of central processing of thermal information in this afferent pathway before it reaches the thalamus and hypothalamus. In cats, rats, and rabbits, thermal inputs from the face are carried by the trigeminal nerve to the caudal trigeminal nucleus in the medulla and projected rostrally to the thalamus–cortex or independently to the hypothalamus (Dickenson et al., 1979). In contrast to the thermal inputs from the trunk and scrotum, the facial inputs are virtually unprocessed between the thermoreceptors and the thalamus: the dynamic and static components of the afferents are preserved. At the lower brainstem there appears to be a functional convergence of thermal inputs. Those from the cold receptors converge at the subcoeruleus area and ascend via the noradrenergic afferents to the hypothalamus to drive heat production, whereas those from the warm receptors converge at the nucleus raphe magnus and ascend via the serotonergic afferents to the hypothalamus to drive heat loss (Bruck and Hinckel, 1982).

5. WEIGHTING OF THERMAL INPUTS FROM DIFFERENT BODY REGIONS

The differential importance of thermal inputs from cutaneous, deep-body, and CNS sites in eliciting thermoregulatory responses appears to be different in mammals and birds. In mammals, the effectiveness of skin inputs versus hypothalamic inputs is greater in smaller animals than it is in larger animals. For instance, the ratio is 3.0 in rats but 0.3 in men (Cabanac, 1975). The lower ratio in the larger animal may reflect the importance of precisely sensing the internal temperature to avoid a large heat deficit or heat surplus due to the larger thermal inertia (Cabanac, 1975). However, there are exceptions; for instance, the cutaneous inputs from the scrotum of the ram (Waites, 1962) and pig (Ingram and Legge, 1972) are much more powerful than the deep-body inputs in modulating heat loss and heat production responses. Within the CNS, the thermal inputs from the POAH and the spinal cord are about equally potent in driving heat production in the dog (Mercer and Jessen, 1980), but in the goat, the POAH inputs are about twice as potent as the spinal inputs (Jessen, 1981). Comparing the thermosensitivity of POAH to increasing heat production in mammals of varying sizes, an inverse relationship seems apparent (Heller, 1983). Whether this is related to the inherent relationship between mass-specific metabolic rate and body size is not known.

In birds, thermal inputs from the skin seem to play only a minor role in thermoregulation (Cabanac, 1975). However, in the pigeon, inputs from feathered skin in the back, wing, and breast (Necker, 1977) as well as from exposed areas such as the face are strongly influential (Schmidt, 1982). As to the CNS inputs, the spinal cord has the highest influence in the pigeon and the Peking duck. The thermosensitivity of the POAH is minimal or even nonhomeostatic, i.e., cooling of the POAH elicits decreased metabolic rate and vasodilatation of the feet, and neither response is homeostatic in raising body temperature (T_b) (Simon, 1974). However, an exception is found in the goose, in which heat production is driven by thermal inputs largely outside of the CNS (Helfmann et al., 1981).

In view of the opposite responses to POAH cooling between mammals and birds even though both possess similar thermosensitive neurons in the POAH, the functional significance of these neurons in their traditional detective and integrative roles on thermoregulation (e.g., Hammel, 1965) has recently been questioned (Necker, 1981; Stitt, 1983). Although thermosensitive neurons have been found in the cortex (Barker and Carpenter, 1970) and dorsal horn of the spinal cord (Necker, 1975), neither area has been shown to exert thermoregulatory influence. In the posterior hypothalamus, which upon cooling gives an opposite response to that of cooling the POAH (Puschmann and Jessen, 1978), the mere presence of thermosensitivity in neurons cannot be construed as being functionally involved in thermoregulation (Stitt, 1983). Transection below the hypothalamus does not totally abolish thermoregulatory ability, and destruction of the POAH (and thus elimination of thermosensitive neurons), while sufficient to abolish most autonomic responses, does not impair behavioral thermoregulation (see Satinoff, 1983). Other observations also suggest the existence of extra-POAH controllers for thermoregulation in the neuraxis. More studies are needed to further clarify the pivotal roles thermosensitive neurons in the POAH play in thermoregulation.

B. CNS Integrator

1. NEURONAL MODEL OF CNS THERMOREGULATION

The similarities between the regulation of temperature in a room and in the body have prompted thermal physiologists to borrow many of the concepts and terminologies used by engineers for feedback control processes. Thus, terms such as input, output, set-point, controller and controller gain, and controlled variable and controlled system have been utilized in the construction of models which would provide a conceptual and practical understanding of how thermoregulation is accomplished. The subject has been reviewed many times in recent years (e.g., Bligh, 1973, 1979; Satinoff, 1978;

South *et al.*, 1978; Stitt, 1980); therefore, only a summary of its current status will be presented here.

a. The "Adjustable Set-Point with Proportional Control" Model First proposed by Hammel *et al.* (1963) and later elaborated by Hammel (1965, 1968, 1983), this model may be summarized by the following equation:

$$R - R_o = -\alpha \, (T_h - T_{set})$$

where $R - R_o$ = the thermoregulatory response to an error signal $(T_h - T_{set})$

   R_o = the basal level of response in a neutral environment

   α = a proportional constant of the thermoregulatory response (the slope or thermosensitivity)

$T_h - T_{set}$ = the difference between the actual hypothalamic temperature and the set-point temperature, equivalent to the error signal of a controller

The major features of this model are (1) the hypothalamic temperature is the controlled variable as well as the feedback element, since it is held constant with only minute fluctuations; (2) the thermosensitive neurons (both high Q_{10} and low Q_{10}) (see also Section I,B,1,c below) in the POAH are responsible for the generation of a "set-point" temperature, the level of which may be functionally adjusted by other thermal or nonthermal (e.g., pyrogen) inputs as well as the arousal state (sleep versus wakefulness); and (3) thermally insensitive neurons activate or inhibit efferent pathways for heat loss or heat conservation/production; the set-point for each response is different and may be independently varied by thermal or nonthermal inputs. However, once initiated, the magnitude of response to temperature change is constant, thus the term *proportional control*. In addition to its simplicity and high experimental predictibility, it also represents a significant conceptual advancement in that it is the hypothalamic thermosensitivity (the adjustable set-point) rather than the controlled variable (hypothalamic temperature) that is being modulated in the maintenance of internal thermal stability against varying thermal or nonthermal perturbations.

b. The "Adjustable Gain" Model Several investigators (Stolwijk and Hardy, 1966; Mitchell *et al.*, 1970; Bruck and Wunnenberg, 1970; Bruck and Hinckel, 1982; Stitt, 1980) have proposed this model, which shares many of the same features as the "adjustable set-point with proportional control" model; for example, the concept of set-point or "threshold" (i.e., a range of

T_b within which no thermoregulatory response is required), error signal, and the modulation of a central integrator by peripheral thermal and nonthermal inputs. However, some differences exist: (1) the controlled variable is several local temperatures rather than a single representative one; (2) the threshold temperature is an averaged T_b rather than a single local temperature; and (3) the thermoregulatory response (e.g., shivering, nonshivering thermogenesis) is not of constant magnitude but depends on the thermal inputs from peripheral (skin) and CNS thermodetectors. Mathematically, this model can be represented as follows:

$$R - R_o = k \, (T_{ho} - T_h) \, (T_{so} - T_s)$$

where $R - R_o$ = net thermoregulatory response beyond basal, R_o

T_{ho} and T_{so} = respective reference (not set-point) temperatures for hypothalamus and skin

T_h and T_s = actual temperatures of hypothalamus and skin

k = a proportional constant for the response

Since the thermoregulatory response results from error signals generated from two sets of thermodetectors, the effect is multiplicative, instead of additive as in Hammel's model. The proportional constant, k, which reflects the thermosensitivity of the hypothalamic controller, is a variable, thus the "adjustable gain" connotation. The gain (or thermosensitivity) of the controller depends on the level of mean skin temperature; the lower the mean skin temperature, the greater the hypothalamic thermosensitivity (Stitt, 1980). However, associated with the change in gain of the controller is also the change in threshold (set-point) temperature for initiating the response: the lower the mean skin temperature, the higher the hypothalamic temperature for increased heat production (Bruck and Schwennicke, 1971; Stitt, 1980).

c. A Combined "Constant Gain and Adjustable Gain" Model Several lines of evidence both at the cellular (Hellon, 1972; Necker, 1981) and at the whole-animal level (Stitt *et al.*, 1966; Hammel *et al.*, 1977) suggest that both additive and multiplicative controls may coexist in the central controller. For instance, thermosensitive neurons in the POAH may show changes in threshold only, gain only, or both threshold and gain in their spontaneous firing rate versus hypothalamic temperature curve in response to skin temperature change (Necker, 1981). Animal experiments have also shown changes in either threshold (Hellström and Hammel, 1967) or gain (Hammel *et al.*, 1977) or both (Heller and Henderson, 1976) in response to different skin inputs. These observations are in support of the model proposed by Boulant (1974), in which differential patterns of spontaneous firing of the POAH warm- and cold-sensitive neurons and their differential response to peripheral thermal influences (shift in threshold versus gain) have been incorporated. Recently, a new mathematical model has been proposed by Simon (1981) incorporating not only the POAH thermosensors but also the recognition of differential temperature sensitivity for intrahypothalamic synaptic transmission of peripheral thermal signals: high Q_{10} for cold and low Q_{10} for warm signals. The strength of this new model lies in its inclusiveness of both the adjustable set-point and the adjustable gain models, plus its possible general applicability to thermoregulation in both birds and mammals.

2. NEUROCHEMICAL MODEL OF CNS THERMOREGULATION

The discoveries that relatively high concentrations of norepinephrine (NE) (Vogt, 1954) and 5-hydroxytryptamine (5-HT) (Amin et al., 1954) exist in the hypothalamus in relation to other parts of the brain led to the speculation that these monoamines may be involved in the central control of T_b (von Euler, 1961). However, not until Feldberg and Myers (1963) injected adrenalin, NE, and 5-HT into the lateral ventricle (i.c.v.) of cats and observed a decrease of T_b by catecholamines and an increase by 5-HT was the role of monoamines in thermoregulation ascertained. Based on these observations, Feldberg and Myers (1964) proposed that the control of T_b depends on the balance between the release of endogenous catecholamines and 5-HT in the anterior hypothalamus which lies close to the ventral wall of the third ventricle. Subsequent studies (Feldberg and Myers, 1965) employing microinjections into discrete loci of brain confirmed that only the anterior hypothalamus is sensitive to manipulations by monoamines. Subsequent studies aiming to elucidate the neurochemical nature of central regulation of T_b include the use of various amines and their agonists and antagonists, ions, prostaglandins, pyrogens, anesthetics, peptides, and many others. These studies have been recently reviewed and summarized (e.g., Myers, 1974, 1976, 1980; Bligh, 1973, 1979; Hellon, 1975; Lomax and Schonbaum, 1979; Clark and Lipton, 1983). Two models are presented here: the first invokes the concept of a central set-point, whose anatomical site of origin and chemical specificity have been identified (Myers, 1980); and the second does not incorporate the concept of a central set-point (Bligh, 1979).

Having reviewed extensively the experimental evidence based on the manipulation of specific hypothalamic sites (i.e., much more anatomically discrete than the i.c.v. approach), Myers (1976, 1980) proposed a neurochemical scheme depicting the cellular mechanisms underlying CNS regulation of T_b in primates and cats and possibly in other mammals. The scheme envisages thermosensitive neurons in the anterior hypothalamus whose firing rates are modulatable by either 5-HT or NE (Beckman and Eisenman, 1970). The 5-HT-sensitive neurons are part of the pathway for heat conservation/production, whereas the NE-sensitive neurons are part of the pathway for heat loss. In the posterior hypothalamus, a set-point is maintained by an inherent ratio in the concentrations of sodium to calcium ions. An increase in the sodium-to-calcium ratio results in an elevation of the set-point, which is followed by increased heat production and T_b. A decrease in the sodium-to-calcium ratio has the opposite effect. Peripheral thermoreceptors send inputs via monoaminergic pathways (Hellon, 1975; Bruck and Hinckel, 1982) through mesencephalic projections to the anterior hypothalamus—the cold receptors to the 5-HT neurons for activation of heat conservation/production and the warm receptors to the NE neurons for activation of heat loss. The efferent pathways from the 5-HT neurons in the anterior hypothalamus to the posterior hypothalamus are cholinergic (ACh), as are the mesencephalic efferents from the posterior hypothalamus to the effectors. Other thermal (temperature of blood) and nonthermal (e.g., pyrogen, prostaglandins, anesthetics, hormones) inputs impinge upon the 5-HT or NE neurons to exert their influences, either through changes in rates of heat production or heat loss, or by altering the set-point.

The neurochemical model proposed by Bligh (1979) is based primarily on results from i.c.v. injections of putative neurotransmitters and their antagonists and agonists in sheep, goats, and rabbits. Similar to that proposed by Myers, this model concurs that there are thermosensitive neurons in the POAH which receive warm and cold inputs from the skin and spinal cord in addition to local thermal sensing. However, major differences exist: (1) The concept of set-point is unnecessary; the dynamic balance between the overlapping, bell-shaped activity–temperature curves of warm and cold sensors in POAH can accomplish the same purpose as a set-point through reciprocal inhibition between heat loss and heat production efferents. (2) 5-HT elicits heat loss and suppressed heat production, opposite to that described in Myers' model; the effect of NE is generally inhibitive of the prevailing thermoregulatory response at a particular ambient temperature; at high ambient temperatures, it inhibits evaporative heat loss resulting in increased T_b, and at low ambient temperatures, it inhibits heat production resulting in decreased T_b. However, the action of ACh is similar in both models; it increases heat production and decreases heat loss. (3) Other putative transmitters such as histamine, dopamine, GABA, glycine, aspartic acid, glutamic acid, and taurine are also found to be synaptically active, suggesting multisynaptic arrangements in efferent pathways. The strength of Bligh's model is that it is based on simple neuronal format and fundamental neurological principles (dynamic balance of sensors and reciprocal inhibition) and, more importantly, it is consistently able to accurately predict the experimental results. Its major weakness, however, is the inability of i.c.v. administration to decipher specific anatomical sites where putative neurotransmitters and synaptically active substances exert their effects.

In summary, no consensus on a neurochemical model for thermoregulation is currently available. The scheme proposed by Myers (1980) provides convincing evidence of an ionic set-point mechanism in the posterior hypothalamus, but how impinging thermal and nonthermal inputs onto the POAH are translated into ionic changes and set-point adjustments remains to be elucidated. The scheme proposed by Bligh (1979), although anatomically diffuse, appears to be experimentally predictive and resolving. The apparent species difference to central monoamines between primates and cats versus sheep, goats, and rabbits cannot be explained until more anatomically precise manipulations are available. Furthermore, in view of the thermoregulatory effects of many endogeneous peptides (see Section II,A,2), it is likely that the neurochemical models presented here are only some of the possible mechanisms which exist.

3. CNS REGULATION OF BEHAVIORAL THERMOREGULATION

Utilizing symmetrical operant responses to determine the specificity of a thermoregulatory behavior (i.e., increased bar pressing for heat and decreased bar pressing for cold), and in combination with specific CNS lesions, there is now ample evidence to show that the neural controls for operant and autonomic thermoregulation are separate (see reviews by Satinoff and Hendersen, 1977; Satinoff, 1978, 1983). For example, in rats, lesions of the preoptic area damage autonomic but not operant responses for maintenance of normal T_b (Lipton, 1968; Carlisle, 1969), and lesions of the lateral hypothalamus disrupt thermoregulatory operant but not autonomic regulation (Satinoff and Shan, 1971). Local temperature changes in the posterior hypothalamus do not elicit autonomic but do alter operant thermoregulation in the squirrel monkey (Adair, 1974) and the rat (Refinetti and Carlisle, 1986). Thus, not only are the operant and autonomic pathways separate, but one can also compensate for the deficit in the other (Adair, 1976; Satinoff and Hendersen, 1977). In view of the early evolutionary appearance of behavioral thermoregulation among the vertebrates (e.g., fish: Hammel et al., 1969; frog: Duclaux et al., 1973; and lizards: Hammel et al., 1967), it is apparent that behavioral thermoregulation has been highly adaptive to warrant its competitiveness against other evolving behavioral patterns in evolution (Satinoff, 1983).

4. NEURAL ORGANIZATION OF CNS INTEGRATOR

An important consideration of CNS regulation of T_b is whether a single central controller with multiple inputs and outputs is sufficient or whether many controllers (as many as there are thermoregulatory responses) are needed to explain the thermoregulatory functions observed. Based on the observation that pharmacological suppression of hypothalamic sensitivity to warmth with capsaicin eliminates all defenses against hyperthermia but has no effect on defense for cold (Jancso-Gabor and Szolcsanyi, 1970), Cabanac (1975) concluded that independent networks for cold and warm defense exist all the way from sensors to effectors. Based on the anatomically discrete control for autonomic and behavioral responses (see above), the varying degree of precision for T_b control at different levels of the CNS (Satinoff, 1983), and the conservative nature of evolution in preserving an existing system for shared functions (e.g., vasomotor control for blood-pressure regulation as well as for heat loss/conservation), Satinoff (1978, 1982, 1983) proposed a parallel, multicontroller model for Jacksonian hierarchical control in the neuraxis for thermoregulation. The essence of Satinoff's model (1982) may be summarized as follows: "1) there are separate channels from thermal detector to motor effector for every thermoregulatory response in an animal's repertoire; 2) these miniature thermoregulatory systems are located at several levels of the neuraxis from spinal cord to cortex; and 3) they are not independent of each other, rather, the activity of lower structures is facilitated and inhibited by those above." This model has gained wide acceptance by researchers in the field of thermoregulation (e.g., Simon et al., 1983; Heller, 1983; Hammel, 1983).

C. Effectors for Thermoregulation

Efferents from the CNS controllers activate the appropriate pathways involved in heat conservation/production and heat loss. These pathways are controlled by both somatic and autonomic nervous systems and are subject to neuroendocrine regulation.

1. EFFECTORS FOR THERMOGENESIS

Two types of thermogenesis are recognized (Himms-Hagen, 1981): (1) obligatory, which includes basal metabolic rate and the thermic effect of food (specific dynamic action), and 2) facultative, which includes shivering or physical activity and cold-induced and diet-induced nonshivering thermogenesis (NST). For thermoregulatory purposes, only facultative thermogenesis is adjustable.

a. Shivering Thermogenesis Shivering is involuntary. It is manifested by rhythmic, random, and uncoordinated but simultaneous contractions of the flexors and extensors. Shivering takes place in almost all muscles of the body with the exception of the facial, perineal, extraocular, and middle-ear muscles (Kawamura *et al.,* 1953). The masseters appear to be the first to shiver in the cold, followed by muscles of the neck and trunk and, lastly, those of the legs. Since no mechanical work is done during shivering, all energies released during muscle contraction (including those involved in excitation–contraction coupling and maintenance of ionic gradients) appear as heat. However, due to increased heat loss of maintaining perfusion to peripheral shivering muscles and the disruption of the boundary insulating air layer by muscle movements, the net heat gain by shivering is estimated at only 48% (Hardy, 1961). The primary center for shivering is located in the dorsomedial area of hypothalamus (Hemingway, 1963); from here, efferents pass through the mid- and hindbrain down the ventrolateral columns of the spinal cord and to the muscles. Shivering can be inhibited by inputs from the cerebral cortex, anterior hypothalamus, and ventromedial hypothalamus and can be inhibited or facilitated by inputs from the septal area (Hemingway, 1963). Based on studies in the guinea pig, the activation of shivering appears to be determined by both the skin and the spinal cord temperature in a multiplicative manner (Bruck and Wunnenberg, 1970). There is evidence to show that sympathetic influence may affect shivering through induced depolarization in skeletal muscle cells (Teskey *et al.,* 1975) and modification of neuromuscular transmission (Moravec and Vyskocil, 1976).

Heat produced during work substitutes the heat generated by shivering since both are produced by the same mechanism, namely muscle contraction. However, in large mammals (dogs, humans, horses), the increase in metabolism by exercise can be to 20 times that of the basal level (see Wang and Abbotts, 1981, for a review), whereas in humans, the increase in metabolism by shivering in intense cold is only about 5 times above basal (Glickman *et al.,* 1967). The precise reason for this discrepancy is presently unknown.

b. Nonshivering Thermogenesis Nonshivering thermogenesis is a heat-producing mechanism which does not involve muscle contraction. Cold-induced NST is a major contributor to thermoregulatory heat production in newborn, cold-acclimated, winter-acclimatized, and hibernating animals (Jansky, 1973). Diet-induced thermogenesis, which is stimulated by overeating (Rothwell and Stock, 1979), shares many of the neuroendocrine and biochemical mechanisms governing cold-induced NST. Its major contribution is to act as an energy buffer for the disposal of excess caloric intake due to overeating and thus to prevent obesity (Himms-Hagen, 1983). Therefore, the subsequent discussion on facultative NST for thermoregulation refers only to cold-induced NST.

In mammals, NST is mediated by NE (Hsieh and Carlson, 1957), and the major site for NST is brown adipose tissue (Foster and Frydman, 1978). In cold-acclimated rats, blood flow measured by radiolabeled microsphere indicates that at maximum thermogenesis more than a third of total cardiac output is directed to the brown fat even though the tissue is less than 2% of body weight (Foster and Frydman, 1978). Furthermore, with each passage through the brown fat, oxygen in the arterial blood is completely depleted (Foster and Frydman, 1978). Consequently, the brown fat can account for 65–80% of the total thermogenesis observed in cold-acclimated rats exposed to cold versus about 30% for warm-acclimated rats (Foster and Frydman, 1979). Among primates, NE-stimulated NST is greatest in cold-acclimated monkeys, who possess the largest amount of brown fat, and least in heat-acclimated monkeys, who lack brown fat (Chaffee and Allen, 1973). In adult humans who typically do not have brown fat, the NE-stimulated NST is very limited (7.5% increase above resting), even in Korean women divers (Ama) who experience severe daily cold-water stress in the winter (Kang et al., 1970). The disproportionally high oxidative capability of brown fat owes to its biochemical uniqueness which allows extensive uncoupling of oxidative phosphorylation and the production of heat instead of ATP. A brown fat-specific mitochondrial protein, thermogenin, which has M_r 32,000 and has specific binding sites for purine nucleotides (e.g., GDP, GTP, ADP, ATP) has been isolated (for reviews, see Cannon et al., 1981; Cannon and Nedergaard, 1983). Thermogenin forms a part of the natural protonophore of the inner mitochondrial membrane; when bound by GDP, the channel is impermeable to proton flow and the proton gradient generated by the electron transport chain provides the energy needed for ATP synthesis by oxidative phosphorylation (the coupled state of respiration). When stimulated by the sympathetic discharge, NE stimulates intracellular cAMP formation via the β-receptor + adenylate cyclase system. The activation of lipase by cAMP leads to lipolysis and the formation of free fatty acids. Free fatty acids competitively bind to thermogenin and displace GDP. Consequently, the proton channel becomes permeable to proton flow and a proton short-circuit is formed which dissipates the energy gradient into heat instead of ATP synthesis (the uncoupled state of respiration). The metabolism of fatty acids provides the substrate needed to sustain this uncoupled state. There is evidence that the amount of thermogenin present in the brown adipose tissue increases dramatically during the 3 weeks when cold acclimation takes place (Desautels *et al.*, 1978; Sundin and Cannon, 1980), and the elevated amount of thermogenin is maintained as long as the animal is in the cold. This is in agreement with the time course of increased thermogenic capabilities of the whole animal during cold acclimation.

Other mechanisms for cold-induced NST which have received major attention involve the role of (Na^+,K^+)-ATPase. Since the membrane permeability to Na^+ and K^+ in the brown fat and skeletal muscle is increased by NE or thyroid hormone (Horwitz, 1978), the compensatory increase in (Na^+,K^+)-ATPase activity in restoring the transmembrane ionic balance necessitates increased energy expenditure and thus heat generation. As will be discussed later (Section II,B,1), the quantitative contributions from these thermogenic mechanisms may be of only minor importance. In contrast to mammals, birds rely primarily on shivering thermogenesis to combat cold; an equivalent NE- or other hormone-mediated NST has not been demonstrated in birds despite exhaustive searches (Calder and King, 1974). Peripheral injection of NE results in hypothermia and suppression of metabolic rate (Hissa *et al.*, 1975; Koban and Feist, 1982). Although there are some indications that thyroid-mediated NST might exist in neonatal fowls (Freeman, 1977), this is still tentative.

2. EFFECTORS FOR THERMOLYSIS

a. Sweating   In the ox, horse, donkey, and llama, sweating in response to thermal stress is continuous (Allen and Bligh, 1969), but in sheep and goat (Allen and Bligh, 1969) and in man and infrahuman primates (Elizondo and Johnson, 1981), sweating is not continuous but cyclic and synchronous in discharge. The synchronous nature of sweat secretion over different parts of the body surface indicates that the efferent pathways for sweating are from a single central controller in the hypothalamus (Elizondo and Johnson, 1981). In horse, man, and rat, the innervation of the atrichial glands is sympathetic but cholinergic (Weiner and Hellman, 1960). In the sheep, goat, ox, horse, and llama, however, innervation appears to be adrenergic, since intravenous adrenalin elicits sweating in all these species (Bligh, 1973). The regulation of sweating by core and skin temperatures has been studied (e.g., Nadal *et al.*,

1971; Smiles *et al.*, 1976; Elizondo and Johnson, 1981). In the rhesus monkey, the rate of sweating is linearly related to the level of hypothalamic temperature when the skin temperature is held constant. At lower skin temperature, the rate of sweating remains constant but the threshold for sweating is shifted upward (Smiles *et al.*, 1976). These observations suggest that the brain and skin temperatures interact additively (linearly) in determining the threshold brain temperature for sweating and that the sensitivity of the sweat gland apparatus itself remains constant (Elizondo and Johnson, 1981).

To have maximal cooling effect, secreted sweat must be evaporated as quickly as it is formed. Thus, the ambient vapor-pressure deficit for water near the evaporating surface becomes a critical consideration for the thermolytic effectiveness of sweating. Sweating fatigue has been thought to be due to the accumulation of a liquid film on the skin. By wiping the skin dry, the rate of sweating is increased, but in dry air, sweating fatigue can be prevented (Cabanac, 1975). The rate of sweating can be affected by many other factors in addition to ambient water vapor-pressure deficit. For instance, dehydration lowers the rate of sweating (Greenleaf and Castle, 1971) but heat acclimation increases the rate, capacity, and sensitivity of the central controller for sweating and decreases the threshold temperatures (both skin and brain) for the onset of sweating (Elizondo and Johnson, 1981).

b. Panting   Panting in its most general definition is a raised respiratory ventilation across the moist surface of the upper respiratory tract in response to thermoregulatory requirements (Bligh, 1973; Hales, 1981). During exposure to heat, panting has been observed in reptiles (Templeton, 1970), birds (Dawson and Hudson, 1970), and most mammals. Since birds do not sweat, panting is a key thermolytic strategy; many birds show increased breathing frequency with increased ambient temperature; others, for example the rock dove and ostrich, pant at a fixed resonant frequency similar to that observed in dogs to reduce the energy cost of panting (Dawson and Hudson, 1970). In mammals under mild heat stress, panting is characterized by a rapid, shallow respiratory pattern. When T_b is elevated during severe heat stress, the rapid, shallow panting changes into a slower, deeper pattern called "second-phase breathing" (Hales, 1981). With second-phase breathing, a marked respiratory alkalosis develops, indicating an imbalance between thermoregulatory and respiratory demands on the respiratory system. The exact control of panting is currently uncertain (see Hales, 1981), but the combined thermal inputs from skin, core, and midbrain to the hypothalamus and the efferents to the bulbo-pontine respiratory center may be responsible for determining the rapid, shallow, slow, or deep patterns of panting (Whittow, 1971).

c. Other Strategies for Evaporative Cooling Thermoregulatory saliva secretion with or without spreading has been observed in dogs (Sharp *et al.*, 1969), rats (Hainsworth and Stricker, 1971), and in some marsupial mammals (Bartholomew, 1956). Wallowing in mud or its own urine has been found in the domestic pig, which does not sweat (Ingram, 1965). Gular fluttering, a rapid movement of air over the gular area (floor of mouth and anterior part of esophagus), which is thin and highly vascularized, is used for evaporative cooling in many birds, including owls, pelicans, and doves (Dawson and Hudson, 1970).

D. Evolution of Homeothermy in the Vertebrates

The deduction of a physiological trait which cannot be preserved and substantiated by fossil records requires cautious interpretation. Deliberation on the evolution of homeothermy has attracted much interest (Hammel, 1983; Bligh, 1973; Satinoff, 1978; Crompton *et al.*, 1978; Crawshaw, 1980; Hulbert, 1981); there appears to be general consensus on the major steps leading from ecto- and poikilothermy to endo- and homeothermy, but some disagreement on the chronology of these steps has occurred (e.g., see Hulbert, 1981). The relationship between body size and the scaling of metabolism among animals and plants and how this relationship might contribute to the evolution of endo- and homeothermy have been discussed extensively by McNab (1983) and, therefore, will not be elaborated further here.

The deployment of resistive (short-term, phasic) and/or capacitive (longer-term, acclimatory) adjustments to thermal challenges depends to a large extent on whether the animal can effectively dissociate itself from environmental constraints. Biochemical and physiological manifestations consequent to thermal acclimation or acclimatization are widespread in the animal kingdom (Hochachka and Somero, 1973; Somero, 1978; Cossins, 1982; Chaffee and Roberts, 1971), and they represent capacitive adjustments at the membrane and enzymatic levels which most likely involve neuroendocrine regulations. But the chronic and diffuse nature of these compensatory changes makes them more difficult to visualize than the resistive, phasic responses in formulating a possible evolutionary scheme on thermoregulation. Although only the latter aspects will be presented herein, it is important to recognize that the capacitive adjustments are very much part of a repertoire in the autonomic responses to thermal challenges in all animals.

The demonstrations that (1) fish select preferred water temperature when given the choice (e.g., Fry and Hochachka, 1970) but preoptic lesions abolish this response (Nelson and Prosser, 1979); (2) heating or cooling the forebrain (Hammel *et al.*, 1969) or stimulating it with biogenic amines (Green and Lomax, 1976) modulates the latency of escape from aversively warm or cold water; and (3) behavioral fever can be induced by pyrogens (Reynolds *et al.*, 1976) indicate that all necessary neural elements for sensory and motor responses already exist in the most primitive of the vertebrates. In the frog, behavioral thermoregulation (Lillywhite, 1970) and its modulation by thermal manipulation of the spinal cord and by pyrogen (Myhre *et al.*, 1977) have also been demonstrated. In reptiles, the shuttling behavior for regulation of T_b within a preferred range is well known (Templeton, 1970), and alteration of escape T_b in cold and heat has also been demonstrated by thermal manipulation of the forebrain (Hammel *et al.*, 1967) and by pyrogens (Bernheim and Kluger, 1976). In fish and amphibians, the ability to counter thermal challenges by autonomic means is very limited. The exceptional ability of certain fishes to maintain a thermal gradient between body and ambient up to 21°C (e.g., tuna and lamnid sharks; Carey and Teal, 1969a,b) is mainly through trapping of heat, a by-product of constant muscle contractions (swimming), by a countercurrent vascular arrangement (Carey and Teal, 1966). In reptiles, autonomic responses such as vasomotor adjustments have also been employed to facilitate the rate of T_b warming during basking and to reduce the rate of cooling while foraging underwater (Bartholomew and Lasiewski, 1965). Panting has been observed in some lizards as a means of evaporative cooling (Templeton, 1970), but to be effective, it requires vasomotor coordination. Only in the Indian python during brooding is a reptile capable of maintaining a relatively constant brood temperature, 7°C above ambient, through metabolic efforts (Hutchison *et al.*, 1966). One may thus deduce that the next important steps would be the acquisition of insulation for heat preservation and the ability to produce heat by oxidative metabolism.

Evidence from the fossil record suggests that the first mammals (small insectivores) exploited a vacant nocturnal niche in the 120 million years during the Jurassic and Cretaceous periods (Crompton *et al.*, 1978). These animals may have had a T_b some 10°C lower than the typical 37°C due to the cool nights and the animals' limited insulative and thermogenic capacities. Major radiation of mammals occurred at the beginning of the Tertiary period; the number of families increased from approximately 14 in the late Cretaceous period to at least 40 in the Palaeocene epoch of the Tertiary period (Crompton *et al.*, 1978). Many of the new species invaded diurnal niches, vacated by the extinction of the ruling reptiles. Since the midlatitude temperature during the early Tertiary period was some 10 to 15°C warmer than that of today (see Hammel, 1983), early diurnal mammals may have evolved more effective insulations than their nocturnal counterparts to limit heat influx in their attempt to combat intense solar radiation. However, tachymetabolism might not have evolved because of its high demand for energy intake and the danger of being overheated during activity (Hammel, 1983). The level of T_b in these diurnal forms might have been between 37 and 40°C, high enough to allow passive heat dissipation without invoking evaporative cooling, which is metabolically costly. Between the Eocene and Pliocene epochs, the earth's climatic temperature began to decline, culminating in the arrival of the Quaternary Ice Age. Since mammals already possessed insulation and shivering thermogenesis, they needed only to evolve tachymetabolism to extend their aerobic capacity for defense of their T_b against cooling and for physical exertion in the cold. However, it is not clear how a 4- to 5-fold increase in resting metabolism beyond that of the reptilian level could be accomplished, although the greater size of organs and the greater number of mitochondria and amount of mitochondrial membrane surface area for each organ found in present-day mammals might explain some of the differences in energetics (Hulbert, 1981). Finally, as to the cellular mechanisms for tachymetabolism, both shivering and NST are involved. It is apparent that all living homeotherms shiver, but only certain mammalian lineages have evolved NST. For example, none of the monotremes, marsupials, edentates, or pangolins have brown fat (Hulbert, 1981), the primary site of NST (see Section I,C,1,b above). In other placental mammals, brown fat is either present in the neonate or persists throughout adulthood, as in the hibernators. Since extant birds do not have brown fat, a dichotomic separation in utilizing this tissue for NST must have occurred very early in the evolution of homeothermy.

II. Hormonal Interactions on Temperature Regulation

A. Hormones and CNS Regulation of Body Temperature

I. BIOGENIC AMINES

Studies on the thermoregulatory effects of centrally administered monoamines, their agonists and antagonists, and drugs which influence their synthesis, degradation, uptake, and depletion have been numerous. These have been reviewed recently (Bligh, 1973, 1979; Myers, 1974, 1980, 1981; Lomax and Schonbaum, 1979) and a representative summary will be presented here.

a. Norepinephrine (NE) At neutral ambient temperature (20–25°C), NE given i.c.v. causes a fall in T_b in many species (e.g., cats, dogs, monkeys, and mice) but a rise in many others (e.g., sheep, goats, rabbits, and ground squirrels) and both a fall and a rise in at least one species, the rat. The NE effect is apparently via the α-receptors since phentolamine, an α-blocker, abolishes the NE response, whereas propranolol, a β-blocker, has no effect. When NE is injected into the POAH, a decrease in T_b has been observed in cats, monkeys, and hamsters, but both an increase and a decrease occurs in rats and rabbits. In cats and monkeys, the NE-induced hypothermia is due to both suppression of heat production (metabolic rate) and activation of heat loss (vasodilatation), indicating that both pathways are invoked by NE (Feldberg and Myers, 1965; Myers, 1980). In sheep, goats, and rabbits, i.c.v. NE causes hyperthermia at neutral and high ambient temperatures (20–40°C) by inhibiting heat loss; at low ambient temperature (5–10°C), however, NE causes hypothermia by inhibiting heat production. This also indicates that NE has dual roles in activating both heat-production and heat-loss pathways. Measurements of NE turnover in the whole brain, hypothalamus, and brainstem in rats indicate an increase of NE turnover regardless of whether the rats were at high (32–40°C) or low (−3–9°C) ambient temperatures (Bruinvels, 1979). Although this further suggests activation of the noradrenergic pathways under both heat and cold, the possibility that this could be a nonthermal stress response cannot be excluded. In fact, NE has recently been demonstrated to suppress both thermosensitive and insensitive neurons of explant cultures from the rat's POAH in a nonspecific way (Baldino, 1986).

Thermal inputs from the skin, spinal cord, and medulla are carried to the hypothalamus via noradrenergic afferents (Bruck and Hinckel, 1980). Fine adrenergic fibers are present in the vicinity of the third ventricle; depletion of their neurotransmitter by capsaicin abolishes the rat's ability to thermoregulate at high ambient temperatures (Jansco-Gabor et al., 1970), indicating a noradrenergic link between skin warmth sensors and heat-loss pathways in the rostral hypothalamus (Bruinvels, 1979). However, chemical lesioning of noradrenergic neurons in the POAH by 6-hydroxydopamine in rats (Van Zoeren and Stricker, 1977; Myers and Ruwe, 1978), sheep (Bligh et al., 1977), and guinea pigs (Behr et al., 1983) results in no impairment of thermoregulation, indicating the nonessential nature of adrenergic neurons in relaying thermal inputs from the periphery. Furthermore, in rats which had POAH lesions, the hypothermic effect of i.c.v. NE was enhanced, indicating that the POAH is not directly involved in mediating the NE-induced hypothermia (Cantor and Satinoff, 1976). Taken together, it may be tentatively concluded that the thermoregulatory responses elicited by NE are largely through inhibition of a prevailing efferent pathway depending on the ambient temperature, and that its action may not be mediated exclusively through the thermosensitive neurons in the rostral hypothalamus.

b. Dopamine (DA) The role of DA in CNS thermoregulation has been reviewed recently (Cox, 1979; Lee *et al.*, 1985b). The typical response of i.c.v. injection of DA is hypothermia, as observed in monkeys (Myers, 1966), mice (Brittain and Handley, 1967), rats (Bruinvels, 1970), cats (Burks, 1972), goats (De Roij *et al.*, 1978), guinea pigs (Amico *et al.*, 1976), pigeons (Hissa and Rantenberg, 1975), and hens (Scott and Van Tienhoven, 1974). However, in rabbits, DA elicits hyperthermia (Jacob and Girault, 1974). Administered directly to the POAH, DA or its agonists, such as apomorphine, also elicit hypothermia in cats (Ruwe and Myers, 1978), rats (Cox *et al.*, 1978), and fowls (Marley and Nistico, 1972). The hypothermic effect of DA is brought about by peripheral vasodilation, increased evaporative cooling, and suppression of heat production (Lin and Su, 1979).

Since DA may be converted metabolically to NE by hydroxylation, there exists the possibility that the thermoregulatory effect of DA may be via NE. This does not appear to be the case because (1) the anatomical specificity of DA- and NE-sensitive sites is discrete in rats (Cox and Lee, 1977) and cats (Ruwe and Myers, 1978); (2) in cats, push–pull perfusion of sensitive sites in POAH previously loaded with radioactive DA and NE releases only [^{14}C]DA but not [^{3}H]NE during heat stress, indicating selective activation of the DA pathway by heat (Ruwe and Myers, 1978); and (3) blocking of central DA receptors (Cox and Lee, 1977) abolishes or significantly reduces the hypothermic effects of systemically injected apormorphine, but blocking of NE receptors has no effect on apomorphine-induced hypothermia (Cox, 1979). Further, DA depresses the firing rates of cold-sensitive neurons in the cat's hypothalamus (Sweatman and Jell, 1977) and decreases the firing rate of cold-sensitive but increases that of warm-sensitive neurons in the perfused hypothalamic slice (Scott and Boulant, 1984), indicating differential selectivity of DA on CNS thermosensitive neurons. Taken together, there appears to be accumulating evidence that DA may have a physiological role in CNS thermoregulation, in particular, in the mediation of heat loss by altering the activities of thermosensitive neurons in the POAH areas.

c. 5-Hydroxytryptamine (5-HT; Serotonin) Although it was first shown by Canal and Ornesi in 1961 (see Jacob and Girault, 1979) that intracisternal injection of 5-HT (1 to 1.5 mg) causes hyperthermia in rabbits, early systematic efforts by Feldberg and Myers (1964) led to current understanding of the role of 5-HT in CNS thermoregulation (see reviews by Bligh, 1979; Jacob and

Girault, 1979; Myers, 1981). When injected i.c.v., 5-HT elicits hyperthermia in cats, dogs, and monkeys, but hypothermia in goats, oxen, mice, pigeons, and chickens, and both hyper- and hypothermia in sheep, rabbits, and rats. When injected intrahypothalamically, 5-HT causes hyperthermia in cats, monkeys, and rats, but hypothermia in rabbits and chickens (Jacob and Girault, 1979). It is presently not clear why there exist such apparent species differences in responses to 5-HT. Recently, Myers (1981) suggested that both the anatomical specificity in site application and the dosage of injection could be responsible for many of the species variations observed (see below).

Since 5-HT injected into the POAH shows the greatest sensitivity for eliciting thermoregulatory responses, it is possible that 5-HT might be modifying the activity of thermosensitive neurons situated in this region. However, tests of the effects of 5-HT on firing rates of thermosensitive neurons have been inconclusive. For instance, in cats and dogs, 5-HT either has no effect or inhibits both cold- and warm-sensitive neurons (Cunningham et al., 1967; Beckman and Eisenman, 1970; Jell, 1973, 1974). In rabbits, 5-HT increases the firing rates of warm-sensitive neurons and decreases the rates of cold-sensitive neurons in the POAH region; however, the opposite effects are seen in thermosensitive neurons located in the midbrain (Hori and Nakayama, 1973). Thus, no definitive cellular mechanisms can be attributed to the action of 5-HT on thermoregulation. The limited number of studies on the turnover rate of brain 5-HT in response to thermal challenges indicate that, in rats, 5-HT turnover is increased with exposure to heat (Reid et al., 1968; Weiss and Aghajanian, 1971; Simmonds, 1970), suggesting the serotonergic nature of the heat-loss pathways in this species. In contrast, the release of 5-HT in the hypothalamus and other brain areas in the monkey is increased during cold exposure (Myers and Beleslin, 1970, 1971), suggesting the serotonergic nature of the heat-production pathways in this species. It is apparent that further studies are needed to ascertain whether the role of 5-HT on thermoregulation is indeed species specific.

The effects of ambient temperature on thermoregulatory responses elicited by 5-HT have been reviewed by Jacob and Girault (1979). In general, the same response to 5-HT is observed at neutral temperatures (20–22°C) or lower, but at higher temperatures (30–40°C) of the 5-HT response is abolished.

In an attempt to resolve the apparent conflicting results the 5-HT response among different species, Myers (1981) proposed that many of the discrepancies may be explained by route and dosage of injection. In cats, which normally respond to POAH injection of 5-HT (1.25–2.5 μg) by a rise in T_b, a fall in T_b can be elicited by a larger dose (3.5 μg). An injection of NE (3.5

μg) into the same site also causes a similar fall in T_b. Furthermore, pretreatment with phentolamine, an α-adrenergic antagonist, abolishes the hypothermic effect of 5-HT. This is interpreted to mean that the hypothermic effect of 5-HT is mediated by the α-adrenergic receptors normally activated by NE; however, 5-HT at high doses may compete for such receptors and activate the heat-loss pathway associated with NE and cause hypothermia. Similar experiments using 5-HT (10.5 μg) and a dopaminergic antagonist, butaclamol, have also demonstrated that hypothermia evoked by 5-HT can be abolished by butaclamol pretreatment, again suggesting the competition of dopaminergic receptors by 5-HT and the activation of dopaminergic pathways for heat loss. However, as so many neurons and nerve terminals are included within the diffusion area after microinjection, the possibility that the hypothermic effect of 5-HT is mediated by a downstream α-adrenergic and/or dopaminergic system cannot be excluded. Therefore, careful investigations, including ones on the interactions between 5-HT and other monoamines, notably NE and DA, must be conducted before a definitive conclusion can be drawn as to the species specificity of 5-HT action on CNS thermoregulation.

 d. Acetylcholine (ACh)　The role of ACh in CNS thermoregulation has been recently reviewed by Crawshaw (1979) and Myers (1980). Injected by the i.c.v. route, ACh or carbachol elicits hyperthermia in sheep and goats, hypothermia in rabbits, rats, mice, and the echidna, and both hyper- and hypothermia in cats and pigeons (see review by Crawshaw, 1979). Since ACh has both nicotinic and muscarinic actions, the use of specific blockers, D-tubocurarine for nicotinic and atropine for muscarinic receptors, has further clarified the action of ACh. The hyperthermic action of ACh or carbachol in monkeys (Hall and Myers, 1972) and hamsters (Reigle and Wolf, 1975) appears to be via muscarinic activation of vasoconstriction and shivering. In young guinea pigs, the hypothermic action of ACh or carbachol on suppression of NST is nicotinic (Zeisberger and Bruck, 1973). In rats, the coordinated heat loss responses by ACh or carbachol (Crawshaw, 1979) appear to be muscarinic (Poole and Stephenson, 1979), and in cats, although both hypo- and hyperthermia have been observed by ACh (Rudy and Wolf, 1972), the hyperthermic effect is blocked by atropine (Baird and Lang, 1973). In rabbits, ACh has little effect on thermoregulation (Cooper *et al.*, 1965), but i.c.v. atropine causes hypothermia by suppression of oxygen consumption and shivering (Preston, 1974; Cooper *et al.*, 1976). Thus, both muscarinic and nicotinic mechanisms are involved in mediating heat conservation/production or heat loss depending on the species.

 In monkeys, the release of ACh from the POAH area is markedly enhanced during cold exposure and markedly reduced during warm exposure, suggesting cholinergic linking between the POAH and effectors for heat conservation and heat production (Myers and Waller, 1973). This is further substantiated by the increased release of ACh from the diencephalic and mesencephalic heat-production pathways following POAH injection of 5-HT, which is thermogenic in monkeys, and the decreased release of ACh following NE, which is thermolytic (Myers and Waller, 1975). A cholinergic pathway has also been proposed (Myers, 1976) which links the POAH to the caudal hypothalamus for alteration of the $Ca^{2+} : Na^+$ ratio and the level of set-point for T_b regulation.

The effects of ACh on thermosensitive neurons in the POAH have been varied and inconclusive. In cats, iontophoretically applied ACh exerts both excitatory and inhibitory influences (Jell, 1973, 1974), and in rats, both inhibitory (Beckman and Eisenman, 1970; Murakami, 1973) and excitatory (Murakami, 1973; Knox *et al.*, 1973) effects have also been observed. Thus, as is the case for 5-HT, no cellular mechanism may be provided presently for the action of ACh on CNS thermoregulation.

2. BRAIN PEPTIDES

Studies on thermoregulatory effects of brain peptides are recent events consequent to the recognition that many of these may act as central neurotransmitters or neuromodulators (see Snyder, 1980, for a review). An extensive survey on peptides and thermoregulation has been compiled by Clark (1979a; Clark and Lipton, 1985), and about 40 peptides from more than 200 studies have been included. In addition, Yehuda and Kastin (1980) and Clark and Lipton (1983) have also provided recent extensive reviews on peptides and thermoregulation. Therefore, only a selective list of peptides is included here, representing those which have been studied more systematically and those which may have possible physiological significance in thermoregulation.

a. Thyrotropin-Releasing Hormone (TRH) The CNS distribution of TRH and its neuropharmacological aspects have been reviewed recently (Nagawa, 1980). With respect to thermoregulation, Metcalf (1974) was the first to demonstrate that when injected i.c.v. in nanogram quantities TRH elicits hypothermia in the cat. However, the hypothermic response is preceded by tachypnea, salivation, growling, vomiting, and vasodilatation (Metcalf and Myers, 1976). Further studies in the cat indicate that neither the anterior hypothalamus nor the posterior hypothalamus, areas which are known to mediate physiological responses for thermoregulation, is sensitive to TRH. The only loci which are responsive to TRH are in the mesencephalon; upon stimulation, tachypnea, profuse salivation, vasodilatation, and consequently, profound hypothermia are observed (Myers *et al.*, 1977). It is apparent that, in cats, the TRH-induced hypothermia is due primarily to hyperventilation, an autonomic response triggered by TRH, but has little physiological significance in mediating thermoregulation.

Injected i.c.v. or into discrete brain sites, TRH has also been observed to elicit hyperthermia in rats (Cohn *et al.*, 1976; Brown *et al.*, 1977; Kalivas and Horita, 1980). The TRH-induced hyperthermia persists in hypophysectomized and thyroidectomized rats, indicating that the peptide's action is central; further, it is not mediated by monoamines but possibly by prostaglandins since adrenergic antagonists have no effect, yet indomethacin or acetylsalicylate completely abolishes the TRH-induced hyperthermia (Cohn *et al.*, 1980). Moreover, the inhibitory effect of TRH on the firing rate of POAH warm-sensitive neurons is much greater than it is to the cold-sensitive neurons, which show a rebound of increased activity following cessation of TRH application (Salzman and Beckman, 1981). Since warm-sensitive neurons presumably drive heat-loss responses and cold-sensitive neurons drive heat production, the hyperthermic effect of TRH may be due to the inhibition of heat loss but little or even no activation of the heat-production pathways. Thus, in rats, TRH may exert a physiological role in CNS thermoregulation.

Microinjected into the hippocampus, TRH (0.1–100 ng) elicits a dose-related decrease in T_b, accompanied by a fall in metabolic rate and quiescence in the euthermic ground squirrel; in the hibernating squirrel, however, 0.1 ng of TRH administered to the same site triggers arousal from hibernation (Stanton *et al.*, 1980). It is suggested that TRH may exert its thermoregulatory effect either through modification of hippocampal influence on POAH thermosensitive neurons or by alteration of CNS excitability (Stanton *et al.*, 1980). In this regard, the reversal of hypothermia induced by barbiturates, alcohol, β-endorphin, neurotensin, and bombesin by TRH (Yehuda and Kastin, 1980) may also be related to the analeptic actions of TRH.

b. Opioid Peptides The pentapeptides Met-enkephalin and Leu-enkephalin, and the 31-amino-acid peptide β-endorphin are endogenous opioids derived from two precursors, proenkephalin and the 91-amino-acid peptide β-lipotropin, which is a fragment of the larger proopiomelanocortin (POMC) (see also Chapter 8, this volume). Their distributions in the CNS are discrete; regions of high enkephalins are usually associated with low endorphin. High levels of enkephalins are found in the medulla, brainstem, limbic system, and hypothalamus except in the anterior and basal hypothalamus, whereas the endorphin concentration is highest in the pituitary but also high in the anterior and basal hypothalamus, septal area, and pons (Snyder, 1980). There are at least three opioid receptors known to date (Snyder, 1980; Clark, 1981): (1) the μ-receptor, with distinct preference for morphine and blocked by naloxone, (2) the δ-receptor, with selectivity for the enkephalins and their derivatives such as [d-Alanine²-d-Leucine⁵] enkephalin and D-Ala²-methionine-enkephalinamide, and which is relatively insensitive to naloxone blockade, and (3) the κ-receptor, whose agonists are ketocyclazocine, pentazocine, and nalorphine, and which, like the δ-receptor, is insensitive to naloxone blockade. In an attempt to evaluate the thermoregulatory role of each opioid receptor type, Geller et al. (1982, 1986) recorded the body temperature changes of the rat after i.c.v. injection of various opioids and opiates. From the results of these studies, they postulated a two-receptor model, with one (possibly μ) mediating an increase in body temperature and the other (possibly κ) being responsible for hypothermic responses. However, this two-receptor model can not fully explain previous findings that pretreatment with naloxone significantly reverses the hyperthermic effect of β-endorphin but is less effective on the hyperthermic effect of enkephalin (Clark, 1979b). It is thus evident that the exact thermoregulatory role of each opioid receptor type is far from being well delineated and requires further investigations using specific opioid receptor antagonists.

When given i.c.v., and in low doses, β-endorphin and the Leu- and Met-enkephalins, like morphine, elicit concerted thermogenic responses in the cat (Clark, 1979b), rat (Tseng et al., 1980; Thornhill and Wilfong, 1982), and mouse (Huidobro-Toro et al., 1979), resulting in elevated T_b in both hot and cold environment; however, larger doses elicit a depression of T_b (Clark, 1981) which may be related to general stress such as restraint (Clark and Lipton, 1983). The anatomical sites at which these peptides act are not fully identified, but the POAH (Martin and Bacino, 1979; Thornhill and Wilfong, 1982), nucleus accumbens, periaqueductal gray, and caudate nucleus are sensitive to the peptides (Tseng et al., 1980). Finally, whether there is a physiological role for endogenous opioid peptides in thermoregulation is currently unknown. Morphine is known to cause a central decrease in brain NE levels, an increase DA levels, and an increase in 5-HT turnover and ACh release (Clark, 1979b); the thermic effects of β-endorphin may be mediated through 5-HT or prostaglandin (Lin and Su, 1979; Martin et al., 1981). It is possible that by modulating the activities of these neurotransmitters, opioids could exert thermoregulatory influences under physiological conditions. However, since naloxone or naltrexone does not alter the level of resting T_b nor its level after pyrogen-induced fever, it is unlikely that endogenous opioids exert tonic or phasic influences on thermoregulation under normal and febrile conditions (Clark, 1981). But when given subcutaneously to rats under acute cold or heat exposure, naloxone or naltrexone decreases or increases their T_b, respectively (Holaday et al., 1978; Thornhill et al., 1980), suggesting a possible role of endogenous opioids in adapting to acute thermal stresses. The future availability of specific δ- and κ-receptor antagonists should help investigators to further understand the physiological role of endogenous opioids in thermoregulation.

c. Bombesin   Bombesin is a 14-amino-acid peptide originally isolated from the skin of the European frog, *Bombina bombina*. Mammalian bombesin has been isolated from the stomach of mammals and the brains of sheep and rats; high concentrations have been found in the hypothalamus and the medulla oblongata (Brown, 1981). In rats, bombesin administered i.c.v. or into the POAH elicits a dose-dependent decrease in T_b in the cold (Wunder et al., 1980; Pittman et al., 1980; Lin and Lin, 1986) due to the suppression of heat production and possibly the elevation of heat loss. The hypothermia may be related to the suppression of TSH secretion by bombesin (Brown et al., 1977). In a warm environment, bombesin elicits an increase in T_b, but the mechanism is not clear (Brown, 1981). The temperature-dependent change in thermoregulatory responses to bombesin leads to the suggestion that this peptide might produce poikilothermia in animals (Tache et al., 1980). However, recent studies demonstrate that bomebesin does not impair thermoregulatory function of the animal but interferes with both regulatory heat production and heat loss to elicit its hypothermic effect (Brown, 1982). Furthermore, the animal tends to choose the cooler environment (Kavaliers and Hawkins, 1981) or increase bar-pressing to escape radiant heat after administration of bombesin (Avery et al., 1981). These results indicate that bombesin may lower the central set-point to produce its thermoregulatory effect. In support of this suggestion, recent findings indeed demonstrate that bomebesin elicits its thermolytic effect by lowering the threshold body temperature (Jansky et al., 1986).

d. Somatostatin   Somatostatin is a 14-amino-acid cyclic peptide (SS-14) found both in the hypothalamus and in the gastrointestinal tract and the pancreas. It inhibits the release of pituitary growth hormone, TSH, and prolactin and the secretion of glucagon, insulin, and gastrin (Synder, 1980). Another 28-amino-acid somatostatin (SS-28) has been isolated from the bovine hypothalamus which has actions that are qualitatively different from the SS-14 and is not the precursor of SS-14 (Brown, 1981). This indicates that there may be several somatostatins which may act through various somatostatin receptors to elicit different biological effects (Brown, 1981). When SS-14, SS-28, or a somatostatin analog, Des AA1,2,3,4,5,12,13[D-Trp^8]-SS (ODT^8-SS) is administered into the rat brain, hyperthermia results due to the elevation of oxygen consumption (Brown, 1982). The site of action appears to be the anterior preoptic region (see Brown, 1981). The potency of ODT^8-SS in inducing hyperthermia is about 10 times greater than that of PGE$_2$, and indomethacin, a prostaglandin-synthesis blocker, has no effect on ODT^8-SS-induced hyperthermia, indicating the action is not mediated via the prostaglandins (Brown, 1981). The somatostatins, in particular ODT^8-SS, completely reverse the hypothermia produced in rats by bombesin, neurotensin, carbachol (ACh agonist), and apormorphin (DA agonist), but have no effect on the hypothermia induced by β-endorphin (Brown and Vale, 1980). The significance of these observations is presently unclear. Similarly, whether somatostatins are involved in the physiological mediation of thermogenic responses remains to be elucidated.

e. Arginine Vasopressin (AVP) Arginine vasopressin is a nonapeptide with antidiuretic properties in systemic action. It is synthesized by neurons whose cell bodies are located in the supraoptic nuclei with axons projecting into the neurophysis for release into the blood. In addition, AVP-containing neurons have been identified to project from hypothalamic nuclei to various autonomic centers in the brainstem and the spinal cord, for instance, the nucleus coeruleus, various vagal nuclei, the septum, the subfornical organ, the amygdala, and other areas (Synder, 1980). These widespread projections strongly suggest that AVP may serve as a putative neurotransmitter or neuromodulator in the CNS. Current evidence suggests that AVP may act as an endogenous antipyretic to decrease or limit the maximal extent of pyrogen-induced fever in sheep, guinea pigs, and rats and to suppress the development of fever in pregnant sheep and guinea pigs just before and after parturition and in the postnatal young (Veale *et al.*, 1981). Evidence supporting such a view is as follows: (1) AVP perfused through the septal area but not elsewhere in the hypothalamus suppresses the endotoxin-induced fever in the nonpregnant ewe in a dose-dependent manner but does not decrease the T$_b$ of nonfebrile animals (Cooper *et al.*, 1979); (2) the perfusion of anti-AVP in the septal area enhances fever in sheep (Veale *et al.*, 1981); (3) AVP is released from the septal area during fever in sheep (Cooper *et al.*, 1979); (4) fever is suppressed 4 days before and 30 hours after parturition in the ewe, during which time an inverse correlation between magnitude of fever and concentration of plasma AVP can be seen (Cooper *et al.*, 1979); (5) there is a significant reduction of pyrogen-induced fever in the pregnant guinea pig 7 days before and 10 hours after parturition and up to 2 days in the newborn (Zeisberger *et al.*, 1981); and (6) an increase in AVP immunoreactive materials has been observed in neuronal projections between hypothalamic–septal and hypothalamic–amygdaloid pathways in the pregnant guinea pig just before and after parturition (Merker *et al.*, 1980), indicating activation of vasopressinergic pathways concurrent with reduction of pyrogen-induced fever. Taken together, there is strong evidence that AVP may exert a physiological role in the negative modulation of fever in these species. However, i.c.v. infusion of AVP exerts no antipyretic effect in the febrile rabbit (Bernardini *et al.*, 1983) and monkey (Lee *et al.*, 1985a). Besides species difference, it is possible that the discrepancy may be due to different routes of administration. But the recent finding that an intraseptal injection of AVP in the febrile rabbit fails to attenuate fever (Bernardini *et al.*, 1983) may represent a caveat against the antipyretic role of endogenous AVP. Moreover, endotoxin fails to elicit fever in the homozygous Brattleboro rat deficient in AVP, but the febrile response can be restored by treatment of the animal with a replacement regimen of AVP (Veale *et al.*, 1982). As a result of this finding, AVP can also be suggested as a central mediator of fever.

f. Adrenocorticotropin (ACTH) and α-Melanotropin (α-MSH) The presence of ACTH and α-MSH in the POAH, septum, and other parts of the brain (see Murphy *et al.*, 1983) in addition to their original sites of discovery in the anterior and intermediate lobes of the pituitary, raises the question of whether these substances may have central effects on physiological functions, in particular thermoregulation (Lipton *et al.*, 1981). Structurally, ACTH and α-MSH are closely related; ACTH contains 39 amino acids, whereas α-MSH contains 13 but shares the same amino-acid sequence of ACTH (1-13). Injected i.c.v., ACTH (1-24) and α-MSH, at 1.25–5 μg, elicit dose-dependent hypothermia in rabbits at $T_a = 10 - 23°C$ by inhibiting both heat conservation and heat production (Lipton and Glyn, 1980; Lipton *et al.*, 1981). However, at a lower dose (0.25 μg) which has no hypothermic effect on normal T_b, both peptides reduce the magnitude of leukocyte pyrogen-induced fever in rabbits (Lipton *et al.*, 1981), squirrel monkeys (Lipton *et al.*, 1984), and guinea pigs

(Kandasamy and Williams, 1984). The defervescence produced by ACTH is a central action since it persists in adrenalectomized rabbits (Lipton *et al.*, 1981). More recently, it has been shown that both peripherally injected α-MSH (Murphy and Lipton, 1982) and α-MSH injected into the septal area (Glyn-Ballinger *et al.*, 1983) reduce fever, and the concentration of α-MSH is observed to increase in the septum but to decrease in the arcuate nucleus during fever (Samson *et al.*, 1981), suggesting an axoplasmic transport of α-MSH from cell bodies in the arcuate nucleus to fibers in the septum in response to fever. Furthermore, α-MSH injected i.c.v. in rabbits is found to be 2500 times more potent by weight than acetaminophen in reducing fever (Murphy *et al.*, 1983). Recently, the involvement of endogenous α-MSH has been indicated by the finding that i.c.v. pretreatment of the rabbit with α-MSH antiserum augments the febrile response to interleukin-1 (Shih *et al.*, 1986). Taken together, these observations strongly suggest that both ACTH (1-24) and α-MSH may act as endogenous antipyretics in limiting the magnitude of fever under pathological conditions.

g. Other Brain Peptides   A 13-amino-acid peptide, neurotensin (NT), is present throughout the brain with high concentrations in the hypothalamus, basal ganglia, and interpeduncular nucleus (Synder, 1980). The thermoregulatory effect of NT has not been extensively studied; results to date indicate that when introduced i.c.v. it elicits hypothermia in the monkey, rat, mouse, gerbil, guinea pig, and golden hamster but has no effect on the ground squirrel, woodchuck, rabbit, or pigeon (Metcalf *et al.*, 1980; Nemeroff *et al.*, 1980). The hypothermic effect of NT may involve the inhibition of heat production and this may be reversed by TRH (Nemeroff *et al.*, 1980; Brown and Vale, 1980). Recently, evidence has been accumulated to suggest that NT may act as a poikilothermic agent to induce temperature changes. After i.c.v. injection of NT, the body temperature of the rat follows the ambient in direct proportion to the magnitude of heat or cold exposure (Chandra *et al.*, 1981; Lee and Myers, 1983). Furthermore, NT has been shown to suppress the firing rate of both warm- and cold-sensitive neurons in the POAH of the rat (Hori *et al.*, 1986). Thus, whether NT plays a physiological role in thermoregulation is currently uncertain.

The manifestations of hot flush, or flash, in postmenopausal women include peripheral vasodilation which begins in the thorax and progresses to the head and neck, followed by sweating and a decrease in core temperature (Tataryn *et al.*, 1980). Since these changes appear to be coordinated thermoregulatory measures, it is possible that the hormonal imbalance after menopause might be the underlying trigger for hot flush. In human subjects, plasma follicle-stimulating hormone (FSH) concentration does not correlate with the onset of flush, but that of the luteinizing hormone (LH) does (Tataryn et al., 1980). Subsequent studies in rats show that peripheral vasodilation and decrease in core temperature are elicited when LH-releasing hormone (LHRH; GnRH) is injected into the POAH but not in other sites in the hypothalamus (Lomax et al., 1980). These preliminary studies suggest that LHRH may have a role in lowering the central set-point for thermoregulation.

3. STEROID HORMONES

The cyclic variation in basal T_b during the menstrual cycle is well known; the high-T_b interval in the luteal phase is associated with high plasma progesterone, while the low-T_b interval in the follicular phase shows a high estrogen/progesterone ratio (Vande Wiele et al., 1970). Using a behavioral test (Cunningham and Cabanac, 1971), women in their luteal phase show a T_{set} which is approximately 0.4°C higher than that in their follicular phase, indicating the elevation of T_{set} by progesterone. In other mammalian species, administration of progesterone results in a rise in T_b, whereas with estrogen administration, a fall in T_b is shown (Marrone et al., 1976). In ovariectomized rats, estradiol therapy lowers the T_{set} (Wilkinson et al., 1980) while both estradiol and progesterone increase heat loss in the cold (Carlisle et al., 1979), resulting in greater operant heat requirement. In lactating rats, T_b is elevated in the first two weeks postpartum and tolerance to heat stress is decreased (Knecht et al., 1980), possibly due to an elevated set-point for evaporative cooling by progesterone whose secretion is stimulated by prolactin (Woodside et al., 1981). Finally, in the guppy, the preferred T_b in adult females and juveniles is approximately 4°C lower than that chosen by adult males; however, testosterone treatment in the former group abolishes this difference (Johansen and Cross, 1980). Taken together, the sex hormones appear to exert thermoregulatory effects in diverse vertebrate groups both through changes in central T_{set} and in effector responses.

B. Hormones and Thermogenesis

I. SYMPATHETICO−ADRENO−MEDULLARY SYSTEM (SAM)

The contribution of the SAM to thermogenesis has been reviewed many times (Himms-Hagen, 1967, 1975, 1983; Gale, 1973; Wunder, 1979;

Young and Landsberg, 1979). The major source of NE is the sympathetic noradrenergic fiber and it acts predominantly as a neurotransmitter rather than as a circulating hormone. A minor source of NE is the adrenal medulla, which also provides all the circulating epinephrine (E). In goats (Andersson *et al.*, 1963), pigs (Baldwin *et al.*, 1969), and baboons (Proppe and Gale, 1970), cooling of the POAH elicits increased catecholamine (CA) excretion at neutral or warm ambient temperatures, indicating the activation of the SAM via central thermosensitive neurons even in the absence of a peripheral cold stimulus. During cold exposure, the sympathetic nervous system appears to be activated first, with increasing adrenomedullary secretion as cold persists (Leduc, 1961). This latter contribution is not mandatory since demedullated rats can be successfully acclimated to cold (Kiang-Ulrich and Horvath, 1979). The general activation of the SAM is indicated by increased CA excretion and synthesis (Shum *et al.*, 1969) and increased tissue NE turnover rates in both birds (Koban and Feist, 1982) and mammals (Young and Landsberg, 1979; Young *et al.*, 1982). The overall contributions of CA in thermogenesis can be summarized to include the following five aspects: (1) muscular and cardiovascular effects for heat conservation and redistribution of blood flow, (2) metabolic effects largely due to enhancement of substrate mobilization, (3) effects of CA on secretion of other hormones which act on the same targets as CA, (4) direct effect of CA on target cells, and (5) regulation of NE-stimulated NST.

The muscular and cardiovascular effects of CA for heat conservation are well known and will not be elaborated further. Suffice it to say that piloerection, reduced blood flow to splanchnic, renal, and peripheral vascular beds, and increased perfusion of heart, brain, skeletal muscle, liver, and brown fat (if present) are typical CA effects during cold exposure. The timely supply of substrates to metabolizing cells has been demonstrated to be the most critical limiting factor in fully eliciting maximum thermogenesis and preventing hypothermia in rats (Wang, 1980, 1981; Wang and Anholt, 1982). In this respect, CA exerts both local and systemic influences. The stimulation of glycogenolysis in cardiac muscle is a local effect since the end product is utilized *in situ*. On the other hand, glycogenolysis in liver and lipolysis in white adipose tissue are systemic effects since the fuels are released into circulation and utilized globally. The lipolytic effect of NE in brown fat may be both local and systemic (Nedergaard, 1982). The CA-stimulated glycogenolysis in liver has long been thought to be mediated by the α-adrenergic–adenyl cyclase–cAMP system; however, recent evidence suggests that an α-adrenergic activation of glycogen phosphorylase also exists which involves changes of Ca influx but is independent of the adenyl cyclase–cAMP system (Assimacopoulos-Jennet *et al.*, 1977). Thus, α- and β-adrenergic effects may act in synchrony rather than in opposition in the regulation of intermediary metabolism (Young and Landsberg, 1979). Glycogenolysis in skeletal muscle as stimulated by CA releases glucose for local use only when muscles lack glucose-6-phosphatase. However, glycolysis of the glucose-6-phosphate in muscle produces lactate, which enters circulation and can be used by the heart or as a precursor for hepatic gluconeogenesis via the Cori cycle. Gluconeogenesis is stimulated by CA via both the formation of lactate, glycerol from lipolysis, and three-carbon amino acids from protein degradation, and a direct stimulation on the hepatocyte via an α-adrenergic mechanism which is independent of the adenyl cyclase–cyclic AMP system (Tolbert *et al.*, 1973). The lipolysis stimulated by CA is a β_1-adrenergic response in both the white and brown adipose tissues (Frisk-Holmberg and Ostman, 1977; Bukowiecki *et al.*, 1980) and α-adrenergic stimulation is inhibitory on lipolysis (Rosenqvist, 1972). *In vivo* kinetic studies show that CA increases the turnover rates of glucose and free fatty acids, indicating increased substrate mobilization and utilization (Young and Landsberg, 1979). However, in demonstrating both the glycogenolytic and lipolytic effects of CA, the effective doses used *in vivo* and *in vitro* are several orders of magnitude greater than the normal circulating levels of CA; this indicates that the NE locally released by the sympathetic nerves may play a more prominent role than the circulating CA in eliciting these metabolic responses. In this regard, adrenergic innervations, in addition to blood vessels, also extend to cells of white and brown adipose tissues and to liver, kidney, and endocrine cells such as the thyroid follicles, pancreatic islets, and juxtaglomerular apparatus, suggesting direct sympathetic regulation of metabolic functions subserved by these structures (Young and Landsberg, 1979. As to the effects of CA on the secretion of other hormones, Young and Landsberg (1979) have provided a detailed review. To summarize, CA stimulates the secretion of glucagon via a β-receptor mechanism but inhibits the secretion of insulin by an α-receptor mechanism; acting together, a maximum hepatic glucose output can be realized. Via the β-receptor mechanism, CA also stimulates the release of thyroid hormones, parathormone, calcitonin, renin, and gastrin, thereby affecting basal metabolism, calcium regulation, water and electrolyte balance and hemodynamics, and gastric secretion, respectively. In addition, recent observations also suggest that insulin may stimulate the central sympathetic outflow (via the ventromedial hypothalamus) and the satiety center, since overeating (increased plasma glucose and insulin) stimulates, while fasting (decreased glucose and insulin) depresses, sympathetic activity (Landsberg and Young, 1981). This is further supported by the observation that overfeeding alone does not improve NST in diabetic rats, but a single supplement of insulin elevates their NST to the same level as observed in normal, overfed rats (Rothwell and Stock, 1981).

The direct action of CA, in particular that of NE on thermogenesis, may involve the increase of membrane permeability to Na^+ and K^+ in the brown adipocyte and skeletal muscle (Horwitz, 1978; Clausen and Flatman, 1977). Since an ionic imbalance across the cell membrane activates the (Na^+, K^+)-ATPase for restoration of resting membrane potential, heat production is increased and respiration is stimulated due to ATP hydrolysis. Increased ouabain-sensitive respiration in the liver and muscle (Guernsey and Stevens, 1977; Horwitz and Eaton, 1977) and increased (Na^+,K^+)-ATPase activity in the liver (Videla et al., 1975) and muscle (see Himms-Hagen, 1978) have been observed in cold-acclimated rodents. However, the quantitative significance of this thermogenic mechanism in the intact animal is currently uncertain (Himms-Hagen, 1983). This is because ouabain inhibits not only (Na^+,K^+)-ATPase but also other cellular processes involved in mediating NST, consequent to the disturbance of intracellular Na^+, K^+, and Ca^{2+} concentrations. Thus, the inhibition of lipolysis in the white adipocytes (Fain and Rosenthal, 1971), of NE-stimulated adenyl cyclase in the brown fat (Fain et al., 1973), and of α-adrenergic-stimulated gluconeogenesis in the kidney (Saggerson and Carpenter, 1979) have all been observed to be causes by ouabain. Furthermore, since the expression of NST does not require visceral organs, including the liver (Depocas, 1958), and the contribution of NST from skeletal muscle is generally low in cold-acclimated (Grubb and Folk, 1976) or winter-acclimatized (Wickler, 1981) animals, it may be tentatively concluded that the direct effect of NE on thermogenesis via altered membrane permeability is perhaps not substantial.

The role of sympathetic influence or NE in mediating NST in cold-acclimated animals was discussed earlier (see Section I,C,1,b). Recent evidence indicates that during cold acclimation, hypertrophy of brown fat (increased total DNA, protein, and cytochrome oxidase contents) (Bukowiecki et al., 1982), increased synthesis of M_r 32,000 thermogenin (Cannon and Nedergaard, 1983), and increased GDP binding sites on thermogenin (Desautels et al., 1978) are typical. It has been demonstrated that these fundamental trophic events underlying NST are all mediated by the sympathetic outflow during cold acclimation (Mory et al., 1982).

2. THYROID HORMONES [THYROXINE (T_4) AND TRIIODOTHYRONINE (T_3)]

The role thyroid hormones play in thermogenesis has long been recognized as important and critical. This view stems from earlier observations that thyroidectomized or hypothyroid rats are unable to survive in the cold due to impaired heat-production capability (Hsieh, 1962; Sellers *et al.*, 1971), which is possibly due to reduced lipolysis (Rosenqvist, 1972). Later, Ismail-Beigi and Edelman (1970, 1971) demonstrated that the increased thermogenesis in thyroid hormone-treated animals is due to increased (Na^+,K^+)-ATPase activity since ouabain significantly blocks the T_3-induced increase in tissue respiration. This leads to the suggestion that thyroid hormones mediate part or all of cold-induced NST (Edelman, 1974). Further evidence suggesting the role of the thyroid in thermogenesis includes (1) the elevated plasma T_3 level in cold-acclimated animals, which is primarily due to accelerated conversion from T_4 to T_3 (Balsam and Sexton, 1975), (2) increased state 4 and 3 mitochondrial respirations in T_4-treated rats (Shear and Bronk, 1979), and (3) increased mitochondrial mass after T_4 treatment (Wooten and Cascarano, 1980).

Evidence against the idea of thyroid-mediated NST of cold-acclimated animals stems from the realization that it is the brown fat, rather than the general body tissues, which contributes up to 80% of total thermogenesis in the cold (Foster and Frydman, 1979). Although there is some evidence suggesting that an increased conversion of T_4 to T_3 can be elicited by NE in brown fat (Silva and Larsen, 1983) and that the activation of the (Na^+,K^+)-ATPase in brown fat may be involved in NE-stimulated NST, particularly in hyperthyroid or overfed rats (Rothwell *et al.*, 1982), the generally accepted view is that the mechanism of NST in brown fat is the short-circuiting of the proton conductance channel in the inner mitochondrial membrane (Nicholls, 1979). Thus, the quantitative importance of thyroid hormone-mediated NST is challenged. The following is a corroboration of old and new observations (Wunder, 1979; Himms-Hagen, 1983). For example, (1) rats acclimated to cold and fed on a T_4-free diet exhibit a normal T_4 turnover rate without developing hyperthyroidism; (2) thyroidectomized rats survive well in the cold if provided with a low maintenance dose of T_4 (Sellers *et al.*, 1974); (3) thyroidectomized cold-acclimated rats require only a T_4 dose similar to that given to room temperature rats to maintain NST (Hsieh, 1962); (4) thyroidectomized rats treated with a maintenance dose of T_4 grow normal brown fat during cold-acclimation despite normal blood T_3 level (Trian-dafillou *et al.*, 1982); and (5) chronic treatment with thyroid hormone not only does not mimic the NE-stimulated NST found in cold-acclimated rats (LeBlanc and Villemaire, 1970), but the excess amount of hormone actually prevents the normal development of brown fat during cold acclimation (Sundin, 1981; Mory *et al.*, 1981). Taken together, these observations indicate that thyroid hormone has only a permissive role in cold-induced NST, most probably through its well-known potentiating effect on the catecholamines (Fregly *et al.*, 1979). The inhibitory effect of excess thyroid hormone on proper development of brown fat and NST in cold-acclimated rats could be due to reduced sympathetic activity in hyperthyroidism (Gibson, 1981). Since NE from the sympathetic innervation in brown fat exerts a trophic effect on adaptive changes for enhanced NST in cold-acclimated animals (Barnard *et al.*, 1980; Himms-Hagen, 1983), a reduced sympathetic activity could result in reduced NST in animals treated with excess thyroid hormone.

3. ADRENAL CORTICOSTEROIDS

The function of glucocorticoids in thermogenesis has been reviewed (Wunder, 1979; Deavers and Musacchia, 1979; Musacchia and Deavers, 1981). During acute cold exposure, the plasma concentration of corticosteroids increases significantly (Maickel *et al.*, 1961; Wang and Peter, 1977) consequent to the activation of the hypothalamo–pituitary–adrenal axis by the cold (Chowers *et al.*, 1972). In adrenalectomized rats maintained on saline and 5% glucose (Maickel *et al.*, 1967a,b) thermoregulation is intact at $T_a = 20°C$, but they become profoundly hypothermic ($T_b = 29.8°C$) when exposed to 4°C for 3 hours. The major deficits appear to be the inability of the adrenalectomized animals to piloerect, vasoconstrict, and elevate their plasma levels of glucose and free fatty acids (FFA)—fuels essential for shivering (Himms-Hagen, 1975). Treatment with CA alone does not improve these deficiencies. However, pretreatment with cortisone 4 hours prior to cold exposure results in pronounced peripheral vasoconstriction, piloerection and shivering, significantly elevated levels of plasma glucose and FFA, and a near-normal T_b (34.1°C). Administering catecholamines in cortisone pretreated rats further elevates plasma glucose and FFA levels during cold exposure. In adrenalectomized dogs, E-induced elevation of plasma FFA is abolished, but replacement therapy with cortisone restores the response (Shafrir and Steinberg, 1960). The substrate-mobilizing effects of glucocorticoids appear to be both direct but slow (Fain *et al.*, 1965; Divakaran and Friedmann, 1976) and indirect but fast (via glucagon, epinephrine, growth hormone) (Goodman, 1970).

In the rabbit aorta (Kalsner, 1969) and in the conjunctival vessels of rabbit and man (Reis, 1960; Lepri and Cristiani, 1964), glucocorticoids potentiate the pressor effect of CA. Taken together, the glucocorticoids appear to exert an important but permissive role on other metabolic hormones for both heat conservation (piloerection and vasoconstriction) and heat production (substrate mobilization) during cold exposure.

4. OTHER HORMONES

In rats, plasma glucagon levels are increased with cold-acclimation and decreased with heat-acclimation—changes which correlate positively with plasma FFA levels (Kuroshima et al., 1981; Doi et al., 1982). The calorigenic action of glucagon on isolated epididymal adipocytes is enhanced in cold-acclimated rats, while it is reduced in heat-acclimated rats. A single injection of glucagon elevates the brown fat temperature by nearly 3°C in cold-acclimated rats but only 1.4°C in warm control rats. This suggests an enhanced NST response to glucagon after cold acclimation. Furthermore, chronic treatment with glucagon (50μg/100 g, s.c., twice daily for 4 weeks) improves NST and, consequently, causes greater cold resistance than in untreated controls (Kuroshima et al., 1981). It is presently unclear whether glucagon exerts its thermogenic effect mainly through its role on substrate mobilization or whether it has potentiating effects on other metabolic hormones involved in thermogenesis.

In birds, glucagon is glycogenolytic, lipolytic, and gluconeogenic (Heald et al., 1965), but its role in thermoregulation appears to be species specific. In pigeons (Hohtola et al., 1977) and 2-day-old chicks (Palokangas et al., 1972), glucagon suppresses oxygen consumption and causes hypothermia in the cold due to the inhibition of shivering, even though plasma glucose and FFA levels are elevated (Hohtola et al., 1977). In the 3-month old Japanese quail, however, glucagon causes a significant increase in oxygen consumption (Krimphove and Opitz, 1975). It is presently unclear as to why there is a species difference. Also, the mechanism through which glucagon inhibits shivering in the pigeon has yet to be understood at this time.

Aside from its regulation of seasonal reproduction in mammals (see Reiter, 1980, for a review) and circadian activity rhythm in birds (Gaston and Menaker, 1968; Gwinner, 1978), the pineal and its hormone, melatonin, have recently been implicated in thermoregulation in vertebrates (see Ralph et al., 1979; Ralph, 1984; Heldmaier and Lynch, 1986, for a review). In the goldfish (Kavaliers, 1982a) and white sucker (Kavaliers, 1982b), pinealectomy results in the preferment and selection of a higher T_b. In salamanders, i.p. melatonin or chlorpromaxine, which blocks the breakdown of endogenous melatonin, results in the selection of a lower preferred T_b and the abolishment of the diel T_b rhythm (Hutchison et al., 1979). In lizards, whose pineal complex consists of an extracranial photoreceptive parietal eye and an intracranial pineal organ which is homologous to the pineal gland of birds and mammals, parietalectomy also results in the selection of a higher preferred T_b (Hutchison and Kosh, 1974) and the lowering of threshold T_b for panting (Firth and Heatwole, 1976); pinealectomy, however, results in the selection of a lower T_b (Firth et al., 1980). In turtles (Erskine and Hutchison, 1981), melatonin injection lowers the selected preferred T_b. The reasons for these divergent results among the reptiles are presently unclear. In the house sparrow, pinealectomy abolishes the typical diel rhythm of T_b and the minimum nocturnal T_b is significantly elevated (Binkley et al., 1971). In pigeons, pinealectomy results in an elevated T_b during both day and night, and melatonin treatment reverses this (George, 1982). In the sheep and rabbit, pinealectomy results in hyperthermia (Ralph et al., 1979); in rats, however, pinealectomy lowers the T_b but has no effect on the diel T_b rhythm (Fioretti et al., 1974; Spencer et al., 1976). Taken together, these observations seem to suggest, with the exception of laboratory-reared rats, that the pineal or parietal eye exerts a protective role against an excessive rise in resting T_b in vertebrates (Ralph et al., 1979; Ralph, 1984), possibly by influencing the level of set-point commensurate with a specific physiological state. Whether or not this is related to the antithyrotropic action of melatonin on the inhibition of TRH secretion (Vriend, 1981), which affects general metabolism, is currently unknown.

In the white-footed mouse (Lynch et al., 1978) and the Djungarian hamster (Heldmaier et al., 1981), NST capability can be modulated by photoperiods independent of the cold: a short photoperiod enhances NST and a long photoperiod depresses NST. However, an enhanced NST capacity similar to that observed under a short photoperiod can be elicited in hamsters kept under a long photoperiod if they are treated chronically with melatonin (Heldmaier et al., 1981). Similar observations on brown-fat hypertrophy or improved NST by melatonin have been reported in the white-footed mouse (Lynch and Epstein, 1976) and the 13-lined ground squirrel (Sinnamon and Pivorun, 1981b). Since melatonin levels in the pineal and blood are highest in the dark, these studies suggest that melatonin may serve as a physiological mediator linking environmental stimuli (a declining photoperiod and perhaps temperature) and the proper development of thermogenic effectors in preparation for predictable seasonal events, e.g., winter. Precisely how melatonin exerts its effects on enhanced NST is presently unknown; this is particularly enigmatic since the pineal typically exerts an inhibitory effect on both thyroid and adrenal cortex functions (Ralph *et al.*, 1979), both of which are known to have permissive roles in improving NST in cold-acclimation (see discussion above).

C. Hormones and Heat Exposure

Compared to thermogenesis, changes in hormonal functions during heat exposure have not been as extensively studied. The earlier studies were reviewed by Collins and Weiner (1968) and more recently by Chaffee and Roberts (1971) and Gale (1973). The following discussion includes both short-term exposure to heat and long-term acclimation or acclimatization to heat.

I. SYMPATHETICO−ADRENO−MEDULLARY SYSTEM (SAM)

Using the urinary excretion rate of free NE and E as an index for SAM activity, local warming of the POAH has been found to decrease SAM activity in goats (Andersson, 1970), pigs (Baldwin *et al.*, 1969), and baboons (Proppe and Gale, 1970). In nonacclimated men, exercising in 50°C for 90 minutes significantly increases urinary NE excretion and T_b; however, both responses are significantly depressed after heat-acclimation (Maher *et al.*, 1972). In the rhesus monkey, short-term heat exposure to raise Tb by 1 and 2°C results in corresponding increases in plasma NE of 100 and 850% and in plasma E of 100 and 525%, respectively (Elizondo and Johnson, 1981); the E response is abolished after ketamine anesthesia, suggesting that its release from the adrenal medulla is not thermally related. In men, short-term exposure to dry heat (sauna, 70−80°C) for 20−40 minutes significantly increases urinary excretion of NE and E, indicating an activation of the SAM possibly related to vascular adjustments (Huikko *et al.*, 1966). Taken together, these studies seem to suggest that short-term heat exposure is sufficient to cause a rise in T_b which stimulates the SAM, but that part of the response is nonthermal and represents a general response to stress. With heat-acclimation, the magnitude of SAM activation is attenuated, but the differential changes of the thermal versus nonthermal components remain to be elucidated.

The neurohumoral control of sweating induced by heat and by exercise appears to be different. In the sheep, goat, donkey, pig, ox, and wild bovids, thermal sweating is mediated by adrenergic nerves and does not involve E from the adrenal medulla; in cats and anthroid primates, including man, thermal sweating is controlled by sympathetic cholinergic fibers (Hales,

1974). In the donkey, thermal sweating is abolished if sympathetic nerves to the sweat glands are cut, but exercise-induced sweating in these glands persists; bilateral denervation of the adrenal medulla to eliminate circulating E, however, abolishes exercise-induced sweating in denervated sweat glands (Robertshaw and Taylor, 1969). In the stump-tailed monkey, the sweating rate is 50% greater during exercise than the maximum observed rate during heat exposure alone, and this increase could be attributed to the increased plasma E during exercise (Robertshaw et al., 1973). Therefore it appears that, during exercise, hormonal influence can either independently regulate or further enhance a sweating response that is neurally regulated under heat exposure alone (Taylor, 1974). An exception to this dual regulation of sweating activity seems to be the horse: both thermal and exercise-induced sweating are controlled by plasma E; increased plasma E during exercise increases sweating (Evans and Smith, 1956).

The physiological manifestations after heat-acclimation in primates include a decreased resting T_b, an increased maximum sweating capacity, a decreased central threshold T_b (set-point) for the initiation of sweating, and an increased sensitivity (gain) for sweating with changes of central T_b (Bruck, 1981; Elizondo and Johnson, 1981). The neurohumoral mechanisms responsible for these adjustments remain largely unknown. However, in animal models (guinea pig), peripheral NE infusion can modulate central threshold T_bs for shivering and heat-induced polynea (Bruck, 1981), suggesting a possible interaction between peripheral SAM activity and modifications of central thermoregulatory functions in long-term thermal acclimation. In humans, heat-acclimation results in lowered T_b in both males and females, but the higher resting T_b of females typically found in the unacclimated state is not altered by heat-acclimation (Shapiro et al., 1980). This indicates that the differential effect of progesterone-mediated elevation in set-point for T_b in females (Cunningham and Cabanac, 1971) persists irrespective of other neurohumoral influences which lead to an overall decrease in set-point T_b in heat-acclimation.

2. THYROID HORMONES

Local warming of the POAH in goats results in a decrease in thyroid activity as determined by plasma PBI-131 level and the radioactivity retained in the thyroid (Andersson, 1970). This suggests that a rise in T_b under heat stress could suppress thyroid function. In a variety of animals (e.g., rabbits, calves, and humans), acclimating to heat (27–34°C) results in depressed thyroid function and, more often than not, reduced basal metabolism (Collins and Weiner, 1968). In the hamster, acclimation to 35°C results in hypoactive thyroid and adrenal medulla (Cassuto and Amit, 1968), manifested by a general decrease in thermogenesis and mitochondrial respiration (Cassuto and Chaffee, 1966). In the desert-dwelling round-tailed ground squirrel, which remains active throughout the day, thyroid activity is extremely low; its basal metabolism is 40% below the predicted value based on weight. When exposed to heat (47°C for 2 hours), this species tolerated a T_b of 42.2°C without mortality, whereas the antelope ground squirrel, a sympatric species with euthyroid function and which is active only during the cooler parts of the day, died at a $T_a = 44$°C and a T_b of 41.6°C (Hudson and Wang, 1969). This illustrates the physiological advantage of having low thyroid activity and low endogenous basal heat production when living under chronic heat stress. Similar principles have also been illustrated by the increased heat tolerance in heat-acclimated rhesus monkeys (Chaffee and Allen, 1973) and in rams and rabbits after thyroidectomy (Collins and Weiner, 1968).

3. OTHER HORMONES

In rats, sheep, cattle and humans, heat-acclimation or seasonal increases in temperature result in reduced glucocorticoid activity based on decreased urinary excretion of 17-OH corticosteroids (Collins and Weiner, 1968). This conclusion has recently been substantiated by the observed decrease in plasma cortisol level in humans after heat-acclimation (Davies et al., 1981). Taken together, the pituitary–adrenal axis appears to be depressed consequent to heat acclimation. The plasma levels of aldosterone and renin activity are relatively constant after heat-acclimation in humans, although both can be significantly increased by acute heat stress and exercise (Davies et al., 1981). This suggests that the increased Na^+ conservation typically observed after heat-acclimation is likely to be a normal response to heat and exercise rather than to a negative Na^+ balance (Davies et al., 1981). Other hormonal adjustments consequent to altered water metabolism in acute and prolonged heat exposure are beyond the scope of this chapter.

As described earlier (Section II,B,4), the pineal and its hormone, melatonin, appear to defend against a debilitating rise in T_b (Ralph et al., 1979; Ralph, 1984). This is further supported by the observation that in the warm-acclimated (25°C) pigeon, pinealectomy during the winter results in an elevation of T_b (George, 1982), an effect which is seasonal (not seen in summer birds), not permanent (lasts less than 1 year), not mediated by thyroid hormones, and reversible by melatonin therapy (George, 1982). In rats, acclimation to heat adversely affects reproductive function (Sod-Moriah et al.,

1974). Male rats pinealectomizid at weanling and acclimated to 35°C for 30 days showed no decrease in resting T_b or serum level of corticosterone but significant depressions of serum LH and testosterone levels and pituitary LH content (Megal *et al.*, 1981). These results suggest that the pineal effect on T_b may be both seasonal and age dependent and that the pineal may moderate the adverse effect of high T_a on reproduction.

III. Temperature Regulation under Modified Physiological States

A. Sleep

Compared to wakefulness, significant modifications of the thermoregulatory responses occur during sleep in practically all mammals and birds thus far studied, and the nature of these modifications is also sleep-state dependent (Heller and Glotzbach, 1977; Heller *et al.*, 1983; Parmeggiani, 1981). During synchronized sleep (slow-wave sleep), there is a downward shift in both the level of T_b that is being regulated and the hypothalamic threshold temperature for a variety of autonomic responses (e.g., shivering, thermal tachypnea, vasomotor response, sweating). The gain (rate of response versus temperature change) of the responses may be decreased (e.g., heat production in the kangaroo rat; Glotzbach and Heller, 1976) or increased (e.g., tachypnea and shivering in cats; Parmeggiani, 1981). During desynchronized sleep (rapid-eye-movement or paradoxical sleep), however, the thermoregulatory system appears to be inactivated; no autonomic thermoregulatory responses can be elicited following drastic alterations of the brain temperature (up to 5°C; Glotzbach and Heller, 1976) and there is a loss of muscle tonus and postural control for behavioral thermoregulation (Parmeggiani, 1981). These observations indicate that the central thermocontrollers are temporarily overridden, perhaps by inhibitory influences from other parts of the CNS, resulting in a temporary suspension of homeothermic regulation during desynchronized sleep (Heller and Glotzbach, 1977; Parmeggiani, 1981). Since the occurrence of desynchronized sleep is greatest at thermoneutral temperatures (Schmidek *et al.*, 1972) and brief skin-temperature changes toward thermoneutrality trigger desynchronized sleep (Szymusiak and Satinoff, 1980), it is possible that peripheral thermal inputs play an important role in activating desynchronized sleep. In humans, cooling disrupts sleep states more effectively than heating and more frequently during desynchronized sleep (Candas *et al.*, 1982), suggesting the maintenance or perhaps even enhancement of peripheral thermosensitivity during desynchronized sleep. Recently, a sleep factor has been isolated from human urine and identified as a glycopeptide (Krueger *et al.*, 1982). When administered i.c.v., it induces slow-wave sleep in the rabbit. Whether similar thermoregulatory changes occur with sleep induced in this manner is presently unknown.

The functional significance of a tuned-down thermoregulatory state during sleep has been related to energy conservation (Heller and Glotzbach, 1977; Heller *et al.*, 1978; Walker and Berger, 1980). As slow-wave sleep is present only in the endotherms, depressed metabolism and T_b during this state serve to reduce the high energy costs associated with homeothermy. In humans, metabolic rate during slow-wave sleep may be 8–40% below the resting waking minimum (Walker and Berger, 1980); in the kangaroo rat and marmot the reductions are 12 and 10%, respectively (Heller *et al.*, 1978), and in the pigeon, approximately 20% (Heller *et al.*, 1983). Furthermore, since electrophysiological evidence has revealed that hibernation in mammals is an extension of slow-wave sleep (Heller and Glotzbach, 1977; Heller *et al.*, 1978; Walker and Berger, 1980) and that up to 88% of energy savings may be realized by exhibition of hibernation (Wang, 1979), a functional homology between sleep, shallow torpor, and hibernation in energy conservation has thus been proposed (Heller and Glotzbach, 1977; Heller *et al.*, 1978; Walker and Berger, 1980). The existence of slow-wave sleep, shallow torpor, and hibernation in the most primitive mammal, the echidna (Allison and Van Twyver, 1972; Allison *et al.*, 1972), suggests the early occurrence of slow-wave sleep in mammalian evolutionary history, since reptiles, based on current evidence, do not exhibit slow-wave sleep. However, in view of the polyphyletic origin of shallow torpor and hibernation in mammals and birds (Lyman, 1982), it is likely that physiological mechanisms underlying slow-wave sleep could have provided the suitable substrate, based on which deeper forms of torpidity were further evolved in response to demands for energy conservation under divergent ecological niches.

B. Fever

The subject of fever has been in recent years one of the most active areas of research in thermal physiology. Several monographs have been published on the biochemical and physiological mechanisms of fever, the CNS effects of pyrogens and their mediators, antipyresis, and the evolutionary adaptive significance of fever (e.g., Kluger, 1979; Lipton, 1980; Milton,

1982a). Only a brief review will be presented here to provide an overview of the subject.

Fever is manifested by an elevation in thermoregulatory set-point, invoked by the invasion of exogenous antigens (virus, bacteria, etc.),which by an unknown mechanism activate the phagocytes (monocytes, granulocytes, Kupffer cells) for the synthesis and release of an endogenous pyrogen. The endogenous pyrogen acts centrally, activating both the autonomic and behavioral effectors in raising the T_b of the febrile animal to the new set-point (Kluger, 1981; Dinarello, 1980). The endogenous pyrogen is a polypeptide, ranging from 14,000 to 60,000 in molecular weight among different species (Dinarello, 1980); in man it is approximately 15,000 dalton (Dinarello et al., 1974). Human endogenous pyrogen can cause fever in rabbits, suggesting the lack of species specificity, but the active sites of the endogenous pyrogen remain unclear (Dinarello, 1980). Recently, evidence has been accumulated to suggest that the endogenous pyrogen may either get into the brain tissue through the organum vasculosum of the lamina terminalis (OVLT) to elicit a secondary cascade event (Blatteis et al., 1983; Stitt, 1985). Many intermediary substances have thus been proposed which may mediate the febrile effect of endogenous pyrogen, for instance, the prostaglandins (Milton and Wendlandt, 1970; Bernheim et al., 1980), many hormones and neurotransmitters (cAMP, NE, DA; see Hellon, 1975, for a review), changing Na^+/Ca^{2+} ratio in the posterior hypothalamus (Myers and Tytell, 1972), and a protein (Siegert et al., 1976). All of these possibilities have been examined in detail (see Milton, 1982a, for a review). To summarize, the supposition that prostaglandins (PGE_2, in particular; Milton, 1982b; Stitt, 1986) or other metabolites of arachidonic acid (precursor of prostaglandins; Hellon et al., 1980) are the mediators of pyrogen fever appears still tenable, despite several reports to the contrary (for reviews, see Cooper, 1987). The recent study in rabbits in which pretreatment with cycloheximide (an inhibitor of protein synthesis) blocks the pyrogen-induced and PGE_2-induced fever equally (Hellon et al., 1980) further suggests that a common, protein-synthesis pathway is shared by both pyretics in their fever-generating action. However, the precise involvement of a protein mediator in pyrogen fever is presently unclear (Milton and Sawhney, 1982). The raising of set-point in pyrogen fever due to an elevation of the diencephalic Na^+/Ca^{2+} ratio has been supported (Myers, 1982), but the involvement of the cyclic nucleotides (Dascombe, 1982, 1985) and monoamines (Cox and Lee, 1982; Dascombe, 1985) in fever mediation is presently conjectural.

With the release of endogenous pyrogen from phagocytes, many other associated physiological changes occur in addition to the onset of fever.

These include activation of the host immune-defense system and a change of plasma trace-metal concentrations, e.g., increased plasma copper but decreased plasma zinc and iron (Kluger, 1981). The reduction in plasma trace metals is apparently independent of T_b changes since antipyretics, which block the rise in T_b in endogenous pyrogen-induced fever, do not affect the decrease of plasma trace metals (Tocco et al., 1983). The significance of altered plasma trace metals in fever remains to be elucidated; however, the decreased plasma iron concentration appears to be of functional importance. The in vitro survival rate of fever-causing bacteria is reduced only if cultured at the febrile temperature and at the reduced iron concentration found during fever but not at the normal T_b regardless of iron concentration (Kluger and Rothenburg, 1979). This suggests that a synergistic effect might exist between elevated T_b and decreased plasma iron in curtailing bacterial growth and thus limiting the extent of bacterial insult (Kluger, 1981). Similarly, since copper is known to be toxic to prokaryotes such as bacteria, the increased serum copper concentration during fever may also have an inhibitory effect on bacterial growth and the extent of fever (Reynolds et al., 1980).

The mechanisms via which pyrogen fever is terminated are not precisely known. In general, antipyretics are known to exert their effects centrally, possibly either by competing against pyrogens for receptor sites or by inhibiting prostaglandin synthesis (Clark, 1980). The competition for similar receptor sites has been demonstrated by the parallel right-hand shift of the dose–response curve for leukocyte pyrogen by antipyretics such as sodium salicylate, acetaminophen (Clark and Coldwell, 1972), and indomethacin (Clark and Cumby, 1975), whereas the inhibition of prostaglandin synthesis by antipyretics has been amply demonstrated (e.g., Milton, 1982b). A new mechanism for antipyresis has been proposed recently which involves the facilitated transport of endogenous pyrogen (assuming it can enter the CNS) or prostaglandin out of the CNS via the L transport system (Lipton, 1980). Other mechanisms for antipyresis include the roles of AVP, ACTH, and α-MSH, which have already been discussed (see Section II,A,2,e and f).

One of the central questions in fever research has been the adaptive significance of fever. Since fever (either behavioral and/or autonomic) can be elicited in all vertebrates as well as in some invertebrates by similar microorganisms, and since defervescence can be induced by similar antipyretics (Reynolds et al., 1980; Kluger, 1981), it is likely that the evolutionary origin of fever is ancient. To evaluate the survival value of fever in ectotherms (fish to reptiles), infected animals can be prevented from developing a behavioral fever simply by removing the external heat source. In such cases, the survival rate of infected animals is significantly lower than that of febrile animals (e.g., in fish: Covert and Reynolds, 1977; amphibians: Kluger, 1977; reptiles: Kluger et al., 1975). In endotherms, reduction of the magnitude of fever from 1.56 to 0.72°C by antipyretic (sodium salicylate) results in significantly higher mortality rates in the infected rabbit (see Kluger, 1981). In newborn mammals (dog, mouse, pig) whose T_b may be easily modified by T_a, viral infections result in much greater rates of mortality if the newborns are kept at lower T_b than at higher (febrile) T_b (see Kluger, 1981). Thus, it is indicative that fever following a bacterial infection is indeed beneficial to the host and that the rise in T_b plus other associated changes (e.g., immune-defense, trace-metal concentration) with the onset of fever can be construed as a specific defense response in the repertoire of an animal's homeostatic regulatory mechanisms.

C. Hibernation

Hibernation in endotherms is characterized by a profound reduction of physiological functions, with T_b approximating T_a for a period of a few days to a few weeks and ending in a spontaneous arousal, within which the normal T_b and physiological functions are restored using heat generated exclusively from within the animal. The biochemical, physiological, and ecological adaptations associated with hibernation were recently reviewed (e.g., Wang and Hudson, 1978; Lyman et al., 1982; Wang, 1985, 1987a), as were the neuroendocrine (Hudson and Wang, 1979; Wang, 1982, 1986) and neurophysiological aspects of hibernation (Beckman, 1978; Beckman and Stanton, 1982; Heller, 1979). There is little doubt that hibernation in endotherms represents an advanced form of thermoregulation—an extension of biochemical and physiological functions from the typical euthermic level down to a T_b just a few degrees C above freezing. Due to space constraint, only selected coverage of hormones and hibernation will be presented here.

Although exceptions exist, polyglandular involution appears to be prevalent prior to seasonal hibernation (Wang, 1982), with the involution of the pituitary–gonad, pituitary–thyroid, and pituitary–adrenal axes being the most conspicuous. During the first half of the hibernation season endocrine functions are generally depressed; this is followed by a reactivation of all functions sometimes after the midhibernation season, and peak activities are reached at the time of or shortly after the spring emergence. Evidence supporting such a scheme has been based largely on changes in gland weight and histological aspects and, to a lesser extent, on measurements of hormone levels in glands and/or blood. Kinetic studies on turnover rates of hormones at different phases of the annual hibernation cycle have been very scarce. It is, therefore, apparent that our current understanding of the relationship between endocrine function and hibernation has been correlative and indirect. A further complication is the apparent endogenous, circannual nature of seasonal hibernation, i.e., repetitive yearly hibernation cycles persist despite the constant photoperiod and temperature regimes under which the hibernators are maintained (Pengelley and Asmundson, 1974). Since the regulation of T_b requires concerted neuroendocrine coordination (Gale, 1973), the shift from the nonhibernating to the hibernating phase with the ensuing drastic reduction in Tb during hibernation suggests that neuroendocrine adjustments necessary for hibernation are likely to occur entirely endogenously, independent of entraining environmental zeitgebers. Thus, aside from the usual problems associated with neuroendocrine studies, e.g., stress on animals due to sampling techniques, removal of endocrine glands, replacement therapy, and the limited functional usefulness based on changes in histological indices and plasma hormone concentrations, the study of neuroendocrine regulation of hibernation is further complicated by (1) the necessity of following annual changes in endocrine functions so that specific endocrine-modulated metabolic adjustments in preparation for hibernation (e.g., fattening) may be identified, (2) the combined effects of suppression of endocrine function due to depressed T_b during hibernation versus depression that is independent of T_b changes, and (3) the polyphyletic origin of hibernation and, thus, the lack of a generalized endocrine scheme for hibernators living under different ecological niches.

I. HYPOTHALAMO−HYPOPHYSEAL INTERACTIONS

Studies on the influence of hypothalamic neurosecretory functions in hibernation have been few in number and fragmentary in coverage. Most of the available studies have employed morphological criteria in deciphering the functional state of the hypothalamo–hypophyseal system. In ground squirrels (Polenov and Yurisova, 1975; Yurisova and Polenov, 1979) and bats (Jasinski, 1970), a general reduction in the synthesis, transport, and release of neurosecretory materials is evident prior to hibernation: the volume of the nuclei and nucleoli of the hypothalamic secretory nuclei is small, the amount of neurosecretory material in the fibers of the hypothalamic nuclei and the hypothalamo–hypophyseal tract in the median eminence is scarce, and the vascular activity (presence of blood cells) in the capillaries of the hypothalamo–hypophyseal system and that of the mantel plexus is low. This is followed by a marked reduction of transport and release of neurosecretory

materials during hibernation, resulting in their accumulation in the hypothalamo–hypophyseal system and the posterior pituitary. In the hedgehog,
using immunocytochemical techniques (Nurnberger, 1983), the AVP- and
oxytocin-containing neurons are less active, whereas those of the
somatostatin- and Met-enkephalin-containing neurons are more active during hibernation than in the active state. Based on the number of synaptoid
and neurohemal contacts in the infundibulum and portal plexus, respectively,
as indices for secretory activity (Fleischhauer and Wittkowski, 1976), the
secretion of hypothalamic releasing and inhibiting hormones (e.g., LHRH,
GHIH or somatostatin, CRH, and TRH) is thought to have decreased
during hibernation. However, in view of the control of the release of hypothalamic releasing factors by aminergic innervations in the neurosecretory
cells (Fuxe *et al.*, 1974b), it is conceivable that the release of hypothalamic
hormones could be accomplished by neural stimulations without substantial
changes in morphology of the synaptoid and neurohemal contact zones. In
the hibernating woodchuck (Young *et al.*, 1979a), TRH concentration is
significantly higher in the hypothalamus, septum, and striatum than their
respective prehibernation values; this suggests local accumulation and correlates well with the greatly reduced thyrotroph activity in the pituitary (Frink
et al., 1980) and thyroid during hibernation (Krupp *et al.*, 1977). During
periodic arousals from hibernation, there are temporary increases in hormonal release from storage (Legait *et al.*, 1970; Yurisova and Polenov, 1979), for
instance, ADH from the neurophysis to facilitate urine formation (Burlet *et
al.*, 1974). In the ground squirrel (Yurisova and Polenov, 1979), the testicular
and adrenal weights are found to increase steadily throughout the hibernation season; it is not known however, whether the hypothalamo–pituitary–
adrenal or pituitary–gonadal axes are active only during periodical arousals
when T_b is high or constantly, even under depressed T_b.

2. BRAIN MONOAMINES

When administered i.c.v., 5-HT elicits hypothermia in the golden hamster (Jansky and Novotona, 1976). In the Richardson's ground squirrel, i.c.v.
5-HT differentially increases heat loss and suppresses heat production in the
cold, when testing is done during the animal's hibernating phase. This results
in a relatively large depression in T_b (1.6°C) in comparison to that (0.4°C)
with the same injection during the nonhibernating phase (Glass and Wang,
1979a). This implies that there might be a seasonal reorganization of the 5-
HT-mediated thermoregulatory functions commensurate with the hibernating phase of the animal to facilitate the onset of hibernation. During arousal

from hibernation, central injection of 5-HT at T_b = 10, 20, and 30°C results in increased heat loss, decreased heat production, and a slower rate of rewarming (Glass and Wang, 1979b), which confirms the thermolytic effect of 5-HT even at depressed T_b. In the arctic ground squirrel, i.p. injection of 5-hydroxytryptophan, a precursor for 5-HT, 1 to 1.5 months before hibernation results in decreased T_b, decreased locomotor activity, increased slow-wave sleep time, and a preference for cold (Pastukhov *et al.*, 1981)—changes which are characteristic during preparation for hibernation.

In addition to their roles in the regulation of T_b, brain monoamines are involved in the regulation of neuroendocrine functions (Fuxe *et al.*, 1974a). For instance, DA is inhibitory to LHRH secretion but NE is stimulatory, and 5-HT stimulates LH, FSH, and prolactin (Fuxe *et al.*, 1974b) but inhibits ACTH (Telegdy, 1977). Although systematic studies of brain 5-HT and its possible neuroendocrine regulation of hibernation are currently lacking, there is increasing evidence to show that it is functionally involved (see, e.g., Beckman, 1978; Jansky, 1978; Jansky *et al.*, 1981; Novotona and Civin, 1979; Glass and Wang, 1979a,b; Popova and Voitenko, 1981). The general consensus is that increased brain serotonergic activity facilitates entry into hibernation since the 5-HT level is lower during hibernation than it is prior to entry into hibernation, and 5-hydroxyindolacetic acid, a metabolite of 5-HT, is higher during hibernation (Novotona *et al.*, 1975; Canguilhem *et al.*, 1977; Duncan and Tricklebank, 1978). Furthermore, injection (i.p.) of parachlorophenylalanine or lesioning of the medial raphe nucleus, both of which serve to reduce the endogenous synthesis of brain 5-HT, results in an inability of the animal to enter hibernation (Spafford and Pengelley, 1971). However, in a recent study on the European hamster, only electrolytic lesions of a small area in the anterior part of the medial raphe nucleus prevented the onset of hibernation, whereas electrolytic lesions of other parts of the raphe nuclei or a general reduction of brain 5-HT by 5,7-dihydroxytryptamine have no effect (Canguilhem *et al.*, 1986). This study indicates that it is not the absolute amount of brain 5-HT, per se, that is important but the intactness of certain serotonergic pathways that is critical for the occurrence of hibernation. In agreement with the suggestion that 5-HT is involved in hibernation, feeding of a tryptophan-rich diet facilitates the occurrence of hibernation in the golden hamster, presumably through an increase in substrates for 5-HT synthesis in the brain (Jansky, 1978). In addition, the turnover rate of brain 5-HT increases some 18 times in prehibernation and 24 times during hibernation in the Syrian hamster (Novotona *et al.*, 1975). These drastically increased 5-HT turnover rates have been interpreted to mean enhanced suppression of the pituitary–adrenal axis, thereby facilitating hibernation (Novotona and Civin, 1979). Recent studies have shown that adrenalectomy actually prevents, whereas chronic adrenal corticosteroids enhance, hibernation in this species (Jansky et al., 1981). Thus, the functional significance of increased brain 5-HT turnover in hibernation remains unclear.

Taken together, the alteration of central 5-HT metabolism appears to be related to hibernation. Based on the differential thermoregulatory effects of 5-HT in animals under different physiological states (Glass and Wang, 1979a), it may be speculated that the increased turnover of brain 5-HT (Novotona et al., 1975) may lead to a significant suppression of heat production, facilitating entry into and maintenance of hibernation, whereas the release of 5-HT suppression of heat production may lead to spontaneous arousal from hibernation. The release of 5-HT suppression could be due to a local shortage of the neurotransmitter if synthesis and axonal transport of 5-HT to specific loci are limited while its utilization continues at a high rate during hibernation. Obviously, further studies are needed to test the validity of this speculation.

In contrast with 5-HT, centrally applied NE increases heat production and T_b in the euthermic Richardson's ground squirrel, a response which does not show seasonal variation (Glass and Wang, 1979b). With regard to hibernation, there appears to be a complete cessation of central NE turnover some 10 hours prior to the onset of hibernation in the 13-lined ground squirrel (Draskoczy and Lyman, 1967). During hibernation, the NE turnover is essentially nil in both the 13-lined ground squirrel (Draskoczy and Lyman, 1967) and the European hedgehog (Sauerbier and Lemmer, 1977). No turnover studies have been made during arousal from hibernation; however, central injection of NE at T_b = 10, 20, and 30°C in the arousing Richardson's ground squirrel results in greater rates of heat production and faster rates of rewarming (Glass and Wang, 1979b), confirming the thermogenic effect of NE even at depressed T_b.

In a series of studies summarized by Beckman (1978), microinjections of NE, 5-HT, and ACh into the POAH trigger arousal in the hibernating golden-mantled ground squirrel, but when injected into the midbrain reticular formation, only ACh triggers arousal. These observations suggest that aminergic and cholinergic neurons in these areas may be involved with mechanisms which trigger arousal from hibernation, but how they interact is presently unknown. Based on the neural model for the regulation of hibernation (Heller, 1979; Beckman and Stanton, 1982), increased inhibitory influence on midbrain reticular formation from the hippocampus facilitates sleep and possibly the maintenance of hibernation, whereas removal of hippocampal inhibition or activation of midbrain reticular formation would reinstate the wakeful state or arousal from hibernation. Further neurochemical characterization of these pathways including turnover studies and seasonal variations may provide the needed information in discerning the regulatory roles of monoamines on hibernation.

3. PITUITARY

In a variety of species, e.g., ground squirrels (Hoffman and Zarrow, 1958a), hamsters (see Kayser, 1961), and bats (Nunez and Gershon, 1982), a histological examination of the pituitary cell types, pituitary weight, and target glands and their hormones shows an annual cycle in pituitary function: the gland is most active in April-May after emergence from hibernation and least active in December-January during early hibernation, and there is a rapid increase in activity between January and March in the latter half of hibernation (see Wang, 1982, for a review). This annual cycle, however, is independent of the occurrence of hibernation, since the cycle exists regardless of whether the animals are maintained at 2°C to facilitate hibernation or at 22–25°C to discourage hibernation (Hoffman and Zarrow, 1958a).

In hedgehogs, cold exposure during the fall does not elicit the typical alarm reaction of an increased plasma ACTH concentration; this is interpreted to mean that a blockage of the pituitary–adrenal axis is essential in the preparation for hibernation (Hoo-Paris, 1971). During periodic arousal, however, the level of plasma ACTH increases significantly above the hibernating level, indicating that the pituitary–adrenal axis can be reactivated during the hibernating season. In the red-cheeked suslik (a ground squirrel; Krass and Khabibov, 1975), the blood level of 11-hydroxycorticosteroids shows a seasonal change similar to that of pituitary ACTH, indicating a functional relationship of the pituitary–adrenal axis throughout the hibernating season. In euthermic golden-mantled ground squirrels maintained under 23°C and 14L : 10D, monthly blood LH concentrations in males (Licht *et al.*, 1982) and ovariectomized, but not intact, females (Zucker and Licht, 1983) showed a circannual rhythm with increasing values in March and peaks in April—times corresponding to late hibernation season and spring emergence. These observations suggest that the hypothalamo–pituitary control of reproduction in seasonal hibernators may be largely endogenous, irrespective of the occurrence of hibernation.

4. THYROID

The role of the thyroid gland in hibernation has been extensively reviewed recently (Hudson and Wang, 1979; Hudson, 1981; Wang, 1982). Among the ground squirrels and woodchucks which show an inactive thyroid prior to hibernation, histological observations (e.g., Hoffman and Zarrow, 1958b; Krupp et al., 1977; Winston and Henderson, 1981) show that plasma T_4 and T_3 are high during the spring, decreasing during the summer and fall, and high and increasing during hibernation (Wenberg and Holland, 1973a; Young et al., 1979b; Demeneix and Henderson, 1978a). The high levels of plasma T_4 and T_3 observed during hibernation are best explained as due to a lack of peripheral utilization of the hormones rather than increased secretions from the thyroid. Evidence supporting such a view includes the following. (1) Cooling of the hypothalamus, which typically activates the pituitary–thyroid axis (Gale, 1973), does not invoke increased release of ^{125}I from thyroid during the hibernating season (Hulbert and Hudson, 1976). (2) High uptake of inorganic iodide has been observed during the hibernating season (Wenberg and Holland, 1973a; Hulbert and Hudson, 1976), but accumulation of colloid in follicles, decreased cell height (Hoffman and Zarrow, 1958b), and a very low (Hulbert and Hudson, 1976) to nil (Hudson and Wang, 1969) organic iodine release rate suggest a low thyroid secretory activity probably due to lack of TSH. This is supported by the inactivity of the pituitary thyrotroph during hibernation (Frink et al., 1980). (3) Exogenous TSH and T_4 restore to normal levels the depressed basal metabolic rate found during the hibernating season but have no effect on the occurrence of hibernation (Hudson, 1981), indicating that both the thyroid and its target tissues remain responsive during the hibernating season and that a normally active thyroid does not prevent hibernation. (4) Although the plasma T_4 and T_3 levels are high during hibernation, the free/total hormone ratios are extremely low (Demeneix and Henderson, 1978b; Young et al., 1979b). Since the binding of hormones to their plasma protein carriers is increased at low T_b and decreased at high T_b, the available plasma free T_4 and T_3 for tissue metabolism can be increased simply by increasing T_b during arousal without invoking an activation of the pituitary–thyroid axis (Young et al., 1979b). Thus the high plasma T_4 and T_3 levels found during hibernation may represent storage rather than active secretion by thyroid. (5) In the Richardson's ground squirrel, which has high plasma T_4 and T_3 (Demeneix and Henderson, 1978b; Henderson and Demeneix, 1981) but markedly decreased thyroid synthetic, reabsorptive, and secretory activities (Winston and Henderson, 1981) during hibernation, thyroidectomy abolishes the high plasma T_4 and T_3 levels but does not affect the pattern of hibernation (Henderson and Demeneix, 1981). Taken together, these studies have demonstrated convincingly that in ground squirrels and woodchucks the thyroid is inactive either prior to or during hibernation and that high circulating T_4 or T_3 concentrations may act as storages for thyroid hormones, but these high levels are not necessary for the occurrence of hibernation. Other species which also show an inactive thyroid and low plasma T_4 level during hibernation include the European hedgehog (Augee et al., 1979).

In contrast with the above group, various indices used to assess thyroid function (e.g., TSH level: Hudson, 1981; organic ^{125}I release: Hudson, 1980; radioiodide uptake and conversion ratio and disappearance of radiolabeled T_4: Tashima, 1965; Bauman et al., 1969) have indicated that the thyroid remains active during hibernation in many species of hamster (e.g., Syrian: Tashima, 1965; Bauman et al., 1969; Turkish: Hudson, 1981; European: Canguilhem, 1970) and in the chipmunk (Hudson, 1980). The importance of the thyroid for the occurrence of hibernation is further demonstrated by the reduction of frequency (Jansky et al., 1981) or absence (Canguilhem, 1970) of hibernation following thyroidectomy and the restoration of a normal hibernation pattern following T_4 treatment (Canguilhem, 1970).

Why there is such a strong dichotomy in thyroid status with regard to hibernation among different species is not entirely clear. It has been proposed that this may be related to the different natural habitats and activity patterns of the individual species and to the different modes of energy storage among them (Hudson, 1981). In ground squirrels and woodchucks, the danger of overheating due to intense summer solar radiation and diurnal activity in open fields would favor a reduced endogenous heat load consequent to reduced thyroid activity, whereas the summer–fall fattening in these species prior to hibernation would also be facilitated by low thyroid activity. In contrast, the nocturnal activity pattern of hamsters and the diurnal but forest habitat of chipmunks may not be subjected to the selection for depressed endogenous heat production, whereas the food-hoarding behavior of hamsters and chipmunks may require thyroid hormones for energy assimilation when they feed between hibernation bouts.

The possible role of reduced thyroid activity in increasing membrane fluidity in preparation for hibernation has been proposed (Hulbert et al., 1976; Hulbert, 1978; Augee et al., 1979). However, in view of the fact that both hamsters and chipmunks can tolerate similar T_b during hibernation as do ground squirrels and hedgehogs, while possessing active thyroid functions during hibernation, much more work is needed before this supposition can be accepted as a generalized mechanism for biochemical adaptation in hibernation.

5. CALCITONIN AND PARATHORMONE

The role of these hormones in hibernation has been assessed by the changes in blood and tissue calcium levels and bone histology. During hibernation, hypercalcemia has been observed in the hedgehog, the European hamster, and the Syrian hamster; increased calcium in the heart and skeletal muscle of the Syrian hamster and the heart and liver of the ground squirrel was also observed (see Wang, 1982). The significance of a high calcium level in blood and tissues during hibernation is presently conjectural but may be related to the maintenance of contractility and irritability in the heart and skeletal muscles (Ferren et al., 1971), the increased membrane fluidity at low T_b (Aloia and Pengelley, 1979), and the sustained mitochondrial energy metabolism at low T_b (Pehowich and Wang, 1981).

Since dietary calcium is absent in species which do not feed between hibernation bouts, calcium must be supplied from bone deposits to replenish excretory loss. Osteoporosis, or the loss of bone material, has been observed during hibernation in the bat (Whalen et al., 1971), the arctic ground squirrel (Mayer and Bernick, 1963), the 13-lined ground squirrel (Haller and Zimny, 1977), and the Syrian hamster (Steinberg et al., 1981), even though this last species does feed upon arousal. Since bone resorption requires parathormone, the observed increase in synthetic and secretory activities of the parathyroid gland during prehibernation and hibernation (Kayser, 1961; Nunez et al., 1972), is consistent with this physiological need. The parafollicular cells of the thyroid, which secrete calcitonin, show maximum activity during prehibernation, a progressive decrease in activity throughout hibernation, and a return to high activity in late hibernation (Nunez et al., 1967). The thyroid calcitonin content in bats is high during prehibernation but low during hibernation (Haymovits et al., 1976). In the 13-lined ground squirrel (Kenny and Musacchia, 1977), significantly lower thyroid calcitonin content has been observed in November and December than in September and February. However, such differences are not related to hibernation per se since both hibernating and euthermic animals show the same trend. Taken together, the low winter activity of the parafollicular cells coupled with the high activity of the parathyroid may explain the observed hypercalcemia and osteoporosis during hibernation. This is further supported by the observation that exogenous calcitonin reverses the osteoporosis found in the hibernating bat (Krook *et al.*, 1977).

6. ENDOCRINE PANCREAS

During hibernation, the primary fuel for metabolism is fat (see Willis, 1982, for a review). In seasonal hibernators, fall fattening is associated with hyperinsulinemia, hypertriacylglycerolemia, peripheral insulin resistance, and increased lipoprotein lipase activity in the white adipose tissue (see Wang, 1987a). In addition to fat, glucose utilization continues during hibernation especially by the CNS in areas which receive thermal afferents, e.g., the paratrigeminal nucleus (Kilduff *et al.*, 1983). During arousal, glucose utilization is increased, being supported by either gluconeogenesis (Galster and Morrison, 1975; Yacoe, 1983) or glycogenolysis (Musacchia and Deavers, 1981; Musacchia, 1984).

In species which do not feed between hibernation bouts (e.g., ground squirrel, woodchuck, hedgehog, dormouse), there is generally a reduction of blood glucose level with progression of time in a hibernation bout. In species which do feed between hibernation bouts (e.g., Syrian and Turkish hamsters), the blood glucose level during hibernation is similar to that observed in euthermia through glycogenolysis from the liver (see Musacchia, 1984, for a review). However, there are exceptions; for instance, in the golden-mantled (Zimmerman, 1982) and the 13-lined ground squirrels (Agid *et al.*, 1978), the blood glucose level is relatively constant throughout the hibernation bout.

In the hedgehog, the plasma level of immunoreactive insulin during hibernation is undetectable, but the pancreatic insulin content is high; intravenous glucose challenge results in hyperglycemia without invoking insulin release (Johanssen and Senturia, 1972). Furthermore, exogenous insulin causes no hypoglycemia and exogenous insulin antiserum no hyperglycemia in the hibernating hedgehog; however, both effects become prominent during arousal from hibernation, when T_b exceeds 20 and 15°C, respectively (Hoo-Paris and Sutter, 1980a). Taken together, these observations suggest that, during hibernation, the secretion of insulin from the pancreas as well as the binding of insulin or insulin antiserum to their respective sites are impaired probably due to the severe depression of T_b, since both functions are restored with the increase of T_b. During arousal, glucose challenge does not invoke additional insulin release despite the elevation in T_b, but the resultant hyperglycemia may be prevented if hedgehogs are pretreated with phentolamine, an α-adrenergic blocker (Hoo-Paris and Sutter, 1980b). This indicates that the strong sympathetic drive during arousal from hibernation has a profound inhibitory effect on insulin secretion from the pancreas. In the 13-lined ground squirrel, plasma-immunoreactive insulin is decreased during hibernation, whereas that of the garden dormouse stays unchanged (Agid *et al.*, 1978). The functional significance of these changes remains to be elucidated. In the dormouse (Castex and Sutter, 1979), the efficacy of insulin-induced glucose oxidation in white adipocytes is greatest during the fall, less during the winter, and least in spring and summer. This seasonal difference in insulin sensitivity cannot be explained by differences in insulin-receptor binding affinity, which is constant year round, or the number of high- or low-affinity binding sites, which are 8 and 15 times greater, respectively, in summer than in winter (Castex and Sutter, 1981). The seasonal difference in insulin sensitivity is due to greater glucose oxidation via the pentose phosphate pathway than the glycolytic pathway; the activity of glucose-6-phosphate dehydrogenase, a key enzyme in the pentose phosphate pathway, is 6 and 4 times greater during the fall and winter, respectively, than it is during spring and summer (Castex and Sutter, 1981). Since an increased pentose phosphate pathway activity favors lipogenesis, the seasonal difference in insulin-induced glucose oxidation is in tune with the prehibernation fat accumulation in this species. In the red-cheeked suslik (Daudova and Soliternova, 1972), the insulin-stimulated glucose transport in adipose tissue is essentially the same between active and hibernating animals, suggesting no seasonal change in tissue sensitivity to insulin in this function.

The changes of glucagon during hibernation are similar to those of insulin. In the hedgehog (Hoo-Paris *et al.*, 1982) and the dormouse (Hoo-Paris *et al.*, 1985), the plasma glucagon level is markedly reduced during hibernation. The A cells are inactive during hibernation and glucagon is probably stored during hibernation (Raths and Kulzer, 1976). Exogenous glucagon fails to induce hyperglycemia during hibernation but does so above a T_b of 25°C during arousal. It appears that a depressed T_b during hibernation has a similar effect on glucagon secretion and binding as it does on insulin as described above. However, during arousal from hibernation, in addition to the increase in T_b, the increased sympathetic activity would facilitate the secretion of glucagon from the A cells (Hoo-Paris *et al.*, 1982). In contrast to the above findings in the hedgehog, both the A and B cells remain functional, even during hibernation, in the marmot (Florant *et al.*, 1986), suggesting the existence of species differences in endocrine pancreas function.

7. PINEAL

The role of the pineal in hibernation is presently unclear (Wang, 1982; Heldmaier and Lynch, 1986). But since gonadal involution typically precedes hibernation and gonadal recrudescence precedes spring emergence, the pineal gland could be involved in the regulation of the pituitary–gonad axis in the annual hibernation cycle. In highly photosensitive hibernators, such as the golden hamster, exposure to a 2L : 22D photoperiod at 8°C results in atrophy of the gonads and accessory sexual glands (Hoffman and Reiter, 1965) and increased incidence of hibernation (Jansky et al., 1981), but pinealectomy results in the maintenance of sexual activity and decreased incidence of hibernation (Smit-Vis, 1972; Jansky et al., 1981). Pinealectomy reduces the incidence of torpor in the white-footed mouse (Lynch et al., 1980) but has no effect in adult golden-mantled or Richardson's ground squirrels (Harlow et al., 1980). However, in juvenile golden-mantled ground squirrels which are born and raised in the laboratory, pinealectomy at 10 days of age increases the length of the hibernation season (Ralph et al., 1982). The duration of individual hibernation bouts is also not affected by pinealectomy in the 13-lined (Sinnamon and Pivorun, 1982) and Richardson's ground squirrels (Harlow et al., 1980). In the golden-mantled ground squirrel, however, there appears to be a transient lengthening of the individual hibernation bouts in the first 30 days of monitoring following pinealectomy, but they return to normal thereafter (Harlow et al., 1980). Pinealectomized golden-mantled ground squirrels show a normal circannual rhythm for hibernation in the first year but a temporally compressed hibernation cycle in the second year: terminal arousal and testicular recrudescence are approximately 6 weeks earlier than normal (Phillips and Harlow, 1982). Thus, in this species, the pineal is somewhat involved in the expression of the annual hibernation cycle, but how it acts remains unknown.

Administration of the pineal hormone, melatonin, has been shown to increase the incidence of daily torpor in the white-footed mouse (Lynch et al., 1980) and of hibernation in the golden-mantled ground squirrel (Palmer and Riedesel, 1976) and the golden hamster if this latter species is kept in constant light, but not in constant darkness (Jansky et al., 1981). Melatonin implants, either in beeswax or silastic tubing, or daily subcutaneous injections late in the photophase have no effect on the duration of hibernation bouts in the Richardson's (Ralph et al., 1982) or the 13-lined ground squirrel (Sinnamon and Pivorun, 1981a,b, 1982), intact or pinealectomized. Reduction of endogenous melatonin titers by melatonin immunogen (Sinnamon and Pivorun, 1982) also fails to alter the duration of individual hibernation bouts. Thus, except in one incidence (Palmer and Riedesel, 1976), melatonin treatment has little effect in altering the incidence or duration of hibernation in ground squirrels.

In the marmot, the plasma melatonin level during hibernation is similar to the lowest level found in euthermia despite the 30°C or more decrease in T_b (Florant and Tamarkin, 1984). In the hibernating golden-mantled ground squirrel (Ralph *et al.*, 1982), the plasma melatonin levels are similar among the intact, sham-pinealectomized, and pinealectomized groups; these levels are also similar to those found in the active state prior to surgery. In periodically aroused animals, plasma melatonin levels are also similar among the three groups, but the absolute levels are decreased to approximately 52–65% of those found during hibernation. The most striking finding is that even after pinealectomy, plasma melatonin titer remains high, substantiating earlier findings that melatonin may be provided from extrapineal sources, e.g., the retina and the Harderian gland (Cardinali and Wurtman, 1972). Thus, the effects of pinealectomy on the seasonal cycle of hibernation may or may not be related to circulating levels of melatonin, and other products from the pineal (e.g., peptides and related substances; Benson and Ebels, 1981) must also be considered (Ralph *et al.*, 1982).

In the male Uinta ground squirrel (Ellis and Balph, 1976), the activity of the enzyme, hydroxyindole-O-methyltransferase, which converts N-acetylserotonin to melatonin in the pineal, is highest just prior to hibernation, suggesting that an increased synthesis of melatonin may occur prior to or during hibernation. Consonant with this possibility, the pinealocytes of hibernating 13-lined ground squirrels are characterized by enlarged nuclei and nucleoli and the presence of fine granular materials in membranes of the endoplasmic reticulum, both of which suggest heightened synthetic activity (McNulty and Dombrowski, 1980; McNulty *et al.*, 1980). A similar enlargement of the nuclei and nucleoli of pinealocytes has also been observed in the dormouse (Legait *et al.*, 1975). On the other hand, based on the number and morphology of Golgi complexes, the secretory activity of the pineal during the hibernating season is uncertain, but it appears to be either increased throughout (McNulty and Dombrowski, 1980) or at least during periodic arousals (Frink *et al.*, 1978). In the hibernating red-cheeked suslik (Popova *et al.*, 1975), the serotonin content of the pineal is reduced to approximately 38%

that of the summer value, and other morphological characteristics are also indicative of an inactive pineal. Since no systematic measurements on pineal and plasma levels of melatonin have been made with the progression of hibernation, it is not possible to ascertain whether the status of the pineal changes throughout the hibernation season. It may be speculated, however, that there could be an initial high synthetic activity for pineal melatonin and polypeptides prior to and during the early phase of hibernation followed by a moderate to low rate of their secretion throughout the hibernation season— events which are not inconsistent with the observed relatively constant plasma melatonin level, the biochemical and histological changes, and the low serotonin content of the pineal. Furthermore, a relatively stable blood level of melatonin could exert a "counter antigonadotropic effect" (Reiter et al., 1974), possibly by releasing the inhibitory effect of melatonin on gonadotropins and prolactin secretion (Vaughan, 1981) through the down-regulation of brain melatonin receptor binding sites (Cardinali, 1981). If this were true, a physiological mechanism could be envisaged by which spontaneous gonadal recrudescence is allowed to occur during hibernation even though in nature the animals are still in total darkness and have a low, stable burrow temperature (Wang, 1979).

8. GONADS

In seasonal hibernators such as the ground squirrel, marmot, and hedgehog, involution of the gonads occurs in the summer and the animals enter hibernation with atrophied gonads. By spring emergence, however, the gonads are fully active, ready for breeding. The reactivation of the gonads, therefore, must have taken place during the hibernation season even though the animals spend much of their time with a depressed T_b and in total darkness in the burrow (Wang, 1982). Measurements on circannual changes in plasma testosterone levels in a variety of species (e.g., European hedgehog: Saboureau and Boissin, 1978; Saboureau, 1986; ground squirrels: Licht et al., 1982; Ellis et al., 1983; chipmunk: Scott et al., 1981; garden dormouse: Ambid and Berges, 1981; edible dormouse: Jallageas and Assenmacher, 1983) indicate very low values during the hibernating season, but these may increase drastically either late in the hibernation season or shortly after spring emergence. How this spontaneous gonadal recrudescence occurs is currently unknown, but a pineal-mediated regulation (see above) or altered hypothalamic, pituitary, and testicular 5-HT metabolism resulting in removal of inhibition of testicular development (Ellis et al., 1983) are some of the possibilities. Despite this gap in knowledge, a fundamental question can nevertheless be raised, namely, whether the involution of gonads is a prerequisite for hibernation and whether gonadal recrudescence is the cause for spring emergence. Recent studies on the golden-mantled ground squirrel (Kenagy, 1980) and the chipmunk (Kenagy, 1981) indicate that reproduction and hibernation are based on interrelated but independent circannual rhythms; the circannual testicular cycle persists even though the animals have been discouraged from hibernating when maintained at a relatively high T_a (23°C). This confirms similar earlier observations as summarized by Kayser (1961) and provides further support for the study of Pengelley and Asmundson (1974), that in male and female golden-mantled ground squirrels, castration does not affect the circannual periodicity of hibernation. Even more convincingly, a study on the antelope ground squirrel, a closely related desert species which does not hibernate (Kenagy, 1981), has revealed that the circannual testicular rhythm persists under a constant photoperiod and temperature even though this species lacks a typical circannual body-weight rhythm which is characteristic of nearly all seasonal hibernators. It therefore appears that in some seasonal hibernators, gonadal activities have little regulatory influence on the occurrence and termination of hibernation.

In contrast, the gonadal influence on occurrence of hibernation is substantial in some other seasonal (e.g., hedgehog) and nonseasonal (e.g., hamsters) hibernators. In the hedgehog (Saboureau, 1986), castration prolongs the hibernation season by 2 months in the male but has no effect in the female. Administration of testosterone to either intact or castrated males significantly reduced the incidence of hibernation. In the hamster, exposure to a short photoperiod and/or cold results typically in gonadal atrophy after 3–4 weeks. Although involution of the testis is not a prerequisite for hibernation in either the golden (Smit-Vis and Smit, 1970) or the Turkish hamster (Hall and Goldman, 1980), gonadal atrophy (Smit-Vis, 1972) or castration (Hall and Goldman, 1980; Jansky et al., 1981) significantly improves the occurrence of hibernation. In the Turkish hamster (Hall and Goldman, 1980), castration prolongs the hibernation season from the typical 5–6 months to more than 18 months. Chronic implants of testosterone inhibit hibernation in a dose-dependent manner in the male golden hamster (Jansky et al., 1980) and in both sexes of Turkish hamsters (Hall and Goldman, 1980). Thus, in hamsters, gonadal activity appears to exert a regulatory role in the onset and termination of hibernation, but the mechanisms are currently unknown. As a possibility, the known effect of testosterone in elevating the set-point for T_b regulation (Johansen and Cross, 1980) may be a contributing factor.

The gradual increase in testicular weight throughout the hibernation season in the ground squirrel, European hedgehog, European hamster, and garden dormouse (see Wang, 1982) indicates that growth continues during this period. Whether cell proliferation and/or differentiation occur during depressed T_b or only during euthermia between hibernation bouts is not clear. Equally unclear are whether the secretion of gonadotropins occurs during periodical arousal and whether the binding of gonadotropins to target cells is also subject to seasonal as well as temperature modulations. In a recent study of the golden-mantled ground squirrel (Barnes *et al.*, 1986), testicular functions (*in vitro* testosterone production as stimulated by LH) were inhibited by low temperature and no major testicular growth occurred during hibernation. The authors proposed that significant testicular growth only occurs in the 30 days after the animal has regained permanent euthermia but remains in the burrow prior to spring emergence. It is apparent that more studies are needed to elucidate both the time course and the endocrine mechanism for the regulation of spontaneous gonadal recrudescence in seasonal hibernators in the cold and dark.

9. ADRENAL GLUCOCORTICOIDS

The role of the adrenals in hibernation is presently unclear (see Wang, 1982). In the ground squirrel (Popovic, 1960), adrenalectomy prevents hibernation but grafting of cortical tissue into the anterior eye chamber or injection of corticosteroids restores hibernation. In the Syrian hamster, adrenalectomy prevents hibernation whereas subcutaneous implants of corticosteroids enhance hibernation (Jansky *et al.*, 1981). In view of the wide-ranging metabolic and thermogenic effects of glucocorticoids (Musacchia and Deavers, 1981), it is difficult to identify the precise mechanisms by which corticosteroids may exert their prohibernation action.

The status of adrenal activity in relation to hibernation appears to be different in different species, and it ranges from inactive and unchanged to highly active. In the red-cheeked suslik (Popova and Koryakina, 1981), the activity of the pituitary-adrenal axis is highest in the spring and lowest in the fall; the plasma level of glucocorticoids is lowest during hibernation. However, no seasonal difference in adrenal sensitivity to ACTH stimulation is apparent; the net increase of plasma corticoids by exogenous ACTH remains constant in the spring, summer, and fall. In contrast, the adrenal from hibernating or newly aroused 13-lined ground squirrels cannot be stimulated by ACTH even at 37°C, whereas the adrenal from summer active squirrels shows a 30-fold increase in secretion by the same treatment (Huibregtse *et al.*, 1971). This suggests a seasonal tissue refractoriness in the adrenal of the 13-lined ground squirrel.

In the garden dormouse and European hedgehog (Boulouard, 1972), the plasma glucocorticoid level during hibernation is similar to that found in the euthermic state following arousal. However, adrenals sampled during hibernation are incapable of secreting steroids at 8°C even with ACTH present, but secretion can be increased by incubating at 37°C and can be fully restored if ACTH is present. This indicates that the relatively constant glucocorticoid level between the hibernating and euthermic states is likely due to diminished secretion and utilization during hibernation and increased secretion and utilization during the arousal process. A diminished peripheral utilization of cortisol during hibernation has been demonstrated in the hedgehog (Saboureau *et al.*, 1980). In the woodchuck, the plasma cortisol level measured through remote blood sampling techniques is similar between spring and fall (Florant and Weitzman, 1980) and is similar to that found in the marmot during hibernation (Kastner *et al.*, 1978), indicating no seasonal variation in basal cortisol level. However, the daily urinary excretion of 17-hydroxysteroids is lowest during the hibernating season and highest in the spring (Wenberg and Holland, 1973b). Thus, a combined decrease in rates of secretion, utilization, and excretion during the hibernating season may be responsible for the relatively constant seasonal plasma cortisol levels.

In the little brown bat (Gustafson and Belt, 1981), the plasma cortisol level is higher during the hibernating season than it is during the nonhibernating season; morphological indices also indicate an increased secretion during winter. This seasonal difference in plasma cortisol level suggests alteration in the feedback regulation of the brain–pituitary–adrenal axis similar to that observed in the red-cheeked suslik (Popova and Koryakina, 1981). The significance of an active adrenal during hibernation in this bat is unknown, but in view of the drastic increase in plasma glucose (3.5 times normal) in the hibernating big brown bat (Hinkley and Burton, 1970), which is a close relative, an altered carbohydrate metabolism may be involved in hibernation in these animals.

10. ADRENAL MINERALOCORTICOIDS

There appears to be evidence that both the secretion and the utilization of aldosterone are maintained or even increased during hibernation. For instance, in the Syrian hamster (Raths and Kulzer, 1976), the width of the zona glomerulosa and the width and size of cell nuclei contained therein are increased during hibernation. In the European hamster (Bloch and Can-guilhem, 1966), both adrenal aldosterone content and urinary aldosterone increase with the duration of hibernation. In the marmot (Kastner *et al.*, 1978), plasma aldosterone and renin concentrations increase with the progression of time in a hibernation bout, and normal plasma renin activity is retained in the hibernating 13-lined ground squirrel (Edmonson, 1976). In the red-cheeked suslik (Kolpakov and Samsonenko, 1970), *in vitro* secretion of aldosterone, as stimulated by bovine angiotensin II, is twice as great in adrenals from hibernating animals as in those from active animals, indicating greater tissue responsiveness during hibernation. In view of the lack of feeding, and thus salt intake, during the hibernation season in these seasonal hibernators and the disturbance of Na^+/K^+ distribution across the cell membrane after prolonged hibernation (Willis, 1982), the normal or increased aldosterone secretion and utilization during hibernation is consistent with the physiological needs of electrolyte conservation.

II. ADRENAL MEDULLA

The adrenal medulla is apparently not necessary for hibernation in the ground squirrel since demedullated animals can enter and arouse from hibernation (Popovic, 1960). However, the role of CA in mediating calorigenesis is well known (see Section II,B,1), and the activation of the SAM system during arousal from hibernation is well documented (Lyman *et al.*, 1982). In the woodchuck (Florant *et al.*, 1982), significant decreases in plasma CA levels have been observed during entry into, and in deep hibernation. During arousal, however, NE and E increase more than 30-fold. These very significant changes attest to the very substantial contributions the adrenal medulla makes in mediating the cardiovascular and calorigenic requirements associated with periodic arousal.

In the hedgehog, the adrenal content of CA increases prior to and during hibernation, mostly due to increased E (Helle *et al.*, 1980), resulting in a peak E to NE ratio (Johansson, 1978). This is reflected in the increase, during the autumn, of the activity of the enzyme phenylethanolamine-*N*-methyltransferase, which catalyzes the conversion of NE to E. During hibernation, NE is not detectable in the medulla but may be synthesized by a rapid switching-on of the enzyme, dopamine-B-hydroxylase, which catalyzes DA to NE (Helle *et al.*, 1980). In the European ground squirrel (Petrovic *et al.*, 1978), a similar increase in total CA and a higher E to NE ratio have also been observed during hibernation; these changes are preceded by an increase in the activity of tyrosine hydroxylase, which converts L-tyrosine to DOPA, the rate-limiting step in the biosynthesis of CA (Petrovic *et al.*, 1978). Enzymes involved in the catabolism of CA, monoamine oxidase, and catechol-O-methyltransferase have highest activities in the winter and spring and lowest in the summer and fall (Johansson, 1978). This is probably an indication of the seasonal differences in sympathetic activities, as these are high in the winter during periodic arousals and in the spring, when emerged animals are confronted by cold.

12. ANTIMETABOLIC PEPTIDES, HIBERNATION INDUCTION TRIGGER, AND ENDOGENOUS OPIOIDS

The possible existence of an antimetabolic peptide in the brain of hibernating ground squirrels has been reviewed recently (Swan, 1981). Although earlier results indicated that a brain extract from the hibernating 13-lined ground squirrel, when injected intravenously, is capable of inducing a state of torpor in the white rat with decreases in oxygen consumption by 35% and T_b by 3°C (Swan and Schatte, 1977), more recent studies employing an i.c.v. injection have failed to demonstrate a similar antimetabolic effect (Swan, 1981).

Dawe and Spurrier (1969) were the first to show that by transfusing blood from a hibernating ground squirrel to a recipient of the same species, hibernation can be induced in the recipient during the summer, when hibernation is normally absent. Subsequent studies have shown that the blood-borne "hibernation induction trigger" is a heat-labile peptide less than 12,000 MW and that no species specificity is involved in its action, i.e., serum from a hibernating woodchuck can induce hibernation in the 13-lined ground squirrel (Dawe, 1978). However, attempts to demonstrate a similar effect using hibernators other than the 13-lined ground squirrel as the recipient have been unsuccessful (Abbotts *et al.*, 1979; Galster, 1978; Minor *et al.*, 1978). It is not presently clear why there exists such a discrepancy.

Biochemical efforts aiming to isolate the hibernation induction trigger from the plasma of hibernating woodchucks have been summarized recently (Oeltgen and Spurrier, 1981). No pure preparation is currently available but the biologically active component is tightly associated with the serum albumin fraction. A bioassay using summer hibernation in the 13-lined ground squirrel as the endpoint indicates that induction of hibernation occurs in 2 days to 5 weeks following intravenous injection of lyophilized albumin fraction. Another bioassay using macaque monkeys has shown that bradycardia, hypothermia, behavioral depression and aphagia (Myers *et al.*, 1981), and reduced renal creatinine clearance (Oeltgen *et al.*, 1982) are typical following i.c.v. administration of the lyophilized albumin fraction. More interestingly, some of these effects may be retarded or reversed by opiate antagonists (naloxone and naltrexone), suggesting that the hibernation induction trigger may be an endogenous opioidlike peptide (Oeltgen *et al.*, 1982).

The possible involvement of endogenous opioids in hibernation has been receiving increasing attention (see Beckman, 1986). In the red-cheeked suslik (Kramarova *et al.*, 1983), significant increases in brain Met- and Leu-enkephalins have been observed during hibernation, which is consistent with the increase in Met-enkephalin immunoreactivity in the hypothalamic neurons of the hibernating hedgehog (Nurnberger, 1983). In the hibernating Turkish hamster, naloxone increases heart rate and triggers arousal, but it has no cardioaccelerating effect in the euthermic animal (Margulis *et al.*, 1979). In the garden dormouse, naltrexone decreases the incidence of deep hibernation (Kromer, 1980). Both observations seem to suggest increased opioid activity during hibernation. Furthermore, in the golden-mantled ground squirrel, physical dependence on morphine fails to develop during hibernation but does so during euthermia (Beckman *et al.*, 1981), suggesting greater occupancy of opiate receptors by endogenous opioids during hibernation. This is further supported by the observation that naloxone, when given i.c.v., results in a dose-dependent decrease in the duration of hibernation bout (Llados-Eckman and Beckman, 1983; Beckman, 1986). It therefore appears that increased central opioid activity might be involved in hibernation and that this increase may be manifested by the presence of opioids in the peripheral circulation and appear as the hibernation induction trigger.

IV. Summary

The quest of achieving an optimal internal thermal state to maximize biochemical efficiency encompasses both behavioral and autonomic efforts. Although superficially divergent in internal thermal state among the vertebrates, both ectotherms and endotherms share many common neural elements in sensing, deciphering, and responding to environmental thermal challenges. Since in fish behavioral selection of preferred T_b and the alteration of such behavior by brain lesions (Nelson and Prosser, 1979), thermal manipulation (Hammel *et al.*, 1969), monoamines (Green and Lomax, 1976), peptides (Kavaliers, 1982a; Kavaliers and Hawkins, 1981), steroids (Johansen and Cross, 1980), melatonin (Kavaliers, 1982a), and pyrogens (Covert and Reynolds, 1977) has been demonstrated, it is apparent that all necessary neural elements for thermoregulation and responsiveness to these agents already exist even in the most primitive of the vertebrates.

In the endotherms, autonomic capabilities allow the maintenance of a relatively constant internal thermal state independent of environmental fluctuations. Numerous studies have been devoted to characterizations of the thermal detectors (Hensel, 1974), afferents (Bruck and Hinckel, 1982), CNS integrators (Myers, 1980; Bligh, 1979), and efferents for thermogenesis (Jansky, 1973) and thermolysis (Hales, 1974) and to the susceptibility of the individual elements to neuroendocrine influence. The bulk of our current knowledge appears to be in two areas: neuroendocrine effects on (1) the CNS controller/integrator and (2) on the effectors; relatively little is known on the afferent side. In the CNS, biogenic amines (Myers, 1980; Bligh, 1979), ions (Myers, 1982), peptides (Clark and Lipton, 1983), and steroids (Wilkinson et al., 1980; Carlisle et al., 1979) act either as neurotransmitters or as neuromodulators to activate or inhibit effector pathways, be they for heat conservation/production or heat loss or to modify the set-point for thermoregulation. Specific examples for each have been presented to illustrate the points but, except in a very few cases, the physiological significance of these neuroendocrine influences remains unclear.

On the effector side, significant advancement has been made in understanding NST and brown adipose tissue in combating cold in certain mammals (Foster and Frydman, 1978, 1979; Nicholls, 1979; Cannon et al., 1981), especially with regard to the roles of the SAM (Gale, 1973; Himms-Hagen, 1975; Young and Landsberg, 1979), thyroid hormones (Edelman, 1974; Himms-Hagen, 1983), and corticosteroids (Deavers and Musacchia, 1979; Musacchia and Deavers, 1981). Here both the dynamic (Himms-Hagen, 1981; Cannon et al., 1981) and the trophic (Barnard et al., 1980) roles of NE on NST and brown adipose tissue and the permissive role of thyroid hormones (Fregly et al., 1979; Himms-Hagen, 1983) and corticosterioids (Musacchia and Deavers, 1981) on catecholamine actions have been delineated. A more recent development in a similar vein is the role of the pineal and its hormone, melatonin, in mediating the effect of photoperiod on winter acclimatization and capacity for NST (Lynch and Epstein, 1976; Heldmaier et al., 1981). Taken together, the physiological importance of these neuroendocrine modulations on NST and survival in the cold has been amply demonstrated.

With regard to thermoregulation under modified physiological states, there appear to be downward shifts in CNS set-point in sleep (Heller and Glotzbach, 1977) and hibernation (Heller, 1979) and upward shifts in fever (Kluger, 1981). The neuroendocrine influence on thermoregulation during sleep is largely unknown, but in view of the recent discovery of a "sleep factor," a glycopeptide (Krueger et al., 1982) which induces slow-wave sleep in rabbits when given i.c.v., the arousal state of the CNS and thus the associated thermoregulatory changes are likely to be modifiable by neuroendocrine influences. In fever, the elevated CNS set-point is thought to be mediated by prostaglandins (Milton, 1982b) with possible involvement of new protein synthesis (Milton and Sawhney, 1982). Antipyresis has been demonstrated possible by at least three endogenous neuropeptides, AVP (Veale et al., 1981), α-MSH, and ACTH (Lipton et al., 1981), although the mechanisms of their actions are not known. For hibernation, although the drastic changes in metabolism and other physiological functions suggest strong neuroendocrine regulation in amplifying the range of T_b within which biochemical and physiological homeostasis can be sustained, a cause–effect relationship has yet to be established (Wang, 1982). But the strong seasonal connotation in the preparation for hibernation, e.g., rapid fattening prior to hibernation, has already provided some molecular insights about the biochemical coupling between insulin receptors and carbohydrate and lipid metabolism in adipocytes during this special phase (Castex and Sutter, 1981). Similarly, the role of brain monoamines in regulating hypothalamo–hypophyseal functions in addition to their thermoregulatory effects have also been receiving increasing attention (Jansky, 1978; Novotona and Civin, 1979). More recently, the involvement of endogenous opioids in hibernation has been gaining interest (Margulis et al., 1979; Oeltgen et al., 1982; Llados–Eckman and Beckman, 1983). Whether there is a functional equivalence between the opioids and the hibernation "trigger" (Dawe and Spurrier, 1969) remains to be elucidated. It is apparent that seasonal hibernators such as ground squirrels and hedgehogs offer rich opportunities for neuroendocrine studies on temporal changes of receptor–ligand interactions both at the CNS and at the effector level.

Acknowledgment

Literature survey was aided by a Natural Science and Engineering Research Council of Canada Operating Grant No. A6455 to L. Wang.

References

Abbotts, B., Wang, L. C. H., and Glass, J. D. (1979). Absence of evidence for a hibernation "trigger" in blood dialyzate of Richardson's ground squirrel. *Cryobiology* **16**, 179–183.

Adair, E. (1974). Hypothalamic control of thermoregulatory behavior: Preoptic-posterior hypothalamic interaction. In "Recent Studies of Hypothalamic Function" (K. Lederis and K. E. Cooper, eds.), pp. 341–358. Karger, Basel.

Adair, E. (1976). Autonomic thermoregulation in squirrel monkey when behavioral regulation is limited. J. Appl. Physiol. 40, 694–700.

Agid, R., Ambid, L., Sable-Amplis, R., and Sicart, R. (1978). Aspects of metabolic and endocrine changes in hibernation. In "Strategies in Cold: Natural Torpidity and Thermogenesis" (L. C. H. Wang and J. W. Hudson, eds.), pp. 499–540. Academic Press, New York.

Allen, T. E., and Bligh, J. (1969). A comparative study of the temporal patterns of cutaneous water vapor loss from some domesticated mammals with epitrichial sweat glands. Comp. Biochem. Physiol. 31, 347–363.

Allison, T., and Van Twyver, H. (1972). Electrophysiological studies of the echidna Tachyglossus aculeatus. II. Dormancy and hibernation. Arch. Ital. Biol. 110, 184–194.

Allison, T., Van Twyver, H., and Goff, W. R. (1972). Electrophysiological studies of the echidna Tachyglossus aculeatus. I. Waking and sleeping. Arch. Ital. Biol. 110, 145–183.

Aloia, R. C., and Pengelley, E. T. (1979). Lipid composition of cellular membranes of hibernating mammals. Chem. Zool. 11, 1–47.

Ambid, L., and Berges, R. (1981). Seasonal rhythm in plasma testosterone levels and gonadal activity in the hibernating garden dormouse. Cryobiology 18, 88.

Amin, A. H., Crawford, T. B. B., and Gaddum, J. H. (1954). The distribution of substance P and 5-hydroxytryptamine in the central nervous system of the dog. J. Physiol. (London) 126, 596–618.

Amico, D. J. D., Calne, B., and Klawans, H. L. (1976). Altered hypothermic responsiveness to (+) amphetamine. J. Pharm. Pharmacol. 28, 154–156.

Andersson, B. (1970). Central nervous and hormonal interaction in temperature regulation of the goat. In "Physiological and Behavioral Temperature Regulation" (J. D. Hardy, A. P. Gagge, and J. A. J. Stolwijk, eds.), pp. 634–647. Thomas, Springfield, Illinois.

Andersson, B., Ekman, L., Gale, C. C., and Sundsten, J. W. (1963). Control of thyrotrophic hormone (TSH) secretion by the "heat loss center." Acta Physiol. Scand. 59, 12–33.

Assimacopoulos-Jennet, F. D., Blackmore, P. F., and Exton, J. H. (1977). Studies on α-adrenergic activation of hepatic glucose output: Studies on role of calcium in α-adrenergic activation of phosphorylase. J. Biol. Chem. 252, 2662–2669.

Augee, M. L., Raison, J. K., and Hulbert, A. J. (1979). Seasonal changes in membrane lipid transitions and thyroid function in the hedgehog. Am. J. Physiol. 236, E589–E593.

Avery, D. D., Hawkins, M. F., and Wunder, B. (1981). The effects of injections of bombesin into the cerebral ventricles on behavioral thermoregulation. Neuropharmacology 20, 23–27.

Baird, J., and Lang, W. J. (1973). Temperature responses in the rat and cat to cholinomimetic drugs injected into the cerebral ventricles. *Eur. J. Pharmacol.* **21,** 203–311.

Baldino, F., Jr. (1986). Norepinephrine suppression of neuronal thermosensitivity. *In* "Homeostasis and Thermal Stress: Experimental and Therapeutic Advances" (K. Cooper, P. Lomax, E. Schonbaum, and W. L. Veale, eds.), pp. 99–102. Karger, Basel.

Baldwin, B. A., Ingram, D. L., and LeBlanc, J. A. (1969). The effects of environmental temperature and hypothalamic temperature on excretion of catecholamines in the urine of the pig. *Brain Res.* **16,** 511–515.

Balsam, A., and Sexton, F. C. (1975). Increased metabolism of iodothyronine in the rat after short-term cold adaptation. *Endocrinology (Baltimore)* **97,** 385–391.

Barker, J. L., and Carpenter, D. O. (1970). Thermosensitivity of neurons in the sensorimotor cortex of the cat. *Science* **169,** 191–194.

Barnard, T., Mory, G., and Nechad, M. (1980). Biogenic amines and the trophic response of brown adipose tissue. *In* "Biogenic Amines in Development" (H. Pervez and S. Parvez, eds.), pp. 391–439. Elsevier, Amsterdam.

Barnes, B. M., Kretzmann, M., Licht, P., and Zucker, I. (1986). Reproductive development in hibernating ground squirrels. *In* "Living in The Cold" (H. C. Heller, X. J. Musacchia, and L. C. H. Wang, eds.), pp. 245–251. Elsevier, New York.

Bartholomew, G. A. (1956). Temperature regulation in the macropod marsupial *Setonix brachyurus. Physiol. Zool.* **29,** 26–40.

Bartholomew, G. A., and Lasiewski, R. C. (1965). Heating and cooling rates in marine iguana. *Comp. Biochem. Physiol.* **16,** 573–582.

Bauman, T. R., Anderson, R. R., and Turner, C. W. (1969). Thyroid hormone secretion rates and food consumption of the hamster (*Mesocricetus auratus*) at 22.5 and 4.5C. *Gen. Comp. Endocrinol.* **10,** 92–98.

Beckman, A. L. (1978). Hypothalamic and midbrain function during hibernation. *In* "Current Studies of Hypothalamic Function" (W. L. Veale and K. Lederis, eds.), Vol. 2, pp. 29–43. Karger, Basel.

Beckman, A. L. (1986). Functional aspects of brain opioid peptide systems in hibernation. *In* "Living in the Cold: Physiological and Biochemical Adaptations" (C. H. Heller, X. J. Musacchia, and L. C. H. Wang, eds.), pp. 225–234. Am. Elsevier, New York.

Beckman, A. L., and Eisenman, J. S. (1970). Microelectrophoresis of biogenic amines on hypothalamic thermosensitive cells. *Science* **170,** 334–336.

Beckman, A. L., and Stanton, T. L. (1982). Properties of the CNS during the state of hibernation. *In* "The Neural Basis of Behavior" (A. L. Beckman, ed.), pp. 19–45. Spectrum, New York.

Beckman, A. L., Llados-Eckman, C., Stanton, T. L., and Adler, M. W. (1981). Physical dependence on morphine fails to develop during the hibernating state. *Science* **212,** 1527–1529.

Behr, R., Zeisberger, E., and Merker, G. (1983). Response of the guinea-pig (*Cavia aperea porcellus*) to external cooling after aminergic denervation of the anterior hypothalamus. *J. Therm. Biol.* **8**, 125–128.

Benson, B., and Ebels, I. (1981). Other pineal peptides and related substances -physiological implications for reproductive biology. In "The Pineal Gland" (R. J. Reiter, ed.), Vol. 2, pp. 165–187. CRC Press, Boca Raton, Florida.

Bernardini, G. L., Lipton, J. M., and Clark, W. G. (1983). Intracerebroventricular and septal injections of arginine vasopressin are not antipyretic in the rabbit. *Peptides (N.Y.)* **4**, 195–198.

Bernheim, H. A., and Kluger, M. J. (1976). Fever: Effect of drug-induced antipyresis on survival. *Science* **193**, 237–239.

Bernheim, H. A., Gilbert, T. M., and Stitt, J. T. (1980). Prostaglandin E levels in third ventricular cerebrospinal fluid of rabbits during fever and changes in body temperature. *J. Physiol. (London)* **301**, 69–78.

Binkley, S., Kluth, E., and Menaker, M. (1971). Pineal function in sparrows: Circadian rhythms and body temperature. *Science* **174**, 311–314.

Blatteis, C. M., Bealer, S. L., Hunter, W. S., Llanos, Q. J., Ahokas, R. A., and Mashburn, T. A., Jr. (1983). Suppression of fever after lesions of the anteroventral third ventricle in guinea pigs. *Brain Res. Bull.* **2**, 529–537.

Bligh, J. (1973). "Temperature Regulation in Mammals and Other Vertebrates." Elsevier, Amsterdam.

Bligh, J. (1979). The central neurology of mammalian thermoregulation. *Neuroscience* **4**, 1213–1236.

Bligh, J., Davis, A. J., Sharman, D. F., and Smith, C. A. (1977). Unimpaired thermoregulation in the sheep after depletion of hypothalamic noreadrenaline by 6-hydroxydopamine. *J. Physiol. (London)* **265**, 51P.

Bloch, R., and Canguilhem, B. (1966). Cycle saisonnier délimination urinaire de l'aldosterone chez un hibernant, Cricetus cricetus. Influence de la température. *C. R. Seances Soc. Biol. Ses Fil.* **160**, 1500–1502.

Boulant, J. A. (1974). The effect of firing rate on preoptic neuronal thermosensitivity. *J. Physiol. (London)* **240**, 661–669.

Boulouard, R. (1972). Adrenocortical function in two hibernators: The garden dormouse and the hedgehog. *Proc. Int. Symp. Environ. Physiol. (Bioenergetics) FASEB* pp. 108–112.

Brittain, R. T., and Handley, S. L. (1967). Temperature changes produced by the injection of catecholamines and 5-hydroxytryptamine into the cerebral ventricles of the conscious mouse. *J. Physiol. (London)* **192**, 805–813.

Brown, M. R. (1981). Bombesin, somatostatin, and related peptides: Actions on thermoregulation. *Fed. Proc., Fed. Am. Soc. Exp. Biol.* **40**, 2765–2768.

Brown, M. R. (1982). Bombesin and somatostatin related peptides: Effects on oxygen consumption. *Brain Res.* **242**, 243–246.

Brown, M. R., and Vale, W. (1980). Peptides and thermoregulation. In "Thermoregulatory Mechanisms and Their Therapeutic Implications" (B. Cox, P. Lomax, A. S. Milton, and E. Schonbaum, eds.), pp. 186–194. Karger, Basel.

Brown, M. R., Rivier, J., and Vale, W. (1977). Actions of bombesin, thyrotropin releasing factor, prostaglandin E2 and naloxone on thermoregulation in the rat. *Life Sci.* **20**, 1681–1687.

Bruck, K. (1981). Basic mechanisms in longtime thermal adaptation. *Adv. Physiol. Sci.* **32**, 263–273.

Bruck, K., and Hinckel, P. (1980). Thermoregulatory noradrenergic and serotonergic pathways to hypothalamic units. *J. Physiol. (London)* **304**, 193–202.

Bruck, K., and Hinckel, P. (1982). Thermoafferent systems and their adaptive modifications. *Pharmacol. Ther.* **17**, 357–381.

Bruck, K., and Schwennicke, H. P. (1971). Interaction of superficial and hypothalamic thermosensitive structures in the control of nonshivering thermogenesis. *Int. J. Biometeorol.* **15**, 156–161.

Bruck, K., and Wunnenberg, W. (1970). Meshed control of two effector systems: Nonshivering and shivering thermogenesis. *In* "Physiological and Behavioral Temperature Regulation" (J. D. Hardy, A. P. Gagge, and J. A. J. Stolwijk, eds.), pp. 562–580. Thomas, Springfield, Illinois.

Bruinvels, J. (1970). Effect of noradrenaline, dopamine and 5-hydroxytryptamine on body temperature in the rat after intracisternal administration. *Neuropharmacology* **9**, 277–282.

Bruinvels, J. (1979). Norepinephrine. *In* "Body Temperature: Regulation, Drug Effects, and Therapeutic Implications" (P. Lomax and E. Schonbaum, eds.), pp. 257–288. Dekker, New York.

Bukowiecki, L., Follea, N., Paradis, A., and Collet, A. J. (1980). Stereospecific stimulation of brown adipose adipocyte respiration by catecholamines via β1-adrenoreceptors. *Am. J. Physiol.* **238**, E552–E563.

Bukowiecki, L., Collet, A. J., Follea, N., Guay, G., and Jahjah, L. (1982). Brown adipose tissue hyperplasia: A fundamental mechanism of adaptation to cold and hyperphagia. *Am. J. Physiol.* **242**, E353–E359.

Burks, T. F. (1972). Central alpha-adrenergic receptors in thermoregulation. *Neuropharmacology* **11**, 615–624.

Burlet, C., Robert, J., and Legait, E. (1974). Double tag study (Cesium 131, Cr-51-tagged red blood cells) of capillary exchanges in the neurohypophysis of a lerot (Eliomys quercinus L.) during his awakening from hibernation. *In* "Neuroscretion: The Final Neuroendocrine Pathway" (F. Knowles and L. Vollrath, eds.), p. 298. Springer, Berlin.

Cabanac, M. (1975). Temperature regulation. *Annu. Rev. Physiol.* **37**, 415–439.

Calder, C., and King, J. R. (1974). Thermal and caloric relations of birds. *In* "Avian Biology" (D. S. Farner and J. R. King, eds.), Vol. 4, pp. 259–413. Academic Press, New York.

Candas, V., Lobert, J. P., and Muzet, A. (1982). Heating and cooling stimulations during SWS and REM sleep in man. *J. Therm. Biol.* **7**, 155–158.

Canguilhem, B. (1970). Effets de la radiothyroidectomie et des injections d'hormone thyroidenne sur l'entrée en hibernation du Hamster d'Europe (*Cricetus cricetus*). *C. R. Seances Soc. Biol. Ses Fil.* **164,** 1366–1369.

Canguilhem, B., Kempf, E., Mack, G., and Schmitt, P. (1977). Regional studies of brain serotonin and norepinephrine in the hibernating awakening or active European hamster, *Cricetus cricetus,* during winter. *Comp. Biochem. Physiol.* **57C,** 175–179.

Canguilhem, B., Miro, J. L., Kempf, E., and Schmitt, P. (1986). Does serotonin play a role in entrance into hibernation? *Am. J. Physiol.* **251,** R755–761.

Cannon, B., and Nedergaard, J. (1983). Biochemical aspects of acclimation to cold. *J. Therm. Biol.* **8,** 85–90.

Cannon, B., Nedergaard, J., and Sundin, U. (1981). Thermogenesis, brown fat and thermogenin. *In* "Survival in the Cold" (X. J. Musacchia and L. Jansky, eds.), pp. 99–120. Am. Elsevier, New York.

Cantor, A., and Satinoff, E. (1976). Thermoregulatory responses to intraventricular norepinephrine in normal and hypothalamic-damaged rats. *Brain Res.* **108,** 125–141.

Cardinali, D. P. (1981). Melatonin: A mammalian pineal hormone. *Endocr. Rev.* **2,** 327–346.

Cardinali, D. P., and Wurtman, R. J. (1972). Hydroxyindol-O-methyl transferases in rat pineal, retina and Harderian gland. *Endocrinology (Baltimore)* **91,** 247–252.

Carey, F. G., and Teal, J. M. (1966). Heat conservation in tuna fish muscle. *Proc. Natl. Acad. Sci. U.S.A.* **56,** 1464–1469.

Carey, F. G., and Teal, J. M. (1969a). Mako and probeagle: Warm-bodied sharks. *Comp. Biochem. Physiol.* **28,** 199–204.

Carey, F. G., and Teal, J. M. (1969b). Regulation of body temperature by bluefin tuna. *Comp. Biochem. Physiol.* **28,** 205–213.

Carlisle, H. J. (1969). The effects of preoptic and anterior hypothalamic lesions on behavioral thermoregulation in the cold. *J. Comp. Physiol. Psychol.* **69,** 391–402.

Carlisle, H. J., Wilinson, C. W., Laudenslager, M. L., and Keith, L. D. (1979). Diurnal variation of heat intake in ovariectomized steroid treated rats. *Horm. Behav.* **12,** 232–242.

Carpenter, D. O. (1981). Ionic and metabolic bases of neuronal thermosensitivity. *Fed. Proc., Fed. Am. Soc. Exp. Biol.* **40,** 2808–2813.

Cassuto, Y., and Amit, Y. (1968). Thyroxine and norepinephrine effects on the metabolic rates of heat-acclimated hamster. *Endocrinology (Baltimore)* **82,** 17–20.

Cassuto, Y., and Chaffee, R. R. J. (1966). Effects of prolonged heat exposure on cellular metabolism of hamster. *Am. J. Physiol.* **210,** 423–426.

Castex, C., and Sutter, B. C. J. (1979). Seasonal variations of insulin sensitivity in edible dormouse (*Glis glis*) adipocytes. *Gen. Comp. Endocrinol.* **38,** 365–369.

Castex, C., and Sutter, B. C. J. (1981). Insulin binding and glucose oxidation in edible dormouse (*Glis glis*) adipose tissue: Seasonal variations. *Gen. Comp. Endocrinol.* **45,** 273–278.

Chaffee, R. R. J., and Allen, J. R. (1973). Effects of ambient temperature on the resting metabolic rate of cold- and heat-acclimated *Macaca mulatta*. *Comp. Biochem. Physiol. A* **44A**, 1215–1225.

Chaffee, R. R. J., and Roberts, J. C. (1971). Temperature acclimation birds and mammals. *Annu. Rev. Physiol.* **33**, 155–202.

Chandra, A., Chou, H. C., Chang, C., and Lin, M. T. (1981). Effects of intraventricular administration of neurotensin and somatostatin on thermoregulation in the rat. *Neuropharmacology* **20**, 715–718.

Chowers, I., Conforti, N., and Feldman, S. (1972). Body temperature and adrenal function in cold-exposed hypothalamic-disconnected rats. *Am. J. Physiol.* **223**, 341–345.

Clark, W. G. (1979a). Changes in body temperature after administration of amino acids, peptides, dopamine, neuroleptics and related agents. *Neurosci. Behav. Rev.* **3**, 179–231.

Clark, W. G. (1979b). Influence of opioids on central thermoregulatory mechanisms. *Pharmacol., Biochem. Behav.* **10**, 609–613.

Clark, W. G. (1980). Antipyresis: Mechanisms of action. *In* "Fever" (J. M. Lipton, ed.), pp. 131–140. Raven Press, New York.

Clark, W. G. (1981). Effects of opioid peptides on thermoregulation. *Fed. Proc., Fed. Am. Soc. Exp. Biol.* **40**, 2754–2759.

Clark, W. G., and Coldwell, B. A. (1972). Competitive antagonism of leukocytic pyrogen by sodium salicylate and acetaminophen. *Proc. Soc. Exp. Biol. Med.* **141**, 669–672.

Clark, W. G., and Cumby, H. R. (1975). The antipyretic effect of indomethacin. *J. Physiol. (London)* **248**, 625–638.

Clark, W. G., and Lipton, J. M. (1983). Brain and pituitary peptides in thermoregulation. *Pharmacol. Ther.* **22**, 249–297.

Clark, W. G., and Lipton, J. M. (1985). Changes in body temperature after administration of amino acids, peptides, dopamine, neuroleptics and related agents. II. *Neurosci. Biobehav. Rev.* **9**, 299–371.

Clausen, T., and Flatman, J. A. (1977). The effects of catecholamines on Na-K transport and membrane potential in rat soleus muscle. *J. Physiol. (London)* **270**, 383–414.

Cohn, M. L., Cohn, M., Krzysik, B. A., and Taylor, F. H. (1976). Regulation of behavioral events by thyrotropin releasing factor and cyclic AMP. *Pharmacol., Biochem Behav.* **5**, Suppl. 1, 129–133.

Cohn, M. L., Cohn, M., and Taube, D. (1980). Thyrotropin releasing hormone induced hyperthermia in the rat inhibited by lysine acetylsalicylate and indomethacin. *In* "Thermoregulatory Mechanisms and Their Therapeutic Implications" (B. Cox, P. Lomax, A. S. Milton, and E. Schonbaum, eds.), pp. 198–201. Karger, Basel.

Collins, K. J., and Weiner, J. S. (1968). Endocrinological aspects of exposure to high environmental temperatures. *Physiol. Rev.* **48**, 785–839.

Cooper, K. E. (1987). The neurobiology of fever: Thoughts on recent developments. *Annu. Rev. Neurosci.* **10**, 297–324.

Cooper, K. E., Cranston, W. I., and Honour, A. J. (1965). Effects of intraventricular and intrahypothalamic injections of noradrenaline and 5-hydroxytryptamine on body temperature in conscious rabbits. *J. Physiol. (London)* **181**, 852–864.

Cooper, K. E., Preston, E., and Veale, W. L. (1976). Effects of atropine, injected into a lateral cerebral ventricle of the rabbit, on fevers due to intravenous leucocyte pyrogen and hypothalamic intraventricular injections of prostaglandin E1. *J. Physiol. (London)* **254**, 729–741.

Cooper, K. E., Kasting, N. W., Lederis, K., and Veale, W. L. (1979). Evidence supporting a role for vasopressin in natural suppression of fever in the sheep. *J. Physiol. (London)* **295**, 33–45.

Cossins, A. R. (1982). The adaptation of membrane dynamic structure to temperature. *In* "Effects of Low Temperature on Biological Membranes" (G. J. Morris and A. Clarke, eds.), pp. 83–106. Academic Press, London.

Covert, J. B., and Reynolds, W. W. (1977). Survival value of fever in fish. *Nature (London)* **267**, 43–45.

Cox, B. (1979). Dopamine. *In* "Body Temperature: Regulation, Drug Effects, and Therapeutic Implications" (P. Lomax and E. Schonbaum, eds.), pp. 231–255. Dekker, New York.

Cox, B., and Lee, T. F. (1977). Do central dopamine receptors have a physiological role in thermoregulation? *Br. J. Pharmacol.* **61**, 83–86.

Cox, B., and Lee, T. F. (1982). Role of central neurotransmitters in fever. *Handb. Exp. Pharmacol.* **60**, 125–150.

Cox, B., Kerwin, R., and Lee, T. F. (1978). Dopamine receptors in the central thermoregulatory pathways of the rat. *J. Physiol. (London)* **282**, 471–483.

Crawshaw, L. I. (1979). Acetylcholine. *In* "Body Temperature: Regulation, Drug Effects and Therapeutic Implications" (P. Lomax and E. Schonbaum, eds.), pp. 305–335. Dekker, New York.

Crawshaw, L. I. (1980). Temperature regulation in vertebrates. *Annu. Rev. Physiol.* **42**, 473–491.

Crompton, A. W., Taylor, C. R., and Jagger, J. A. (1978). Evolution of homeothermy in mammals. *Nature (London)* **272**, 333–336.

Cunningham, D. J., and Cabanac, M. (1971). Evidence from behavioral thermoregulatory responses of a shift in setpoint temperature related to the menstrual cycle. *J. Physiol. (Paris)* **63**, 236–238.

Cunningham, D. J., Stolwijk, J. A. J., Murakami, N., and Hardy, J. D. (1967). Responses of neurons in the proptic area to temperature, serotonin and epinephrine. *Am. J. Physiol.* **213**, 1570–1581.

Dascombe, M. J. (1982). Cyclic nucleotides and fever. *In* "Handbook of Experimental Pharmacology, Vol. 60, Pyretics and Antipyretics" (A. S. Milton, ed.), pp. 219–255. Springer-Verlag, Berlin.

Dascombe, M. J. (1985). The pharmacology of fever. *Prog. Neurobiol.* **25,** 327–373.

Daudova, G. M., and Soliternova, I. B. (1972). Influence of insulin on the absorption of glucose by adipose tissue of the ground squirrel Citellus suslicus during hibernation and arousal. *J. Evol. Biochem. Physiol. (Engl. Transl.)* **8,** 399–401.

Davies, J. A., Harrison, M. H., Cochrane, L. A., Edwards, R. J., and Gibson, T. M. (1981). Effect of saline loading during heat acclimatization on adrenocortical hormone levels. *J. Appl. Physiol.* **50,** 605–612.

Dawe, A. R. (1978). Hibernation trigger research updated. *In* "Strategies in Cold: Natural Torpidity and Thermogenesis" (L. C. H. Wang and J. W. Hudson, eds.), pp. 541–563. Academic Press, New York.

Dawe, A. R. and Spurrier, W. A. (1969). Hibernation induced in ground squirrels by blood transfusion. *Science* **163,** 298–299.

Dawson, W. R., and Hudson, J. W. (1970). Birds. *In* "Comparative Physiology of Thermoregulation" (G. C. Whittow, ed.), Vol. 1, pp. 223–310. Academic Press, New York.

Deavers, D. R., and Musacchia, X. J. (1979). The function of glucocorticoids in thermogenesis. *Fed. Proc., Fed. Am. Soc. Exp. Biol.* **38,** 2177–2181.

Demeneix, B. A., and Henderson, N. E. (1978a). Serum T4 and T3 in active and torpid ground squirrel, *Spermophilus richardsoni. Gen. Comp. Endocrinol.* **35,** 77–85.

Demeneix, B. A., and Henderson, N. E. (1978b). Thyroxine metabolism in active and torpid ground squirrels. *Gen. Comp. Endocrinol.* **35,** 86–92.

Depocas, F. (1958). Chemical thermogenesis in the functionally eviscerated cold-acclimated rat. *Can. J. Biochem. Physiol.* **36,** 691–699.

De Roij, Bligh, T. A. J. M., Smith, C. A., and Frens, J. (1978). Comparison of the thermoregulatory responses to intracerebro-ventricularly injected dopamine and noradrenaline in the sheep. *Naunyn-Schmiedeberg's Arch. Pharmacol.* **303,** 263–269.

Desautels, M., Zaror-Behrens, H., and Himms-Hagen, J. (1978). Increased purine nucleotide binding, altered polypeptide composition and thermogenesis in brown adipose tissue mitochondria of cold-acclimated rats. *Can. J. Biochem.* **56,** 378–383.

Dickenson, A. H., Hellon, R. F., and Taylor, C. C. M. (1979). Facial thermal input to the trigeminal spinal nucleus of rabbits and rats. *J. Comp. Neurol.* **185,** 203–210.

Dinarello, C. A. (1980). Endogenous pyrogens. *In* "Fever" (J. M. Lipton, ed.), pp. 1–9. Raven Press, New York.

Dinarello, C. A., Goldin, N. P., and Wolff, S. M. (1974). Demonstration and characterization of two distinct human leukocytic pyrogens. *J. Exp. Med.* **139,** 1369–1381.

Divakaran, P., and Friedmann, N. (1976). A fast in vitro effect of glucocorticoids on hepatic lipolysis. *Endocrinology (Baltimore)* **98,** 1550–1553.

Doi, K., Ohno, T., and Kuroshima, A. (1982). Role of endocrine pancreas in temperature acclimation. *Life Sci.* **30,** 2253–2259.

Draskoczy, P. R., and Lyman, C. P. (1967). Turnover of catecholamines in active and hibernating ground squirrels. *J. Pharmacol. Exp. Ther.* **155,** 101–111.

Duclaux, R., Fantino, M., and Cabanac, M. (1973). Comportement thermoregulateur chez Rana esculenta. Influence du rechauffement spinal. *Pfluegers Arch.* **342,** 347–358.

Duncan, R. J. S., and Tricklebank, M. D. (1978). On the stimulation of the rate of hydroxylation of tryptophan in the brain of hamsters during hibernation. *J. Neurochem.* **31,** 553–556.

Edelman, I. S. (1974). Thyroid thermogenesis. *N. Engl. J. Med.* **290,** 1303–1308.

Edmonson, E. J. (1976). Plasma renin activity and plasma electrolyte concentration in a hibernator. *Fed. Proc., Fed. Am. Soc. Exp. Biol.* **35,** 705.

Eisenman, J. S. (1972). Unit activity studies of thermoresponsive neurons. *In* "Essays on Temperature Regulation" (J. Bligh and R. E. Moore, eds.), pp. 55–69. Elsevier/North-Holland, Amsterdam.

Elizondo, R. S., and Johnson, G. S. (1981). Peripheral effector mechanisms of temperature regulation: The regulation of seating activities in primates. *Adv. Physiol. Sci.* **32,** 397–408.

Ellis, L. C., and Balph, D. F. (1976). Age and seasonal differences in the synthesis and metabolism of testosterone by testicular tissue and pineal hydroxindole-O-methyl transferase activity of Uinta ground squirrels, *Spermophilus armatus*. *Gen. Comp. Endocrinol.* **28,** 42–51.

Ellis, L. C., Palmer, R. A., and Balph, D. F. (1983). The reproductive cycle of male Uinta ground squirrels: Some anatomical and biochemical correlates. *Comp. Biochem. Physiol. A.* **74A,** 239–245.

Erskine, D. J., and Hutchinson, V. H. (1981). Melatonin and behavioral thermoregulation in the turtle, *Terrapene carolina triunguis*. *Physiol. Behav.* **26,** 991–994.

Evans, C. L., and Smith, D. F. G. (1956). Sweating responses in the horse. *Proc. R. Soc. London, Ser. B* **145,** 61–83.

Fain, J. N., and Rosenthal, J. W. (1971). Calorigenic action of triiodothyronine on white fat cells: Effects of ouabain, oligomycine, and catecholamines. *Endrocrinology (Baltimore)* **89,** 1205–1211.

Fain, J. N., Kovacev, V. P., and Scow, R. O. (1965). Effect of growth hormone and dexamethasone on lipolysis and metabolism in isolated fat cells in the rat. *J. Biol. Chem.* **240,** 3522–3529.

Fain, J. N., Jacobs, M. D., and Clement-Cormier, Y. C. (1973). Interrelationship of cyclic AMP, lipolysis, and respiration in brown fat cells. *Am. J. Physiol.* **224,** 346–351.

Feldberg, W., and Myers, R. D. (1963). A new concept of temperature regulation by amines in the hypothalamus. *Nature (London)* **200,** 1325.

Feldberg, W., and Myers, R. D. (1964). Effects on temperature of amines injected into the cerebral ventricles. A new concept of temperature regulation. *J. Physiol. (London)* **173**, 226–237.

Feldberg, W., and Myers, R. D. (1965). Changes in temperature produced by microinjections of amines into the anterior hypothalamus of cats. *J. Physiol. (London)* **177**, 239–245.

Ferren, L. G., South, F. E., and Jacobs, H. K. (1971). Calcium and magnesium levels in tissues and serum of hibernating and cold-acclimated hamsters. *Cryobiology* **8**, 506–508.

Fioretti, M. C., Riccardi, C., Menconi, E., and Martini, L. (1974). Control of the circadian rhythm of the body temperature in the rat. *Life Sci.* **14**, 211–219.

Firth, B. T., and Heatwole, H. (1976). Panting thresholds of lizards: The role of the pineal complex in panting responses in an agamid, *Amphibolurus muricatus*. *Gen. Comp. Endocrinol.* **29**, 388–401.

Firth, B. T., Ralph, C. L., and Boardman, T. J. (1980). Independent effects of the pineal and a bacterial pyrogen in behavioural thermoregulation in lizards. *Nature (London)* **285**, 399–400.

Fleischhauer, K., and Wittkowski, W. (1976). Morphological aspects of the formation, transport and secretion of releasing and inhibiting hormones. *Acta Endocrinol. (Copenhagen), Suppl.* **202**, 11–12.

Florant, G. L., and Tamarkin, L. (1984). Plasma melatonin rhythms in euthermic marmots (*Marmota flaviventris*). *Biol. Reprod.* **30**, 332–337.

Florant, G. L., and Weitzman, E. D. (1980). Diurnal and episodic pattern of plasma cortisol during fall and spring in young and old woodchucks (*Marmota monax*). *Comp. Biochem. Physiol. A* **66A**, 575–581.

Florant, G. L., Weitzman, E. D., Jayant, A., and Cote, L. J. (1982). Plasma catecholamine levels during cold adaptation and hibernation in woodchucks (*Marmota monax*). *J. Therm. Biol.* **7**, 143–146.

Florant, G. L., Hoo-Paris, R., Caster, Ch., Bauman, W. A., and Sutter, B. Ch. J. (1986). Pancreatic alpha and beta cell stimulation in euthermic and hibernating marmots, *Marmota flaviventris:* effects of glucose and arginine administration. *J. Comp. Physiol.* **156B**, 309–314.

Foster, D. O., and Frydman, M. L. (1978). Nonshivering thermogenesis in the rat. II. Measurements of blood flow with microsphres point to brown adipose tissue as the dominant site of the calorigenesis induced by noradrenaline. *Can. J. Physiol. Pharmacol.* **56**, 110–122.

Foster, D. O., and Frydman, M. L. (1979). Tissue distribution of cold-induced thermogenesis in conscious warm- or cold-acclimated rats reevaluated from changes in tissue blood flow: The dominant role of brown adipose tissue in the replacement of shivering by nonshivering thermogenesis. *Can. J. Physiol. Pharmacol.* **57**, 257–270.

Freeman, M. B. (1977). Lipolysis and its significance in the response to cold of the neonatal fowl, *Gallus domesticus*. *J. Therm. Biol.* **2**, 145–149.

Fregly, M. J., Field, F. P., Katovich, M. J., and Barney, C. C. (1979). Catecholamine-thyroid hormone interaction in cold-acclimated rats. *Fed. Proc., Fed. Am. Soc. Exp. Biol.* **38**, 2162–2169.

Frink, R., Krupp, P. P., and Young, R. A. (1978). Seasonal ultrastructural variations in pinealocytes of the woodchuck, *Marmota monax*. *J. Morphol.* **158**, 91–108.

Frink, R., Young, R. A., and Krupp, P. P. (1980). Seasonal changes in the ultrastructure of the thyrotroph in the woodchuck (*Marmota monax*) adenohypophysis. *J. Anat.* **130**, 499–505.

Frisk-Holmberg, M., and Ostman, J. (1977). Differential inhibition of lipolysis in human adipose tissue by adrenergic beta receptor blocking drugs. *J. Pharmacol. Exp. Ther.* **200**, 598–605.

Fry, F. E. J., and Hochachka, P. W. (1970). Fish. *In* "Comparative Physiology of Thermoregulation" (G. C. Whittow, ed.), Vol. 1, pp. 79–134. Academic Press, New York.

Fuxe, K., Goldstein, M., Hokfelt, T., Jonsson, G., and Lofstrom, A. (1974a). New aspects of the catecholamine innervation of the hypothalamus and the limbic system. *In* "Neurosecretion: The Final Neuroendocrine Pathway" (F. Knowles and V. Vollrath, eds.), pp. 223–228. Springer, Berlin.

Fuxe, K., Hokfelt, T., Jonsson, G., and Lofstrom, A. (1974b). Aminergic mechanisms in neuroendocrine control. *In* "Neurosecretion: The Final Neuroendocrine Pathway" (F. Knowles and V. Vollrath, eds.), pp. 269–275. Springer, Berlin.

Gale, C. C. (1973). Neuroendocrine aspects of thermoregulation. *Annu. Rev. Physiol.* **35**, 391–430.

Gallego, R., Eyzaguirre, C., and Monti-Bloch, L. (1979). Thermal and osmotic responses of arterial receptors. *J. Neurophysiol.* **42**, 665–689.

Galster, W. A. (1978). Failure to initiate hibernation with blood from the hibernating arctic ground squirrel, *Citellus undulatus*, and eastern woodchuck, *Marmota monax*. *J. Therm. Biol.* **3**, 93.

Galster, W. A., and Morrison, P. R. (1975). Gluconeogenesis in arctic ground squirrels between periods of hibernation. *Am. J. Physiol.* **228**, 325–330.

Gaston, S., and Menaker, M. (1968). Pineal function: The biological click in the sparrow? *Science* **160**, 1125–1127.

Geller, E. B., Hawk, C., Tallarida, R. J., and Alder, M. W. (1982). Postulated thermoregulatory roles for different opiate receptors in rats. *Life Sci.* **31**, 2241–2244.

Geller, E. B., Rowan, C. H., and Alder, M. W. (1986). Body temperature effects of opioids in rats: Intracerebroventricular administration. *Pharmacol. Biochem. Behav.* **24**, 1761–1765.

George, J. C. (1982). Thermogenesis in the avian body and the role of the pineal in thermoregulation. *In* "The Pineal and its Hormones" (R. J. Reiter, ed.), pp. 217–231. Liss, New York.

Gibson, A. (1981). The influence of endocrine hormones on the autonomic nervous system. *J. Auton. Pharmacol.* **1**, 331–358.

Glass, J. D., and Wang, L. C. H. (1979a). Thermoregulatory effects of intracerebroventricular injection of serotonin and a nonoamine oxidase inhibitor in a hibernator, *Spermophilus richardsonii*. *J. Therm. Biol.* **4**, 149–156.

Glass, J. D., and Wang, L. C. H. (1979b). Effects of central injection of biogenic amines during arousal from hibernation. *Am. J. Physiol.* **236**, R162–R167.

Glickman, N., Mitchell, H. H., Keeton, R. W., and Lambert, E. H. (1967). Shivering and heat production in men exposed to intense cold. *J. Appl. Physiol.* **22**, 1–8.

Glotzbach, S. F., and Heller, H. C. (1976). Central nervous regulation of body temperature during sleep. *Science* **194**, 537–539.

Glyn-Ballinger, J. R., Bernardini, G. L., and Lipton, J. M. (1983). α-MSH injected into the septal region reduces fever in rabbits. *Peptides (N.Y.)* **4**, 199–203.

Goodman, H. M. (1970). Permissive effects of hormones on lipolysis. *Endocrinology (Baltimore)* **86**, 1064–1074.

Green, M. D., and Lomax, P. (1976). Behavioral thermoregulation and neuramines in fish, *Chromus chromus*. *J. Therm. Biol.* **1**, 237–240.

Greenleaf, J. E., and Castle, B. L. (1971). Exercise temperature regulation in man during hypohydration and hyperhydration. *J. Appl. Physiol.* **30**, 847–853.

Grubb, B., and Folk, G. E., Jr. (1976). Effect of cold acclimation on norepinephrine stimulated oxygen consumption in muscle. *J. Comp. Physiol. B* **110B**, 217–226.

Guernsey, D. L., and Stevens, E. D. (1977). The cell membrane sodium pump as a mechanism for increasing thermogenesis during cold acclimation in rats. *Science* **196**, 908–910.

Gustafson, A. E., and Belt, W. D. (1981). The adrenal cortex during activity and hibernation in the male little brown bat, Myotis lucifugus: Annual rhythm of plasma cortisol levels. *Gen. Comp. Endocrinol.* **44**, 269–278.

Gwinner, E. (1978). Effects of pinealectomy on circadian locomotor activity rhythm in European starlin, *Sturnus vulgaris*. *J. Comp. Physiol.* **126**, 123–129.

Hainsworth, F. R., and Stricker, E. M. (1971). Relationship between body temperature and salivary secretion by rats in the heat. *J. Physiol. (Paris)* **63**, 257–259.

Hales, J. R. S. (1974). Physiological responses to heat. *MTP Int. Rev. Sci.* **7**, 107–162.

Hales, J. R. S. (1981). Peripheral effector mechanisms of thermoregulation: Regulation of panting. *Adv. Physiol. Sci.* **32**, 421–426.

Hall, G. H., and Myers, R. D. (1972). Temperature changes produced by nicotine injected into the hypothalamus of the conscious monkey. *Brain Res.* **37**, 241–251.

Hall, V., and Goldman, B. (1980). Effects of gonadal steroid hormones on hibernation in the Turkish hamster (*Mesocricetus brandti*). *J. Comp. Physiol. B* **135B**, 107–114.

Haller, A. C., and Zimny, M. L. (1977). Effects of hibernation on interradicular alveolar bone. *J. Dent. Res.* **56**, 1552–1557.

Hammel, H. T. (1965). Neurons and temperature regulation. In "Physiological Controls and Regulation" (W. S. Yamamoto and J. R. Brobeck, eds.), pp. 71–97. Saunders, Philadelphia, Pennsylvania.

Hammel, H. T. (1968). Regulation of internal body temperature. *Annu. Rev. Physiol.* **30**, 641–710.

Hammel, H. T. (1983). Phylogeny of regulatory mechanisms in temperature regulation. *J. Therm. Biol.* **8**, 37–42.

Hammel, H. T., Jackson, D. C., Stolwijk, J. A. J., Hardy, J. D., and Stromme, S. B. (1963). Temperature regulation by hypothalamic proportional control with an adjustable set point. *J. Appl. Physiol.* **18**, 1146–1154.

Hammel, H. T., Caldwell, F. T., and Abrams, R. M. (1967). Regulation of body temperature in the blue-tongued lizard. *Science* **156**, 1260–1263.

Hammel, H. T., Stromme, S. B., and Myhre, K. (1969). Forebrain temperature activates behavioral thermoregulatory response in arctic sculpins. *Science* **165**, 83–85.

Hammel, H. T., Elsner, R. W., Heller, H. C., Maggert, J. A., and Bainton, C. R. (1977). Thermoregulatory responses to altering hypothalamic temperature in the harbor seal. *Am. J. Physiol.* **232**, R18–R26.

Hardy, J. D. (1961). Physiology of temperature regulation. *Physiol. Rev.* **41**, 521–606.

Harlow, H. J., Phillips, J. A., and Ralph, C. L. (1980). The effect of pinealectomy on hibernation in two species of seasonal hibernators, *Citellus lateralis* and *C. richardsonii*. *J. Exp. Zool.* **213**, 301–303.

Haymovits, A., Gershon, M. D., and Nunez, E. A. (1976). Calcitonin, serotonin, and parafollicular cell granules during the hibernation activity cycle in the bat. *Proc. Soc. Exp. Biol. Med.* **153**, 383–387.

Heald, P. J., McLachlan, D. M., and Rookledge, K. A. (1965). The effects of insulin, glucagon and adrenocorticotrophic hormone on the plasma glucose and free fatty acids in the domestic fowl. *J. Endocrinol.* **38**, 83–95.

Heldmaier, G., and Lynch, G. R. (1986). Pineal involvement in thermoregulation and acclimatization. *Pineal Res. Rev.* **4**, 97–139.

Heldmaier, G., Steinlechner, S., Rafael, J., and Vsiansky, P. (1981). Photoperiodic control and effects of melatonin on nonshivering thermogenesis and brown adipose tissue. *Science* **212**, 917–919.

Helfmann, W., Jannes, P., and Jessen, C. (1981). Total body thermosensitivity and its spinal and supraspinal fractions in the conscious goose. *Pfluegers Arch.* **391**, 60–67.

Helle, K. B., Bolstad, G., Pihl, K. E., and Knudsen, R. (1980). Catecholamines, ATP and dopamine-B-hydroxyglase in the adrenal medulla of the hedgehog in the prehibernating state and during hibernation. *Cryobiology* **17**, 74–92.

Heller, H. C. (1979). Hibernation: Neural aspects. *Annu. Rev. Physiol.* **41**, 305–321.

Heller, H. C. (1983). Central nervous mechanisms regulating body temperature. In "Microwaves and Thermoregulation" (E. R. Adair, ed.), pp. 161–190. Academic Press, New York.

Heller, H. C., and Glotzbach, S. F. (1977). Thermoregulation during sleep and hibernation. *Int. Rev. Physiol.* **15**, 147–188.

Heller, H. C., and Henderson, J. A. (1976). Hypothalamic thermosensitivity and regulation of heat storage behavior in a day-active desert rodent, *Ammospermophilus nelsoni. J. Comp. Physiol.* **108**, 255.

Heller, H. C., Walker, J. M., Florant, G. L., Glotzbach, S. F., and Berger, R. J. (1978). Sleep and hibernation: Electrophysiological and thermoregulatory homologies. *In* "Strategies in Cold: Natural Torpidity and Thermogenesis" (L. C. H. Wang and J. W. Hudson, eds.), pp. 225–265. Academic Press, New York.

Heller, H. C., Graf, R., and Rautenberg, W. (1983). Circadian and arousal state influences on thermoregulation in the pigeon. *Am. J. Physiol.* **14**, R321–R328.

Hellon, R. F. (1972). Temperature-sensitive neurons in the brain stem: Their responses to brain temperature at different ambient temperatures. *Pfluegers Arch.* **335**, 323–334.

Hellon, R. F. (1975). Monoamines, pyrogens and cations: Their actions on central control of body temperature. *Pharmacol. Rev.* **26**, 289–321.

Hellon, R. F. (1983). Central projections and processing of skin-temperature signals. *J. Therm. Biol.* **8**, 7–8.

Hellon, R. F., and Mitchell, D. (1975). Convergence in a thermal afferent pathway in the rat. *J. Physiol. (London)* **248**, 359–376.

Hellon, R. F., and Misra, N. K. (1973). Neurons in the ventrobasal complex of the rat thalamus responding to scrotal skin temperature changes. *J. Physiol. (London)* **232**, 389–400.

Hellon, R. F., Cranston, W. I., Townsend, Y., Mitchell, D., Dawson, N. J., and Duff, G. W. (1980). Some tests of the prostaglandin hypothesis of fever. *In* "Fever" (J. M. Lipton, ed.), pp. 159–164. Raven Press, New York.

Hellström, B., and Hammel, H. T. (1967). Some characteristics of temperature regulation in the unanesthetized dog. *Am. J. Physiol.* **213**, 547–556.

Hemingway, A. (1963). Shivering. *Physiol. Rev.* **43**, 397–422.

Henderson, N. E., and Demeneix, B. A. (1981). Hibernation in thyroidectomized ground squirrels, *Spermophilus richardsonii. Gen. Comp. Endocrinol.* **43**, 543–548.

Hennessey, T. M., Saimi, Y., and Kung, C. (1983). A heat-induced depolarization of paramecium and its relationship to thermal avoidance behavior. *J. Comp. Physiol.* **153**, 39–46.

Hensel, H. (1974). Thermoreceptors. *Annu. Rev. Physiol.* **36**, 233–249.

Hensel, H. (1983). Recent advances in thermoreceptor physiology. *J. Therm. Biol.* **8**, 306.

Hensel, H., and Schafer, K. (1974). Effects of calcium on warm and cold receptors. *Pfluegers Arch.* **352**, 87–90.

Hensel, H., Andres, K. H., and von During, M. (1974). Structure and function of cold receptors. *Pfluegers Arch.* **352**, 1–10.

Himms-Hagen, J. (1967). Sympathetic regulation of metabolism. *Pharmacol. Rev.* **19**, 367–461.

Himms-Hagen, J. (1975). Role of the adrenal medulla in adaptation to cold. *In* "Adrenal Gland" (H. Blaschko, G. Sayers, and A. D. Smith, eds.), pp. 637–665. *Am. Physiol. Soc.,* Washington, D. C.

Himms-Hagen, J. (1978). Biochemical aspects of nonshivering thermogenesis. *In* "Strategies in Cold: Natural Torpidity and Thermogenesis" (L. C. H. Wang and J. W. Hudson, eds.), pp. 595–617.

Himms-Hagen, J. (1981). Nonshivering thermogenesis, brown adipose tissue and obesity. *In* "Nutritional Factors: Modulating Effects on Metabolic Processes" (R. F. Beers, Jr. and E. G. Bassett, eds.), pp. 85–99. Raven Press, New York.

Himms-Hagen, J. (1983). Thyroid hormones and thermogenesis. *In* "Mammalian Thermogenesis" (M. J. Stock and L. Girardier, eds.), pp. 141–177. Chapman & Hall, London.

Hinckel, P., and Schroder-Rosenstock, K. (1981). Response to potine units to skin-temperature changes in the guinea-pig. *J. Physiol. (London)* **314**, 189–194.

Hinkley, R. E., and Burton, P. A. (1970). Fine structure of the pancreatic islet cells of normal and alloxan treated bats (*Eptesicus fuscus*). *Anat. Rec.* **166**, 67–86.

Hissa, R., and Rautenberg, W. (1975). Thermoregulatory effects of introhypothalamic injections of neurotransmitters and their inhibitors in pigeons. *Comp. Biochem. Physiol. A* **51A**, 319–326.

Hissa, R., Pyornila, A., and Saarela, S. (1975). Effects of peripheral noradrenaline on thermoregulation in temperature acclimated pigeons. *Comp. Biochem. Physiol. C* **51C**, 243–247.

Hochachka, P. W., and Somero, G. N. (1973). "Strategies of Biochemical Adaptation." Saunders, Philadelphia, Pennsylvania.

Hoffman, R. A., and Reiter, R. J. (1965). Pineal gland: Influence on gonads of male hamsters. *Science* **148**, 1609–1611.

Hoffman, R. A., and Zarrow, M. X. (1958a). Seasonal changes in the basophillic cells of the pituitary gland of the ground squirrel (*Citellus tridecemlineatus*). *Anat. Rec.* **131**, 727–735.

Hoffman, R. A., and Zarrow, M. X. (1958b). A comparison of seasonal changes and the effect of cold on the thyroid gland of the male rat and ground squirrel (*Citellus tridecemlineatus*). *Acta Endocrinol. (Copenhagen)* **27**, 77–84.

Hohtola, E., Hissa, R., and Saarela, S. (1977). Effect of glucagon on thermogenesis in the pigeon. *Am. J. Physiol.* **232**, E451–E455.

Holaday, J. W., Wei, E., Loh, H. H., and Li, C. H. (1978). Endorphins may function in heat adaptation. *Proc. Natl. Acad. Sci. U.S.A.* **75**, 2923–2927.

Hoo-Paris, R. (1971). Hibernation and ACTH in the hedgehog, Erinaceus europaeus. *Ann. Endocrinol.* **32**, 743–752.

Hoo-Paris, R., and Sutter, B. C. J. (1980a). Blood glucose control by insulin in the lethargic and arousing hedgehog (*Erinaceus europaeus*). *Comp. Biochem. Physiol. A* **66A**, 141–143.

Hoo-Paris, R., and Sutter, B. C. J. (1980b). Role of glucose and catecholamines in the regulation of insulin secretion in the hibernating hedgehog (*Erinaceus europaeus*). *Gen. Comp. Endocrinol.* **41,** 62–65.

Hoo-Paris, R., Hamsany, M., Sutter, B. C. J., Assan, R., and Boillol, J. (1982). Plasma glucose and glucagon concentrations in the hibernating hedgehog. *Gen. Comp. Endocrinol.* **46,** 246–254.

Hoo-Paris, R., Castex, Ch., Hamsany, M., Thari, A., and Sutter, B. (1985). Glucagon secretion in the hibernating edible dormouse (*Glis glis*). *Comp. Biochem. Physiol.* **81A,** 277–281.

Hori, T., and Nakayama, T. (1973). Effects of biogenic amines on central thermoresponsive neurons in the rabbit. *J. Physiol. (London)* **232,** 71–85.

Hori, T., Yamasaki, M., Kiyohara, T., and Shibata, M. (1986). Responses of preoptic thermosensitive neurons to poikilothermia-inducing peptides—bombesin and neurotensin. *Pfluegers Arch.* **407,** 558–560.

Horwitz, B. A. (1978). Neurohumoral regulation of nonshivering thermogenesis in mammals. *In* "Strategies in Cold: Natural Torpidity and Thermogenesis" (L. C. H. Wang and J. W. Hudson, eds.), pp. 619–653. Academic Press, New York.

Horwitz, B. A., and Eaton, M. (1977). Ouabain-sensitive liver and diaphragm respiration in cold-acclimated hamsters. *J. Appl. Physiol.* **42,** 150–153.

Hsieh, A. C. L. (1962). The role of the thyroid in rats exposed to cold. *J. Physiol. (London)* **161,** 175–188.

Hsieh, A. C. L., and Carlson, L. D. (1957). Role of adrenaline and noradrenaline in chemical regulation of heat production. *Am. J. Physiol.* **190,** 243–246.

Hudson, J. W. (1980). The thyroid gland and temperature regulation in the prairie vole, *Microtus ochragaster* and the chipmunk, *Tamias striatus*. *Comp. Biochem. Physiol. A* **65A,** 173–179.

Hudson, J. W. (1981). Role of the endocrine glands in hibernation with special reference to the thyroid gland. *In* "Survival in the Cold" (X. J. Musacchia and L. Jansky, eds.), pp. 33–54. Am. Elsevier, Amsterdam.

Hudson, J. W., and Wang, L. C. H. (1969). Thyroid function in desert ground squirrels. *In* "Physiological Systems in Semiarid Environments" (C. C. Hoff and M. L. Riedesel, eds.), pp. 17–33. Univ. of New Mexico Press, Albuquerque.

Hudson, J. W., and Wang, L. C. H. (1979). Hibernation: Endocrinologic aspects. *Annu. Rev. Physiol.* **41,** 287–303.

Huibregtse, W. H., Gunville, R., and Ungar, F. (1971). Secretion of corticosterone in vitro by normothermic and hibernating ground squirrels. *Comp. Biochem. Physiol. A* **38A,** 763–768.

Huidobro-Toto, J. P., Anderson, H. H., and Way, E. L. (1979). Pharmacological characterization of the hyperthermia produced by β-endorphin in mice. *Fed. Proc., Fed. Am. Soc. Exp. Biol.* **38,** 364.

Huikko, M., Jouppila, P., and Karki, N. T. (1966). Effect of Finnish bath (sauna) on the urinary excretion of noradrenaline, adrenaline, and 3-methoxy-4-hydroxy-mandelic acid. *Acta Physiol. Scand.* **66**, 316–321.

Hulbert, A. J. (1978). The thyroid hormones: A thesis concerning their action. *J. Theor. Biol.* **73**, 81–100.

Hulbert, A. J. (1981). Evolution from ectothermia towards endothermia. *Adv. Physiol. Sci.* **32**, 237–247.

Hulbert, A. J., and Hudson, J. W. (1976). Thyroid function in a hibernator, *Spermophilus tridecemlineatus. Am. J. Physiol.* **230**, 1138–1146.

Hulbert, A. J., Augee, M. L., and Raison, J. K. (1976). The influence of thyroid hormones on the structure and function of mitochondrial membranes. *Biochim. Biophys. Acta* **455**, 597–601.

Hutchison, V. H., Dowling, H. G., and Vinegar, A. (1966). Thermal regulation in brooding Indian python. *Science* **151**, 694–696.

Hutchison, V. H., and Kosh, R. J. (1974). Thermoregulatory function of the parietal eye in the lizard *Anolis carolinensis. Oecologia* **16**, 173–177.

Hutchison, V. H., Black, J. J., and Erskine, D. (1979). Melatonin and chlorpromazine: Thermal selection in the mudpuppy, *Necturus maculosus. Life Sci.* **25**, 527–530.

Ingram, D. L. (1965). Evaporative cooling in the pig. *Nature (London)* **207**, 415–416.

Ingram, D. L., and Legge, K. F. (1972). The influence of deep body and skin temperatures on thermoregulatory responses to heating of the scrotum in pigs. *J. Physiol. (London)* **224**, 477–487.

Ismail-Beigi, F., and Edelman, I. S. (1970). Mechanism of thyroid calorigenesis: Role of active sodium transport. *Proc. Natl. Acad. Sci. U.S.A.* **67**, 1071–1078.

Ismail-Beigi, F., and Edelman, I. S. (1971). The mechanism of the calorigenic action of thyroid hormone. *J. Gen. Physiol.* **57**, 710–722.

Jacob, J. J., and Girault, J. M. T. (1974). The influence of cyproheptadine and d-lysergamide on the rise in temperature induced by intracerebroventricular 5-hydroxytryptamine and dopamine in conscious rabbits. *Eur. J. Pharmacol.* **27**, 59–67.

Jacob, J. J., and Girault, J. M. T. (1979). 5-hydroxytryptamine. *In* "Body Temperature: Regulation, Drug Effects, and Therapeutic Implications" (P. Lomax and E. Schonbaum, eds.), pp. 183–230. Dekker, New York.

Jallageas, M., and Assenmacher, I. (1983). Annual plasma testosterone and thyroxine cycles in relation to hibernation in the edible dormouse *Glis glis. Gen. Comp. Endocrinol.* **50**, 452–462.

Jancso-Gabor, A., and Szolcsanyi, J. (1970). Stimulation and desensitization of the hypothalamic heat-sensitive structures by capsaicin in rats. *J. Physiol. (London)* **208**, 449–459.

Jancso-Gabor, A., Szolcsanyi, J., and Jansco, N. (1970). Irreversible impairment of thermoregulation induced by capsaicin and similar pungent substances in rats and guinea-pigs. *J. Physiol. (London)* **206,** 495–507.

Jansky, L. (1973). Non-shivering thermogenesis and its thermoregulatory significance. *Biol. Rev. Cambridge Philos. Soc.* **48,** 85–132.

Jansky, L. (1978). Time sequence of physiological changes during hibernation: The significance of serotonergic pathways. *In* "Strategies in Cold: Natural Torpidity and Thermogenesis" (L. C. Wang and J. W. Hudson, eds.), pp. 299–326. Academic Press, New York.

Jansky, L., and Novotona, R. (1976). The role of central aminergic transmission in thermoregulation and hibernation. *In* "Regulation of Depressed Metabolism and Thermogenesis" (L. Jansky and X. J. Musacchia, eds.), pp. 64–80. Thomas, Springfield, Illinois.

Jansky, L., Kahlerova, Z., Nedoma, J., and Andrews, J. F. (1981). Humoral control of hibernation in golden hamsters. *In* "Survival in the Cold" (X. J. Musacchia and L. Jansky, eds.), pp. 13–32. Elsevier, Amsterdam.

Jansky, L., Vybiral, S., Moravec, J., Nachazel, J., Riedel, W., Simon, E., and Simon-Oppermann, C. (1986). Neuropeptides and temperature regulation. *J. Therm. Biol.* **11,** 79–83.

Jasinski, A. (1970). Hypothalamic neurosecretion in the bat, *Myotis myotis* Borkhausen, during the period of hibernation and activity. *In* "Aspects of Neuroendocrinology" (W. Bargmann and B. Scharrer, eds.), pp. 301–309. Springer, Berlin.

Jell, R. M. (1973). Responses of hypothalamic neurons to local temperature and to acetylcholine, noradrenaline and 5-hydroxytryptamine. *Brain Res.* **55,** 123–134.

Jell, R. M. (1974). Responses of rostral hypothalamic neurons to peripheral temperature and to amines. *J. Physiol. (London)* **240,** 295–307.

Jessen, C. (1981). Independent clamps of peripheral and central temperatures and their effects on heat production in the goat. *J. Physiol. (London)* **311,** 11–22.

Jessen, C. (1985). Thermal afferents in the control of body temperature. *Pharmacol. Ther.* **28,** 107–134.

Jessen, C., Feistkorn, G., and Nagel, A. (1983). Effects on heat loss of central-leg cooling in the conscious goat. *J. Thermal, Biol.* **8,** 65–67.

Johansen, P. H., and Cross, J. A. (1980). Effects of sexual maturation and sex steroid hormone treatment on the temperature preference of the guppy *Poecilia reticulata. Can. J. Zool.* **58,** 586–588.

Johansson, B. W. (1978). Seasonal variations in the endocrine system of hibernators. *In* "Environmental Endocrinology" (I. Assenmacher and D. S. Farner, eds.), pp. 103–110. Springer, Berlin.

Johansson, B. W., and Senturia, J. B. (1972). Seasonal variations in the physiology and biochemistry of the European hedgehog (*Erinaceus europaeus*) including comparisons with nonhibernators, guinea pig and man. *Acta Physiol. Scand., Suppl.* **380.**

Kalivas, P. W., and Horita, A. (1980). Thyrotropin-releasing hormone: Neurogenesis of actions in the pentobarbital narcotized rat. *J. Pharmacol. Exp. Ther.* **212**, 203–210.

Kalsner, S. (1969). Mechanism of hydrocortisone potentiation of responses to epinephrine and norepinephrine in rabbit aorta. *Circ. Res.* **24**, 383–395.

Kandasamy, S. B., and Williams, B. A. (1984). Hypothermic and antipyretic effect of ACTH (1-24) and α-mealnotropin in guinea-pigs. *Neuropharmacology* **23**, 49–53.

Kang, B. S., Han, D. S., Paik, K. S., Park, Y. S., Kim, J. K., Kim, C. S., Rennie, D. W., and Hong, S. K. (1970). Calorigenic action of norepinephrine in the Korean women divers. *J. Appl. Physiol.* **29**, 6–9.

Kastner, P. R., Zatzman, M. L., South, F. E., and Johnson, J. A. (1978). Renin-angiotensin-aldosterone system of the hibernating marmot. *Am. J. Physiol.* **234**, R178–R182.

Kavaliers, M. (1982a). Peptides, the pineal gland and thermoregulation. *In* "The Pineal and its Hormones" (R. J. Reiter, ed.), pp. 207–215. Liss, New York.

Kavaliers, M. (1982b). Seasonal and circannual rhythms in behavioral thermoregulation and their modifications by pinealectomy in the white sucker, *Catostomus commersoni*. *J. Comp. Physiol.* **146**, 235–243.

Kavaliers, M., and Hawkins, M. F. (1981). Bombesin alters behavioral thermoregulation in fish. *Life Sci.* **28**, 1361–1364.

Kawamura, Y., Kishi, K., and Fujimoto, J. (1953). Electromyographic analysis of the shivering movement. *Jpn. J. Physiol.* **15**, 676–686.

Kayser, C. (1961). "The Physiology of Natural Hibernation." Pergamon, New York.

Kenagy, G. J. (1980). Interrelation of endogenous annual rhythms of reproduction and hibernation in the golden-mantled ground squirrel. *J. Comp. Physiol. A* **135A**, 333–339.

Kenagy, G. J. (1981). Endogenous annual rhythm of reproductive function in the non-hibernating desert ground squirrel *Ammospermophilus leucurus*. *J. Comp. Physiol. A* **142A**, 251–258.

Kenny, A. D., and Musacchia, X. J. (1977). Influence of season and hibernation on thyroid calcitonin content and plasma electrolytes in the ground squirrel. *Comp. Biochem. Physiol. A* **57A**, 485–489.

Kiang-Ulrich, M., and Horvath, S. M. (1979). Successful cold acclimation following bilateral adrenodemedullation in rats. *Proc. Soc. Exp. Biol. Med.* **162**, 449–453.

Kilduff, T. S., Sharp, F. R., and Heller, H. C. (1983). Relative 2-deoxyglucose uptake of the paratrigeminal nucleus increases during hibernation. *Brain Res.* **262**, 117–123.

Kluger, M. J. (1977). Fever in the frog *Hyla cinerea*. *J. Therm. Biol.* **2**, 79–81.

Kluger, M. J. (1979). "Fever: Its Biology, Evolution and Function." Princeton Univ. Press, Princeton, New Jersey.

Kluger, M. J. (1981). Is fever a nonspecific host defense response? *In* "Infection: The

Physiological and Metabolic Responses of the Host" (M. C. Powanda and P. G. Canonico, eds.), pp. 75–95. Elsevier, Amsterdam.

Kluger, M. J., and Rothenburg, B. A. (1979). Fever and reduced iron: Their interaction as a host defense response to bacterial infection. *Science* **203**, 374–376.

Kluger, M. J., Ringler, D. H., and Anver, M. R. (1975). Fever and survival. *Science* **188**, 166–168.

Knecht, E. A., Toraason, M. A., and Wright, G. L. (1980). Thermoregulatory ability of female rats during pregnancy and lactation. *Am. J. Physiol.* **239**, R470–R475.

Knox, G. V., Campbell, C., and Lomax, P. (1973). The effects of acetylcholine and nicotine on unit activity in the hypothalamic thermoregulatory centers of the rat. *Brain Res.* **51**, 215–223.

Koban, M., and Feist, D. D. (1982). The effect of cold on norepinephrine turnover in tissues of seasonally acclimatized redpolls (*Carduelis flammea*). *J. Comp. Physiol.* **146**, 137–144.

Kolpakov, M. G., and Samsonenko, R. A. (1970). Reactivity of the adrenals of the red-cheeked suslik at various periods of vital activity. *Dokl. Akad. Nauk SSSR, Ser. Biol.* **191**, 1424–1426.

Kramarova, L. I., Kolaeva, S. H., Yukhananov, R. Yu., and Rozhanets, V. V. (1983). Content of DSIP, enkephalins and ACTH in some tissues of active and hibernating ground squirrels (*Citellus suslicus*). *Comp. Biochem. Physiol. C* **74C**, 31–33.

Krass, P. M., and Khabibov, B. (1975). Seasonal rhythm of functional activity of the adrenocorticotropic function of the pituitary in the red-cheeked suslik. *Dokl. Biol. Sci. (Engl. Transl.)* **225**, 474–476.

Krimphove, M., and Opitz, K. (1975). Untersuchungen der calorigenen Wirkung von Glucagon. *Arch. Int. Pharmacodyn. Ther.* **216**, 328–350.

Kromer, W. (1980). Naltrexone influence on hibernation. *Experientia* **36**, 581–582.

Krook, L., Wimsatt, W. A., Whalen, J. P., MacIntyre, I., and Nunez, E. A. (1977). Calcitonin and hibernation bone loss in the bat (*Myotis lucifugus*). *Cornell Vet.* **67**, 265–271.

Krueger, J. M., Pappenheimer, J. R., and Karnovsky, M. L. (1982). The composition of sleep-promoting factor isolated from human urine. *J. Biol. Chem.* **257**, 1664–1669.

Krupp, P. P., Young, R. A., and Frink, R. (1977). The thyroid gland of the woodchuck, *Marmota monax:* A morphological study of seasonal variations in the follicular cells. *Anat. Rec.* **187**, 495–513.

Kuroshima, A., Doi, K., Yahata, T., Jurahashi, M., and Ohno, T. (1981). Glucagon and temperature acclimation. *Adv. Physiol. Sci.* **32**, 305–307.

Landsberg, L., and Young, J. B. (1981). Diet-induced changes in sympathoadrenal activity: Implications for thermogenesis. *Life Sci.* **28**, 1801–1819.

LeBlanc, J., and Villemaire, A. (1970). Thyroxine and noradrenaline on noradrenaline sensitivity, cold resistance, and brown fat. *Am. J. Physiol.* **218**, 1742–1745.

Leduc, J. (1961). Catecholamine production and release in exposure and acclimation to cold. *Acta Physiol. Scand.*, **53**, Suppl. 183, 1–101.

Lee, T. F., and Myers, R. D. (1983). Analysis of the thermolytic action of ICV neurotensin in the rat at different ambient temperatures. *Brain Res. Bull.* **10**, 661–665.

Lee, T. F., Mora, F., and Myers, R. D. (1985a). Effect of intracerebroventricular vasopressin in body temperature and endotoxin fever of macaque monkey. *Am. J. Physiol.* **248**, R674–678.

Lee, T. F., Mora, F., and Myers, R. D. (1985b). Dopamine and thermoregulation: An evaluation with special reference to dopaminergic pathways. *Neurosci. Behav. Rev.* **9**, 589–598.

Legait, E., Burlet, C., and Marchetti, J. (1970). Contribution to the study of the hypothalamo-neurohypophyseal system during hibernation. *In* "Aspects of Neuroendocrinology" (W. Bargmann and B. Scharrer, eds.), pp. 310–321. Springer, Berlin.

Legait, H., Roux, M., Dussart, G., Richoux, J. P., and Contet-Audonneau, J. L. (1975). Données morphometriques sur la glande pinéale du loir (*Glis glis*) et du lerot (*Eliomys quercinus*) au cours du cycle annuel. *C. R. Seances Soc. Biol. Ses Fil.* **169**, 132–136.

Lepri, G., and Cristiani, R. (1964). Ability of certain adrenocortical hormones to potentiate the vasoconstrictor action of noradrenaline on the conjunctival vessels in the rabbit and in man. *Br. J. Opthalmol.* **48**, 205–208.

Licht, P., Zucker, I., Hubbard, G., and Boshes, M. (1982). Circannual rhythms of plasma testosterone and luteinizing hormone levels in golden-mantled ground squirrels (*Spermophilus lateralis*). *Biol. Reprod.* **27**, 411–418.

Lillywhite, H. B. (1970). Behavioral temperature regulation in the bullfrog, *Rana catesbeiana. Copeia* pp. 158–168.

Lin, K. S., and Lin, M. T. (1986). Effects of bombesin on thermoregulatory responses and hypothalamic neuronal activities in the rat. *Am. J. Physiol.* **251**, R303–309.

Lin, M. T., and Su, C. Y. (1979). Metabolic, respiratory, vasomotor and body temperature responses to beta-endorphin and morphine in rabbits. *J. Physiol. (London)* **295**, 179–189.

Lipton, J. M. (1968). Effects of preoptic lesions on heat-escape responding and colonic temperature in the rat. *Physiol. Behav.* **3**, 165–169.

Lipton, J. M., ed. (1980). "Fever." Raven Press, New York.

Lipton, J. M., and Glyn, J. R. (1980). Central administration of peptides alters thermoregulation in the rabbit. *Peptides (N.Y.)* **1**, 15–18.

Lipton, J. M., Glyn, J. R., and Zimmer, J. A. (1981). ACTH and α-melanotropin in central temperature control. *Fed. Proc., Fed. Am. Soc. Exp. Biol.* **40**, 2760–2764.

Lipton, J. M., Glyn-Ballinger, J. R., Murphy, M. T., Zimmer, J. A., Barnardini, G., and Samson, W. K. (1984). The central neuropeptides ACTH and α-MSH in fever control. *J. Therm. Biol.* **9**, 139–143.

Llados-Eckman, C., and Beckman, A. L. (1983). Reduction of hibernation bout duration by icv infusion of naloxone in *Citellus lateralis*. *Proc. Soc. Neurosci.* **9,** 796.

Lomax, P., and Schonbaum, E., eds. (1979). "Body Temperature: Regulation, Drug Effects, and Therapeutic Implications." Dekker, New York.

Lomax, P., Bajorek, J. G., Chesarek, W., and Tatryn, I. V. (1980). Thermoregulatory effects of luteinizing hormone releasing hormone in the rat. *In* "Thermoregulatory Mechanisms and Their Therapeutic Implications" (B. Cox, P. Lomax, A. S. Milton, and E. Schonbaum, eds.), pp. 208–211. Karger, Basel.

Lyman, C. P. (1982). Who is who among the hibernators. *In* "Hibernation and Torpor in Mammals and Birds" (C. P. Lyman, J. S. Willis, A. Malan, and L. C. H. Wang, eds.), pp. 12–36. Academic Press, New York.

Lyman, C. P., Willis, J. S., Malan, A., and Wang, L. C. H., eds. (1982). "Hibernation and Torpor in Mammals and Birds." Academic Press, New York.

Lynch, G. R., and Epstein, A. L. (1976). Melatonin-induced changes in gonads, pelage and thermogenic characters in the white-footed mouse, *Peromyscus leucopus*. *Comp. Biochem. Physiol. C* **53C,** 67–68.

Lynch, G. R., White, S. E., Grundel, R., and Berger, M. S. (1978). Effects of photoperiod, melatonin administration and thyroid block on spontaneous daily torpor and temperature regulation in the white-footed mouse, *Peromyscus leucopus*. *J. Comp. Physiol.* **125,** 157–163.

Lynch, G. R., Sullivan, J. K., and Gendler, S. L. (1980). Temperature regulation in the mouse, *Peromyscus leucopus*: Effects of various photoperiods, pinealectomy and melatonin administration. *Int. J. Biometeorol.* **24,** 49–55.

Maher, J. T., Bass, D. E., Heistad, D. D., Angelakos, E. T., and Hartley, L. H. (1972). Effect of posture on heat acclimatization in man. *J. Appl. Physiol.* **33,** 8–13.

Maickel, R. P., Westermann, E. O., and Brodie, B. B. (1961). Effects of reserpine and cold exposure on pituitary-adrenocortical function in rats. *J. Pharmacol. Exp. Ther.* **134,** 167–175.

Maickel, R. P., Matussek, N., Stern, D. N., and Brodie, B. B. (1967a). The sympathetic nervous system as a homeostatic mechanism. I. Absolute need for sympathetic nervous function in body temperature maintenance of cold exposed rats. *J. Pharmacol. Exp. Ther.* **157,** 103–110.

Maickel, R. P., Stern, D. N., Takabatake, E., and Brodie, B. B. (1976b). The sympathetic nervous system as a homeostatic mechanism. II. Effect of adrenocortical hormones on body temperature maintenance of cold-exposed adrenalectomized rats. *J. Pharmacol. Exp. Ther.* **157,** 111–116.

Margulis, D. L., Goldman, B., and Finck, A. (1979). Hibernation: An opioid-dependent state? *Brain Res. Bull.* **4,** 721–724.

Marley, E., and Nistico, G. (1972). Effects of catecholamines and adenosine derivatives given into the brains of fowls. *Br. J. Pharmacol.* **46,** 619–636.

Marrone, B. L., Gentry, R. T., and Wade, G. N. (1976). Gonadal hormones and body temperature in rats: Effects of estrous cycles, castration and steroid replacement. *Physiol. Behav.* **17,** 419–425.

Martin, G. E., and Bacino, C. B. (1979). Action of intracerebrally injected β-endorphin on the rat's core temperature. *Eur. J. Pharmacol.* **59,** 227–236.

Martin, G. E., Bacino, C. B., and Papp, N. L. (1981). Action of selected serotonin antagonists on hyperthermia evoked by intracerebrally injected β-endorphin. *Peptides (N.Y.)* **2,** 213–217.

Mayer, W. V., and Bernick, S. (1963). Effect of hibernation on tooth structure and dental caries. *In* "Mechanisms of Hard Tissue Destruction," Publ. No. 75, pp. 285–296. Am. Assoc. Adv. Sci., Washington, D. C.

McNab, B. K. (1983). Energetics, body size, and the limits to endothermy. *J. Zool.* **199,** 1–29.

McNulty, J. A., and Dombrowski, T. A. (1980). Ultrastructural evidence for seasonal changes in pinealocytes of the 13-lined ground squirrel, *Spermophilus tridecemlineatus:* A qualitative and quantitative study. *Anat. Rec.* **196,** 387–400.

McNulty, J. A., Dombrowski, T. A., and Spurrier, W. A. (1980). A seasonal study of pinealocytes in the 13-lined ground squirrel, *Spermophilus tridecemlineatus. Reprod. Nutr. Dev.* **20,** 665–672.

Megal, E., Kaplanski, J., Sod-Moriah, U. A., Hirschmann, N., and Nir, I. (1981). Role of the pineal gland in male rats chronically exposed to increased temperature. *J. Neural Transm.* **50,** 267–273.

Mense, S. (1978). Effects of temperature on the discharges of muscle spindles and tendon organs. *Pfluegers Arch.* **374,** 159–166.

Mercer, J. B., and Jessen, C. (1978). Central thermosensitivity in conscious goats: Hypothalamus and spinal cord versus residual inner body. *Pfluegers Arch.* **374,** 179–186.

Mercer, J. B., and Jessen, C. (1980). Thermal control of respiratory evaporative heat loss in exercising dogs. *J. Appl. Physiol.* **49,** 979–984.

Mercer, J. B., and Simon, E. (1983). A comparison between total body thermosensitivity and local thermosensitivity in the guinea-pig (*Cavia porcellus*). *J. Therm. Biol.* **8,** 43–45.

Merker, G., Blahser, S., and Zeisberger, E. (1980). The reactivity pattern of vasopressin-containing neurons and its relation to the antipyretic reaction in the pregnant guinea pig. *Cell Tissue Res.* **212,** 47–61.

Metcalf, G. (1974). TRH: A possible mediator of thermoregulation. *Nature (London)* **252,** 310–311.

Metcalf, G., and Myers, R. D. (1976). A comparison between the hypothermia induced by intraventricular injections of othyrotropin releasing hormone, noradrenaline or calcium ions in unanesthetized cats. *Br. J. Pharmacol.* **58,** 489–495.

Metcalf, G., Dettmar, P. W., and Watson, T. (1980). The role of neuropeptides in thermoregulation. *In* "Thermoregulatory Mechanisms and Their Therapeutic Implications" (B. Cox, P. Lomax, A. S. Milton, and E. Schonbaum, eds.), pp. 175–179. Karger, Basel.

Milton, A. S., ed. (1982a). "Handbook of Experimental Pharmacology," Vol. 60. Springer-Verlag, Berlin.

Milton, A. S. (1982b). Prostaglandins in fever and the mode of action of antipyretic drugs. *Handb. Exp. Pharmacol.* **60,** 248–303.

Milton, A. S., and Sawhney, V. K. (1982). Protein synthesis and fever. *Handb. Exp. Pharmacol.* **60,** 305–315.

Milton, A. S., and Wendlandt, S. (1970). A possible role for prostaglandin E_1 as a modulator for temperature regulation in the central nervous system of the cat. *J. Physiol. (London)* **207,** 76–77.

Minor, J. D., Bishop, D. A., and Badger, C. R. (1978). The golden hamster and the blood-borne hibernation trigger. *Cryobiology* **15,** 557–562.

Mitchell, D., Snellen, J. W., and Atkins, A. R. (1970). Thermoregulation during fever: Changes in set-point or change in gain. *Pfluegers Arch.* **321,** 293–302.

Moravec, J., and Vyskocil, F. (1976). Neuromuscular transmission in a hibernator. *In* "Regulation of Depressed Metabolism and Thermogenesis" (L. Jansky and X. J. Musacchia, eds.), pp. 81–92. Thomas, Springfield, Illinois.

Mory, G., Ricquier, D., Presquies, P., and Hemon, P. (1981). Effects of hypothyroidism on the brown adipose tissue of adult rats: Comparison with the effects of adaptation to cold. *J. Endocrinol.* **91,** 515–527.

Mory, G., Ricquier, D., Nechad, M., and Hemon, P. (1982). Impairment of trophic response of brown fat to cold in guanethidine-treated rats. *Am. J. Physiol.* **242,** C159–C165.

Murakami, N. (1973). Effects of iotophoretic application of 5-hydroxytryptamine, noradrenaline and acetylcholine upon hypothalamic temperature-sensitive neurons in rats. *Jpn. J. Physiol.* **23,** 435–447.

Murphy, M. T., and Lipton, J. M. (1982). Peripheral administration of α-MSH reduces fever in older and younger rabbits. *Peptides (N.Y.)* **3,** 775–779.

Murphy, M. T., Richards, D. B., and Lipton, J. M. (1983). Antipyretic potency of centrally administered α-melanocyte stimulating hormone. *Science* **221,** 192–193.

Musacchia, X. J. (1984). Comparative physiological and biochemical aspects of hypothermia as a model for hibernation. *Cryobiology* **21,** 583–592.

Musacchia, X. J., and Deavers, D. R. (1981). The regulation of carbohydrate metabolism in hibernators. *In* "Survival in the Cold" (J. Musacchia and L. Jansky, eds.), pp. 55–75. Elsevier, Amsterdam.

Myers, R. D. (1966). Temperature regulation in the conscious monkey: Chemical mechanisms in the hypothalamus. *Int. Biometeorol. Congr. Proc. p. 125.*

Myers, R. D. (1974). Temperature regulation. *In* "Handbook of Drugs and Chemical Stimulation of the Brain," pp. 237–285. Van Nostrand-Reinhold, New York.

Myers, R. D. (1976). Chemical control of body temperature by the hypothalamus: A model and some mysteries. *Proc. Aust. Physiol. Pharmacol. Soc.* **7,** 15–32.

Myers, R. D. (1980). Hypothalamic control of thermoregulation: Neurochemical mechanisms. *In* "Handbook of Hypothalamus" (P. Morgane and J. Panksepp, eds.), pp. 83–210. Dekker, New York.

Myers, R. D. (1981). Serotonin and thermoregulation: Old and new views. *Journal of Physiol. (Paris)* **77**, 505–513.

Myers, R. D. (1982). The role of ions in thermoregulation and fever. *Handb. Exp. Pharmacol.* **60**, 151–186.

Myers, R. D., and Beleslin, D. B. (1970). The spontaneous release of 5-hydroxytryptamine and acetylcholine within the diencephalon of the unanesthetized rhesus monkey. *Exp. Brain Res.* **11**, 539–552.

Myers, R. D., and Beleslin, D. B. (1971). Changes in serotonin release in hypothalamus during cooling or warming the monkey. *Am. J. Physiol.* **220**, 1746–1754.

Myers, R. D., and Ruwe, W. E. (1978). Thermoregulation in the rat: Deficits following 6-OHDA injections in the hypothalamus. *Pharmacol., Biochem. Behav.* **8**, 377–385.

Myers, R. D., and Tytell, M. (1972). Fever: Reciprocal shift in brain sodium to calcium ratio as the set-point temperature rises. *Science* **178**, 765–767.

Myers, R. D., and Waller, M. B. (1973). Differential release of acetylcholine from the hypothalamus and mesencephalon of the monkey during thermoregulation. *J. Physiol. (London)* **230**, 273–293.

Myers, R. D., and Waller, M. B. (1975). 5-HT- and norepinephrine-induced release of ACh from the thalamus and mesencepahlon of the monkey during thermoregulation. *Brain Res.* **84**, 47–61.

Myers, R. D., Metcalf, G., and Rice, J. C. (1977). Identification by microinjection of TRH-sensitive sites in the cat's brain stem that mediate respiratory, temperature and other autonomic changes. *Brain Res.* **126**, 105–115.

Myers, R. D., Oeltgen, P. R., and Spurrier, W. A. (1981). Hibernation "trigger" injected in brain induces hypothermia and hypophagia in the monkey. *Brain Res. Bull.* **7**, 691–695.

Myhre, K., Cabanac, M., and Myhre, G. (1977). Fever and behavioral temperature regulation in the frog,*Rana esculenta. Acta Physiol. Scand.* **101**, 219–229.

Nadal, E. R., Bullard, R. W., and Stolwijk, J. A. J. (1971). Importance of skin temperature in the regulation of sweating. *J. Appl. Physiol.* **31**, 80–87.

Nagawa, Y. (1980). Pharmacology of the central nervous system: Effects of thyrotropin releasing hormone. *J. Takeda Res. Lab.* **39**, 151–191.

Nakayama, T., Ishikawa, Y., and Tsurutani, T. (1979). Projection of scrotal thermal afferents to the preoptic and hypothalamic neurons in rats. *Pfluegers Arch.* **380**, 59–64.

Necker, R. (1975). Temperature-sensitive ascending neurons in the spinal cord of pigeons. *Pfluegers Arch.* **353**, 275–286.

Necker, R. (1977). Thermal sensitivity of different skin areas in pigeons. *J. Comp. Physiol.* **116**, 239–246.

Necker, R. (1981). Thermoreception and temperature regulation in homeothermic vertebrates. *Prog. Sens. Physiol.* **2**, 1–47.

Nedergaard, J. (1982). Catecholamine sensitivity in brown fat cells from cold-acclimated hamsters and rats. *Am. J. Physiol.* **242**, C250–C257.

Nelson, D. O., and Prosser, C. L. (1979). Effects of preoptic lesions on behavioral thermoregulation of green sunfish, *Lepomis cyanellus*, and of goldfish, *Carassius auratus. J. Comp. Physiol.* **129**, 193–197.

Nemeroff, C. B., Bissette, G., Manberg, P. J., Moore, S. I., Ervin, G. N., Osbahr, A. J., and Prange, A. J. (1980). Hypothermic responses to neurotensin in vertebrates. *In* "Thermoregulatory Mechanisms and Their Therapeutic Implications" (B. Cox, P. Lomax, A. S. Milton, and E. Schonbaum, eds.), pp. 180–185. Karger, Basel.

Nicholls, D. J. (1979). Brown adipose tissue mitochondria. *Biochim. Biophys. Acta* **549**, 1–29.

Novotona, R., and Civin, J. (1979). The relationship between tryptophan pyrrolase activity, 5-HT metabolism in the brain and induction of hibernation. *Physiol. Bohemosl.* **28**, 339–346.

Novotona, R., Jansky, L., and Drahota, Z. (1975). Effect of hibernation on turnover of serotonin in the brain stem of golden hamster (*Mesocricetus auratus*). *Gen. Pharmacol.* **6**, 23–26.

Nunez, E. A., and Gershon, M. D. (1982). Appearance and disappearance of tubular paracrystalline structures in somatotophs and lactotrophs during the annual life cycle of the bat (*Myotis lucifugus*). *Am. J. Anat.* **165**, 101–111.

Nunez, E. A., Gould, R. P., Hamilton, D. W., Hayward, J. S., and Holt, S. J. (1967). Seasonal changes in the fine structure of the basal granular cells of the bat thyroid. *J. Cell Sci.* **2**, 401–410.

Nunez, E. A., Whalen, J. P., and Krook, L. (1972). An ultrastructural study of the natural secretory cycle of the parathyroid gland of the bat. *Am. J. Anat.* **134**, 459–480.

Nurnberger, F. (1983). Der Hypothalamus des Igels (*Erinaceus europaeus* L.) unter besonderer Beruecksichtigung des Winterschlafes. Cytoarchitektonische und immunocytochemische Studien. Ph.D. Dissertation, University of Marburg, FRG.

Oeltgen, P. R., and Spurrier, W. A. (1981). Characterization of a hibernation trigger. *In* "Survival in the Cold" (X. J. Musacchia and L. Jansky, eds.), pp. 139–157. Elsevier, Amsterdam.

Oeltgen, P. R., Walsh, J. W., Hamann, S. R., Randall, D. C., Spurrier, W. A., and Myers, R. D. (1982). Hibernation "trigger": Opioid-like inhibitory action on brain function of the monkey. *Pharmacol. Biochem. Behav.* **17**, 1271–1274.

Palmer, D. L., and Riedesel, M. L. (1976). Responses of whole-animal and isolated hearts of ground squirrels, *Citellus lateralis*, to melatonin. *Comp. Biochem. Physiol. C* **53C**, 69–72.

Palokangas, R., Vihko, V., and Nuuja, I. (1972). The effects of cold and glucagon on lipolysis, glycogenolysis and oxygen consumption in young chicks. *Comp. Biochem. Physiol. A* **45A,** 489–495.

Parmeggiani, P. L. (1982). Sleep and temperature regulation. *Physiol. Sci.* **32,** 207–215.

Pastukhov, Yu. F., Chepkasov, I. E., Slovikov, B. I., Kolaeva, S. G., Lutsenko, N. D., and Vain-Rib, M. A. (1981). The role of serotonin in preparing the ground squirrels for hibernation. *Cryobiology* **18,** 91.

Pehowich, D. J., and Wang, L. C. H. (1981). Temperature dependence of mitochondrial Ca2+ transport in a hibernating and nonhibernating ground squirrel. *Acta Univ. Carol., Biol.* pp. 291–293.

Pengelley, E. T., and Asmundson, S. J. (1974). Circannual rhythmicity in hibernating mammals. *In* "Circannual Clocks, Annual Biological Rhythms" (E. T. Pengelley, ed.), pp. 95–160. Academic Press, New York.

Petrovic, V. M., Janic-Sibalic, V., Aminot, A., and Roffi, J. (1978). Adrenal tyrosine hydroxyglase activity in the ground squirrel—effect of cold and arousal from hibernation. *Comp. Bioch. Physiol. C.* **61C,** 99–101.

Phillips, J. A., and Harlow, H. J. (1982). Long-term effects of pinealectomy on the annual cycle of golden-mantled ground squirrel, *Spermophilus lateralis. J. Comp. Physiol. B* **146B,** 501–505.

Pittman, Q. J., Tache, Y., and Brown, M. (1980). Bombesin acts in the preoptic area to produce hypothermia in rats. *Life Sci.* **26,** 725–730.

Polenov, A. L., and Yurisova, M. N. (1975). The hypothalamo-hypophyseal system in the ground squirrels, *Citellus erythrogenys* Brandt and *Citellus undulatus* Pallas. I. Microanatomy and cytomorphology of the Gormori-positive neurosecretory system with special reference to its state during hibernation. *Z. Mikrosk.-Anat. Forsch.* **89,** 991–1014.

Poole, S., and Stephenson, J. D. (1979). Effects of noradrenaline and carbachol on temperature regulation of cold-stressed and cold-acclimated rats. *Br. J. Pharmacol.* **65,** 43–51.

Popova, N. K., and Koryakina, L. A. (1981). Seasonal changes in pituitary-adrenal reactivity in hibernating spermophiles. *Endocrinol. Exp.* **15,** 269–276.

Popova, N. K., and Voitenko, N. N. (1981). Brain serotonin metabolism in hibernation. *Pharmacol. Biochem. Behav.* **14,** 773–777.

Popova, N. K., Kolaeva, S. G., and Dianova, I. I. (1975). State of the pineal gland during hibernation. *Bull. Exp. Biol. Med. (Engl. Transl.)* **79,** 116–117.

Popovic, V. (1960). Endocrines in hibernation. *Bull. Mus. Comp. Zool.* **124,** 104–130.

Preston, E. (1974). Central effects of cholinergic-receptor blocking drugs on the conscious rabbit's thermoregulation against body cooling. *J. Pharmacol. Exp. Ther.* **188,** 400–409.

Proppe, D. W., and Gale, C. C. (1970). Endocrine thermoregulatory responses to local hypothalamic warming in unanesthetized baboons. *Am. J. Physiol.* **219,** 202–207.

Puschmann, S., and Jessen, C. (1978). Anterior and posterior hypothalamus: Effects of independent temperature displacements on heat production in conscious goats. *Pfluegers Arch.* **373,** 59–68.

Ralph, C. L. (1984). Pineal bodies and thermoregulation. *In* "Pineal Gland" (R. J. Reiter, ed.), pp. 193–219. Raven Press, New York.

Ralph, C. L., Firth, B. T., Gern, W. A., and Owens, D. W. (1979). The pineal complex and thermoregulation. *Biol. Rev. Cambridge Philos. Soc.* **54,** 41–72.

Ralph, C. L., Harlow, H. J., and Phillips, J. A. (1982). Delayed effect of pinealectomy on hibernation of the golden-mantled ground squirrel. *Int. J. Biometeorol.* **26,** 311–328.

Raths, P., and Kulzer, E. (1976). Physiology of hibernation and related lethargic states in mammals and birds. *Bonn. Zool. Monogr.* **9,** 1–93.

Rawson, R. O., and Quick, K. P. (1972). Localization of intra-abdominal thermoreceptors in the ewe. *J. Physiol. (London)* **222,** 665–677.

Refinetti, R., and Carlisle, H. J. (1986). Effects of anterior and posterior hypothalamic temperature changes on thermoregulation in the rat. *Physiol. Behav.* **36,** 1099–1103.

Reid, W. D., Volicer, L., Smookler, H., Beaven, M. A., and Brodie, B. B. (1968). Brain amines and temperature regulation. *Pharmacology* **1,** 329–344.

Reigle, T. G., and Wolf, H. H. (1975). A study of potential cholinergic mechanisms involved in the central control of body temperature in golden hamsters. *Neuropharmacology* **14,** 67–74.

Reis, D. J. (1960). Potentiation of the vasoconstrictor action of topical norepinephrine on the human bulbar conjunctival vessels after topical application of certain adrenocortical hormones. *J. Clin. Endocrinol. Metab.* **20,** 446–456.

Reiter, R. J. (1980). The pineal and its hormones in the control of reproduction in mammals. *Endocr. Rev.* **1,** 109–131.

Reiter, R. J., Vaughan, M. K., Balsk, D. E., and Johnson, L. Y. (1974). Melatonin: Its inhibition of pineal antigonadotrophic activity in male hamsters. *Science* **185,** 1169–1171.

Reynolds, W. W., Casterlin, M. E., and Covert, J. B. (1976). Behavioral fever in teleost fishes. *Nature (London)* **259,** 41–42.

Reynolds, W. W., Casterlin, M. E., and Covert, J. B. (1980). Behaviorally mediated fever in aquatic ectotherms. *In* "Fever" (J. M. Lipton, ed.), pp. 207–212. Raven Press, New York.

Riedel, W. (1976). Warm receptors in the dorsal abdominal wall of the rabbit. *Pfluegers Arch.* **361,** 205–206.

Riedel, W., Siaplauras, S., and Simon, E. (1973). Intra-abdominal thermosensitivity in the rabbit as compared with spinal thermosensitivity. *Pfluegers Arch.* **340,** 59–70.

Robertshaw, D., and Taylor, C. R. (1969). Sweat gland function of the donkey (*Equus asinus*). *J. Physiol. (London)* **205,** 79.

Robertshaw, D., Taylor, C. R., and Mazzia, L. M. (1973). Sweating in primates: Secretion by adrenal medulla during exercise. *Am. J. Physiol.* **244,** 678–681.

Rosenqvist, U. (1972). Adrenergic receptor response in hypothyroidism. *Acta Med. Scand., Suppl.* **532,** 1–28.

Rothwell, N. J., and Stock, M. J. (1979). A role for brown adipose tissue in diet-induced thermogenesis. *Nature (London)* **281,** 31–35.

Rothwell, N. J., and Stock, M. J. (1981). A role for insulin in the diet induced thermogenesis of cafeteria fed rats. *Metab. Clin. Exp.* **30,** 673–678.

Rothwell, N. J., Saville, M. E., Stock, M. J., and Wyllie, M. G. (1982). Catecholamine and thyroid hormone influence on brown fat ($Na+$, $K+$)-ATPase activity and thermogenesis in the rat. *Horm. Metab. Res.* **14,** 261–265.

Rudy, T. A., and Wolf, H. H. (1972). Effects of intracerebral injections of carbamylcholine and acetylcholine on temperature regulation in the cat. *Brain Res.* **38,** 117–130.

Ruwe, W. D., and Myers, R. D. (1978). Dopoamine in the hypothalamus of the cat: Pharmacological characterization and push-pull perfusion analysis of sites mediating hypothermia. *Pharmacol. Biochem. Behav.* **9,** 65–80.

Saboureau, M. (1986). Hibernation in the hedgehog: influence of external and internal factors. *In* "Living in The Cold" (H. C. Heller, X. J. Musacchia, and L. C. H. Wang, eds.), pp. 253–263. Elsevier, New York.

Saboureau, M., and Boissin, J. (1978). Seasonal changes and environmental control of testicular function in the hedgehog, *Erinaceus europaeus* L. *In* "Environmental Endocrinology" (I. Assenmacher and D. S. Farner, eds.), pp. 111–112. Springer, Berlin.

Saboureau, M., Bobet, J. P., and Boissin, J. (1980). Activité cyclique de la fonction corticosurrenalienne et variations saisonnieres du métabolisme périphérique du cortisol chez un mammifère hibernant, le hérisson (*Erinaceus europaeus* L.). *J. Physiol. (Paris)* **76,** 617–629.

Saggerson, E. D., and Carpenter, C. A. (1979). Ouabain and $K+$ removal blocks α-adrenergic stimulation of gluconeogenesis in bubule fragments from fed rats. *FEBS Lett.* **106,** 189–192.

Salzman, S. K., and Beckman, A. L. (1981). Effects of thyrotropin releasing hormone on hypothalamic thermosensitive neurons of the rat. *Brain Res. Bull.* **7,** 325–332.

Samson, W. K., Lipton, J. M., Zimmer, J. A., and Glyn, J. R. (1981). The effect of fever on central α-MSH concentrations in the rabbit. *Peptides (N.Y.)* **2,** 419–423.

Satinoff, E. (1978). Neural organization and evolution of thermal regulation in mammals. *Science* **201,** 16–22.

Satinoff, E. (1982). Are there similarities between thermoregulation and sexual behavior? *In* "The Physiological Mechanism of Motivation" (D. W. Pfaff, ed.), pp. 217–251. Springer, Berlin.

Satinoff, E. (1983). A reevaluation of the concept of the homeostatic organization of temperature regulation. *Handb. Behav. Neurobiol.* **6,** 443–472.

Satinoff, E., and Hendersen, R. (1977). Thermoregulatory behavior. *In* "Handbook of Operant Behavior" (W. K. Honig and J. E. R. Staddon, eds.), pp. 153–173. Prentice-Hall, Englewood Cliffs, New Jersey.

Satinoff, E., and Shan, S. (1971). Loss of behavioral thermoregulation after lateral hypothalamic lesions in rats. *J. Comp. Physiol. Psychol.* **77,** 302–312.

Sauerbier, I., and Lemmer, B. (1977). Seasonal variations in the turnover of noradrenaline of active and hibernating hedgehogs (*Erinaceus europaeus*). *Comp. Biochem. Physiol. C* **57C,** 61–63.

Schafer, K., Braun, H. A., and Hensel, H. (1982). Static and dynamic activity of cold receptors at various calcium levels. *J. Neurophysiol.* **47,** 1017–1028.

Schingnitz, G., and Werner, J. (1983). Thalamic neurons in the rat responding to thermal and noxious stimulation at various sites. *J. Therm. Biol.* **8,** 23–25.

Schmidek, W. R., Hoshino, K., Schmidek, M., and Timo-Iaria, C. (1972). Influence of environmental temperature on the sleep-wakefulness cycle in the rat. *Physiol. Behav.* **8,** 363–371.

Schmidt, I. (1982). Thermal stimulation of exposed skin area influences behavioral thermoregulation in pigeons. *J. Comp. Physiol.* **146,** 201–206.

Scott, I. M., and Boulant, J. A. (1984). Dopamine effects on thermosensitive neurons in hypothalamic tissue slices. *Brain Res.* **306,** 157–163.

Scott, I. M., D'Agostino, G., Cervona, T., Becker, L., and Giovinazzo, L. (1981). Seasonal testosterone in the eastern chipmunk, *Tamias striatus. Cryobiology* **18,** 89.

Scott, N. R., and Van Tienhoven, A. (1974). Biogenic amines and body temperature in the hen Gallus domesticus. *Am. J. Physiol.* **227,** 1399–1405.

Sellers, E. A., Flattery, K. V., Shum, A., and Johnson, G. E. (1971). Thyroid status in relation to catecholamines in cold and warm environment. *Can. J. Physiol. Pharmacol.* **49,** 268–275.

Sellers, E. A., Flattery, K. V., and Steiner, G. (1974). Cold acclimation of hypothyroid rats. *Am. J. Physiol.* **226,** 290–294.

Shafrir, E., and Steinberg, D. (1960). The essential role of the adrenal cortex in the response of plasma free fatty acids, cholesterol and phospholipids to epinephrine injection. *J. Clin. Invest.* **39,** 310–319.

Shapiro, Y., Pandolf, K. B., and Goldman, R. F. (1980). Sex differences in acclimation to a hot-dry environment. *Ergonomics* **23,** 635–642.

Sharp, F., Smith, D., Thompson, M., and Hammel, H. T. (1969). Thermoregulatory salivation proportional to hypothalamic temperature above threshold in the dog. *Life Sci.* **8,** 1069–1076.

Shear, S. B., and Bronk, J. R. (1979). The influence of thyroxine administered in vivo on the transmembrane protonic electrochemical potential difference in rat liver mitochondria. *Biochem. J.* **178,** 505–507.

Shih, S. T., Khorram, O., Lipton, J. M., and McCann, S. M. (1986). Central administration of α-MSH antiserum augments fever in the rabbit. *Am. J. Physiol.* **250,** R803–R806.

Shum, A., Johnson, G. E., and Flattery, K. V. (1969). Influence of ambient temperature on excretion of catecholamines and metabolites. *Am. J. Physiol.* **216,** 1164–1169.

Siegert, R., Phillipp-Dormston, W. K., Radsak, K., and Menzel, H. (1976). Mechanisms of fever induction in rabbits. *Infect. Immun.* **14,** 1130–1137.

Silva, J. E., and Larsen, P. R. (1983). Adrenergic activation of triiodothyronine production in brown adipose tissue. *Nature (London)* **305,** 712–714.

Simmonds, M. A. (1970). Effect of environmental temperature on the turnover of 5-hydroxytryptamine in various areas of rat brain. *J. Physiol. (London)* **211,** 93–108.

Simon, E. (1974). Temperature regulation: The spinal cord as a site of extrahypothalamic thermoregulatory functions. *Rev. Physiol. Biochem. Pharmacol.* **71,** 1–76.

Simon, E. (1981). Effects of CNS temperature on generation and transmission of temperature signals in homeotherms: A common concept for mammalian and avian thermoregulation. *Pfluegers Arch.* **392,** 79–88.

Simon, E., and Simon-Oppermann, C. (1979). Metabolic thermoregulatory responses to CNS cooling and to general hypothermia in the conscious Peking duck. *Pfluegers Arch.,* **382,** Suppl. R27.

Simon, E., Mercer, J. B., and Inomoto, T. (1983). Temperature-dependent synapses and primary thermosensors in the thermoregulatory central nervous network. *J. Therm. Biol.* **8,** 137–139.

Sinnamon, W. B., and Pivorun, E. G. (1981a). Effect of chronic melatonin administration on the duration of hibernation in *Spermophilus tridecemlineatus*. *Comp. Biochem. Physiol. A* **70A,** 435–437.

Sinnamon, W. B., and Pivorun, E. G. (1981b). Melatonin induces hypertrophy of brown adipose tissue in *Spermophilus tridencemlineatus*. *Cryobiology* **18,** 603–607.

Sinnamon, W. B., and Pivorun, E. G. (1982). Effects of pinealectomy, melatonin injections and melatonin antibody production on the mean duration of individual hibernation bouts in *Spermophilus tridecemlineatus*. *J. Therm. Biol.* **7,** 243–249.

Smiles, K. A., Elizondo, R. S., and Barney, C. C. (1976). Sweating responses during changes of hypothalamic temperature in the rhesus monkey. *J. Appl. Physiol.* **40,** 653–657.

Smit-Vis, J. H. (1972). The effect of pinealectomy and of testosterone administration on the occurrence of hibernation in adult male golden hamsters. *Acta Morphol. Neerl.-Scand.* **10,** 269–281.

Smit-Vis, J. H., and Smit, G. J. (1970). Hibernation and testis activity in the golden hamster. *Neth. J. Zool.* **20,** 502–506.

Snyder, S. H. (1980). Brain peptides as neurotransmitters. *Science* **209,** 976–983.

Sod-Moriah, U. A., Goldberg, G. M., and Bedrak, E. (1974). Intrascrotal temperature, testicular histology and fertility of heat-acclimatized rats. *J. Reprod. Fertil.* **37,** 263–268.

Somero, G. N. (1978). Temperature adaptation of enzymes: Biological optimization through structure-function compromises. *Annu. Rev. Ecol. Syst.* **9**, 1–29.

South, F. E., Miller, V. M., and Hartner, W. C. (1978). Neuronal models of temperature regulation in euthermic and hibernating mammals: An alternate model for hibernation. In "Strategies in Cold: Natural Torpidity and Thermogenesis" (L. C. H. Wang and J. W. Hudson, eds.), pp. 187–224. Academic Press, New York.

Spafford, D. C., and Pengelley, E. T. (1971). The influence of the neurohumor serotonin on hibernation in the golden-mantled ground squirrel, *Citellus lateralis. Comp. Biochem. Physiol. A* **38A**, 239–250.

Spencer, F., Shirer, H. W., and Yochim, J. M. (1976). Core temperature in the female rat: Effect of pinealectomy or altered lighting. *Am. J. Physiol.* **231**, 355–360.

Stanton, T. L., Winokur, A., and Beckman, A. L. (1980). Reversal of natural CNS depression by TRH action in the hippocampus. *Brain Res.* **181**, 470–475.

Steinberg, B., Singh, I. J., and Mitchell, O. G. (1981). The effects of cold-stress, hibernation, and prolonged inactivity on bone dynamics in the golden hamster, *Mesocricetus auratus. J. Morphol.* **167**, 43–51.

Stitt, J. T. (1980). Variable open-loop gain in the control of thermogenesis in cold exposed rabbits. *J. Appl. Physiol.* **48**, 494–499.

Stitt, J. T. (1983). Hypothalamic generation of effector signals. *J. Therm. Biol.* **8**, 113–117.

Stitt, J. T. (1985). Evidence for the involvement of the organism vasculosum laminae terminalis in the febrile response of rabbits and rats. *J. Physiol. (London)* **368**, 501–511.

Stitt, J. T. (1986). Prostaglandin E as the neural mediator of the febrile response. *Yale J. Biol. Med.* **59**, 137–149.

Stitt, J. T., Hardy, J. D., and Stolwijk, J. A. J. (1966). PGE fever: Its effect on thermoregulation at different low ambient temperatures. *Am. J. Physiol.* **227**, 622–629.

Stolwijk, J. A. J., and Hardy, J. D. (1966). Temperature regulation in man—a theoretical study. *Pfluegers Arch.* **291**, 129–162.

Sundin, U. (1981). GDP binding to rat brown fat mitochondria: Effects of thyroxine at different ambient temperatures. *Am. J. Physiol.* **241**, C134–C139.

Sundin, U., and Cannon, B. (1980). GDP-binding to the brown fat mitochondria of developing and cold-adapted rats. *Comp. Biochem. Physiol. B* **65B**, 463–471.

Swan, H. (1981). Neuroendocrine aspects of hibernation. In "Survival in the Cold" (X. J. Musacchia and L. Jansky, eds.), pp. 121–138. Elsevier, Amsterdam.

Swan, H., and Schatte, C. (1977). Antimetabolic extract from the brain of the hibernating ground squirrel *Citellus tridecemlineatus. Science* **195**, 84–85.

Sweatman, P., and Jell, R. M. (1977). Dopamine and histamine sensitivity of rostral hypothalamic neurons in the cat: Possible involvement in thermoregulation. *Brain Res.* **127**, 173–178.

Szymusiak, R., and Satinoff, E. (1980). Brief skin temperature changes toward thermoneutrality trigger REM sleep in rats. *Physiol. Behav.* **25,** 305–311.

Tache, Y., Pittman, Q., and Brown, M. R. (1980). Bombesin-induced poikilothermy in rats. *Brain Res.* **188,** 525–530.

Tashima, L. S. (1965). The effects of cold exposure and hibernation on the thyroidal activity of Mesocricetus auratus. *Gen. Comp. Endocrinol.* **5,** 267–277.

Tataryn, I. V., Meldrum, D. R., Frumar, A. M., Lu, K. H., Judd, H. L., Bajorek, J. G., Chesarek, W., and Lomax, P. (1980). The hormonal and thermoregulatory changes in postmenopausal hot flushes. *In* "Thermoregulatory Mechanisms and Their Therapeutic Implications" (B. Cox, P. Lomax, A. S. Milton, and E. Schonbaum, eds.), pp. 202–207. Karger, Basel.

Taylor, C. R. (1974). Exercise and thermoregulation. *MTP Int. Rev. Sci.* **7,** 163–184.

Telegdy, G. (1977). Brain serotonin and pituitary-adrenal function. *Proc. Int. Union Physiol. Sci., Int.* **12,** 541.

Templeton, J. R. (1970). Reptiles. *In* "Comparative Physiology of Thermoregulation" (G. C. Whittow, ed.), Vol. 1, pp. 167–221. Academic Press, New York.

Taskey, N., Horwitz, B., and Horowitz, J. (1975). Norepinephrine-induced depolarization of skeletal muscle cells. *Eur. J. Pharmacol.* **30,** 352–355.

Thompson, F. J., and Barnes, C. D. (1969). Evidence for thermosensitive receptors in the femoral vein. *Fed. Proc., Fed. Am. Soc. Exp. Biol.* **28,** 722.

Thornhill, J. A., and Wilfong, A. (1982). Lateral cerebral ventricle and preoptic-anterior hypothalamic area infusion and perfusion of β-endorphin and ACTH to unrestrained rats: Core and surface temperature responses. *Can. J. Physiol. Pharmacol.* **60,** 1267–1274.

Thornhill, J. A., Cooper, K. E., and Veale, W. L. (1980). Core temperature changes following administration of naloxone and naltrexone to rats exposed to hot and cold ambient temperatures. Evidence for the physiological role of endorphin in hot and cold. *J. Pharm. Pharmacol.* **32,** 427–430.

Tocco, R. J., Kahn, L. L., Kluger, M. J., and Vander, A. J. (1983). Relationship of trace metals to fever during infection: Are prostaglandins involved? *Am. J. Physiol.* **244,** R368–R373.

Tolbert, M. E. M., Butcher, F. R., and Fain, J. N. (1973). Lack of correlation between catecholamine effects on cyclic adenosine 3′, 5′-monophosphate and gluconeogenesis in isolated rat liver cells. *J. Biol. Chem.* **248,** 5686–5692.

Triandafillou, J., Gwilliam, C., and Himms-Hagen, J. (1982). Role of thyroid hormone in cold-induced changes in rat brown adipose tissue mitochondria. *Can. J. Biochem.* **60,** 530–537.

Tseng, L. F., Wei, E. T., Loh, H. H., and Li, C. H. (1980). β-endorphin: Central sites of analgesia, catalepsy and body temperature changes in rats. *J. Pharmacol. Exp. Ther.* **214,** 328–332.

Vande Wiele, R. L., Bogumil, J., Dyrenfurth, I., Ferin, M., Jewelewicz, R., Warren, M., Rizkallah, T., and Mikhail, G. (1970). Mechanisms regulating the menstrual cycle in women. *Recent Prog. Horm. Res.* **26,** 63–103.

Van Zoeren, J. G., and Stricker, E. M. (1977). Effects of preoptic, lateral hypothalamic, or dopamine-depleting lesions on behavioral thermoregulation in rats exposed to the cold. *J. Comp. Physiol. Psychol.* **91**, 989–999.

Vaughan, M. K. (1981). The pineal gland—a survey of its antogonadotrophic substances and their actions. *Int. Rev. Physiol.* **24**, 41–95.

Veale, W. L., Kasting, N. W., and Cooper, K. E. (1981). Arginine vasopressin and endogenous antipyresis: Evidence and significance. *Fed. Proc., Fed. Am. Soc. Exp. Biol.* **40**, 2750–2753.

Veale, W. L., Eagan, P. C., and Cooper, K. E. (1982). Abnormality of the febrile response of the Brattleboro rat. *Ann. N. Y. Acad. Sci.* **394**, 776–779.

Videla, L., Flattery, K. V., Sellers, E. A., and Israel, Y. (1975). Ethanol metabolism and liver oxidative capacity in cold acclimation. *J. Pharmacol. Exp. Ther.* **192**, 575–582.

Vogt, M. (1954). The concentration of sympathin in different parts of the central nervous system under normal conditions and after the administration of drugs. *J. Physiol. (London)* **123**, 451–481.

von Euler, C. (1961). Physiology and pharmacology of temperature regulation. *Pharmacol. Rev.* **13**, 361–398.

Vriend, J. (1981). The pineal and melatonin in the regulation of pituitary-thyroid axis. *Life Sci.* **29**, 1929–1936.

Waites, G. M. H. (1962). The effect of heating the scrotum of the ram on respiration and body temperature. *Q. J. Exp. Physiol. Cogn. Med. Sci.* **47**, 314–323.

Walker, J. M., and Berger, R. J. 91980). Sleep as an adaptation for energy conservation functionally related to hibernation and shallow torpor. *Brain Res.* **53**, 255–278.

Wang, L. C. H. (1979). time patterns and metabolic rates of natural torpor in the Richardson's ground squirrel. *Can. J. Zool.* **57**, 149–155.

Wang, L. C. H. (1980). Modulation of maximum thermogenesis by feeding in the white rat. *J. Appl. Physiol.* **49**, 975–978.

Wang, L. C. H. (1981). Effects of feeding on aminophylline induced supra-maximal thermogenesis. *Life Sci.* **29**, 2459–2466.

Wang, L. C. H. (1982). Hibernation and the endocrine. *In* "Hibernation and Torpor in Mammals and Birds" (C. P. Lyman, J. S. Willis, A. Malan, and L. C. H. Wang, eds.), pp. 206–236. Academic Press, New York.

Wang, L. C. H. (1985). Life at low temperature: Mammalian hibernation. *Cryo-Letters* **6**, 257–274.

Wang, L. C. H. (1986). Some neuroendocrine aspects of mammalian hibernation. *In* "Endocrine Regulations as Adaptive Mechanisms to the Environment" (I. Assenmacher and J. Boissin, eds.), pp. 341–349. C.N.R.S., Paris.

Wang, L. C. H. (1987). Mammalian hibernation. *In* "The Effects of Low Temperatures on Biological Systems" (G. J. Morris and B. Grout, eds.), pp. 349–386. Arnold, London.

Wang, L. C. H., and Abbotts, B. (1981). Maximum thermogenesis in hibernators:

Magnitudes and seasonal variations. *In* "Survival in the Cold" (X. J. Musacchia and L. Jansky, eds.), pp. 77–97. Elsevier, Amsterdam.

Wang, L. C. H., and Anholt, E. C. (1982). Elicitation of supramaximal thermogenesis by aminophylline in the rat. *J. Appl. Physiol.* **53,** 16–20.

Wang, L. C. H., and Hudson, J. W., eds. (1978). "Strategies in Cold: Natural Torpidity and Thermogenesis." Academic Press, New York.

Wang, L. C. H., and Peter, R. E. (1977). Changes in plasma glucose, FFA, corticosterone, and thyroxine in He-O2-induced hypothermia. *J. Appl. Physiol.* **42,** 694–698.

Weiner, J. S., and Hellman, K. (1960). The sweat glands. *Biol. Rev. Cambridge Philos. Soc.* **35,** 141–186.

Weiss, B. L., and Aghajanian, G. K. (1971). Activation of brain serotonin metabolism by heat: Role of midbrain raphe neurons. *Brain Res.* **26,** 37–48.

Wenberg, G. M., and Holland, J. C. (1973a). The circannual variations of thyroid activity in the woodchuck (*Marmota monax*). *Comp. Biochem. Physiol. A* **44A,** 775–780.

Wenberg, G. M., and Holland, J. C. (1973b). The circannual variations of some of the hormones of the woodchuck (*Marmota monax*). *Comp. Biochem. Physiol.* **46A,** 523–535.

Whalen, J. P., Krook, L., and Nunez, E. A. (1971). Bone resorption in hibernation and its relationship to parafollicular and parathyroid cell structures. *Invest. Radiol.* **6,** 342–343.

Whittow, G. C. (1971). Ungulates. *In* "Comparative Physiology of Thermoregulation" (G. C. Whittow, ed.), Vol. 2, pp. 191–281. Academic Press, New York.

Wickler, S. J. (1981). Nonshivering thermogenesis in skeletal muscle of seasonally acclimatized mice, *Peromyscus*. *Am. J. Physiol.* **241,** R185–R189.

Wilkinson, C. W., Carlisle, H. J., and Reynolds, R. W. (1980). Estrogenic effects on behavioral thermoregulation and body temperature of rats. *Physiol. Behav.* **24,** 337–340.

Willis, J. S. (1982). Intermediary metabolism in hibernation. *In* "Hibernation and Torpor in Mammals and Birds" (C. P. Lyman, J. S. Willis, A. Malan, and L. C. H. Wang, eds.), pp. 124–139. Academic Press, New York.

Winston, B. W., and Henderson, N. E. (1981). Seasonal changes in morphology of the thyroid gland of a hibernator, *Spermophilus richardsoni*. *Can. J. Zool.* **59,** 1022–1031.

Woodside, B., Leon, M., Attard, M., Feder, H. H., Siegel, H. I., and Fischette, C. (1981). Prolactin-steroid influences on the thermal basis for mother-young contact in Norway rats. *J. Comp. Physiol. Psychol.* **95,** 771–780.

Wooten, W. L., and Cascarano, J. (1980). The effect of thyroid hormone on mitochondrial biogenesis and cellular hyperplasia. *J. Bioenerg. Biomembr.* **12,** 1–12.

Wunder, B. A. (1979). Hormonal mechanisms. *In* "Comparative Mechanisms of Cold Adaptation" (L. S. Underwood, L. L. Tieszen, A. B. Callahan, and G. E. Folk, Jr., eds.), pp. 143–158. Academic Press, New York.

Wunder, B. A., Hawkins, M. F., Avery, D. D., and Swan, H. (1980). The effect of bombesin injected into the anterior and posterior hypothalamus on body temperature and oxygen consumption. *Neuropharmacology* **19,** 1095–1097.

Yacoe, M. E. (1983). Protein metabolism in the pectoralis muscle and liver of hibernating bats, *Eptesicus fuscus. J. Comp. Physiol. B* **152B,** 137–144.

Yehuda, S., and Kastin, A. J. (1980). Peptides and thermoregulation. *Neurosci. Behav. Rev.* **4,** 459–471.

Young, J. B., and Landsberg, L. (1979). Catecholamines and the sympathoadrenal system: The regulation of metabolism. *In* "Contemporary Endocrinology" (S. H. Ingbar, ed.), pp. 245–303. Plenum, New York.

Young, J. B., Saville, E., Rothwell, N. J., and Stock, M. J. (1982). Effect of diet and cold exposure on norepinephrine turnover in brown adipose tissue of the rat. *J. Clin. Invest.* **69,** 1061–1071.

Young, R. A., Robinson, D. S., Vagenakis, A. G., Saavedra, J. M., Lovenberg, W., Krupp, P. P., and Danforth, E., Jr. (1979a). Brain TRH, monoamines, tyrosine hydroxyglase, and tryptophan hydroxyglase in the woodchuck, Marmota monax, during the hibernation season. *Comp. Biochem. Physiol. C* **63C,** 319–323.

Young, R. A., Danforth, E., Jr., Vagenakis, A. G., Krupp, P. P., Frink, R., and Sims, E. A. H. (1979b). Seasonal variation and the influence of body temperature on plasma concentrations and binding of thyroxine and triiodothyronine in the woodchuck. *Endocrinology (Baltimore)* **104,** 996–999.

Yurisova, M. N., and Polenov, A. L. (1979). The hypothalamo-hypophysial system in the ground squirrels, *Citellus erythrogenys* Brant. II. Seasonal changes in classical neurosecretory system of a hibernator. *Cell Tissue Res.* **198,** 539–556.

Zeisberger, E., and Bruck, K. (1973). Effects of intrahypothalamically injected noradrenergic and cholinergic agents on thermoregulatory responses. *In* "The Pharmacology of Thermoregulation" (E. Schonbaum and P. Lomax, eds.), pp. 232–243. Karger, Basel.

Zeisberger, E., Merker, G., and Blahser, S. (1981). Fever response in the guinea pig before and after parturition. *Brain Res.* **212,** 379–392.

Zimmerman, M. L. 91982). Carbohydrate and torpor duration in hibernating goldenmantled ground squirrel (*Citellus lateralis*). *J. Comp. Physiol. B* **147B,** 129–135.

Zucker, I., and Licht, P. (1983). Circannual and seasonal variations in plasma luteinizing hormone of avariectomized ground squirrels (*Spermophilus lateralis*). *Biol. Reprod.* **28,** 178–185.

Author Index

Subject Index